Food
Biotechnology

Contents at a Glance

Further Study

The reader should refer to *Food Processing and Preservation* by DS Warris, published by CBS Publishers & Distributors, for the following topics:

- Food flavours and food additives
- Spices, sweetners and vitamins
- Food poisoning, toxicity and safety
- Hazard analysis critical control point (HACCP)
- Food laws and food ethics

Food Biotechnology

GN Foster

CBSPD

CBS Publishers & Distributors Pvt Ltd

New Delhi • Bengaluru • Chennai • Kochi • Kolkata • Lucknow • Mumbai
Hyderabad • Jharkhand • Nagpur • Patna • Pune • Uttarakhand

Food Biotechnology

ISBN: 978-93-89396-34-8

Copyright © Author and Publisher

First Edition: 2020
Reprint: 2023

Published by **Satish Kumar Jain** and produced by **Varun Jain** for
CBS Publishers & Distributors Pvt Ltd
4819/XI Prahlad Street, 24 Ansari Road, Daryaganj, New Delhi 110 002, India
Ph: 011-23289259, 23266861   Website: www.cbspd.com
                                    e-mail: delhi@cbspd.com

Corporate Office: 204 FIE, Industrial Area, Patparganj, Delhi 110 092, India
Ph: 011-4934 4934   Fax: 011-4934 4935   e-mail: publishing@cbspd.com;
                                             publicity@cbspd.com

Branches

- **Bengaluru:** Seema House 2975, 17th Cross, KR Road, Banasankari 2nd Stage, Bengaluru 560 070, Karnataka, India
  Ph: +91-80-26771678/79   Fax: +91-80-26771680   e-mail: bangalore@cbspd.com
- **Chennai:** 7, Subbaraya Street, Shenoy Nagar, Chennai 600 030, Tamil Nadu, India
  Ph: +91-44-26680620, 26681266   Fax: +91-44-42032115   e-mail: chennai@cbspd.com
- **Kochi:** 42/1325, 1326, Power House Road, Opp KSEB, Power House, Ernakulum Kochi 682 018, Kerala, India
  Ph: +91-484-4059061-65,67   Fax: +91-484-4059065   e-mail: kochi@cbspd.com
- **Kolkata:** 147, Hind Ceramics Compound, 1st Floor, Nilgunj Road, Belghoria, Kolkata-700056, West Bengal, India
  Ph: +033-25633055, 033-25633056   e-mail: kolkata@cbspd.com
- **Lucknow:** Basement, Khushnuma Complex, 7 Meerabai Marg (Behind Jawahar Bhawan),Lucknow-226001, UP, India
  Ph: +91-522-4000032   e-mail: tiwari.lucknow@cbspd.com
- **Mumbai:** PWD Shed, Gala no 25/26, Ramchandra Bhatt Marg, Next to JJ Hospital Gate no. 2, Opp. Union Bank of India Noorbaug, Mumbai-400009, Maharashtra, India
  Ph: 022-66661880/89   e-mail: mumbai@cbspd.com

Representatives

• Hyderabad	0-9885175004	• Jharkhand	0-9811541605	• Nagpur	0-9421945513
• Patna	0-9334159340	• Pune	0-9923910676	• Uttarakhand	0-9716462459

Printed at: SRK Graphics, Shahdara, Delhi, India

Preface

For centuries, humans have been selecting, sowing, and harvesting seeds to produce food products that will sustain them. In this present age, global food demand has increased the need for improved crops. Biotechnology offers the needed technology to produce higher crop yields, plants that are naturally protected from disease and insects, and potentially more nutritious and better tasting foods. A general definition of biotechnology is the use of a living organism or its products for commercial purposes. Today, biotechnology involves the use of techniques which take genetic material from one organism and put it into another, thus obtaining desired qualities or products. Crops produced by biotechnology include soyabeans, corn, cotton, canola, papaya, tomatoes and squash. Also, an enzyme used to make cheese and yeast to make bread is commonly produced by biotechnology. Biotechnology can extend advances in cross-breeding, allowing for new food varieties. For example, seedless melons and mini avocadoes. Farmers can also develop food with better flavour and a better nutrient profile.

Modern food biotechnology may help promote public health, providing fruits, vegetables and grains with more nutritional benefits. These include more proteins, vitamins and minerals, or less fat and saturated fat. Already some oils have a better fatty acid profile, less saturated fat and trans fat, and more monounsaturated fat. This can promote heart health. For those with food allergies, biotechnology is seeking ways to reduce allergens in peanuts, wheat and other crops.

Thus, food biotechnology is the application of modern biotechnological techniques to the manufacture and processing of food, for example through fermentation of food (which is the oldest biotechnological process) and food additives, as well as plant and animal cell cultures. New developments in fermentation and enzyme technological processes, molecular thermodynamics, genetic engineering, protein engineering, metabolic engineering, bio-engineering, and processes involving monoclonal antibodies, nanobiotechnology and quorum sensing have introduced exciting new dimensions to food biotechnology, a burgeoning field that transcends many scientific disciplines.

This reference textbook on *Food Biotechnology* is divided into ten sections and contains 31 chapters.

Section I discusses *general considerations and biological/biotechnological aspects*. Chapter 1 is devoted to food biotechnology: An overview. Chapter 2 deals with micro-organims associated with food. Micro-organisms can be used as processing aids in the production of fermented foods. A variety of food chemicals and additives may be produced by fermentation involving select species of micro-organisms. Chapter 3 focuses on applications and impact of biotechnology in food industry. Biotechnology has already benefitted the food industry in a big way. It has given us high quality foods that are tasty, nutritious, wholesome, convenient, shelf stable and safe. Chapter 4 concentrates on biotechnology of fermentation. Fermentation processes utilise micro-organisms to convert solid or liquid substrates into various products. The substrates used vary widely, any material that supports microbial growth being a potential substrate. Similarly, fermentation derived products show tremendous variety.

Section II discusses *genetically modified crops*. Chapter 5 is devoted to feeding the world and eradicating hunger. This chapter discusses some of the key issues pertaining to the appropriate role of biotechnology in addressing hunger and food production in the developing world. Chapter 6 deals with genetically modified crops. There is a scientific consensus that currently available food derived from GM crops poses no greater risk to human health than conventional food, but that each GM food needs to be tested on a case-by-case basis before introduction.

Section III discusses *biotechnology of probiotics and functional foods*. Chapter 7 explains probiotics in food. Probiotics are live micro-organisms thought to be beneficial to the host organism. Probiotics are usually introduced to food, condiments and beverages as a component of fermentation process at appropriate stage. Chapter 8 concentrates on health benefits of probiotics. A number of health effects are associated with usage of probiotics. There are differing degrees of evidence supporting the verification of such effects. Also there are reports showing no clinical effects of certain probiotic strains in specific situations. Chapter 9 focuses on functional foods. Functional foods are foods or dietary components that claim to provide health benefits aside from basic nutrition. These foods contain biologically active substances such as antioxidants that may lower the risks from certain diseases associated with ageing. Chapter 10 discusses health benefits of functional ingredients from marine bioresources. This chapter discusses potential marine bioresources with respect to their extractable biomolecules in detail, while explaining the present and prospective methods of identification and extraction, which are integrated with advanced techniques in modern biotechnology.

Section IV discusses *biotechnology of dairy and milk products*. Chapter 11 is devoted to milk and dairy products. Dairy products include—milk and any of the foods made from milk, including butter, cheese, ice cream, yogurt, and condensed and dried milk. Nutritional and genetic interventions to alter the milk composition for specific health and/or processing opportunities are gaining importance in dairy biotechnology. Chapter 12 deals with genetically modified cheese. Chapter 13 focuses on health benefits of milk and functional dairy products.

Section V discusses *biotechnology of meat, fish and poultry*. Chapter 14 concentrates on biotechnology of meat. Meat is one of the most valuable and demanding food products. Meat is animal flesh that is used as food. Most often, this means the skeletal muscle and associated fat, but it may also describe other edible tissues such as organs, livers, skin, brains, bone marrow, kidneys or lungs. Chapter 15 explains biotechnology of fish. Chapter 16 discusses impact of biotechnology on poultry nutrition. Poultry is defined as economically important birds used for food. This includes chickens, turkeys, quail, ducks, geese, and guineas. Other birds, such as pheasants, partridges and peafowl, can be classified as poultry.

Section VI discusses *genetically modified foods*. Chapter 17 is devoted to fruits and vegetables biotechnology. Biotechnology of fruit and vegetable production are an aid to conventional breeding and its ability to transfer genes between different organisms. Chapter 18 deals with biotechnology of mushrooms. A mushroom is the fleshy, spore-bearing fruiting body of a fungus. Chapter 19 concentrates on genetically modified fruits. Chapter 20 focuses on genetically modified vegetables. Chapter 21 explains genetically modified foods. Genetically modified foods (GM foods), also known as genetically engineered foods (GE foods), or bioengineered foods are foods produced from organisms that have had changes introduced into their DNA using the methods of genetic engineering.

Section VII discusses *production, purification and application of enzymes in food industry.* Chapter 22 is devoted to microbial enzymes, production, purification and isolation. Production of a new microbial enzyme starts with screening of micro-organisms for desirable activity using appropriate selection procedures. Chapter 23 deals with applications of enzymes in food industry.

Section VIII discusses *brewing of yeast, beer and wine industry.* Chapter 24 concentrates on yeast as a versatile tool in biotechnology. Yeasts represent a very diverse group of micro-organisms, and even strains that are classified as the same species often show a high level of genetic divergence. Chapter 25 focuses on biotechnology of brewer's yeast. Chapter 26 explains biotechnology of beer and wine industry. Beer is an undistilled beverage produced from fermentation of barley malt by yeast especially *Sacccharomyces crevisiae* and *S. carsibergenesis.* Winemaking, or vinification, is the production of wine, starting with selection of the grapes or other produce and ending with bottling the finished wine. Although most wine is made from grapes, it may also be made from other fruit or nontoxic plant material.

Section IX discusses *environmental and ecological aspects of food biotechnology.* Chapter 27 is devoted to carbon footprint of food industry. The carbon footprint is a measure of the amount of greenhouse gases (GHG) produced by our activities in relation to carbon dioxide (CO_2) or carbon. The carbon footprint on food is an estimate of all the emissions caused by the production (e.g. farming), manufacture and delivery to the consumer and the disposal of packaging. Chapter 28 deals with utilisation of food wastes for sustainable development. Chapter 29 explains environmental and ecological aspects of genetically modified crops.

Section X discusses *bioethics and intellectual property rights.* Chapter 30 concentrates on bioethics and biotechnology. Biotechnology is at the intersection of science and ethics. Technological developments are shaped by an ethical vision, which in turn is shaped by available technology. Much in biotechnology can be celebrated for how it benefits humanity. The ethical assessment of new technologies, including biotechnology, requires a different approach to ethics. Chapter 31 is devoted to food security and intellectual property rights in developing countries. Food insecurity is a major problem throughout the world. It is a concern at all levels, from individuals to states. At a basic level, food security is about fulfilling each individual's human right to food. Food security is defined as physical and economic access to sufficient, safe and nutritious food by all people to meet their dietary needs and food preferences for an active and healthy life. Meeting food security objectives implies improving access to food which is itself linked to poverty eradication.

Diagrams, figures, tables, glossary and index supplement the text. All topics have been covered in a cogent and lucid style to help the reader grasp the information quickly and easily.

It may not be wrong to hold that the present reference textbook is a complete treatise on this subject. It is an essential reading for BTech (environmental biotechnology/microbiology/ food microbiology/biomedical and biochemical engineering) and students pursuing

BSc/MSc course in biotechnology and microbiology. The book will appeal to professional food scientists as well as graduate and advanced undergraduate students by addressing the latest exciting food biotechnology research in areas such as genetically modified foods, bioenergy, bioplastics, functional foods/nutraceuticals. Besides students, this book will prove useful to industrialists, consultants and researchers in their respective fields.

This reference textbook also caters to the requirement of the syllabus prescribed by various universities for undergraduate and postgraduate courses in the above subjects. It has been prepared with meticulous care, aiming at making the book error-free. Constructive suggestions are always welcome from readers of this book.

GN Foster

Contents

Section VII
PRODUCTION, PURIFICATION AND APPLICATION OF ENZYMES IN FOOD INDUSTRY

Section X
BIOETHICS AND INTELLECTUAL PROPERTY RIGHTS

SECTION I

General Considerations and Biological/Biotechnological Aspects

Chapter 1

Food Biotechnology: An Overview

INTRODUCTION

The term 'biotechnology' was first coined by a Hungarian, Karl Ereky, towards the end of World War I. Ereky used the word to refer to intensive agricultural methods. Since that time biotechnology has been variously defined, but it has nearly always been associated with food production and processing. In particular biotechnology has usually encompassed the traditional manufacture of bread, wine, cheese and other fermented foods. On these grounds, biotechnology can trace its roots back several thousand years to the ancient Sumerians, who brewed beer with naturally occurring yeasts.

Ancient fermentations were not always successful. The microbes that fell into the wine maker's vat could yield the finest vintage or transform the entire product to vinegar. In the 1800s Louis Pasteur laid the foundations of microbiology and identified micro-organisms — bacteria, fungi, algae and protozoa — as the cause of both desirable and undesirable changes in food. The application of Pasteur's research led to safer, more reliable food processing and preservation and helped ensure the consistent high quality of, for example, fine wines and cheeses.

Pasteur asserted that fermentation processes were inextricably linked to the activities of living microbes. Towards the end of the last century it was discovered that cell-free extracts from yeast could also bring about chemical changes without the intervention of the microbes from which they were derived. The active components of such extracts were named enzymes (enzyme means 'in yeast'). Enzymes are proteins, made by all living things, that catalyse specific chemical reactions. Without realising it, the makers of cheese had always used a mixture of natural enzymes—rennet—to transform milk into solid curds and liquid whey. During the 1940s, large-scale fermentation equipment was developed which led to the efficient industrial production by micro-organisms of pure enzymes and additives and other valuable compounds (such as vitamins) for use in food.

Just as different breweries have their own carefully maintained proprietary strains of yeast, enzyme manufacturers culture specially selected strains of their chosen micro-organisms. Over many years great improvements have been made to the efficiency of production and the safety, quality and range of microbial products available. However, much still depends on chance occurrence followed by systematic isolation of organisms with desirable characteristics. With the advent during the 1970s of the ability to

3

make precise changes to genetic material, biotechnology was transformed. The performance of organisms can now be 'fine tuned', and biotechnology has now almost became a synonym for 'genetic modification'. In 1980, an influential British report (the 'Spinks Report') attempted to encapsulate nearly half a century of European and United States thought, defining biotechnology as 'the application of biological organisms, systems or processes to the manufacturing and service industries'.

This broad definition suits our purposes, as it includes the production of food by living organisms, its subsequent processing with the assistance of microbes or enzymes and the assurance of food quality and safety using the tools of molecular biology.

METHODS USED IN BIOTECHNOLOGY

Genetics and Genetic Modification

Chromosomes, genes and DNA

Genes, passed from one generation to the next, determine all inherited characteristics. Genes are made from DNA (deoxyribonucleic acid), most of which is packaged, in fungal (including yeast), plant and animal cells, into chromosomes within the nucleus of the cell. Some genes are also found outside the nucleus: in the mitochondria (which release energy for cellular activities) and within the chloroplasts (sites of photosynthesis) of plant cells. In bacteria, most genes occur on a single circular chromosome, although small rings of DNA called plasmids may also be present. The double helix of DNA can be likened to a twisted rope ladder. The two intertwined helices are chains made from sugar and phosphate molecules linked together alternately. Attached to each sugar molecule is a 'base'. There are four different bases: adenine (A), thymine (T), cytosine (C) and guanine (G). Weak bonds between the bases join the two strands of the double helix together like the rungs of a ladder. A always pairs with T, and C always pairs with G.

This 'base pairing' mechanism ensures identical replication of DNA strands during cell division (Fig. 1.1). A particular gene (a stretch of DNA with a particular sequence) determines the structure of all or part of a specific protein. The sequences of bases in the DNA specify the amino acid residues that are needed to make proteins.

Three bases in a row specify each amino acid, and the sequence that specifies each—the genetic code—is the same in all living organisms. Also encoded within the DNA are instructions to regulate protein production. Although all cells of an organism will contain the same DNA, only certain proteins will be made at any one time or in any particular type of cell, that is, only certain genes will be expressed.

DNA has an identical structure in all living things, and because the genetic code is universal, the possibility is raised that genes can be transferred between completely different species. The process of transferring, removing or altering genetic information by the modification of DNA is commonly called genetic modification (or genetic engineering).

Why alter nature?

In nature, proteins are often made in minute quantities and are therefore difficult or impossible to extract and purify. These proteins include enzymes and a variety of pharmacologically active compounds such as insulin for diabetics, interferons for cancer therapy and vaccines to help prevent diseases. Often large quantities of such valuable proteins are needed. Biotechnologists can achieve this by transferring the relevant genes into microbes that can easily be cultivated in large numbers. The same technology, applied to the production of food, could bring significant benefits.

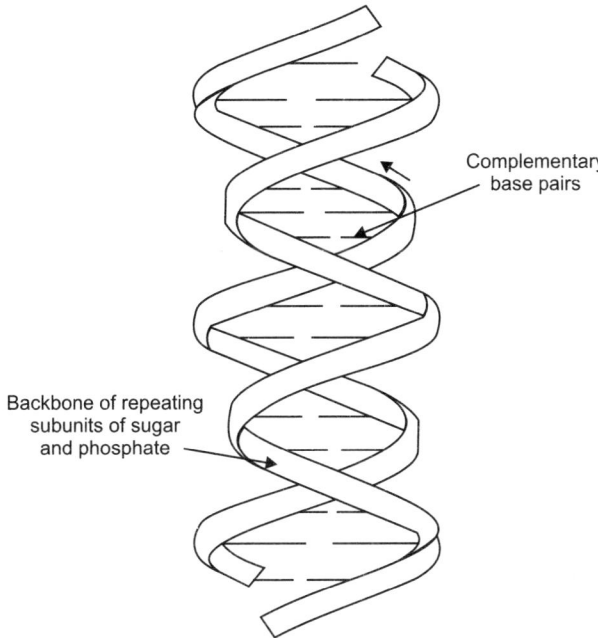

Fig. 1.1: Structure of DNA.

Improved varieties for agriculture: Animal and plant breeders have for centuries selected livestock and plants with desirable characteristics. Breeding from chosen stock is a very slow process that can be set back by the chance recombination of genes in the offspring. A breeder may select a preferred trait, only to find that it is accompanied by an equally undesirable one, which then has to be painstakingly bred out.

Traditional methods of selecting the best plants or animals from which to breed have been greatly aided by modern genetic techniques. Furthermore, it is now possible to make very precise changes to the genetic material. This can help improve the resistance to disease and environmental stress amongst crop plants and farm animals. It can also help boost agricultural productivity and enhance the nutritional status, storage properties and ease of processing of food products.

Food additives and processing aids: Enzymes are specialised proteins that are essential for life. They catalyse all biological processes and thus control metabolism in living organisms. Once extracted from living organisms, these proteins allow certain processes in food production to be conducted. For thousands of years enzymes such as rennet from animals and papain from plants have been used to enhance the flavour, texture and appearance of food. Because of the diversity of micro-organisms, it has been possible to find a wide range of microbial enzymes that are active in the conditions encountered in food processing. With genetic modification a greater range of pure and highly specific enzymes can be produced more efficiently.

These enzymes can be used to make desirable changes to food both rapidly and at relatively low temperatures, with a subsequent reduction in fuel requirements and in the environmental impact of food processing. To the consumer, the direct benefits include better flavour, texture and shelf life of food, often with a reduction in the need for processing and additives.

Modification of genetics

Cutting and pasting DNA: Special enzymes, obtained from bacteria, are an essential tool of the molecular biologist. In nature, these enzymes help bacteria fend off viral attack by precisely dissecting the foreign DNA of invading viruses. In this way, the proliferation of the viruses is restricted. Restriction enzymes (as they are known) recognise and cut DNA molecules at specific locations. Many hundreds of restriction enzymes have been isolated from different microbes and are available commercially. With restriction enzymes almost any section of DNA, and consequently any single gene, can be excised at will. The end of one DNA molecule will readily link to that of another that has been cut with the same enzyme. To join two DNA molecules permanently it is necessary to form chemical bonds along the DNA's sugar-phosphate backbone. An enzyme called DNA ligase can do this job. The function of these 'cut and paste' enzymes in assembling novel DNA molecules is obvious, but the genetic engineer's tool kit would be incomplete without one or two other enzymes. To understand their role it is necessary to appreciate how proteins are made.

A genetic intermediary: The genetic information encoded in DNA lies within the nucleus of the cell. However, proteins are not made in the nucleus, but elsewhere, at special structures called ribosomes. Before a particular protein can be made, a copy of the appropriate instructions must first be transcribed from the DNA and then ferried to the ribosomes. The copied instructions are made from mRNA (messenger ribonucleic acid). This mRNA is virtually a mirror image of the sequence of bases on one DNA strand, according to the basepairing rules. Upon arrival at the ribosomes the base sequence within the mRNA directs the construction of proteins from amino acids. A sequence of three adjacent bases in the mRNA molecule is needed to determine each amino acid in the protein. Cells that are producing a particular protein will have many identical copies of that protein's mRNA inside them. It is often easier to search for genes among the small mRNA molecules rather than along the entire length of the cell's DNA. Once a desired length of mRNA has been isolated, two additional enzymes are needed.

DNA from RNA: The enzyme reverse transcriptase assembles a single strand of complementary DNA alongside a corresponding piece of mRNA. A second enzyme (DNA polymerase) can then be used to construct a double-stranded helix using the first DNA strand as a template. DNA made in this way is called copy or complementary DNA (cDNA). A copy of a gene from a donor cell is used in genetic modification.

Gene synthesis: By the judicious use of restriction and other enzymes, molecular biologists are able to assemble DNA molecules which contain one or more genes of interest. Where a particular piece of DNA is difficult to isolate, it is sometimes possible to make it artificially using a DNA synthesiser. Under computer control, these devices string together the biochemical precursors needed to make short stretches of DNA. Of course, to programme the synthesiser it is necessary to know the sequence of bases present in the desired gene, this too can be determined automatically using a DNA sequencer. It is also possible to copy specific genes using the polymerase chain reaction (PCR). The PCR has been likened to a 'genetic photocopier'. From a very small amount of DNA millions of copies of a specific section of DNA can be made quickly. The PCR lies behind many of the spectacular successes of forensic genetic fingerprinting, where criminals have been identified from the DNA in just a few drops of blood or even a couple of cells on a cigarette butt.

Plasmids: Once a suitable DNA molecule has been constructed, it must be moved into a cell in which it can be expressed and duplicated so that it passes from one cell division to the next. For micro-organisms, one of the most successful methods involves the use of plasmids as a vehicle for transferring genes. Plasmids are rings of DNA that are found in some cells. They carry a limited set of genes and normally constitute only a few percent of a cell's total DNA. During the course of evolution, plasmids carrying genes that help their microbial hosts survive have been selected by nature. Some plasmids confer on their hosts the ability to degrade substances in the environment such as nutrients and antibiotics. Many traditional foods, such as yoghurt, cheese and other fermented dairy products, contain large numbers of living microbes that naturally harbour plasmids.

Like the DNA of chromosomes, that of plasmids can be cut with restriction enzymes and additional DNA pasted into it. The result is a ring of 'recombinant' DNA that can be put into a bacterium. Specialised plasmids can be used to ferry genes from bacteria into yeast cells or even plants. A limitation of plasmids is that they cannot accommodate DNA fragments longer than 15000–20000 base pairs. However, some harmless, specially tailored viruses can package larger DNA molecules. Such viruses have been used to transfer genes into microbes, plants and animals and even to treat human disease.

Genetically modified plants: A vector system that is used for a wide variety of plants is the plant tumour-inducing plasmid (Ti-plasmid) found in the soil bacterium *Agrobacterium tumefaciens*. Through its plasmid, *Agrobacterium* has the ability to naturally engineer plant cells so that they grow tumours that produce compounds which the bacteria need to sustain themselves. Molecular biologists use disarmed (nontumour-inducing) versions of this plasmid to introduce foreign genes of their choice into plants. Because every cell carries a complete copy of all the plant's genes in its chromosomes, it is possible to regrow an entire plant from a single modified cell. Specially modified Ti-plasmids have now been produced which help transfer fairly large genes into plants. Unfortunately, monocotyledons (including the important cereal crops) are resistant to *Agrobacterium*. *Agrobacterium* has proved especially useful when working with trees, which, because they are slow g rowing and large, are difficult to improve by conventional breeding. Apricot, plum, apple and walnut trees have all been genetically modified with *Agrobacterium*.

Gene ballistics: A procedure called ballistic bombardment has achieved success with several crops, including rice, wheat and soya. With this method, the DNA to be introduced into the plant cells is first stuck onto minute tungsten or gold particles. The DNA-coated particles are fired at high velocity into soft plant tissue, usually callus. This introduces functional DNA into the plant cells.

Electroporation: DNA can also be introduced into the thin-walled tubes which develop from pollen grains by subjecting them to microsecond pulses of a strong electric field. This technique, called electroporation, causes pores to appear momentarily in the pollen tubes through which DNA from a surrounding solution can enter. Seeds that develop from ovules fertilised with such pollen carry the introduced genes. Electroporation also works with plant cells from which the cell wall has been removed by enzyme treatment. From these naked plant cells, whole plants can be regenerated by cell culture (see Cell culture). Electroporation has also been used to transfer DNA molecules into a broad spectrum of micro-organisms.

Genetically modified animals: The DNA of animals can also be modified by genetic engineering. It is necessary to introduce genes at an early stage of development if they are to be present in all of the cells of a mature animal and be passed on to its offspring. DNA can be injected into newly fertilised egg cells through a very fine glass pipette. Only a small proportion of such injected eggs take up the new genes. The injected eggs are transferred into the uterus of a suitable foster mother. This is the only method so far that works for cows, pigs, sheep and goats. Microinjection can also be used to introduce new genes into fish eggs, but it is not suitable for the eggs of birds. However, specially modified viruses have been used to introduce, for example, disease resistance into chickens. The viruses, which are made harmless, are inserted through the shell of the egg.

Marker genes and gene probes: Whatever method is used, at best only a small proportion of treated cells take up the introduced DNA. Screening is therefore necessary to discover which cells have done so. As mentioned above, plasmids often carry genes which help the microbes that possess them break down particular antibiotics. These genes can be used as 'markers' to identify those cells which have taken up plasmids, for when the cells are placed in a growth medium which contains an appropriate antibiotic, only those with plasmids will thrive.

Several different types of marker genes exist within the plant kingdom, and these are being developed as alternatives to antibiotic markers. Other methods for identifying transferred genes include the PCR. This method enables the amplification of transplanted DNA sequences, which can then be detected using 'gene probes'. Gene probes are small fragments of singlestranded DNA or RNA which bind to complementary sequences in the DNA that is being sought out. Probes can also be used to detect the DNA of micro-organisms that might contaminate food. They are so sensitive and specific that it is possible to use probes to differentiate between strains of the same species and to determine whether particular micro-organisms are capable of producing toxins.

Genetic switches: To ensure that the recipient cell's biochemical machinery will allow introduced genes to be expressed, sequences of DNA called control regions are required. One well-studied control region switchesm on and off a gene for making β-galactosidase. This enzyme enables bacteria to metabolise lactose, but it is produced only when that sugar is present in the surrounding medium. The 'genetic switch' associated with the gene encoding β-galactosidase can be placed in front of other genes, so that the addition of lactose to broth in which the engineered cells are growing triggers expression of the new genes. There are several other genetic switches, some of which can even be thrown by a simple change in temperature.

Other Biotechnological Methods

Cell culture

Many types of plant cells from a variety of species can be cultivated *in vitro* by supplying them with nutrients and growth substances under strict aseptic conditions. Illumination may also be required. Cultures of individual cells, grown in a fermenter, can be used in preference to whole plants for producing high-value products such as natural food colourings (for example, betanin, the red colour from beetroot) and flavourings (for example, vanilla and mint oil). Problems with this technique arise from the fact that most of these substances are produced only by mature cells that are not dividing. In culture, the cells tend to be actively growing, and so do not yield large quantities of these desirable products.

A recent development with considerable promise is the cultivation of 'hairy root cultures'. By infecting plant tissues with *Agrobacterium rhizogenes* (a relative of *A. tumefaciens*), the production of root-like structures can be triggered. These transformed cells can be grown indefinitely on simple solid or liquid media. It is possible that they could be used to produce natural food flavourings and colourings without having to rely on plants grown in the open, which are subject to variations in availability and quality. However, work on hairy root cultures is still at the experimental stage.

From undifferentiated cells (or callus) grown *in vitro*, whole plants can be regenerated by adjusting the proportions of various growth factors in the surrounding medium. This enables the propagation of many thousands of plants from just a few cells, greatly accelerating the process of plant breeding. Although plants produced by cell culture should be genetically identical, the chromosomes in callus cells frequently undergo considerable rearrangement. Potato cells are especially prone to this. Such variation is a rich source of new strains, often with desirable characteristics. However, this phenomenon can be a serious drawback when great effort has been devoted to cloning a precious new plant variety. Problems of this nature beset a major project to grow oil palms in the mid-1980s. Cell culture is also used for the conservation of those plant varieties which cannot be maintained in a normal seed bank. Genetic changes are in this case highly undesirable. For some species, plantlets raised from cloned cells can be planted directly in farmers' fields. Although expensive, this method is especially suited to crops such as bananas, which do not have seeds. More often, plant cell culture is used to generate uniform, virus - free plants of high value (such as ornamental species) or the plants from which seeds are subsequently obtained. Cauliflower seeds, for example, are often obtained from such stock.

Cell culture can also be used to make 'artificial seeds' — cultured plant embryos that are subsequently coated by a protective layer of hard gel. The cost of producing such artificial seeds currently restricts their use to high value crops. However, the process is highly productive, one estimate suggests that from a 10-litre fermenter sufficient embryos could be produced to satisfy French farmers' entire annual demand for carrot seeds. The walls of plant cells are composed mostly of cellulose. A cellulase enzyme preparation can be used to remove these walls, leaving spherical protoplasts bounded by a thin membrane. The absence of a cell wall makes protoplasts especially amenable to genetic engineering by electroporation or ballistic impregnation. Animal cells may also be cultured *in vitro*, although it is not possible to grow whole animals or anything other than sheets of tissue from them. Their main use is in medical research, where, for instance, they are used to maintain viruses which cannot be cultured other than by infecting living cells.

Cell fusion

Plant cell protoplasts from different species can be fused to create complex 'hybrids' with two or more nuclei. Even protoplasts derived from different genera can be joined, it does not matter that these plants are unable to interbreed. In theory, hybrid plants with characteristics from both donor cells can be regenerated from fused protoplasts. This could provide a simple route for introducing, say, the ability to fix atmospheric nitrogen from legumes into cereals. However, enthusiasm for this technique has waned since the first successful experiments during the 1970s. Protoplasts of many species, fused or not, have proved difficult or impossible to culture, let alone regenerate into entire plants.

Cell fusion of a different type, using animal cells, has been more valuable. Antibodies can be produced by hybridomas grown *in vitro*. A hybridoma is an antibody producing cell (which is normally difficult to culture in isolation) fused with a tumour cell (which is virtually immortal in culture). Large quantities of specific antibodies (called monoclonal antibodies because they a re all identical) can be produced by cultivating hybridomas in a fermenter. These antibodies can be used in diagnostic tests of great sensitivity. For example, aflatoxins are potentially lethal compounds that are produced by fungi which contaminate stored food. Antibodies raised against aflatoxins are now used routinely in inexpensive and simple-to-use diagnostic kits. The same antibody technology is also used to detect the hormones associated with ovulation and pregnancy. Such knowledge brings significant advantages to animal breeding programmes.

FOOD PRODUCTION

Microbial Production of Food, Food Additives and Processing Aids

Fermented foods

Fermented foods are foods in which desirable changes have been produced by the action of microbes or enzymes. Over 3500 different traditional fermented foods have been catalogued. They include the bread, yoghurt and cheese that are familiar in Europe and North America. In Africa, foods made from fermented starch crops (yams, cassava, etc.) are more important, whereas in Asia products derived from fermented soya beans or fish predominate. Fermented beverages include not only the obvious alcoholic drinks, but also tea, coffee and cocoa (fermentation of the leaves or beans occurs after tea, coffee or cocoa have been harvested). Fermentation can make the food more nutritious, tastier or easier to digest or can enhance food safety. Fermentation also helps preserve food and increase its shelf life, reducing the need for additives, refrigeration or other energy-intensive preservation methods. For several thousand years, traditional fermentation (such as brewing) has given people the opportunity to cultivate microbes on a large scale and in a safe manner.

Single-cell protein: In the 1960s, protein from microbial sources (single-cell protein or SCP) was thought to have considerable potential, particularly for Third World countries. Few of the early projects aimed to produce food for humans, aiming instead to provide nutritious, low cost animal feed.

Unfortunately, most of the SCP organisms (yeasts, fungi, and bacteria) used petrochemical derivatives as a source of carbon. Even when oil prices were low, the processes were only marginally economic. Consequently the oil price rises of the 1970s ended most SCP projects. One company had an efficient large-scale process using the bacterium *Methylophilus methylotrophus* that could utilise methanol to produce a partially purified protein for animal feed. However, the cost of soya or fishmeal for animal feed remained considerably lower than that of the bacterial protein. Consequently, production of the protein ceased in the late 1980s.

Of all the SCP projects started in the 1960s, only one has survived to become a commercial success. Fungal protein (in the form of mushrooms and yeasts) has been accepted as food for generations. One company pioneered the production of fungal protein from the mycelium of *Fusarium graminearum*, which is cultivated on glucose made from maize starch. The food is marketed in the United Kingdom, Ireland, Germany, Switzerland and Belgium. By 1995, annual production is expected to reach 14000 T enough for about 28 million meals for a family of four. Unlike soya products, the fungal protein has an excellent texture, attributed to a special linear arrangement of the fungal hyphae which is similar to that of muscle fibres in meat. It has a high protein and fibre content, yet contains almost no fat, which accounts for its popularity with health-conscious consumers. The product has little taste of its own, yet absorbs other flavours readily and can be combined with a wide range of ingredients.

As a form of SCP, algae are of interest in subtropical and tropical regions, where sunlight can be utilised as an energy source. Often the production of algal protein is associated with fish farming. Algae of the genera *Chlorella* and *Scenedesmus* have been used as food in Japan, and *Spirulina* is being produced commercially in several countries, including the United States, Mexico and Israel. Often the product is sold as a high-value 'health food'.

Improved microbial cultures: All SCP processes use naturally occurring micro-organisms that have been carefully selected from wild populations. Production of the fungal protein, for example, uses a strain of *Fusarium graminearum* that was isolated from a soil sample obtained close to a research laboratory. Before that, many thousands of samples from around the globe had been laboriously screened to see whether they contained a fungus with a suitable nutritional profile, pattern of growth and other desirable properties. For the microbial production of a substance such as an amino acid, it is occasionally possible to find a rare natural variant or 'mutant' which lacks a critical step in one of its biochemical pathways and consequently overproduces the desired material. Where such mutants are hard to find, they can sometimes be induced artificially, but such techniques are slow and rely heavily on chance. The goals of genetic modification applied to strain improvement are essentially the same as those of traditional methods, but the techniques are more precise and far quicker. In addition, genetic modification allows the genetic 'solutions' developed by nature to be transferred from one species to another, so that this species can also benefit from certain genetically determined traits, for example, disease resistance.

Genetically modified yeasts: In 1990 the United Kingdom became the first country to permit the use of a live, genetically modified organism in food. This was a special strain of baker's yeast engineered to make bread dough rise faster. Existing genes were placed under the control of stronger, constitutive promoters, which helped the yeast break down sugar maltose faster than usual. Ordinary brewer's yeast (*Saccharomyces cerevisiae*) is able to utilise a variety of carbohydrates as an energy source. These include glucose, sucrose and maltose.

Although sucrose is readily available (as cane or beet sugar), glucose and the other sugars must be prepared by the enzymic breakdown of starch (see *Sweetener production*). Unlike *S. cerevisiae*, the closely related yeast *S. diastaticus* is able to grow on starch and dextrins because it makes an extracellular enzyme, glucoamylase, which catalyses the breakdown of starch. *Saccharomyces diastaticus* cannot be used directly for brewing because it produces a compound which gives beer a spicy flavour. Great interest has therefore focused on transferring the gene for glucoamylase from *S. diastaticus* into *S. cerevisiae*. Such a yeast would be better able to utilise the carbohydrate present in conventional feedstocks, which would increase the yield of alcohol and enable the production of a full-strength, lowcarbohydrate beer without the use of extra enzymes after the beer had been brewed. A modified yeast of this sort, produced by a research foundation, recently received approval for use in beer production in the United Kingdom.

Food yeasts which have been genetically modified to metabolise a wider range of sugars also help reduce the levels of polluting waste in effluent. Sugar beet molasses is widely used as the main raw material in the production of baker's yeast. Beet molasses contains, in addition to sucrose, a small proportion of raffinose. This sugar is not fully broken down and utilised by the yeast, and the unused part (melibiose) is found in waste water from factories. The new strains of yeast utilise raffinose completely, enhancing the yield of baker's yeast and leading to a cleaner effluent.

Improved starter cultures for the dairy industry: When cheese, yoghurt and similar dairy products are made it is important that only the desired micro-organisms be allowed to ferment the milk. Failure to exclude unwanted organisms can lead to poor flavours, low yields and even food poisoning. Scrupulous hygiene is required, and often the milk is heat treated to kill all or some of its microbial flora. Starter cultures of only the desired microbes are then added to the milk in sufficient volumes and in the appropriate conditions to ensure their rapid growth.

When improved starter cultures are developed, it is sometimes possible to choose between long-established, conventional techniques and the modern methods of genetic modification. However, the organisms that are produced may be identical, irrespective of the method chosen. One example where different routes led to the same result is a new yoghurt that keeps its fresh 'home made' taste for several weeks without the risk of turning acidic and bitter.

Normally, the starter culture bacteria in yoghurt turn the milk's sugar, lactose, into lactic acid within days. All that is needed to produce a yoghurt that stays mild is a *Lactobacillus* strain that cannot metabolise lactose. Such variants already exist in nature. One possibility is therefore to search for these organisms in nature. Another conventional approach consists of treating the lactobacilli in various ways to increase their mutation rate in the hope that the desired trait will accidentally be created.

A third option involves modern genetic techniques: identifying the precise gene (or genes) responsible for lactic acid production, isolating and modifying them to inactivate lactic acid production, and then replacing them in the bacteria.

All three techniques have proved successful and led to identical mild-tasting yoghurts. In many other cases, however, only the 'designer gene' approach can deliver the desired results. Although this work is still at the laboratory stage, starter cultures for yoghurt and cheese production have now been altered to provide built-in protection against other microbes that cause food poisoning. The starter cultures have been modified so that they produce a compound which breaks down the cell walls of the potentially lethal food poisoning organism *Listeria monocytogenes*. Modified bacteria could also be put into other foods as a fail-safe mechanism in case the food is stored in the wrong conditions. It might also be possible to design foods that could be protected against a range of other food poisoning organisms, such as *Salmonella*. Other modified dairy starter cultures under development produce flavour compounds for better tasting products and are able to resist attack from viruses that would otherwise ruin the production of yoghurt, cheese and similar products.

Food additives and processing aids

For many years, a wide range of food additives, supplements and processing aids have been obtained from microbial sources. These include amino acids, citric acid, vitamins, natural colourings and gums as well as enzymes. The production micro-organisms have been selected from nature to ensure that, for example, they produce high yields of a good-quality product and/or are easy to grow or that the product is easy to separate and purify. Laborious screening and selection processes are now being augmented by modern genetics, allowing quicker, more precise selection and improvement of existing production strains.

Amino acids: In foods, amino acids are used to enhance flavours and to act as seasonings, nutritional additives and improvers (in flour). They are used in both human food and animal feed. Bacteria or fungi which have been specially selected to overproduce specific amino acids are grown in large fermenters. The acids are secreted into the fermentation medium and harvested. The most important commercial products are glutamic acid, which is used as monosodium glutamate, a flavour enhancer, lysine, cysteine and methionine, which are used as supplements in animal feeds that are usually deficient in these essential amino acids, and phenylalanine, which is used in animal feed and in the manufacture of the sweetener aspartame. Most of the 20 amino acids needed to make proteins are produced by fermentation in hundred- or thousand tonne quantities. In 1990, a company made a batch of tryptophane that was subsequently implicated in a rare degenerative disorder. Concern arose because the tryptophane had been produced by a genetically engineered strain of *Bacillus*. However, it is thought that the illness was caused by insufficient purification of the tryptophane and not by the genetic modification itself.

Gums: Several gums produced by micro-organisms and plants are used widely in the food industry as thickeners, emulsifiers and fillers. Recently a process was developed to turn relatively cheap guar gum (obtained from seeds) into something akin to the more expensive locust bean gum. An enzyme (α-galactosidase) from the guar seeds is responsible for this transformation. The gene encoding the enzyme was inserted into baker's yeast, and α-galactosidase can now be produced in quantity. Production and processing of α-galactosidase by modified organisms brings several other benefits. Bacterial polysaccharides currently occupy only a small fraction of the food ingredients market, but genetically modified cells could produce a wide range of novel gums with improved properties.

Sweeteners: Aspartame, a peptide which is 160 times sweeter than sucrose, is now used in an increasing range of foods and beverages. Aspartame's sweetness was (allegedly) discovered by accident when James Schlatter, working in the United States, licked his fingers to separate a stack of papers. He had previously spilt some aspartame on his hands while working in the laboratory. Chemists at another laboratory in the United Kingdom had independently synthesised aspartame some years before but had failed to notice its sweetness. The original manufacturing process linked the two amino acids phenylalanine and aspartic acid (both products of fermentation) by chemical means. Today a more efficient enzymatic method has been developed.

Other food additives from microbes: Citric, acetic, lactic and ascorbic acids are produced in large volumes for the food industry by microbial fermentation.

Enzymes: A wide range of microbial enzymes are used by the food industry.

Plant Biotechnology

Plant breeding techniques

Since the origins of agriculture some 10000 years ago, farmers have been trying to improve their crops. Initially, people simply replanted some of the seeds from their best plants each year. Crosses between chosen individuals of the same species led to gradual improvements in the stock. Ancient South American civilisations relied on sweet corn (maize), the product of a cross between two dissimilar wild plants, for much of their diet. For thousands of years this species has been totally dependent on human intervention because it has no natural mechanism for dispersing its closely bound seeds. Around the turn of the century the scientific principles which govern inheritance started to be understood, and plant breeding began to be conducted on a more systematic basis. Artificially induced mutation and selection, like that applied to microbial strains, has yielded more productive varieties of the world's major cereal crops. Genetic mapping methods similar to those used to pinpoint human genes have enabled scientists to identify precisely those plants which carry specific desirable genes. Such techniques have and will continue to lead to major improvements in yield, quality and resistance to disease. Despite these refinements to the crude hybridisation methods of former centuries, several problems remain. For instance, although plant breeders would like to introduce specific genes into crops, conventional breeding permits the recombination only of whole sets of chromosomes. Undesirable traits are therefore likely to be inherited alongside the desirable ones. Artificially induced mutation occasionally gives rise to improvements, but more often the mutations have a deleterious effect. Traditional plant breeding remains a tediously slow process , governed mainly by the time it takes plants to grow and set seed. Genetic modification, while it will likely never replace traditional methods, complements them and offers the opportunity to overcome some of their limitations.

About 80% of contemporary research in plant biotechnology is directed towards the improvement of food plants, the remaining 20% is concerned with nonfood products such as cotton, tobacco, ornamental plants and medicines. The initial emphasis in plant biotechnology has generally been directed towards the improvement of agronomic qualities. The second and third generations of genetically modified food plants will bring direct advantages to the consumer and commercial food processor.

Pest resistance

The main emphasis in this area has been the development of crops that produce a bacterial protein from *Bacillus thuringiensis* (*Bt*). Proteins from different strains of *Bt* act on specific pests such as beetles, moths and soil nematodes but do not affect mammals. Over many years, more than 2000 T of *Bt* spores or proteins derived from them have been used as biological alternatives to conventional pesticides. However, *Bt* pesticides are expensive to produce, and because they break down quickly frequent reapplication is necessary. Several genes encoding *Bt* proteins have been inserted into a variety of food plants, including cabbage, mustard, oilseed rape, maize, potato and tomato. The hope is that crops with such built-in resistance will reduce the reliance on conventional pesticides.

Herbicide tolerance

The production of genetically engineered plants that are tolerant of or even resistant to herbicides has generated considerable interest and much controversy in recent years. Critics fear that the availability of herbicidetolerant plants will lead to an increase in the use of herbicides. The true picture is more complex. Farmers currently apply herbicides as a routine preventative measure, usually before weed infestation has progressed too far. It is argued that with the new crops farmers will need to apply chemicals only when they are really needed, confident that this treatment will be effective. This should lead to a reduction in the use of herbicides, which would seem to be against the interests of the agrochemical companies. However, many of the modified plants now under test are tolerant of the safest herbicides available, so that farmers will be able to replace older chemicals with ecologically more favourable ones. Agrochemical companies are interested in selling these more advanced products.

Tolerance of the herbicide glyphosate, for example, has been introduced into soya, maize, oilseed rape and sugar beet. Glyphosate is effective at low concentrations, is not toxic to humans or other mammals and is rapidly degraded by soil micro-organisms. Unfortunately this herbicide cannot be used on conventional crops because it would kill them. When the modified crops become available farmers could use glyphosate in preference to more hazardous chemicals.

Disease resistance

Viral diseases: Plant viruses cause major reductions in yield and adversely affect the quality of many crops. At present there is no effective chemical control for viral diseases of plants. The conventional response has been to select varieties (using classical plant breeding methods) that are naturally resistant to viral attack. The mechanism of natural resistance is not properly understood, and very few of the genes responsible have so far been identified. However, this does not mean that it is impossible to create virus resistant plants. For example, hybrids have been made between virus resistant and virus susceptible potato species. Hybrids cannot usually be created between different species. However, in this case protoplasts from two potato species were fused together, and the resultant cells were cultured on a nutrient medium to regenerate whole plants. These plants were resistant to three major virus diseases. Field trials of this type of potato plant have been carried out in several countries since the mid-1980s.

Antiviral vaccines: The insertion of genes for viral proteins into a plant inhibits subsequent infection by that virus or a closely related one. Potatoes, melons, tomatoes and cucumbers have now been modified to resist viruses in this way.

Antiviral antisense: Another route to engineering virus resistant plants is the suppression of viral genes using antisense technology. Here antisense viral genes are inserted into the plant's DNA. The plant then produces antisense RNA to block the RNA of the invading virus (98% of plant viruses have RNA rather than DNA as their genetic material). This approach is still experimental, but initial results show promise.

Genes fight fungi — and locusts: Fungal diseases, particularly of soft fruits such as tomatoes, are being combated by the transfer of chitinase genes. Unlike the cell walls of plants, those of fungi contain chitin. Therefore the novel enzyme allows the plant to fend off fungal infection without damaging its own cells. Researchers in the United Kingdom, working with Brazilian scientists, recently isolated a chitinase from a tropical cereal that can degrade fungal cell walls. An unusual feature of this protein is that it also inhibits certain digestive enzymes of locusts. This was the first demonstration of a protein that acts as both an enzyme and an enzyme inhibitor. If the gene encoding this protein could be put into major cereal crops, it might offer an attractive means of combating damage of stored grain.

Improved food quality: The antisense route

Adiverse range of food crops with qualities to benefit the food processor, retailer and consumer are currently undergoing field trials. Many of these plants have been modified so that their fruits reach the consumer in peak condition after transportation and storage. Antisense technology has been used to limit or neutralise the action of undesirable genes. The first food plants to be altered in this way were tomatoes.

Ripening research: In the mid-1970s researchers at the University of Nottingham in the United Kingdom started to investigate the complex processes involved in the ripening of fruit. The tomato was selected for this work because it has a relatively small set of genes and is comparatively easy to work with (for example, it can readily be grown by tissue culture). The aim was not to produce a better tomato but to understand the ripening process more fully so that the knowledge acquired could be applied to fruit and vegetables in general. This might allow farmers in tropical countries to benefit from the demand for exotic fruit and out-of-season temperate crops, because the fruit could be transported without refrigeration yet retain its texture and flavour. Food processors might also benefit because they would be able to produce high-quality fruit juices and purees that required less processing and fewer additives.

Pectinase key: Many genes are involved in the development of colour, flavour and texture in fruit. After several years the key to changing texture during tomato ripening was identified, an enzyme called polygalacturonase (PG). As tomatoes ripen, PG breaks down the pectin which holds cell walls together, causing the fruit to soften. A group of industry scientists, working with the Nottingham group, discovered a way to inhibit the PG gene without interfering with other aspects of the tomato's natural ripening mechanism. This meant that the softening process could be slowed down while the tomatoes continued to develop the desirable colour and flavour.

Antisense activity: The PG gene was 'neutralised' by the introduction of an antisense gene into the tomato plant. Although in early experiments some 10% of the PG remained active — sufficient to break down pectin and lead to mushy fruit — the antisense gene was inherited stably, so that cross-pollination between plants with the lowest levels of PG activity produced offspring which had just 1% of the usual enzyme activity. Researchers in the United States made similar use of an antisense PG gene. Their modified tomato is called Flavr Savr™, and has gained approval for sale to US consumers. Pending regulatory approval, European shoppers will have to wait a little longer for a homegrown alternative to watery, flavourless tomatoes.

Other slow-ripening fruit: The gas ethylene is produced by fruit as part of the natural ripening process. An atmosphere of ethylene can be used to ripen fruit artificially. Antisense technology has been used to inhibit the production of ethylene and thereby impede ripening. Among the crops to have benefited from this type of modification are broccoli, raspberries, tomatoes and the 'Euromelon'—the result of a joint project among French, Spanish, British and Greek scientists. All of these fruits can be picked when unripe and will remain in that condition until exposed to an atmosphere of ethylene.

Examples of other improvements to food quality

Potatoes pass the acid test: Human beings synthesise only half of the 20 different amino acids that are needed to make proteins. A healthy diet must therefore contain adequate amounts of the remaining 10 amino acids. The risk of malnutrition is enhanced where there is reliance on a single staple food. The proteins in some crops (for example, many of the legumes that are cultivated in Africa and South America) contain only small amounts of certain essential amino acids. These deficiencies could be overcome by inserting into affected crops genes from other organisms which have substantial amounts of these amino acids. Genetically modified potatoes with an improved amino acid content have already been developed in Israel. This work could prove highly beneficial, since potatoes are the world's fourth most important crop and an important source of nutrition in some countries.

Potato production could also be enhanced following the isolation, by Australian researchers, of a gene which doubles the yield of potato tubers from plants into which it has been inserted. Tests of the nutritional status and cooking properties of the potatoes are now underway. Other plant breeders have introduced a bacterial gene into potato plants that increases the proportion of starch in the tubers while reducing their water content. This means that they absorb less fat on frying.

Altered oils: A combination of conventional plant breeding, genetic modification and protoplast fusion has been used by numerous companies and research groups to alter the properties of oil from oilseed rape. The results range from plants that produce a greater proportion of saturated fats (suitable for margarine production) to others which yield a high-temperature frying oil that contains a low proportion of saturated fat.

Sweeter fruit with no added sugar: Sweeter crops have been produced by transplanting into them the genes for two natural protein sweeteners, monellin and thaumatin. The genes for both proteins came originally from tropical plants. Thaumatin is 3000 times sweeter than sugar.

Removal of antinutritional factors: Many plants produce chemicals to fend off attack by pests and pathogens. Unfortunately, these natural chemicals can act as antinutritional compounds. Most legume seeds, for example, contain chemicals which inhibit the action of digestive enzymes. Many legumes contain relatively high concentrations of lectins, which, if they are not removed by soaking or destroyed during cooking, can cause severe nausea, vomiting and diarrhoea. In cassava and some legumes which form the staples of several African countries, levels of cyanide-generating compounds can lead to death or chronic neurological disease if these foods are not cooked properly. In some countries, all new varieties of potatoes are screened for levels of the toxin solanine. Genetic modification could be used to produce plant varieties with low levels of these antinutritional factors.

Processed products from genetically modified plants: In January 1995 the United Kingdom government announced the food safety clearance of several processed products obtained from three genetically modified crops: oil derived from oilseed rape that had been genetically modified to confer male sterility, thereby preventing self-fertilisation and allowing the production of vigorous high-yielding hybrids, processed products (oil, meal and protein fractions) from soya plants modified to be resistant to the herbicide glyphosate, and tomato paste from tomatoes modified to slow down the process of softening. Unlike the antisense tomatoes described above, these tomatoes had been modified using 'sense' technology. They contain a truncated polygalacturonase gene in the 'sense' orientation, which, for reasons that are not entirely clear, has the same effect as the antisense polygalacturonase gene.

Long-term goals

Nitrogen fixation: The capacity to fix atmospheric nitrogen into a form that can be taken up by plants in the same way they use nitrogen in fertilisers—an ability possessed by *Rhizobium* bacteria that live in close association with leguminous plants—has been a major preoccupation of researchers. Despite decades of intensive study, the goal of a self-fertilising cereal is still elusive. Doubts have now arisen as to whether this aim is realistic and should be a high priority, for two reasons. The first is that to receive ammonia from its associated bacteria, the plant has to provide the bacteria with 'food' so that they can survive in small nodules on its roots. This food is provided at the expense of the plant's productivity. Legumes have had millions of years to evolve a mutually beneficial physiology with their Rhizobia, and it seems unlikely that such relationships could be artificially established with cereals, indeed there is evidence to suggest that nitrogen-fixing cereals could have decreased productivity. The second reason concerns the root nodules: these complex structures would have to be provided on the plant's roots. This is likely to involve the manipulation of very many genes. A better approach might be to transfer nitrogen-fixing abilities into *Agrobacterium*, which infects a wider range of plants. Unfortunately this does not include cereals, which form the majority of food crops. For the foreseeable future it would probably be more profitable to try to improve existing nitrogen-fixing species, especially tropical legumes and their associated bacterial strains. Other readily available options include the traditional improvement of crop management, such as rotation between legumes and cereals.

Drought resistance: Drought-resistant crops would bring obvious advantages to farmers in areas of low rainfall. Numerous organisms are able to withstand dehydration without harm and can be resuscitated simply by adding water, dried yeast is a familiar example. A desert plant, appropriately named the resurrection plant, is similarly able to withstand prolonged periods of drought. Both organisms rely for this remarkable ability on a sugar in their cells called trehalose. In 1993, the first successful attempt was made to place the genes involved in the production of trehalose into potato and tomato plants. The results of this work have yet to be evaluated.

Animal Biotechnology

Animal nutrition

Probiotics: For many years low levels of antibiotics have been added to animal feed, with the result that the animals gain extra weight. This is thought to be because the antibiotics kill detrimental bacteria in the animal's gut. An alternative approach is to add live, nonpathogenic bacteria to the animal feed. These organisms (termed probiotics) are thought to compete with the detrimental species, leading to a similar weight gain. Genetically modified bacteria might one day be used as probiotics to improve the health and efficiency of feed to weight conversion of farm animals.

Rumen bacteria: The nutrition of ruminant animals (for example, sheep and cattle) is highly dependent on the bacteria which live in their stomachs, enabling them to digest plant material effectively. Genetically modified bacteria could be introduced into the rumen by adding them to animal feed, enabling the animals to make better use of a wider range of food plants.

Silage: Silage is grass which has been fermented by naturally occurring bacteria to enhance its nutritional value. Microbial inoculants are now used to improve the formation of silage, but there is a possibility that genetically modified species could be used to further improve the process.

Bovine growth hormone: Bovine somatotropin (BST) is a protein produced by cows that has a variety of physiological effects, including the regulation of milk yield. BST is naturally present in all cow's milk. The gene for BST has been cloned into bacterial cells so that it can be produced in bulk. Injections of BST can be used to enhance milk yield. The hormone also improves growth rate and protein-to-fat ratios in meat. There has been a moratorium on the general use of BST in countries of the European Union for several years, although it is approved elsewhere, including the United States. BST is probably one of the most severely tested substances in the history of the US Food and Drug Administration, but its use has been opposed in some quarters. Such opposition is based partly on uncertainties over its effects on animal health, and on worries that the hormone might contaminate meat or milk. There has also been opposition on political and economic grounds, as a surplus of milk is already produced in many developed countries. The advantage of BST is that it would allow the same volume of milk to be produced by fewer cattle.

Animal breeding and health

Classical animal breeding has done much to improve the productivity and well-being of farmed livestock. Changes in characteristics such as maturity, fecundity and the distribution of muscle tissue are noticeable in many modern breeds compared with their wild ancestors and old domestic breeds. The achievements of traditional breeding are now being augmented by modern genetic analysis.

The majority of the desirable features in livestock seem to be controlled by many genes, each with a small effect, working in concert. The modification of animals by genetic engineering is still in its infancy, so the genes that should be altered to improve animal productivity or health are still difficult to predict. Much of the work so far has concentrated on simple changes, such as the introduction of growth hormone genes. The results, some of which are described below, were not completely foreseen. Most of the transgenic animals created to date are mice which are used in medical research. This lies outside the scope of the current discussion. However, this work is likely to benefit human *and* animal health.

Agriculture in Europe and North America already produces sufficient food for the indigenous population. The real benefits from improved animal production might be seen in the Third World. For example, it may one day be possible to introduce disease resistance into otherwise vulnerable animals. There are well-advanced porcine, ovine and bovine genome projects, which parallel similar efforts to map and sequence the entire human genome. The Bovine Genome Project could result in, for instance, resistance to trypanosomiasis being introduced into more productive breeds of cattle from their naturally resistant African counterparts. However, in the immediate future more benefit is likely to come from the development of new diagnostic agents, vaccines and therapeutic agents than by modifying the animals themselves.

Genetically modified pigs: Genetically modified pigs were first produced in the United States in 1988. Pig embryos were injected with a gene encoding growth hormone. Genetic 'switches' were included so that a change in the animals' diet would greatly increase the levels of hormone circulating in the blood. Contrary to some reports, the result was not giant pigs, but animals with very lean muscle tissue. Unfortunately, the high levels of growth hormone led to gastric ulcers, arthritis, kidney disease and other undesirable conditions. Insufficient attention had been paid to what was already known about growth hormone, namely, that it is generally produced only for limited periods during an animal's life, not continuously.

Classical breeding methods supplemented by modern genetic mapping have produced more desirable results. For example, the genes which control leanness in pigs are linked to those that may cause sudden death during a period of stress. Humans also carry similar genes and may, for example, die inexplicably when undergoing routine surgery, although such a genetic predisposition is very rare in humans. Selection for leanness in pigs over many generations has also increased the frequency of the stress - related genes, but modern breeding methods are allowing such genes to be bred out.

Herman the transgenic bull: The world's first transgenic bull was bred in the Netherlands. Herman, the animal in question, became famous in 1992 when the Dutch parliament had to decide whether or not he would be allowed to mate. Scientists at the company that bred Herman wished to reduce cows susceptibility to mastitis. To do this, they transplanted an extra copy of the gene for a milk protein called lactoferrin into several bovine embryos. Lactoferrin is part of the mammalian (including the human) defence against bacterial infection. Unfortunately, the only embryo to take up the transplanted gene produced not a cow, but a bull. Undeterred, the company hoped that Herman would pass on the gene to his daughters and that they would be less likely to suffer from mastitis. The Dutch parliament eventually gave its consent to the request from the company. Herman has sired several offspring, at least one of which carries the extra lactoferrin gene.

Cloned sheep: Breeders of sheep in the United Kingdom and New Zealand are trying to disentangle the role of genes and the environment in an effort to improve the quality of their livestock. They intend to do this with the help of identical sheep placed in several flocks. The sheep are produced by splitting an embryo when it consists of just eight cells. This happens naturally with the production of identical offspring, such as twins. Each of the eight cells is implanted in a foster mother. Differences between the identical sheep will be due solely to environmental influences (such as diet or exercise). The hope is that this comparison will enable sheep breeders to conduct their work on a more rational basis.

Chickens: Because viral infections are often a problem in intensive poultry production, a lot is known about the viruses which affect chickens. This knowledge has been useful to genetic engineers, because modified versions of the viruses themselves can be used to combat disease. Birds have been given immunity to fatal viruses that proliferate in crowded broiler houses by adding genes for viral proteins to the chicken genome.

Transgenic fish: Compared to mammals and birds, fish are still relatively wild, with little history of improvement through selective breeding. There are very many fish species and consequently a large pool from which genes can be drawn for breeding programmes. Fish, especially when farmed, are prone to infection. Recent genetic research on fish has concentrated on inserting genes that impart resistance to disease. Successful modifications have also been made to genes that influence growth, which could lead to more productive fish for aquaculture. Other possibilities are the introduction of 'antifreeze' genes to extend the range of Atlantic salmon into colder waters and to delay the breeding season of fish so that they gain weight. Concerns have been raised about the wisdom of increasing the geographical range of a predator at the top of a food web. Work in this area must proceed with careful consideration of its possible ecological consequences.

REGULATORY, SAFETY AND SOCIO-ECONOMIC CONSIDERATIONS

To date, relatively few products of genetic modification have reached the supermarket shelves. Genetic modification of farm animals is in its infancy, and work with the major crop plants, especially cereals, has proved more difficult than was at one time anticipated. Outside the laboratory, modified plants and microbes have so far been restricted to closely monitored field trials. Nevertheless, during the coming century the impact of food biotechnology is likely to be both profound and far reaching. For this reason, there has been much debate about the potential social and economic implications of biotechnology. The safety of new developments has been examined with rigour, and legislation has been drafted to protect the interests of the public.

Food Safety

In addition to issues of environmental safety, the implications of modern biotechnology for food safety need to be considered. Recognising the numerous benefits to health, nutrition, food preservation and food production that biotechnology could bring, the Organisation for Economic Cooperation and Development (OECD) established a working group of national experts to consider the safety implications of modern food biotechnology. The intention was to exchange ideas, data and information among experts to enhance international cooperation in the field.

The working group considered numerous examples of how the safety of novel foods and food components had been evaluated in the past and established some concepts and principles that underpin the safety evaluation of foods derived by modern biotechnology. These principles have been widely accepted and are similar to recommendations made by other influential groups such as the World Health

Organisation and the Food and Agriculture Organisation of the United Nations. The major conclusions of the OECD report are summarised below.

Food is considered safe if there is reasonable certainty that no harm will result from its consumption under anticipated conditions. Historically, food prepared and used in traditional ways is considered safe on the basis of long-term experience, even though it may naturally contain harmful substances. In principle, food is presumed to be safe unless a significant hazard has been identified.

Modern biotechnology broadens the range of genetic changes that can be made to food and widens the range of possible sources of food, although this does not inherently lead to food that is less safe than that developed by conventional techniques. Therefore the evaluation of foods derived from modern biotechnology does not require a fundamental change in established principles of food safety, nor does it require a different standard of safety.

Furthermore, modern biotechnology provides precise techniques for the direct and focused assessment of safety (for example, by the detection of minute amounts of contaminating material), which can be usefully applied to foods derived from both modern and traditional methods.

Substantial equivalence

The OECD report said that the most practical method to establish food safety was to consider whether a novel food (or food component) was *substantially equivalent* to an analogous conventional food product, where one existed. Account should be taken of the processing (such as cooking) that the food might undergo, as well as how much food was to be consumed, by whom and the dietary pattern.

To demonstrate substantial equivalence, a number of factors have to be considered, such as:

- The characteristics and composition of the conventional food to which the new one is to be compared
- Knowledge of the component parts of the new product or organism, such as any introduced genes, the method used to introduce the new genetic material and how that new genetic material is expressed.
- The characteristics and composition of the new product or organism compared with the existing food or food component.

If the novel food is judged to be substantially equivalent to an existing food, it is treated in the same manner as its conventional counterpart. Where new classes of foods or food components are introduced it is more difficult to apply the concept of substantial equivalence. Here experience gained in the evaluation of similar materials is taken into account. Where a product is thought not to be substantially equivalent to an existing one, further investigations focusing on the identified differences are required. Totally new foods, where no similar materials have ever been consumed, must be evaluated solely on the basis of their own composition and properties.

Genetically modified potatoes

As an example of the application of the principle of substantial equivalence, potatoes are an established part of the human diet. They can contain toxic alkaloids, but people generally know how to prepare them and avoid eating green potatoes, which contain significant amounts of alkaloids. Potatoes are often infected by naturally occurring viruses, but these do no harm to humans and have a long history of human consumption. If a potato were genetically modified with one of these viruses so that it produced viral protein at levels comparable to those from naturally infested potatoes, it would be considered to be substantially equivalent to the infected potatoes that have a long history of safe use and consumption. This theoretical analysis applies only to viral proteins in the parts of the potato plant that are traditionally consumed. It also assumes that the insertion of the viral coat protein gene does not lead to secondary effects through, for example, interruption of coding sequences within the plant's genome.

Chymosin from genetically modified bacteria

As described in the section on cheese manufacture, the enzyme chymosin can be obtained from genetically modified micro-organisms. Chymosin from the bacterium *Escherichia coli* is obtained from a completely different organism and by a method that is completely different from its traditional counterpart, which comes from an animal. Thus the types of potential impurities differ and the characteristics of the enzymes may differ, too. To determine whether the two preparations were substantially equivalent, the US Food and Drug Administration compared the activities of the two enzymes and evaluated whether impurities in the product from the modified organism affected its safe use. After a battery of tests, both enzymes were found to be substantially equivalent in safety and function.

Safety implications of marker genes

The current generation of genetically modified organisms frequently contains marker genes, some of which encode resistance to antibiotics. In such cases it is important to consider whether food derived from an organism with such a gene would be substantially equivalent to a conventional product. Among the questions that must be asked are whether the marker gene encodes a protein such as an enzyme which catalyses the breakdown of an antibiotic and, if so, what levels of that protein (if any) would be expected in the food. Does the gene encode resistance to a clinically useful antibiotic, and if so, would ingestion of the food while someone was being treated with that antibiotic interfere with the drug's effectiveness? Finally, is it at all likely that the gene could be transferred to other organisms such as microbes in the food or in the intestine of the consumer? For food enzymes produced by genetically modified organisms, the Organisation for Economic Cooperation and Development (OECD) considered that levels of marker genes or their products in food would almost always be biologically insignficant. Would the same be true of plants with marker genes? With these concerns in mind, the Food Safety Unit of the World Health Organisation (WHO) organised a workshop. WHO recognised that there was a need for marker genes, which may have no function in the final product, but that at present it was impractical for such genes to be removed from modified organisms after they had fulfilled their function. The presence of marker genes *per se* in food was not thought to constitute a safety concern. It was also noted that the number of marker genes in plant varieties approaching commercialisation was restricted to two antibiotic resistance markers and to a few herbicide tolerance markers. The workshop concluded that there is no recorded evidence for the transfer of genes from plants to micro-organisms in the gut. This is highly unlikely to occur for theoretical reasons, since the mechanism for controlling gene expression is fundamentally different in bacteria and plants. Unless the gene transferred into the plant were under the control of a bacterial system (and therefore unable to be expressed in the plant), there would be no mechanism for expression of the gene, even if it were transferred to the gut bacteria.

Regulatory approval

European Union: Within the countries of the European Union, regulations have been proposed to cover the use of novel foods or food ingredients. In particular, the regulations will apply to modified or new molecules, to any products that have not previously been eaten by humans to a significant degree, to genetically modified organisms and their products, and to novel processing methods. This proposal is still under discussion by the European Union Council of Ministers.

United States: In the United States, foods are regulated by three bodies. The Food and Drug Administration (FDA) is concerned with food safety relating to new plant varieties, dairy products, seafood, food additives and processing aids, whereas the Department of Agriculture regulates meat and poultry products and field tests all genetically modified plants. The Environmental Protection Agency is responsible for pesticide chemicals, and it may therefore have to approve new plant varieties resistant to attack by pests. Of the three agencies, the policy of the FDA with respect to new plant varieties is most clearly defined at present. The FDA regards the key factors in reviewing safety to be the characteristics of the food and its intended use, rather than the fact that new methods have been used in its production. Novel food products are not subject to regulatory approval if the constituents of the food are the same or substantially similar to substances currently found in other foods (such as proteins, fats, oils and carbohydrates). For example, if a gene from a banana were transferred to a tomato, approval would not ordinarily be required before that food was placed on the market. However, if a sweetening agent that had never been an ingredient of any other food were added to a variety of grapefruit, the novel food would need regulatory approval. The sweetener would be regarded as a food additive and therefore be subject to other, more stringent regulations.

Deliberate release of modified organisms

The conduct of laboratory-based genetic modification of micro-organisms, while requiring care, presents few hazards other than those normally associated with the use of similar micro-organisms. Such risks are readily contained by well-established and tested laboratory procedures. However, when a proposal is made to use a modified organism (microbe or plant) beyond the confines of the laboratory, particularly a robust organism capable of withstanding the rigours of the natural environment, a different set of safety questions arises.

What might the risks be? The deliberate planting or dispersal of genetically modified organisms outside the laboratory could cause unforeseen effects if, for example, modified plants or microbes were able to:

- Avoid limiting factors that regulate naturally occurring populations and thereby change the usual balance between populations (for example, an increased tolerance of drought, high temperatures or high salt concentrations could achieve this).
- Produce new compounds, (for example, insecticides, herbicides).
- Transfer their newly inserted genetic material to other plants.

Note that these changes, should they occur, would not necessarily be harmful, indeed some, such as the production of toxins or improved nitrification, might be the purpose of the modification in the first place. The nature of the potential environmental impact falls into two broad categories: effects which result from the activities of the modified organism itself and those which might arise should genetic material be transferred into other organisms.

Making predictions: Predicting the influence of any organism (genetically modified or not) introduced into a new habitat is difficult. However, it is important to recognise that nearly all agricultural species are introductions which have been cultivated by humans, often for many thousands of years. In this time they have been subjected to enormous random genetic change—selection by nature and by human intervention. It is unlikely that the deliberate and targeted genetic modification of such species would result in hazards any greater than those that have already occurred.

Field trials: More than a thousand field trials of genetically modified crops have been conducted worldwide. Extensive research with the specific aim of determining the ecological effects of modified crops has been carried out in the United Kingdom. The Planned Release of Selected and Modified Organisms research took 3 years, and was funded by European industry and the United Kingdom government. Independent public sector ecologists conducted the work as they saw fit. The results suggested that the genetically modified plants (potatoes, maize, oilseed rape and sugar beet) presented no intrinsic environmental threat. Genes were not readily transferred from the modified plants to their wild relatives, and there was no increase in the ability of the modified crops to invade natural habitats. The ecologists warned that their findings should not be simply extrapolated to other crops, but suggested that a cautious step-by-step approach to planned releases has much to commend it. Similar work involving more than 200 field trials and more than 500 locations and transgenic lines has been carried out in France.

European regulations governing deliberate releases: All European Union nations are party to a European Council directive governing the deliberate release of genetically modified organisms (GMOs) into the environment. This includes provisions for marketing food products containing or made from GMOs. Under this directive all of the member states have agreed to put into place laws and regulations to ensure that common policies on GMOs are adopted throughout the union. The regulations do not apply to organisms obtained by conventional breeding techniques (for example, mutagenesis) that have been applied over a long time. The directive requires that proposed releases should be considered individually. People intending to make a release must first seek permission from the appropriate national authority. The application must contain a technical dossier including information about the personnel concerned and their training and a full environmental risk assessment with details of appropriate safety and emergency responses.

In certain cases interest groups or the public may be consulted about a proposed release. Action has been taken in several countries to ensure that there is public interest representation on the relevant advisory bodies and that the public has access to information about releases. Companies or others may ask for some information to be kept confidential on grounds of commercial competitiveness, but in such cases they must give verifiable justification. Following the destruction of some field trials in the Netherlands, there a re now provisions in Dutch law allowing for the location of field test sites to be kept secret, but normally this is not done.

A general principle is that releases should be conducted a step at a time and that the scale of release should be increased gradually only as it becomes apparent that it is safe to do so. Once a release has been made, a report on it must be sent to the appropriate authorities, with particular reference to any risk to human health or the environment. The directive makes provisions for the exchange of information on releases between member states and for the publication of details. It should be emphasised that the European directive discussed above is based on a precautionary approach. As more experience has been gained, the risks presented by genetically modified plants appear to be smaller than was first anticipated. International research in this area is leading increasingly to the conclusion that there is no greater risk from introducing genetically modified organisms into the environment than is presented by similar introductions of organisms produced by conventional breeding techniques.

Social, Economic and Other Effects

Biotechnology could bring considerable benefits to world agriculture, but there could be accompanying disadvantages. Insufficient space permits only an indication of some of the main issues here. Many of the consequences of modern biotechnology could be similar to those produced by existing trends, such as the shift towards larger farms and more capital intensive farming systems. This tends to favour wealthy farmers in the Northern Hemisphere who can invest in new technologies rather than those in the impoverished South. Biotechnology may reduce developed countries dependency on products imported from developing countries, with adverse effects in the producer nations. Concerns in the Western world about overproduction are unlikely to be shared by countries whose population growth far outstrips the capacity of farmers to provide sufficient food. However, if biotechnology brings greater agricultural productivity to developing nations, there could be major shifts in employment patterns, with adverse effects on the countries' stability.

Some people also question whether developments provided by the new biotechnology are really needed—the so-called fourth hurdle (after the safety, efficacy and quality of a new product have been demonstrated). However, the only place where the market need for a product can be assessed properly is the marketplace. The requirement to demonstrate need in advance could severely restrict innovation.

Biodiversity

Biotechnological methods (including the genetic modification of plants) have already led to plants with greatly improved characteristics compared to the old cultivars produced by conventional methods. Excessive planting of a limited number of cultivars would lead to increased genetic uniformity. This might be accompanied by a greater risk of widespread epidemics of plant diseases, and it could lead to a reduction in biodiversity (in crop plants, weed species, insects and microflora in the fields in question).

Conversely, biotechnology might *encourage* the production of a far wider variety of new crops , *increasing* biodiversity. In the developed nations, the 'green revolution' of the 1960s and 1970s has been so successful in increasing the yield of food that farmers in the United States and Europe are paid to take land out of cultivation, again increasing diversity. The biotechnological revolution may further accelerate this trend.

On a wider scale, biotechnologists depend on the genetic resources of the world for their raw materials, and thus have a vested interest in maintaining biodiversity. The techniques of plant tissue culture might be used to help maintain rare and endangered species. Some argue in favour of patents on living organisms, as they might allow Third World countries to obtain payment for the use of their genetic resources. The need for public involvement Although the potential of genetic engineering in the food sector is often exaggerated, many improvements are possible in the areas of food quality, variety and safety. The economic effects of such developments could be considerable, and there are opportunities to reduce the environmental impact of food processing and production, making a positive contribution towards more sustainable development. Much will depend on the attitudes of consumers. Some of the new foods will be indistinguishable, at least on the supermarket shelf, from those currently available. In other products, genetic engineering will lead to significant and noticeable improvements, such as better flavour. The provision of appropriate information and public understanding are therefore a crucial concern. Although concerns about some aspects of biotechnology are justifiable, it is important that these be judged against the enormous benefit that technologies such as genetic engineering could bring.

In a democratic society public involvement is essential. The strong case for the involvement of the informed lay public on decision-making bodies is reflected in the legislative framework for biotechnology devised in several countries. Some sectors of the food industry are willing to search for applications of genetic modification that are mutually acceptable to consumers, legislators, food producers and processors. It is to be hoped that this monograph will contribute to that discussion.

Chapter 2

Micro-organisms Associated with Food

INTRODUCTION

Food serves as an interacting medium between various living species because it is a source of nutrients for humans, animals as well as micro-organisms. This is a natural consequence of cohabitation. Human and animal food is basically derived from plant and animal sources. Food fit for human consumption is also a medium for the growth and activity of micro-organisms. Hence human food is always associated with a variety of micro-organisms. Since the primary function of micro-organisms is self perpetuation, they use the human or animal food as a source of nutrients for their own growth and activity. Microbial activity in a food can be beneficial in certain cases but in most cases it leads to deterioration of the food and renders it unfit for human consumption. Micro-organisms can be used as processing aids in the production of fermented foods. A variety of food chemicals and additives may be produced by fermentation involving select species of micro-organisms.

Pathogenic micro-organisms grow in the food utilising the nutrients in the food and produce toxins which are detrimental to the health of the consumer when such food is consumed. Food also serves as a vector or medium for certain pathogens that cause food infections and diseases.

The metabolic activity of various micro-organisms not only utilises the nutrients in food but also causes the spoilage of food through undesirable enzymatic changes affecting the quality of the food.

BACTERIA, YEAST AND MOLDS

Thousands of genera and species of micro-organisms have been identified and classified. Several hundreds of these are associated in one way or other with food products. Many of them are of industrial importance as they find use in the production of new foods and food chemicals by fermentation and also in the preservation of food products. Micro-organisms are capable of spoiling food and causing diseases. The micro-organisms which are of importance in food microbiology include bacteria, yeasts and molds.

Bacteria

Bacteria are microscopic single-celled organisms that thrive in diverse environments. They can live within soil, in the ocean and inside the human gut. Humans relationship with bacteria is complex.

Sometimes they lend a helping hand, by curdling milk into yogurt, or helping with our digestion. At other times they are destructive, causing diseases like pneumonia and Methicillin-resistant *Staphylococcus Aureus* (MRSA). Based on the relative complexity of their cells, all living organisms are broadly classified as either prokaryotes or eukaryotes. Bacteria are prokaryotes. The entire organism consists of a single cell with a simple internal structure. Unlike eukaryotic DNA, which is neatly packed into a cellular compartment called the nucleus, bacterial DNA floats free, in a twisted thread-like mass called the nucleoid.

Bacterial cells also contain separate, circular pieces of DNA called plasmids. Bacteria lack membrane-bound organelles, specialised cellular structures that are designed to execute a range of cellular functions from energy production to the transport of proteins. However, both bacterial and eukaryotic cells contain ribosomes. These spherical units are where proteins are assembled from individual amino acids, using the information encoded in a strand of messenger RNA.

Yeasts

Yeasts consist of one cell, and belong to the taxonomic group called fungi, which also contains molds. There are many species of yeasts. The most common yeast known is *Saccharomyces cerevisiae*, which is used in the baking and brewing industry. Yeasts also play an important role in the production of wine, kefir and some other products. Most yeasts used in the food industry are round and divide themselves through budding. This budding is a characteristic used to recognise them through a microscope. During budding the cells appear in 8-shaped forms.

Yeasts need sugar to grow. They produce alcohol and carbon dioxide from sugar. This reaction makes yeast so important for the food industry. Yeasts also produce pleasant aroma components. These aroma compounds play a very important role for the flavour of the end product. In beer the yeast is needed to produce the alcohol and the carbon dioxide for the brim. In the bread industry, both alcohol and carbon dioxide are formed, the alcohol evaporates during baking.

Yeasts can be found everywhere in nature, especially on plants and fruits. After fruits fall off the tree, fruits become rotten through the activity of molds, which form alcohol and carbon dioxide from the sugars in it. Yeasts are grown in the industry in big tanks with sugary water in the presence of oxygen. When the desired amount of yeast is reached the liquid is pumped out, and the yeast is then dried. Nothing else is added in the production of yeast.

Molds

Molds are microscopic fungi that live on plant or animal matter. No one knows how many species of fungi exist, but estimates range from tens of thousands to perhaps 300000 or more. Most are filamentous (threadlike) organisms and the production of spores is characteristic of fungi in general. These spores can be transported by air, water, or insects. Unlike bacteria that are one-celled, molds are made of many cells and can sometimes be seen with the naked eye. Under a microscope, they look like skinny mushrooms. In many molds, the body consists of:

- Root threads that invade the food it lives on.
- A stalk rising above the food.
- Spores that form at the ends of the stalks.

The spores give mold the colour you see. When airborne, the spores spread the mold from place to place like dandelion seeds blowing across a meadow. Molds have branches and roots that are like very thin threads. The roots may be difficult to see when the mold is growing on food and may be very deep in the food. Foods that are moldy may also have invisible bacteria growing along with the mold. Bacteria, yeasts and molds attack virtually all the constituents of foods. Depending on their nature and availability of enzymes some ferment sugars and hydrolyse starches and cellulose while other hydrolyse fats and produce rancidity.

PRIMARY SOURCES OF MICRO-ORGANISMS COMMONLY ASSOCIATED WITH FOOD

The primary source of micro-organisms associated with food is the environment with which the species are associated. The genera and species that cause deterioration of foods are found normally in food products. Each genus has its own particular requirements, and each is affected in predictable ways by the parameters of its environment. Eight environmental sources of micro-organism found in food are given below:

1. Soil and water.
2. Plants and plants products.
3. Food utensils.
4. Intestinal tract of human and animals.
5. Food handlers.
6. Animal feeds.
7. Animal hides.
8. Air and dust.

Soil and water: These two environments are place together because many of the bacteria and fungi that inhibit both share a lot in common. Soil organism may inter the atmosphere by the action of wind and latter enter the water bodies when it rains. They also enter water when rainwater flows over soils into bodies of water. Aquatic organisms can be deposited onto soils through the actions of cloud formation and subsequent rainfall. *Alteromonus* spp. is aquatic forms that requires seawater salinity for growth and would not be expected to persist in soils.

Plant and plant products: Many or most soil and water organisms contaminate plants. Relatively, only a small numbers of organisms find the plant environment suitable to their overall well-being. Among others that are commonly associated with plants are bacterial plant pathogens in the genera *Corenebacterium, Curtobacterium, Pseudomonas* and *Xanthomonas* and fungal pathogens among several genera of molds.

Food utensils: When vegetables are harvested in containers and utensils, one would expect to find some or all of the surface organisms on the products, contact surfaces. In a similar way, the cutting block in a meat market along with cutting knives and grinders are contaminated from initial samples and this process leads to build up of organisms.

Intestinal track of human and animals: This flora becomes a water source when polluted water is used to wash raw food products. The intestinal flora consist of many organisms that do not persist as long in water as do others and notable among these are pathogens such as *Salmonellae*.

Food handlers: The micro flora on the hands and other garments of handlers generally reflects the environment and habits of individuals, and the organisms in question may be those from soils, water, dust, and other environmental sources. Other important sources are those that are common in nasal cavities and the mouth and the skin and those from gastrointestinal track that may enter foods through poor personal hygienic practices.

Animal feeds: This is an important source of *salmonellae* to poultry and other farm animals. Listeria monocytogenes is found in dairy and meat animals. The organisms in dry animal feed are spread throughout the animal environment and may be expected to occur on animal hides.

Animal hides: The types of organisms found in raw milk can be reflection of the flora of the udder when proper procedures are not followed in milking. From both the udder and hide, organisms can contaminate the general environment.

Air and dust: Many organisms are found in air and dust in a food processing operation. Among fungi, a number of molds may be expected to occur in air and dust along with some yeast. In generals, the types of organisms in air and dust would be those that are constantly reseeded to the environment.

FACTORS INFLUENCING MICROBIAL ACTIVITY

The growth and activity of the micro-organisms depend on the nature of the food and its composition. The factors that govern the microbial activity in a food include both intrinsic and extrinsic factors. The intrinsic factors include pH, water activity, oxidation-reduction potential, nutrients content and presence of inhibitors in the food. The extrinsic factors include temperature, relative humidity and the atmosphere surrounding the food. These factors operate individually as well as in combination and affect the growth and activity of micro-organisms in foods. Some intrinsic factors are interlinked with some extrinsic factors. For example, water activity rises with increasing temperature, there is an increase in water activity of 0.03 with a 10°C rise in temperature.

Intrinsic Factors

pH

The intracellular pH of any organism must be maintained above the pH limit that is critical for that organism. The control of intracellular pH is required in order to prevent the denaturation of intracellular proteins. Each organism has a specific requirement and pH tolerance range, some are capable of growth in more acid conditions than others. Most micro-organisms grow best at neutral pH (7.0). Yeasts and molds are typically tolerant of more acidic conditions than bacteria but several species of bacteria will grow down to pH 3.0. These species are typically those that produce acid during their metabolism such as the acetic or lactic acid bacteria.

Bacterial pathogens are usually unable to grow below pH 4.0. The type of microbial growth typically seen in a particular food is partly related to the pH of that product. Fruits are naturally acidic, which inhibits the growth of many bacteria, therefore spoilage of these products is usually with yeasts and molds. Meat and fish however have a natural pH much nearer neutral and they are therefore susceptible to the growth of pathogenic bacteria. Individual strains of a particular species can acquire acid resistance or acid tolerance compared to the normal pH range for that organism. For example acid-adapted *Salmonella* have been reported that are capable of growth at pH 3.8.

There is a broad distinction between high and low acid foods, with low acid foods being those with a pH above 4.6, and high acid, below this. This is because pH 4.6 is the lower limit for the growth of mesophilic *Clostridium botulinum*. Foods with a pH greater than 4.6 must either be chilled, or if ambient stored, undergo a thermal process to destroy *C. botulinum* spores, or have a sufficiently low water activity to prevent its growth.

Different foods tend to spoil in different ways. For example, carbohydrate rich foods often undergo acid hydrolysis when they spoil, this usually reduces the pH, and tends to reduce the risk of pathogen growth. This principle is used in the fermentation of dairy and lactic meat fermentations. In contrast, protein-rich foods tend to increase in pH when they spoil, making them possibly less safe, as the pH rise to the zone where more pathogens can grow. Micro-organisms are able to grow in an environment with a specific pH, as shown in Table 2.1.

Table 2.1: Growth of micro-organisms in an environment.

Micro-organisms	Minimum pH value	Optimum pH value	Maximum pH value
Gram +ve bacteria	4.0	7.0	8.5
Gram –ve bacteria	4.5	7.0	9.0
Yeasts	2.0	4.0–6.0	8.5–9.0
Molds	1.5	7.0	11.0

Some bacteria are:

- Acidophilic bacteria, e.g. Lactic acid bacteria (pH 3.3–7.2) and acetic acid bacteria (pH 2.8–4.3).
- Basophilic bacteria, e.g. *Vibrio parahaemolyticus* (pH 4.8–11.0) and *Enterococcus spp* (pH 4.8– 10.6).
  › Increasing the acidity of foods either through fermentation or the addition of weak acids could be used as a preservative method.

Moisture content and water activity

The water requirements of micro-organisms are generally expressed in terms of water activity of the substrate in which they grow.

Water activity (a_w)

- Water activity is a measure of the water available for micro-organisms to grow or reactions to take place, i.e. measure of the amount of water disposable for the micro-organisms.
- It is a ratio of water vapour pressure of the food substance to the vapour pressure of pure water at the same temperature.
- Water activity is expressed as:
- Water activity (a_w) = P/P_w where P = water vapour pressure of the food substance and P_w = water vapour pressure of pure water (P_w = 1.00).
- The growth of micro-organisms is limited due to minimum water activity values (Table 2.2).

Table 2.2: Growth of micro-organisms due to minimum water activity values.

Micro-organisms	Minimum water activity (a_w) values
Gram +ve bacteria	9.5
Gram –ve bacteria	0.91
Yeasts	0.88
Molds	0.80

Oxidation reduction

Oxidation-reduction potential (O–R potential, Eh): Micro-organisms are sensitive to the oxidation-reduction potential of the substrate. The O–R potential of a substrate is defined as the ease with which the substrate loses or gains electrons and is expressed by the symbol Eh. A substance that loses electrons readily is a good reducing agent while a substance which gains electrons is a good oxidant. When electrons are transferred from one compound to another, a potential difference exists between the two compounds which is expressed in millivolts (mV). A highly oxidised substance has more positive potential while the more reduced substance has a negative potential. When the concentrations of oxidant and reductant are equal the potential difference is zero. Fruit juices are highly oxidised substrates with Eh values in the range of +400 mV. Solid meat is a reducing medium with Eh values in the range of −200 mV while minced meat is an oxidising medium with Eh values in the range of +200 mV. Aerobic micro-organisms such as aerobic bacteria require positive Eh values (oxidised) for their growth while anaerobes such as *Clostridium* species require negative Eh values (reduced). Some aerobic bacteria grow better in slightly reduced conditions and these organisms are referred to as *microaerophiles*. Examples of microaerophilic bacteria include lactobacilli and streptococci. Facultative organisms are capable of growing under aerobic as well as in anaerobic conditions.

The O–R potential of a food is determined by: (i) the characteristic O–R potential of the original food, (ii) the poising capacity, i.e. the resistance to change the potential of the food, (iii) the oxygen tension of the atmosphere surrounding the food and (iv) the access of atmosphere to the food. Most fresh plant and animal foods have a low and well-poised O–R potential in their interior. The presence of reducing sugars and ascorbic acid in plants and fruits and the presence of −SH groups in meat are responsible for the low O–R potential. In living plant and animal cells these reducing substances tend to poise the O–R potential at a low level resisting the effect of oxygen due to respiration and diffusion from the outside. Thus a cut of fresh meat or a whole fruit would have aerobic conditions only close to the surface that supports the growth of aerobic micro-organisms while the interior of the meat or fruit supports the growth of anaerobic bacteria. Processing of foods alters the poising power of food by destroying or altering the reducing or oxidising substances and also may allow more diffusion of oxygen. Thus clear fruit juices do not contain the natural reducing substances which are lost during extraction and filtration and become substrates with more positive Eh values. Micro-organisms affect the Eh values of foods during their growth. Thus aerobes consume oxygen and reduce the Eh values while anaerobes cannot affect the Eh values. In the presence of limited amounts of oxygen the same aerobic and facultative organisms may produce incompletely oxidised products such as organic acids from carbohydrates while in the presence of large amounts of oxygen, complete oxidation of carbohydrates to carbon dioxide and water occur. Similarly, protein decomposition under anaerobic conditions results in putrefaction while under aerobic conditions the products may be less· obnoxious.

Nutrient content

The kinds and proportions of nutrients in the food are all-important in determining what organism is most likely to grow. Consideration must be given to: (i) foods for energy, (ii) foods for growth and (iii) accessory food substances, or vitamins, which may be necessary for energy or growth.

Foods for energy: The carbohydrates, especially the sugars, are most commonly used as an energy source, but other carbon compounds may serve, e.g. esters, alcohols, peptides, amino acids, and organic acids and their salts. Complex carbohydrates, e.g. cellulose, can be utilised by comparatively few organisms, and starch can be hydrolysed by only a limited number of organisms. Micro-organisms differ even in their ability to use some of the simpler soluble sugars. Many organisms cannot use the disaccharide lactose (milk sugar) and therefore do not grow well in milk. Maltose is not attacked by some yeasts. Bacteria often are identified and classified on the basis of their ability or inability to utilise various sugars and alcohols. Most organisms, if they utilise sugars at all, can use glucose.

Foods for growth: Micro-organisms differ in their ability to use various nitrogenous compounds as a source of nitrogen for growth. Many organisms are unable to hydrolyse proteins and hence cannot get nitrogen from them without help from a proteolytic organism. One protein may be a better source of nitrogenous food than another because of different products formed during hydrolysis, especially peptides and amino acids. Peptides, amino acids, urea, ammonia, and other simpler nitrogenous compounds may be available to some organisms but not to others or may be usable under some environmental conditions but not under others.

Accessory food substances, or vitamins: Some micro-organisms are unable to manufacture some or all of the vitamins needed and must have them furnished. Most natural plant and animal foodstuffs contain an array of these vitamins, but some may be low in amount, or lacking. Thus meats are high in B vitamins and fruits are low, but fruits are high in ascorbic acid. Egg white contains biotin but also contains avidin, which ties up biotin, making it unavailable to micro-organisms and eliminating as possible spoilage organisms those which must have biotin supplied. The processing of foods often reduces the vitamin content. Thiamine, pantothenic acid, the folic acid group, and ascorbic acid (in air) are heat-labile, and drying causes a loss in vitamins such as thiamine and ascorbic acid. Even storage of foods for long periods, especially if the storage temperature is elevated, may result in a decrease in the level of some of the accessory growth factors.

Presence of inhibitory substances

Inhibitory substances, originally present in the food, added purposely or accidentally, or developed there by growth of micro-organisms or by processing methods, may prevent growth of all micro-organisms or, more often, may deter certain specific kinds. Examples of inhibitors naturally present are the lactenins and anticoliform factor in freshly drawn milk, lysozyme in egg white, and benzoic acid in cranberries.

A micro-organism growing in a food may produce one or more substances inhibitory to other organisms, products such as acids, alcohols, peroxides, and even antibiotics. Propionic acid produced by the propionibacteria in Swiss cheese is inhibitory to molds, alcohol formed in quantity by wine yeasts inhibits competitors, and nisin produced by certain strains of *Streptococcus lactis* may be useful in inhibiting lactate-fermenting, gasforming clostridia in curing cheese and undesirable in slowing down some of the essential lactic acid *streptococci* during the manufacturing process. There also is the possibility of the destruction of inhibitory compounds in foods by micro-organisms. Certain molds and bacteria are able to destroy some of the phenol compounds that are added to meat or fish by smoking or benzoic acid added to foods, sulphur dioxide is destroyed by yeasts resistant to it, and lactobacilli can inactivate nisin. Heating foods may result in the formation of inhibitory substances. Heating lipids may hasten autoxidation and make them inhibitory, and browning concentrated sugar sirups may result in the production of furfural and hydroxymethyl furfural, which are inhibitory to fermenting organisms. Long storage at warm temperatures may produce similar results.

Extrinsic Factors

For the growth of the micro-organisms certain external factors should be favourable.

Temperature

Micro-organisms grow over a wide range of temperatures. The lowest temperature at which a micro-organism grows is about $-34°C$ and the highest temperatures is about $90°C$. The various micro-organisms are usually grouped on the basis of their temperature requirements for growth. Thus organisms that grow well below 20°C and have optimum temperature in the range of 20 to 30°C are referred to as psychrophiles or psychrotrophs. Organisms that grow well between 20 and 45°C with optimum temperature in the range of 30 and 40°C are called mesophiles. Thermophiles are those organism that grow well at temperatures above 45°C with optimum temperature in the range of 55 to 65°C. At temperatures higher than an organism's optimum growth range, cells die rapidly. Lower temperatures still result in cell death but at a slower rate.

Temperature can therefore be used to eliminate or control the growth of micro-organisms. Heat treatments (pasteurisation or sterilisation) eliminate contaminating micro-organisms via the application of heat for a specific time period (time and temperatures used being dependent upon the target organism). Refrigeration of a food can prevent spoilage by controlling the growth of thermophilic or mesophilic organisms. Most pathogens are capable of growth at refrigeration temperatures and therefore cannot be controlled via refrigeration alone. Some, for example *Listeria*, can grow at very low temperatures.

Relative humidity of environment

The relative humidity of the storage environment must be such that excessive drying of the food or absorption of moisture by the food does not occur. If the growth of micro-organisms in a food is controlled by the water activity of a product then it is very important that the food be stored under relative humidity conditions which will not allow the uptake of moisture from the air, and therefore an increase in water activity. Packaging can be used to limit the migration of moisture into the product.

Atmosphere

It is possible to inhibit surface spoilage of foods by a controlled atmosphere surrounding the food. As all micro-organisms have specific requirements for oxygen and carbon dioxide, by altering the atmosphere within a food package the growth of micro-organisms can be controlled. Vacuum packing food removes available oxygen and thereby prevents the growth of aerobic organisms, it does however still allow the growth of anaerobes such as *C. botulinum*. Modified atmosphere packaging (MAP) allows the food producer to select the atmosphere within the package using varying combinations of oxygen, carbon dioxide and nitrogen, depending upon the product type and target micro-organisms. The majority of MAP foods have varying combinations of carbon dioxide and nitrogen.

MICRO-ORGANISMS IMPORTANT IN FOOD INDUSTRY

In addition to natural microflora determined by type of plant or animal and environmental conditions, every food may be contaminated from outside sources on the way from the field to the processing plant, or during storage, transport and distribution. There are thousands of different types of micro-organisms everywhere in air, soil and water, and consequently on foods, and in the digestive tract of animals and human. Fortunately, the majority of micro-organisms perform useful functions in the environment and also in some branches of of food industry, such as production of wine, beer, bakery products, dairy products, etc. On the other hand unwanted spoilage of foods is generally caused by micro-organisms and contamination of food with pathogens causes food safety problems.

The micro-organisms occurring on and/or in foods are from a practical point of view divided into three groups: molds, yeast and bacteria. Molds are generally concerned in the spoilage of foods, their use in the food industry is limited (e.g. mold ripened cheese). Yeasts are the most widely used micro-organisms in the food industry due to their ability to ferment sugars to ethanol and carbon-dioxide.

Some types of yeast, such as bakers yeasts are grown industrially, and some may be used as protein sources, mainly in animal feed. Bacteria important in food micribiology may be divided into groups according to the product of fermentation, e.g. lactic acid bacteria, acetic acid bacteria, propionic acid bacteria. Bearing in mind the food constituent attacked (used as food for micro-organisms), proteolytic, saccharolytic and lipolytic bacteria may be distinguished. Their systematic classification is based primarily on morphological and physiological properties (e.g. aerobic and anaerobic bacteria, gas forming bacteria, etc.). Lactic acid bacteria are widely used in the dairy industry, and acetic acid bacteria in vinegar production. Many bacteria are known as micro-organisms that cause spoilage and some are pathogens (e.g. *salmonellae*, *staphylococci*, etc.).

Importance of Molds in Foods

Main groups

1. Deuteromycota (fungi imperfecti): This is a class of septate fungi in which a sexual stage (perfect state) of the life cycle has not been observed. The Fungi Imperfecti are essentially a provisional taxonomic grouping, if a sexual stage is observed in a fungus assigned to this group, the organism will then be assigned a new name that reflects its perfect state.
   (a) *Aspergillus:* Very important genus that includes species used in industrial fermentations and the production of enzymes and organic acids. It also includes several spoilage micro-organisms and two species, A. *flavus* and A. *parasiticus*, which produce carcinogenic aflatoxins.
   (b) *Botrytis:* Cause grey mold rot in fruits.
   (c) *Penicillium:* Like *Aspergillus*, this genus includes industrially important species as well as others that are involved in food spoilage and the production of mycotoxins.
2. *Ascomycota:* Septate mycelium, sexual reproduction characterised by the formation of a sac-like structure called an ascus. The ascus eventually ruptures and the enclosed spores are liberated. This is a very large class of fungi that includes the genera *Byssochlamys*, and the perfect states of some fungi formerly classified under *Aspergillus*.
   (a) *Byssochlamys:* Form heat resistant spores that can survive heat treatment and result in the spoilage of high-acid canned foods, especially fruits.
3. *Zygomycota:* A diverse group of lower, filamentous fungi which have aseptate mycelium and are capable of rapid growth. This group has sexual and asexual modes of reproduction. Members include the genus *Rhizopus*, which olds several species that cause bread molds as well as *R. oligosporus*, a species used in oriental soyabean fermentations (tempeh).
   (a) Yeasts: Classified into the same group as molds, most do not live in soil but are instead found in environments with high sugar content such as the nectar of flowers or the surface of fruits.

Ascomycota

Debaromyces: One of the most prevalent genera in dairy products. *D. hansenii* is an important food spoilage species, it can grow in 24% NaCl and at a_w as low as 0.65.

Saccharomyces: Includes the important industrial species, *S. cerevisiae*, used in bread and alcoholic fermentations. *S. bailli* is another important species because it is involved in the spoilage of several types of foods and is resistant to the antifungal preservatives benzoate and sorbate.

Deuteromycota

Candida: Includes the pathogen *C. albicans*, this genus also includes the most common species of yeasts in fresh ground beef and poultry, a few species are involved in industrial fermentations.

Moisture and Air Loving Molds

Probably the best known micro-organisms, molds are widely distributed in nature and grow under a variety of conditions in which air and moisture are present. They are also plants and a part of the fungi family. Nearly everyone has seen mold growth on damp clothing and old shoes. So, many may find it hard to believe that mold is a micro-organism. However, the mold we see with the naked eye is actually a colony of millions of mold cells growing together. Molds vary in appearance. Some are fluffy and filament-like, others are moist and glossy, still others are slimy.

Growth and importance to food industry

Unlike bacteria, molds are made up of more than one cell. Vegetative cells sustain the organism by taking in food substances for energy and the production of new cell material. Reproductive cells produce small 'seed' cells called spores. Unlike bacterial spores, mold spores are the source of new mold organisms. Bacterial spores generally form only when environmental conditions are unfavourable.

Molds produce a stem consisting of several cells. Together, these cells form a 'fruiting body'. The fruiting body produces the spores, which detach and are carried by air currents and deposited to start new mold colonies whenever conditions are favorable. Mold spores are quite abundant in the air. So, any food allowed to stand in the open soon becomes contaminated with mold if adequate moisture is present. Some types of molds are also psychrophiles and can cause spoilage of refrigerated foods.

Molds are important to the food industry. Among their many contributions are the flavour and colour they add to cheeses and the making of soya sauce. They also play a role in the making of such chemicals as citric and lactic acid and many enzymes. Molds can also cause problems in foods. Certain kinds can produce poisons called mycotoxins. Probably the best known use of molds is in the drug industry, where they help produce such antibiotics as penicillin.

IMPORTANCE OF BACTERIA IN FOOD INDUSTRY

Bacteria that play significant roles in foods are often grouped on the basis of their activity in foods without regards to their systematic classifiaction. Bacteria make up the largest group of micro-organisms. People often think of them only as germs and the harm they do. Actually, only a small number of bacteria types are *pathogenic* (disease causing). Most are harmless and many are helpful. Micro-organisms, including bacteria, can also be grouped according to their requirement for oxygen. Some grow only in the presence of oxygen (*aerobes*). Others grow only in the absence of oxygen (*an-aerobes*). Some are able to grow with or without oxygen (*facultative anaerobes*).

Bacterial eating habits: Bacteria and other micro-organisms need food in order to grow and multiply. They vary in their food needs, but nearly everything we consider as food can also be used as food by some type of bacteria. To be used by bacteria, a food substance must pass into the cell where it can be processed into energy and new cell material. Because most foods are too complex to move into a bacterial cell, they must be broken down into simpler substances. *Enzymes* do this by increasing the rate of biochemical reactions. Produced within the bacterial cell, enzymes move through the cell wall to break down the food on the outside into a form bacteria can use.

Growth and Importance in Food Industry

Bacteria reproduce by a process called *binary fission*—one cell divides and becomes two. Some can reproduce at a very rapid rate under proper conditions. If food and moisture are adequate and the temperature is right, certain bacteria can reproduce in as little as 20 minutes. Within 20 minutes, one cell becomes two, in 40 minutes, there will be four, and so on. In only eight hours, the original cell will have multiplied to nearly 17 *million new bacteria*. Of course, conditions don't remain favourable for such a rate of reproduction for long. If they did, we could be buried in bacterial cells.

Lactic acid bacteria their uses in food

Lactic acid bacteria refers to a large group of beneficial bacteria that have similar properties and all produce lactic acid as an end product of the fermentation process. They are widespread in nature and are also found in our digestive systems. Although they are best known for their role in the preparation of fermented dairy products, they are also used for pickling of vegetables, baking, winemaking, curing fish, meats and sausages. Without understanding the scientific basis, people thousands of years ago used lactic acid bacteria to produce cultured foods with improved preservation properties and with characteristic flavours and textures different from the original food.

Similarly today, a wide variety of fermented milk products including liquid drinks such as kefir and semi-solid or firm products like yoghurt and cheese respectively, make good use of these illustrious microbial allies. The manufacture involves a microbial process by which the milk sugar, lactose is converted to lactic acid. As the acid accumulates, the structure of the milk protein changes (curdling) and thus the texture of the product. Other variables such as temperature and the composition of the milk, also contribute to the particular features of different products.

IMPORTANCE OF YEASTS IN FOODS

Saccharomyces species are the most widely used yeasts. The leading species *S. cerevisiae* is used in the manufacture of many foods, with special strains used for the leavening of bread and for the production of ale, wine, alcohol, glycerol and invertase. Top yeasts are active fermenters and grow rapidly at 20°C. The clumping of the cells and the rapid evolution of carbon dioxide sweep the cells to the surface and hence the name top yeasts. Bottom yeasts do not clump and settle to the bottom, hence the name bottom yeasts. They grow more slowly and ferment at lower temperatures of 10–15°C.

S. cerevisiae var ellipsoideus is a high alcohol yielding variety used in the production of industrial alcohol, wine and distilled liquors. *S. uvarum* is a bottom yeast used in beet manufacture. *S. Fragilis* and *S. lactis* have the ability to ferment lactose and hence are important in milk products.

Zygosaccharomyces species are osmophilic and are involved in the spoilage of honey, syrups and molasses. They find use in the fermentation of soya souce and some wines.

Candida species spoil foods high in acid salt content. *C. utilis* is grown for food and *C. krusei* is used with dairy starter cultures to maintain the activity and longevity of lactic acid bacteria. *C. lipolytica* spoils butter and oleomargarine.

Brettanomyces species are involved in the fermentation of Belgian and English beers and in French wines, while *Kloeckera* species are found commonly on fruits and flowers and in soil. *Trichosporon* species (*T. pullulans*) grow at low temperatures and are found in breweries and on chilled beef.

Rhodotorula species cause discolouration of foods by forming coloured spots on meat and sauerkraut.

Sugar-Loving Yeasts

Yeasts are small, single-celled plants. They are members of the family *fungi* (singular, *fungus*), which also includes mushrooms. Fungi differ from other plants in that they have no chlorophyl. Bacteria thrive on many different types of food. But most yeasts can live only on sugars and starches. From these, they produce carbon dioxide gas and alcohol. Thus, they have been useful to man for centuries in the production of certain foods and beverages. They are responsible for the rising of bread dough and the fermentation of wines, whiskey, brandy and beer. They also play the initial role in the production of vinegar.

Growth and Importance to Food Industry

Some yeasts are *psychrophilic* and so they can grow at relatively low temperatures. In fact, the fermentation of wines and beer is often carried out at temperatures near 40°F. Because some kinds are psychrophiles, they can create a spoilage problem in meat and other refrigerated storage areas.

Unlike bacteria, which multiply by binary fission, yeasts reproduce by a method called *budding*. A small knob or bud forms on the parent cell, grows and finally separates to become a new yeast dell. Although this is the most common method of reproduction, yeasts also multiply by the formation of spores. Because they can grow under conditions of high salt or sugar content, they can cause the spoilage of certain foods in which bacteria would not grow. Examples are honey, jellies, maple syrup and sweetened condensed milk. Foods produced by the bacterial fermentation process, such as pickles and sauerkraut, can also be spoiled by yeasts which interfere with the normal fermentative process.

Certain yeasts are pathogenic. However, yeast infections are much less common than are bacterial infections. Today, the impact of yeasts on food and beverage production extends beyond the original and popular notions of bread, beer and wine fermentations by *Saccharomyces cerevisiae* (Table 2.3).

Table 2.3: The commercial and community significance of yeasts in food and beverage production.

Production of fermented foods and beverages
Production of ingredients and additives for food processing
Spoilage of foods and beverages
Biocontrol of spoilage micro-organisms
Probiotic and biotherapeutic agents
Source of food allergens
Source of opportunistic, pathogenic yeasts

In a positive context, they contribute to the fermentation of a broad range of other commodities, where various yeast species may work in concert with bacteria and filamentous fungi. Many valuable food ingredients and processing aids are now derived from yeasts. Some yeasts exhibit strong antifungal activity, enabling them to be exploited as novel agents in the biocontrol of food spoilage. The probiotic activity of some yeasts is another novel property that is attracting increasing interest. Unfortunately, there is also a darker side to yeast activity. Their ability to cause spoilage of many commodities, with major economic loss, is well known in many sectors of the food and beverage industries, while the public health significance of yeasts in foods and beverages is a topic of emerging concern.

APPLICATIONS OF MICRO-ORGANISMS IN FOOD BIOTECHNOLOGY

Micro-organisms have been used for preparing food products like bread, yoghurt or curd, alcoholic beverages, cheese, etc. for a long time without even knowing their involvement in fermentation. Louis

Pasteur showed the role of micro-organisms in spoilage and subsequent elucidation that fermentation also involves micro-organisms. Once this fact was established, the scientists tried to isolate micro-organisms, which were more efficient in producing better products or improvement of processes. Some species are useful for development of flavour unique to certain wines. Thus traditionally certain micro-organisms were used in such fermented foods.

Need for Improved Cultures

When large-scale commercialisation of such products occurred, there was a need to increase the production to meet the increasing demands. Microbial techniques (selection, isolation of pure culture, mutation, protoplast fusion, etc.) well developed by middle of last century, were used advantageously in the maximum output of the desired product with minimum by-product formation. These techniques, however, are slow in developing a micro-organism having the desirable traits and very few undesirable properties. Sometimes when protoplast fusion is carried out, some undesirable properties are transferred, which have to be slowly removed by further mutation. All this takes a long time and the results are not precise as much of the development is being performed empirically.

With the capabilities of modern biotechnology, the scientists can now transfer desirable characteristics or genes, without simultaneous transfer of other undesirable genes. Cut and paste techniques, developed in genetic engineering, can incorporate only the desirable genes. Thus genetically modified organism takes only a few months compared to a few years of laboratory work by traditional methods. This communication presents developments related to foods, wherein a review of genetically modified micro-organisms useful in foods is given. An attempt is also made to point out some desirable traits for commercial cultures, which might prove useful in industrial food production and processing.

Organic Acids by Micro-organisms

Citric acid is the most important organic acid produced by fermentation with an estimated annual production of about half a million tonnes with the value more than half a billion dollars. It is primarily used in foods. Some of the other acids produced in large quantities by fermentation are gluconic acid, lactic acid and ascorbic acid. Citric acid had been prepared from citrus fruits like lemon but now it is mostly produced by fermentation using *Aspergillus niger*, in large corrosion resistant fermenters having stirrers. Some yeasts like *Candida* have also been used to a smaller extent. A smaller amount is also made by older technique with surface fermenters. In submerged culture, when environmental conditions are controlled, organisms grow into small pellets. Sugar from cane molasses is commonly used in the medium, which needs to be controlled for trace metals like iron, copper, etc. Maintenance of very low pH avoids by-products formation. High aeration rate is needed for higher yields. Conversion of glucose to product is high (70–90%) depending on the strain, purity of carbohydrate raw material, and environmental conditions. By-product formation of oxalic or gluconic acid can be reduced by strict control of growth conditions. *A. niger* strains have been developed by mutagenesis and screening, for higher productivity and adaptability to industrial fermenters.

Some studies have been undertaken on parasexual recombination, diploidisation, and heterokaryon formation, etc. Although recombinant DNA technology has been reported for *Aspergillus* species, no reports are available on using this technique for commercial citric acid production. Genes cloned from *A. niger* for pyruvate kinase and phosphofructokinase will tremendously improve the commercial strains producing citric acid. Lactic acid is another important acid produced by fermentation, although an equal amount is also chemically synthesised. The acid is mostly used for the manufacture of emulsifiers and as additives in food industry. It has two enantiomers, L(+) and D(–) lactic acid. The L-lactic acid is involved in normal human metabolism which can selectively be produced by fermentation and this is used in food applications, whereas the chemical synthesis produces DL-lactic acid.

Strains of *Lactobacillus delbruckii, L. casei, L. helveticus* and *L. acidophilus,* employed in commercial fermentation, can ferment a medium containing 12–15% sugar in 2 to 4 days with more than 90% yield. Most lactobacilli cannot use starch. *L. amylophilus* and *L. amylovorus* are able to ferment starch to lactic acid. The production of lactic acid or products like ethanol, acetic acid, etc. depends on the strains as well as the substrate and environmental conditions. Lactobacilli are very fastidious and require many nutrients including nucleotides, amino acids and vitamins provided by yeast extract or peptone. Mutants can be generated spontaneously or by mutagenic agents to give higher conversion or concentration of lactic acid. For commercial and economical production of lactic acid, the further improvements will expand substrate range, improve product tolerance, use of simpler nitrogen, increase disease resistance and control LID isomer ratio.

Gluconic acid is prepared by fermentation using mostly *A. niger* and less commonly *Acetobacter* (*Gluconobacter*) *suboxidans* and some *Penicillium* species. *A. niger* produces acid at high levels (97–99%) with negligible by-products when the trace elements are controlled and sufficient manganese is present. Glucose is converted to glucono delta-lactone by glucose oxidase enzyme. The lactone hydrolyses to gluconate. Fermentation conditions are designed to maintain high levels of this enzyme. Medium contains glucose with low levels of phosphate and nitrogen to prevent excess growth. High aeration, temperature about 30°C and mildly acidic *pH* produces gluconate rapidly. The *pH* is maintained by adding $CaCO_3$ to produce calcium gluconate whereas NaOH gives sodium gluconate. *A. niger* gene for glucose oxidase has been cloned into *S. cerevisiae, A. niger* and *A. nidulans* to yield improved organisms and fermentation.

Microbial Enzymes in Food Industry

Enzymes have been used in foods such as leavening of bread, fermentation of fruit juices or malt, clotting of milk for cheese, etc. Purified enzymes are being used mostly in food industry although some still use live cells as in leavening of bread and alcoholic fermentation. Newer applications for enzymes are in detergents, textiles, paper and pulp, chemical industry, etc. However, the largest application (over 45% of the total enzyme produced), mostly bulk enzymes, is used in foods. Largest market is for rennet (25%) followed by glucoamylase (20%), α-amylase (16%) and glucose isomerase (15%).

Cheese is traditionally prepared using calf rennet, a protease. In 60s and 70s, due to severe shortage of calf rennet, several substitutes from micro-organisms, *Rhizomucor miehei, Endothia parasitica* and *Rhizomucor pusillus,* were chosen by mutation and selection method. Recently, several genetically modified micro-organisms (*E. coli, K. lactis, A. niger),* which contain calf rennet gene, have been developed. Chymosin gene coding for calf rennet was taken from calf stomach cells and inserted into a plasmid, which was inserted into microbial cells. The micro-organisms started producing calf rennet. These rennets by GMOs have been commercially produced since 1980s and, in India, only the microbial rennet is being used since the ban on calf rennet. *Bacillus* spp., mostly extracellular, are important source of stable enzymes for use in food industry. These degrade substrate into smaller molecules, which are easily absorbed by the bacteria. Some industrial enzymes produced by *Bacillus* sp. are Pullulanase by *B. acidopullulyticus,* α-amylase by *B. amyloliquefaciens* and *B. licheniformis,* glucose isomerase by *B. coagulans,* β-glucanase by *B. subtilis,* etc. Amylases and related enzymes are mostly obtained from *Bacillus* species. The α-amylase, that hydrolyse internal 1-4 α-bonds of starch resulting in rapid reduction of viscosity of the substrate, is called liquefying enzyme and produces mostly maltohexaose or maltopentaose. This enzyme from *B. licheniformis* is exceptionally thermostable and is used in starch processing. The enzyme from *B. subtilis* is referred to as saccharifying α-amylase as it predominantly produces maltose and glucose from starch. It shares limited sequence homology with the liquefying enzymes and is less thermostable. The cyclodextrin glucanotransferases catalyse the hydrolysis of amylose to cyclic dextrins, α-, β- and γ-cyclodextrins, which have useful properties in food industry for stabilisation of volatile flavours, enhancers, etc. These are also produced by a number of *Bacillus* sp. While amylases hydrolyse 1-4 linkages, pullulanase hydrolyse 1-6 α-linkages in pullulan and amylopectin. Whereas saccharifying amylases yield mostly maltose, glucoamylase produces glucose. These are commercially produced mostly by fungal cultures but a few *Bacillus* species are also used.

The *bacilli* also produce proteases useful in food industry. Proteases are divided into exo- and endoacting types. The industrially important exopeptidases are classified as per their catalytic mechanisms: serine, cysteine (thiol), metallo- and aspartic (carboxyl) proteases. The serine proteases (pH, 9–11) have no metal ion requirement and are resistant to high temperature and oxidising agents. Such properties are further enhanced by protein engineering making them useful in laundry detergents. They are useful in production of fish-meal and protein hydrolysates from fish. Acid proteases, useful in cheese industry, are found less in bacteria than in molds such as *Mucor* from which commercial microbial rennet is prepared. Similarly thiol proteases like papain are not common in *bacilli*.

Fermentative Production of Amino Acids

Many amino acids are used in food industry, *L-glutamate* as flavour enhancer, glycine as sweetener, lysine and methionine as food and feed additives, aspartate and phenylalanine for aspartame, an low-calorie sweetener, etc. *Corynebacterium glutamicum* is the most versatile organism used commercially to prepare glutamate, lysine, threonine, phenylalanine, etc. *Escherichia, Serratia, Bacillus, Hansenula, Candida,* and *Saccharomyces* are also used in amino acid production, some of them are genetically modified. Bacteria normally do not accumulate large amounts of amino acids because of regulatory control over their synthesis. Mutants have to be prepared by laborious mutagenesis and selecting the mutant producing highest amount. By using recombinant DNA technology, new producers can be developed rapidly by increasing limiting enzyme activities, etc.

The genome analysis of producer strains is now becoming a useful tool. Entire sequence of the chromosomes of *C. glutamicum* and *E. coli* is available. It is possible to compare mutants and identify mutations necessary for overproduction of metabolites. Genetic manipulations including transduction, transformation and conjugation have been used in genetic study of these bacteria. This has led to the understanding of regulatory mechanism for microbial metabolism at genes level. Genetic engineering modification aimed at improving amino acid production by these organisms has not yet resulted in substantial increase in amino acid production. Amplification of genes coding for limiting enzymes might result in increased amino acid production and is carried out by multiple copy plasmids. There are problems of maintaining stability unless selective pressure is exerted by adding antibiotics in the media. A new system with Mu recombinant phages can integrate several copies into the host chromosomes showing stable accumulation without antibiotic selection pressure.

Biotechnology of Dairy Products

Lactic acid bacteria *(Lactobacillus, Leuconostoc, Pediococcus, Bifidobacterium,* and *Lactococcus)* have been used to improve the flavour, texture, preservation and nutritive value of dairy as well as vegetable, cereal and legume fermentation products including yoghurt, buttermilk, cheese, pickled vegetables, idli, etc. In addition, some are even used as probiotics, which contribute to the overall health of the user. In milk, the lactic acid bacteria ferment lactose and other sugars. Some proteases play role in the process along with the sugar metabolising enzymes. Formation of these products as well as compounds affecting flavour and texture gives the typical pleasant aroma, taste and body to the product. The metabolic activity also forms some useful vitamins. Many lactic acid bacteria like *L. acidophilus* and *L. sake* produce antimicrobial bacteriocins, which help in controlling unwanted micro-organisms. Molecular strategies are being studied. Genetically engineered lactics with better fermentation efficiency, better shelf-life, nutritional and sensory properties for the product, etc. will be the target of these studies.

When cheese, yoghurt, etc. are made, undesirable contaminants can lead to poor flavour, low yield and food poisoning. Lactic acid bacteria can be genetically engineered to grow faster than the contaminants, as well as inhibit and destroy the growth of the contaminants including pathogens by producing antimi-385 crobial agents. The starter cultures have been modified to produce an antimicrobial agent, which destroys cell walls of *Listeria monocytogenes*. Similar modification can also be carried out to protect against organisms like *Salmonella*.

Alcoholic Fermentation using Improved Cultures

Yeast strain used in beer brewing is selected on the basis of flavour and aroma, imparted by the strain during fermentation. Flocculation, fermentation rate, ethanol tolerance, osmotolerance, and oxygen requirements are other important factors in considering different strains. For commercial beer production, mostly *S. uvarum* and *S. cerevisiae* have been commonly used. Earlier protoplast fusion, which used to produce a fusion product from *S. uvarum* and *S. diastaticus,* was not only more rapid in fermenting but utilised available sugars more completely. Many of the desirable properties can be incorporated by genetic modification (GMO). Though GMOs are not being used commercially for brewing beer, but with better understanding of genes controlling the properties of brewers yeast and application of genetic engineering, increasing efficiency and productivity at minimum cost without affecting adversely the beer quality will soon become a reality. Some work is also carried out to develop zymocide resistant strain. The studies to improve the distiller's yeast strain include manipulation of alcohol dehydrogenase promoter gene, leading to increased production of uamylase in *S. cerevisiae,* cloning of regulatory genes into *S. cerevisiae* resulting in higher maltase activity thereby improved conversion to ethanol, transforming amylase genes from *S. diastaticus* into *S. cerevisiae* to enable latter to utilise dextrins, incorporating glucoamylase genes from *R. oryzae* and *A. awamori*. Most of the efforts were directed towards faster and more complete conversion of carbohydrates to ethanol.

Miscellaneous Microbial Products

Candida utilis has been used industrially in the production of SCP for food and fodder, waste treatment and the production of fine chemicals used as flavour enhancers. Among the products useful in foods, besides SCP, are 5'-GMP & 5'-IMP, ethanol, ethylacetate, acetylaldehyde, amino acids like serine, histidine, glutamic acid and lysine, xylitol, etc. *C. utilis* does not possess enzymes to hydrolyse starch, cellulose or pectic substrates. Two step dual fermentation can be carried out using *C. utilis* with organisms like *S. fibuliger,* which produces amylases and can be used in starch wastes, and *T. reesei,* which has cellulases and can be used in cellulosic waste. Molecular genetics of *C. utilis* is not adequately studied. Although some transformations have been successfully carried out, no commercial strain has been developed by GMO including protoplast fusion. Since some of the enzymes are lacking in this organisms, incorporation of genes encoding these enzymes would produce a desirable modified organism with application in food industry. *Bacillus* species have provided traditional biotech products such as extracellular enzymes and insect toxins. *B. thuringiensis* strains with toxicities against a variety of pests have been exploited to the extent of getting the genes inserted into food crops for successful development of resistance against these pests. Protein engineering and molecular technologies will slowly replace screening programmes.

Future Applications of Biotechnology in Foods

At present, the large amounts of Genetically Modified (GM) Foods including soya beans, corn, tomatoes, etc. as well as ingredients from GM organisms are being used in most parts of the world. With wide acceptance to biotechnology in food applications, commercial interest would stimulate research in the area. With genetic engineering techniques being used to develop improved cultures, there will be marked improvement in production and quality in addition to many new applications of micro-organisms.

Chapter 3

Applications and Impact of Biotechnology in Food Industry

INTRODUCTION

Staple food constitutes the most indispensable and basic need of man for fulfilling minimal nutritional requirements to sustain human life on earth. Man has been traditionally depending upon agriculture and livestock to meet food demands since times immemorial. It is obligatory on the part of all Governments to provide safe, wholesome and nutritious foods to their citizens belonging to all the sections of society. A healthy diet can play a significant role in creating a healthy mind and healthy society in the country. Adequately nourished and healthy citizens can serve as the work force in building a Nation by boosting the growth, prosperity and productivity. However, the overall quality and safety of food commodities can be considerably influenced by the food processing and packaging to provide optimal nutritive value to the consumers. With the advent of new scientific knowledge and technological innovations, food sector is witnessing a phenomenal growth across the world particularly in developed countries. Although, developing countries like India are the potential markets for variety of such processed foods, their indigenous food processing industry is still in the transition stage to adopt modern and advanced food processing tools to compete with the developed countries. One such powerful technique that can be very promising and highly relevant to food processing industry in countries like India is the Biotechnology. By judiciously applying biotechnological tools and processes, the quality, safety and nutritive value of processed foods can be improved considerably with lot of value addition.

ROLE OF BIOTECHNOLOGY IN FOOD SECTOR

Biotechnology has already benefitted the food industry in a big way. It has given us high quality foods that are tasty, nutritious, wholesome, convenient, shelf stable and safe. As research and development initiatives continue, it seems inevitable that biotechnology will have an increasing impact on the food we eat. It offers huge potential for increasing the range and quality of food available to us, particularly more nutritious and palatable foods. It also seems likely that it will continue to bring advantages to the processing and safety monitoring of food supply due to emergence of new technologies at a faster pace.

Although, traditional biotechnology that makes use of natural microbial fermentations has been playing a vital role in the development of our food supplies such as cheese and yoghurt-making and the use of yeast to leaven bread and ferment alcohol for thousands of years, the second generation food biotechnology is based on initiatives to screen enzymes and micro-organisms in the natural environment and exploit them for useful applications such as food ingredients, microbial fermentation to manufacture several products like lactic acid, citric acid and other flavor enhancers, etc. However, the major focus is now on exploring the modern biotechnology which is based on a combination of molecular genetics, applied enzymology and fermentation technology for value addition to foods. It is the modern biotechnology which is becoming increasingly important part of the over all efforts to improve methods of food production and to increase the variety, quality and safety of foods we eat.

POTENTIAL AREAS IN FOOD PROCESSING FOR BIOTECHNOLOGICAL APPLICATIONS

There are several potential areas in the food industry where the traditional and modern biotechnological tools can be applied during processing for the overall improvement of the nutritional quality, safety and health promoting attributes of the processed foods specifically with regard to the dairy based fermented products. Some of the potential areas of considerable commercial interest in food industry that can be targeted for biotechnological interventions are listed below:

1. Food fermentations
2. Starter cultures technology and genetic manipulation
3. Recombinant Enzymes
4. Biopreservation of foods
5. Functional/Health foods and Nutraceuticals
6. Probiotics, prebiotics and symbiotic foods
7. Genetically modified foods (GM Foods)
8. Milk derived bioactive peptides and other functional ingredients
9. Low calorie foods
10. Food packaging
11. Diagnostic tests for food safety and quality assurance
12. Biosensors

The scope and impact of biotechnological interventions in these areas will be briefly described and highlighted below.

Food Fermentations

Biotechnology as applied to food processing makes use of microbial inoculants to enhance properties such as the taste, aroma, shelf-life, texture and nutritional value of foods. The process whereby micro-organisms and their enzymes bring about these desirable changes in food materials is known as fermentation. Fermentation processing is also widely applied in the production of microbial cultures, enzymes, flavours, fragrances, food additives and a range of other high value-added products. These high value products are increasingly produced in more technologically advanced developing countries for use in their food and non-food processing applications. Fermentation is one of the oldest biotechnological processes traditionally used by man since time immemorial for food preservation. Fermented foods such as bread, beer, wine, vinegar, sauerkraft, pickles, etc. and traditional products like dahi, lassi and shrikhand account for one third of the human diet across the world. Other fermented dairy products such as cheese, yoghurt, kumis, kefir and others like sausages and soya sauce, etc. are now being produced commercially and marketed globally. Food fermentations contribute substantially to food safety and food security particularly during the off season when there is decline in the production of raw material. The fermentation bioprocess is one of the major biotechnological applications in food processing and often constitutes an important step in a sequence of food-processing operations, which may include cleaning, size reduction, soaking and cooking. Fermentation bioprocessing makes use of microbial inoculants for enhancing properties such as the taste, aroma, shelf-life, safety, texture and nutritional value of foods. Microbes associated with the raw food material and the processing environment serve as inoculants in spontaneous fermentations, while inoculants containing high concentrations of live micro-organisms, referred to as starter cultures, are used to initiate and accelerate the rate of fermentation processes in non-spontaneous or controlled fermentation processes. Current literature documents volumes of research reports on the characterisation of microbes associated with the production of traditional fermented foods in developing countries. The development and improvement of microbial cultures has been a driving force for the transformation of traditional food fermentations in developing countries from an 'art' to a science. Microbial culture development has also been a driving force for innovation in the design of equipment suited to the hygienic processing of traditional fermented foods under controlled conditions in many developing countries.

Improvements in the commercially important properties of microbial cultures, together with the improvement and development of bioreactor technology for the control of fermentation processes in developed countries, has played a pivotal role in the production of high-value products such as enzymes, novel microbial cultures, and functional food ingredients. These products are produced in more advanced developing economies, and are increasingly imported by less advanced developing countries, as inputs for their food processing. Although, the fermentation technology has been in vogue for many years in different countries, the output resulting from these technologies is not very high and hence needs optimisation to minimise losses and maximise product recovery to make the technology efficient, commercially viable and cost effective. Microbial cultures can be genetically manipulated using both traditional and molecular approaches to improve their fermentation characteristics for producing better quality fermented foods through enhanced enzymatic activity and flavor development. However, genetic improvement of bacteria, yeasts and moulds has often been the subject of intensive debate because of safety and health concerns likely to be associated with such genetically modified microbes.

Starter Culture Technology in Food Fermentations and their Genetic Manipulation

One of the most important areas relevant to food processing industry is the use of lactic starter cultures. Starter cultures, comprising of Lactic Acid Bacteria (LAB) such as lactococci, lactobacilli, pediococci and propioni bacteria are used in the production of cultured dairy products such as dahi, yogurt and cheese, etc. A starter culture is bound to provide particular characteristics in a more controlled and predictable fermentation. The availability of good starter bacteria is an essential pre-requisite for preparing quality fermented foods. The commercial value of the fermented products is, therefore, chiefly dependent upon the performance of the starter cultures. With the days of spontaneous fermentation and back slopping far behind us, we dwell in an era where the burgeoning fermented milk industry today demands 'multifunctional' or 'tailor made' starters fulfilling technical and metabolic requirements. Utility of starters go beyond imparting preservation and palatability to the final product. They could be selected for accelerated acid, flavor and, bacteriocin production in the fermented food to suppress spoilage and pathogenic bacteria apart from expressing additional health promoting functions. The deliberate use of functional traits within bacteria is supported by knowledge on their phylogenctics, characterisation of genome structure and flexibility, gene regulation and gene functionality particularly in relation to their commercially important traits. The strategies used for genetic manipulation of lactic starters to enhance their commercially important metabolic activities in the fermented foods for value addition have been described previously.

Recent advances in the field of metabolic engineering, genomics, and bioinformatics are expected to contribute to the future development of functional starter cultures to be used for food processing industry. Exploration studies of the natural diversity of wild strains occurring in traditional, artisan foods, fermented dairy products, etc. along with newer approaches such as comparative genomics, microarray analysis, transcriptomics, proteomics, and metabolomics, will generate useful information leading to the generation of new, industrial starter strains with increased diversity, stability, and industrial performance. These techniques will permit rapid, high-throughput screening of promising wild strains with interesting functional properties and lacking negative characteristics, as well as the construction of genetically modified starter cultures with a tailored functionality. Bioinformatics can be used to search genomes for essential components, for instance with regard to flavor development, such as peptidases, amino-transferases, enzymes for biosynthesis of amino acids, and transport systems for peptides and amino acids.

Recombinant Enzymes

Food industry is constantly in search of advanced technologies to meet consumer demand for nutritionally balanced and safe food products. Enzymes are a useful biotechnological processing tool whose action can be controlled in the food matrix to produce high quality products. The emerging area of enzyme engineering addresses the requirement of food processing sector, reducing the investment and the processing cost dramatically. Currently-used food enzymes are extracted from animals and plants (for example, a starch-digesting enzyme, amylase, can be obtained from germinating barley seeds) but most of the enzymes come from beneficial micro-organisms by large scale fermentation through optimisation of temperature, nutrients and air supply and later purified. Fermentation-derived enzymes are now the tools of choice for the innovative food processing industry. Moreover, several of the enzymes used in food processing industries are produced using recombinant micro-organisms. The industrial production of enzymes for use in food processing dates back to 1874 when Danish scientist Christian Hansen extracted rennin (chymosin) from calves stomachs for use in cheese manufacturing. Bovine chymosin was the first enzyme to be produced through biotechnological approaches in *E. coli*. Since then, genetic manipulation has been used to make tailor made enzymes for specific consumer requirement. Now enzymes can be produced through recombinant DNA technology in large quantities for their subsequent application in food industry. Most of these enzymes including bovine, goat or buffalo rennet are now being produced through rDNA technology for commercial applications in the food industry. Although, these recombinant enzymes are highly cost effective, their application in foods require approval from the regulatory agencies (DBT) from safety point of view.

Biopreservation of Foods

Although, recent developments in innovative modern technologies implemented in food processing and more stringent microbiological food-safety standards have reduced the incidences of food borne illnesses and product spoilage, they can not completely rule out the possibility of health risks associated with such foods. Improved food safety has been achieved through drastic physical treatments like high temperatures, high pressure technology as well as chemical preservatives. Toxicity of many of the commonest chemical preservatives (e.g. nitrites, sulphites), the alteration of the organoleptic and nutritional properties of foods, and especially consumers demand for safe and minimally processed foods without chemical additives have necessitated the need for alternative food grade safe biopreservatives. Hence, the food industry is constantly looking for new procedures and methods to produce minimally processed, ready to eat food with intact nutritional, taste, and flavor. Biopreservation of ready to eat processed foods is one such safe approach. It is defined as the extension of shelf life and safety of foods using the natural food grade antimicrobial compounds that are of plant, animal and microbial origin and do not pose any adverse effect on human health. The most common form of biopreservation of food products is through fermentation. The fermentation is brought about by food grade GRAS status lactic acid bacteria (LAB) belonging to *lactococci, lactobacilli, streptococci, pediococci, leuconostocs,* etc. which are being extensively used as starter cultures in the manufacture of dairy, meat and vegetable food products. These bacteria preserve the nutritional qualities along with inhibition of pathogenic and spoilage organisms due to the production of organic acids, hydrogen peroxide as well as proteinaceous metabolites such as bacteriocins.

Functional/Health Foods and Nutraceuticals

Functional/health foods and /nutarceuticals are currently the focus of attention across the world because of their immense health potentials and commercial value. Although, the term 'functional foods' currently lacks a common definition, this category is generally thought to include products that influence specific functions in the body and thereby offer benefits for health, well-being or performance, beyond their regular nutritional value. The concept of functional foods is not new as its origin dates back to prehistoric days. The old practice of using specific foods for some ailments figured prominently in ancient Hindu scriptures like Sushrita. The famous Greek physician Hippocrates also strongly advocated this concept through his tenet 'Let food be thy medicine and medicine be thy food'. However, the relevance of this concept gained sudden momentum during the last few decades due to unprecedented interest evoked amongst the health conscious consumers. This has been largely attributed to radical change in the modern lifestyle and perception of consumers towards their diet beyond nutrition. As a result of shift in the mindset of consumers towards the linkage of diet with their health, the commercial interest in functional food market has grown enormously and there is a boom in the functional foods and nutraceutical products in the market as can be reflected from the availability of variety of health foods in the food counters. The driving forces behind the reemergence of functional food concept in the present context is driven by a number of factors including the increasing life expectancy of people, quest for safe alternative to drugs, self care movement, rising health care costs, overwhelming scientific evidences to link diet with health, advances in food and ingredients technology for product diversification and the greater media coverage given to these high profile foods with novel health claims These products result from technological innovations, such as cholesterol lowering spreads, xylitol sweetened chewing gum and dairy products fermented with specific lactic acid bacteria, or are from a naturally functional food such as soy, oats and grains high in fiber. Functional foods have been developed in most food categories and the global market size is conservatively estimated to exceed that for organic foods. In addition to providing new options for improving health and well-being, the functional foods sector offers potential for new economic opportunities.

Biotechnology has a key role to play in this new industry. Traditionally, the application of biotechnology techniques in the food industry focused on the major energy-providing foods, such as bread, alcohol, fermented starch, yogurt, cheese, vinegar, and others. More recently, there has been increased interest in biologically active non-nutritive ingredients from natural products like herbals or foods. The functional food concept has in recent years moved progressively towards the development of dietary supplements that may affect the intestinal microbial composition and activities and hence may influence the gut health. In this context, dairy based food products which form an integral part of our diet can be very attractive candidate for application as functional or health foods after fortification/supplementation with novel bioactive ingredients which have the ability to trigger general health promoting and specific physiological functions in the host. These bioactive ingredients include Probiotics/prebiotics, bioactive peptides, biotherapeutic proteins, omega 3, CLA to low calorie sugars, PUFA, isoflavones, etc. which can be added to dairy foods to enhance their functionality for protecting health of the consumers against chronic diseases such as gastro-intestinal illnesses, CVD, strokes, hypertension, diabetes, cancers, etc.

Major breakthroughs have occurred and enormous progress has been made in this area of considerable health significance during the past few decades due to new advancements in biotechnological tools particularly with regard to genetic engineering and biotechnology. Nutrigenomics is the new era in the development of third generation of health/functional foods and is expected to revolutionise wellness and disease management across the world. Very soon need based customised health foods with specific bioactive functions intended for the target population will appear at the counters in the super markets and food outlets. This effort, however, requires a strong proactive synergy between Food and Pharmaceutical industry as well as Nutritionists, Biotechnologists and Dietetic and Medical professionals.

Probiotics as Functional Foods

The term 'probiotics' refers to live microorganisms that confer a health benefit to the host when ingested in adequate amounts. They are usually bacteria selected from species found in the intestinal tract. Lactobacilli and bifidobacteria are the two key members of this group used extensively in the production of probiotic food formulations for health applications across the world particularly in the developed countries. Milk and milk products specifically fermented dairy foods are considered as excellent carriers of probiotic strains to express their health promoting functions most optimally. Probiotic microorganisms may be concentrated and added directly to a food or to a milk product in small amounts and allowed to grow. Yoghurt is a classical example of a functional food with probiotics. Yoghurt with probiotics, called bio-yoghurt, should contain living bacterial cultures. Probiotics have been used as dietary supplements and oral agents for intestinal disorders. Probiotics have recently emerged as one of the most valuable bugs on account of expressing a multitude of novel health promoting functions which are highly strain specific. The most notable probiotic functions include immuno-modulation, restoring the balance of disturbed gut flora, strengthening the mucosal barrier function, prevention of lactose intolerance, etc. However, the current focus of attention is to explore probiotics as possible biotherapeutics against chronic inflammatory metabolic disorders such as diabetes, CVD, obesity, irritable bowel disease (IBD) and syndrome (IBS), Ulcerative Colitis (UC), Crohn's disease (CD), acute diarrhea, serum cholesterol reduction, shortening of the duration of respiratory infections, blood pressure control, colon cancer, and urinary tract infection (UTI), etc. Because of their immense health potentials, probiotics are now recognised as the vital health care concept of 21st century. Probiotics are one of the fastest growing food category within functional foods. And, as the list of health benefits accredited to them continues to expand, so does their use in new dairy and functional food applications.

Recent advances in biotechnological tools such as genetic engineering, recombinant DNA technology, PCR and availability of whole genome sequences of common probiotic strains, have improved the prospects of designing novel probiotics with improved functional efficacy and safety for human health applications and new product development. The most notable novel recombinant probiotic at present is a derivative of Lb. johnsonii La1. La1 is a well characterised probiotic strain used extensively in commercial preparation of probiotic foods due to its strong health-related attributes and positive immuno-modulatory effects on the host. Milk fermented with this culture normally produces a racemic mixture of D and L-lactate in the ratio 60:40. Presence of D-lactate in milk fermented with La1 and ability of the strain to produce D-lactate after ingestion does not pose any problem to most of the adult population. But, it can indeed cause D-acidosis and encephalopathy in patients suffering from bowel syndrome and intestinal failures, and in new born infants with immature liver. However, inactivation of the single copy D-lactate dehydrogenase (LdhD) gene of La1 resulted in rerouting of pyruvate mainly to L-lactate with no D-lactate production. This novel strain has the same beneficial properties as the parent probiotic while the absence of D-lactate makes it a safer alternative for specific populations.

Amongst several other possibilities is the design of recombinant strains with novel properties that confer competitive advantage to their survival. One way to accomplish this strategy is by expressing and secreting colicin V, a narrow host range antibacterial bacteriocin produced by E. coli in La1. This strategy has allowed the expression and secretion of the Gram negative antimicrobial in probiotic organisms to extend their inhibitory spectrum to Gram negative enteropathogens too. Established probiotic lactobacilli can also serve as attractive candidates for oral vaccination against HIV, tetanus, Rota virus, E. coli, Salomonella and H. pylori, etc. in view of their long history of safe use, ease of oral administration, low intrinsic immunogenicity and extensive industrial handling experience. Robust genetically engineered probiotic bacteria have also been developed by applying powerful genetic engineering techniques for better survival and stability during the harsh technological processing conditions used in the product development and hostile gut environment. In foods, genetically engineered bacteria have been used to improve the flavor and stability, or to block the formation of unwanted flavors. Metabolic engineering and Genetic engineering should make it possible to strengthen the effects of existing bacterial strains and create new ones. Global gene and protein expression techniques as well as metabolomics are now extensively being used to provide evidence of probiotic adaptations in food products, their survival and host-microbe interactions in the mammalian gut. However, acceptance of genetically engineered probiotics in product development is a subject of intensive debate due to long term safety and public health concerns and requires approval from the regulatory bodies constituted for this purpose in the country.

Genetically Modified and Transgenic Foods

Within the last two decades, the application of recombinant DNA techniques in the production of foods and food ingredients has developed from the level of basic research into a commercial business. From the very beginning, development and use of genetic engineering have been accompanied by strict regulations. These regulatory requirements cover the contained use of genetically modified organisms (GMO), their deliberate release into the environment as well as the placing on the market of products containing or consisting of GMO. So far, there is no report on any adverse effects on humans resulting from the consumption of foods produced by application of recombinant DNA techniques. Nevertheless, the sensitive nature of the subject (ethical issues) and the speed of the developments elicited fear among consumers. Potential hazards of the new technology are automatically projected into risks. In most developed countries, the willingness to accept such (perceived) risks is low, because the first generation of GM foods (herbicide-tolerant, insect-resistant crops and products thereof) were of no obvious advantage to regular consumers. The potential of recombinant DNA techniques in food production goes far beyond the applications reported so far. However, the benefits expected from the next generation of GM crops (improved nutritional value, functional foods) will result in new issues (complex metabolic changes, significant impact on overall nutritional status) and thus will pose new challenges in terms of food safety assessment. Nevertheless, it continues to be a grey area with high commercial stakes and as a result of that, the list of GMO and transgenic foods/crops has been expanding steadilty. While consumers accept the functional food concept readily, the acceptance of novel foods as they are defined according to the EU Regulation, and particularly transgenic food, is controversial.

Complementary to conventional breeding techniques, gene technology allows the transfer of genes between unrelated species. Thereby, breeding targets can also be achieved more quickly both in plant and animal breeding. This is one of the key, but hotly debated technologies of our times. Important topics that need to be addressed vary from legislation, such as labeling requirements, to safety and environmental issues. Concerns about application of this agricultural biotechnology are on the ecological impact of growing genetically modified foods, the impact of these crops on biological diversity, and on the safety of food supply, or the development of resistance by insect pests. However, the potential of the agricultural new biotechnologies is enormous also for developing countries. Therefore, questions about agricultural biotechnology must be addressed for people in both developed and developing countries, as we have to address the issue of food security for a world population of some 9000 million people in the year 2050. Furthermore, genetic engineering is not just a new technology for crop improvement, it is a powerful research tool that is helping to provide fresh and better insights of molecular mechanisms involved in biological processes.

Milk Derived Bioactive Components

Milk and milk products are functional foods. Milk contains bioactive components beyond proteins, minerals and vitamins. These minor elements include immunoglobulins, hormones, growth factors, cytokines, nucleotides, polyamines, enzymes and bioactive peptides. Many bioactive peptides are embedded within milk proteins and remain inactive until released and activated by gastrointestinal digestion or during food processing. Bioactive peptides are naturally found in milk, fermented milk and cheese. Successful commercialisation of milk bioactives is dependent upon developing new technologies for their production, producing innovative food and health ingredients, studying the mechanisms of actions, and conducting clinical studies to verify health effects. These bioactive peptides have been reported to exhibit anti-hypertensive activity, immune-modulatory properties and antimicrobial activity against high risk pathogens and hence can play a very important role in alleviation of gut related diseases such as peptic ulcers. Whey proteins such as lactoferrin, beta-lactoglobulin and alpha-lactalbumin also enhance immune cell function. The metabolic activity of probiotic lactic acid bacteria generates de novo immunoregulatory peptides from milk via enzymatic degradation of parent milk proteins. Opioids refer to natural opiates (opium) and synthetic narcotics (i.e. morphine, heroin) that induce sleep and soothe pain. The opioid agonists in milk peptides, derived from casein or whey protein, have shown morphine-like activity. For example, caseinomorphins prolong gastrointestinal transit time, prevent diarrhea, stimulate secretion of insulin and somatostatin and could play a role in appetite suppression. Opioid antagonists such as casoxins and casoplatelins block the agonist effect of externally administered opioids and enkephalin (the endogenous neurotransmitter) thus affecting the release of pain-reducing endorphins.

Packaging of Processed Foods

The rising sales of convenience processed foods/ready to eat foods and developments in their packaging have been a major issue in innovative packaging to attract the consumers. Packaging techniques have developed to an extent to provide very attractively packed foods and consumers are prepared to pay a premium for quality and safe ready to eat foods. Modern consumers being highly health conscious and with the improvement in economy and rising financial status, demand for bottled water, fruit and vegetable juices, milk drinks and milk products, functional foods, sausages etc along with increasing demand for packaged fresh food products is all time high.

In emerging markets where super and hypermarkets are expanding very rapidly, the demand for high barrier materials, active packaging, intelligent packaging, modified atmosphere packaging (MAP), active packaging with antibacterial activity for increasing shelf life of processed foods, nanotechnology and digital print for packaging is rising extraordinarily. This is precisely due to the growing awareness of health conscious consumers towards safety and demand for processed foods with extended shelf life. Packaging industry in India is one of the fast growing industries which has its influence on all the other industries directly or indirectly.

Recently, nanotechnology has been significantly increasing its impact on the food and beverage packaging industry by altering the structure of the materials on the molecular scale, to give the materials desired properties which can significantly enhance the shelf life, efficiently preserve flavour and colour as well as facilitate transportation and usage. Nanotechnology applications for food contact materials/ matrices (FCM) and food packaging constitute the largest share of the market for applications in the food sector. Nano-structured film can effectively prevent the food from the invasion of microorganisms and ensure the food safety. Sensors can alarm us before the food goes rotten or can inform us the exact nutrition status contained in the contents.

Active FCMs generally incorporate nanoparticles with antimicrobial or oxygen scavenging properties whereas Intelligent food packaging can incorporate nanosensors to monitor and report the condition of the food. Based on the antimicrobial action of nanosilver, a number of active FCMs have been developed that are claimed to preserve the foods for longer period by inhibiting the growth of microorganisms. Examples include 'FresherLonger™ Miracle Food Storage Containers' and 'FresherLonger™ Plastic Storage Bags' from Sharper Image® USA, 'Nano Silver Food Containers' from A-DO Korea and 'Nano Silver Baby Milk Bottle' from Baby Dream® Co. Ltd. (South Korea).

The embedded sensors in a packaging film can detect food-spoilage organisms and trigger a colour change to alert the consumer about the end of the shelf life. One of the examples is Nano Bioswitch/ 'Release-on-Command' system that releases a preservative if food begins to spoil. Nanoscale-sensing devices are also being developed that will enable the food or food ingredients to be traced back to the source of origin.

Low Calorie Foods

The current trend towards a more health- and nutrition-conscious lifestyle has encouraged the development of low calorie foods. The non-nutritive sweetener market has been predicted to reach $500 million by the year 2000. A new class of compounds called taste-active proteins functions as sweeteners and flavor modifiers and includes compounds such as aspartame, thaumatin, and monellin. The gene which codes for the protein thaumatin has been isolated and characterised. Transfer of this gene into bacteria would allow the production of thaumatin via fermentation. If engineered into plants, new and unique foods could be developed. Another application of biotechnology in low calorie food production is the development of low calorie fats and oils. Genetically inducing the production of shorter chain fatty acids in soybean or rapeseed would speed the development of a low calorie vegetable oil. The market for this oil could reach $2 billion a year by the end of the next decade.

Diagnostic Tests for Food Safety and Quality Assurance

Detection and identification of pathogenic bacteria in foods is extremely important for ensuring the safety of food supplies as well as for confirming food-related outbreaks. However, microbial detection and identification is a challenge. First of all, high sensitivities are required for preliminary enrichment and subsequent isolation steps that separate the microorganisms from the foods. Secondly, high specificities are needed for microbial identification to rule out the possibility of false positive and false negative results. The identification step can effectively separate the target pathogens from the background microflora. In this context, the conventional methods based on microbiological culturing and metabolic activation have been traditionally used for analysis of foods in dairy industry across the world. The conventional methods of pathogen identification and confirmation based on culturing on selective medium, biochemical tests and immunological assays are extremely laborious, cumbersome and many times remain inconclusive and results are invariably delayed to make them virtually redundant for any follow up corrective actions. However, with the advancements in Biotechnology and Molecular Biology, new innovative molecular and immunological techniques have also been developed and applied in specialised food laboratories in advanced countries like USA and Europe for assessing the microbiological quality and safety of foods. Some of these rapid assays based on molecular techniques such as PCR and Real Time based diagnostics, Immunological assays, biosensors and enzyme based kits, etc. are considered to be more reliable and rapid for quick detection of pathogens.

Biosensors

Biosensors represent analytical new generation of powerful tools incorporating biologically derived material or biomimic with a physiochemical transducer or transducing microsystem. Biosensors are currently being explored for a wide range of applications in food industry. The techniques based on Biosensors are being developed for rapid direct or indirect detection of foodborne microorganisms, toxins, or undesirable metabolites or other compounds. These systems have a potential application in real-time validation of critical control points. Sensitive, specific and rapid processes have been developed that require minimal culture enrichment and utilise immuno-based biosensors, such as imunomagnetic-electrochemiluminescence to detect pathogenic microorganisms in food systems. Immuno-based biosensors to detect low levels of *E. coli* 0157 and *Salmonella* within 2–8 hr are being used in meat and poultry plants as well as in dairy industry. New technologies such as acoustic wave biosensors and radio frequency identification (RFID) sensor tags promise to greatly improve food safety. Research to develop a single computer chip that will automatically assess food safety at any point from source to consumption is ongoing. The advanced biosensor based systems have the advantage to be integrated into the processing line for monitoring the possible contaminants and pathogens on line during different food processing stages so that follow up action could be taken immediately.

IMPACT OF BIOTECHNOLOGY IN FOOD INDUSTRY

In the backdrop of growing human population at an alarming rate in third world countries including India and the ensuing poverty that continue to daunt the countries, the demand for food and nutritional security has increased dramatically. As a result of this, the role of food industry has become extremely pertinent in producing high quality nutritious and wholesome foods which are safe and cost effective to cater to the needs of their vast respective populations. The application of traditional biotechnology in food industry has been in vogue for quite some time and has made a significant impact on commercial production and processing of foods by improving the fermentation efficiency of the micro-organisms through optimisation of processing parameters to produce the desired quality of food products. The most recent application of modern biotechnology to food industry is the genetic modification (GM) also known as genetic engineering/genetic manipulation/gene technology or recombinant DNA technology. The aims are to increase the range and quality of products available, to reduce their price and to protect the environment. Biotechnology has already made a strong impact on food and dairy industry by improving the nutritional quality, shelf life and safety of processed foods with lot of value addition for different applications including health benefits. By adopting new advancements and innovations in modern biotechnology such as rDNA technology, transgenics, animal/plant cloning, tissue culture and improved bioprocess engineering tools, food industry can benefit immensely through not only improving the yield and quality of the processed foods but also bringing in lot of product diversification by producing novel foods customised for specific consumers. Application of biotechnology in true letter and spirit is likely to revolutionise the concept of food processing in the Indian food industry and hence can make a dramatic influence on our lives and that of future generations if used properly and judiciously. However, preceding the advent of such products onto the market, questions were being raised about their safety, labeling, need and ethics. The use of modern biotechnology (recombinant DNA technology) to produce foods and food ingredients is a subject of heightened discussions and controversy among consumers and public policy makers, and within the scientific community and hence can have lot of impact on the industry and the consumers from the health benefits and safety perspectives.

Biotechnological Interventions in Food Processing

Since proteins and vitamins are often lost in traditional food processing, fermentation processes may offer a way to preserve them. Biotechnology can be used for the upgrading of traditional food processing based on fermentation such as the procedures used to produce high quality, nutritious and wholesome fermented foods such as traditional dairy based products like dahi, lassi, shrikhand and nondairy products like gari, a fermented, gritty and starchy food derived from cassava. Biotechnology can also help to eliminate toxic components, either by genetic engineering or through food processing. In addition to eliminating unwanted components, biotechnology can be used for the inexpensive production of additives that increase the nutritive value of the final product or that improve its flavour, texture or appearance. Present-day applications of biotechnology in food processing are far more advanced than applications in the field of plant genetic engineering. The genetic manipulation of micro-organisms used in food processing is considerably easier than the manipulation of more complex plants. It is, therefore. intriguing that research centers primarily on plant genetic engineering, where there are still many obstacles to overcome, while the chance to improve food processing is largely neglected.

Current Status of Biotechnology in Food Processing - Food Fermentations

Microorganisms are an integral part of the processing system during the production of fermented foods. Microbial cultures can be genetically improved using both traditional and molecular approaches, and improvement of bacteria, yeasts and moulds is the subject of much academic and industrial research. Traits which have been considered for commercial food applications in both developed and developing countries include sensory quality (flavour, aroma, visual appearance, texture and consistency), bacteriophage resistance in the case of dairy fermentations, and the ability to produce antimicrobial compounds (e.g. bacteriocins, hydrogen peroxide) for the inhibition of undesirable microorganisms. In many developing countries, the focus is on the degradation or inactivation of natural toxins (e.g. cyanogenic glucosides in cassava), mycotoxins (in cereal fermentations) and anti-nutritional factors (e.g. phytates).

Biotechnology has also been extensively explored in the production of enzymes for application in raw and processed foods. In the past, enzymes were isolated primarily from plant and animal sources, and thus a relatively limited number of enzymes were available to the food processor at a high cost. Today, bacteria and fungi are exploited and used for the commercial production of a diversity of enzymes. Several strains of microorganisms have been selected or genetically modified to increase the efficiency with which they produce enzymes. In most cases, the modified genes are of microbial origin, although they may also come from different kingdoms. For example, the DNA coding for chymosin, an enzyme found in the stomach of bovine and buffalo calves, that causes milk to curdle during the production of cheese, has been successfully cloned into yeasts (*Kluyveromyces lactis/Pichia pastoris*), bacteria (*Escherichia coli*) and moulds (*Aspergillus niger* var. *awamori*). Chymosin produced by these recombinant microorganisms is currently commercially produced and is widely used in cheese manufacture. Genetic technologies have not only improved the efficiency with which enzymes can be produced, but they have increased their availability, reduced their cost and improved their quality. This has had the beneficial impact of increasing efficiency and streamlining processes which employ the use of enzymes as processing aids in the food industry.

Some Issues Relevant to Developing Countries

Biotechnological research as applied to bioprocessing in the majority of developing countries, targets development and improvement of traditional fermentation processes. In this context, some areas specifically relevant to developing countries as listed below need to be looked into before adopting advanced biotechnological tools in the food processing industry.

Socio-economic and cultural factors

Traditional fermentation processes employed in most developing countries are low input, appropriate food processing technologies with minimal investment requirements. They make use of locally produced raw materials and are an integral part of village life. These processes are, however, often uncontrolled, unhygienic and inefficient and generally result in products of variable quality and short shelf lives. Traditional fermented foods like dahi, srikhand and butter milk, etc. nevertheless, find wide consumer acceptance in developing countries and contribute substantially to food security and nutrition. Applications of biotechnology to fermented foods can have a strong impact on these socio-economic and cultural factors.

Infrastructural and logistical factors

Physical infrastructural requirements for the manufacture, distribution and storage (e.g. by refrigeration) of microbial cultures or enzymes on a continuous basis is generally available in urban areas of many developing countries. However, this is not the case in most rural areas of developing countries. Should research be oriented to ensure that individuals at all levels can benefit from applications of biotechnology in food fermentation processes, i.e. should logistical arrangements for starter culture development be integrated into biotechnological research targeting improvement of traditional fermentations? What is required for the level of fermentation technologies and process controls to be upgraded in order to increase efficiency, yields and the quality and safety of fermented foods in developing countries?

Nutrition and food safety

Fermentation processes enhance the nutritional value of foods through the biosynthesis of vitamins, essential amino acids and proteins, through improving protein and fibre digestibility, enhancing micronutrient bioavailability and degrading antinutritional factors. Many bacteria in fermented foods also exhibit functional properties (probiotics). The safety of fermented food products is enhanced through reduction of toxic compounds, such as mycotoxins and cyanogenic glucosides, and production of antimicrobial factors, such as bacteriocins, carbon dioxide, hydrogen peroxide and ethanol, which facilitate inhibition or elimination of food-borne pathogens. Are the nutritional characteristics (and safety aspects) of fermented foods adequately documented and appreciated in developing countries? Is there a need for consumer education about the benefits of fermented foods?

Intellectual property rights (IPRs)

The processes used in the more advanced areas of agricultural biotechnology tend to be covered by IPRs and these rights tend to be owned by parties in developed countries. This applies also to biotechnology processes used in food processing. On the other hand, many of the traditional fermentation processes applied in developing countries are based on traditional knowledge. In addition to biotechnology processes, microbial strains may also be the object of IPRs. For example, an era of massive private investment in biotechnology was initiated when the United States Supreme Court ruled in 1980 (in the Diamond versus Chakrabarty case) that a live GM bacterium (of the genus *Pseudomonas*, modified to degrade components of crude oil) could be patented. Many of the microorganisms associated with traditional fermentation processes in developing countries are unique. Issues of ownership will become increasingly important as bacterial strains are characterised and starter cultures are developed in developing countries.

COMMERCIAL OPPORTUNITIES

Biotechnological innovations have greatly assisted in industrialising production of certain indigenous fermented foods. Indonesian tempe and Oriental soy sauce are well known examples of indigenous fermented foods that have been industrialised and marketed globally. The results of biotechnology research will lead to fermented foods of improved quality, safety and consistency. Should biotechnology developments in developing countries target commercialisation? Should they target diversification into new value-added products? Should biotechnology development be linked to technological developments in food processing? Can the application of biotechnology to food processing allow farmers in developing countries to add value to their agricultural products (for export or for local consumption) and improve their revenues?

Biotechnology has already made a strong impact on food and dairy industry by improving the nutritional quality, shelf life and safety of processed foods with lot of value addition for different applications including health benefits. By adopting new advancements and innovations in modern biotechnology such as rDNA technology, transgenics, animal/plant cloning, tissue culture and improved bioprocess engineering tools, food industry can benefit immensely though not only improving the yield and quality of the processed foods but also bringing in lot of product diversification by producing novel foods customised for specific consumers. Application of biotechnology in true letter and spirit is likely to revolutionise the concept of food processing in the Indian food industry and hence can make a dramatic influence on our lives and that of future generations if used properly and judiciously.

Biotechnology undoubtedly has a potential role in food processing industry in India and other developing countries and hence can help in meeting the food and nutritional security effectively. Judicious use of modern biotechnology tools and strategies could be extremely valuable not only to increase the food production for the growing population but also can aid in improving the processing quality, taste, nutritional value, texture, shelf life, marketability and added advantages of having medicinal properties for various ailments, thereby, enhancing the commercial value of these foods considerably. The resurgence of concept of functional foods and nutraceuticals for health applications gained momentum at the global level through biotechnological applications. These value added biotech based farm products tailored to processing industries certainly can increase farmers and processors revenue, at the same time satisfying the consumer preferences. Biotechnology has tremendous potential for increasing food production and improving food processing although the real impact will only be felt after a few decades and it will differ from country to country. Nevertheless, biotechnology can have a dramatic impact on the food processing industry in developing countries like India by not only improving the efficiency of food processing but also through value addition and product diversification for catering to the needs of both domestic market and their exports. Additionally, biotechnology interventions in the food chain of agriculture and food processing sectors can generate lot of employment opportunities in the country. By producing safe, high quality, nutritious wholesome and healthy foods within reach of common consumers, biotechnology can help in creating a healthy society and can tremendously boost the growth, productivity and economic status of India at the global level.

Chapter 4

Biotechnology of Fermentation

INTRODUCTION

Fermentation is traditionally a process which enabled to preserve food and as such has been used for centuries until present. However now-a-days, the main purpose of food fermentation is not to preserve, since other preservation techniques are known, but to produce a wide variety of fermentation products with specific taste, flavour, aroma and texture. Using various microbial strains, fermentation conditions (micro-organisms, substrates, temperature, time of fermentation, etc.) and chemical engineering achievements, enable us to manufacture hundreds of types of dairy (cheeses, fermented milk products), vegetable (sauerkraut, pickles, olives), meat (fermented sausages) products, breads, alcoholic beverages (wine, beer, cider), vinegar and other food acids, as well as oils.

Fermentation processes utilise micro-organisms to convert solid or liquid substrates into various products. The substrates used vary widely, any material that supports microbial growth being a potential substrate. Similarly, fermentation derived products show tremendous variety. Commonly consumed fermented products include bread, cheese, sausage, pickled vegetables, cocoa, beer, wine, citric acid, glutamic acid and soya sauce.

Fermentations may require only a single species of micro-organism to effect the desired chemical change. In this case the substrate may be sterilised, to kill unwanted species prior to inoculation with the desired micro-organism. However, most food fermentations are non-sterile. Typically fermentations used in food processing require the participation of several microbial species, acting simultaneously and/or sequentially, to give a product with the desired properties, including appearance, aroma, texture and taste. In non-sterile fermentations, the culture environment may be tailored specifically to favour the desired micro-organisms. For example, the salt content may be high, the pH may be low, or the water activity may be reduced by additives such as salt or sugar.

Factors influencing fermentation: A fermentation is influenced by numerous factors, including temperature, pH, nature and composition of the medium, dissolved O_2, dissolved CO_2, operational system (e.g. batch, fedbatch, continuous), feeding with precursors, mixing (cycling through varying environments), and shear rates in the fermenter. Variations in these factors may affect: the rate of fermentation, the product spectrum and yield, the organoleptic properties of the product (appearance, taste, smell and texture), the

56

generation of toxins, nutritional quality, and other physico-chemical properties. The formulation of the fermentation medium affects the yield, rate and product profile. The medium must provide the necessary amounts of carbon, nitrogen, trace elements and micronutrients (e.g. vitamins). Specific types of carbon and nitrogen sources may be required, and the carbon : nitrogen ratio may have to be controlled. An understanding of fermentation biotechnology is essential for developing a medium with an appropriate formulation. Concentrations of certain nutrients may have to be varied in a specific way during a fermentation to achieve the desired result. Some trace elements may have to be avoided—for example, minute amounts of iron reduce yields in citric acid production by *Aspergillus niger*. Additional factors, such as cost, availability, and batch-to-batch variability also affect the choice of medium.

TYPES OF FERMENTATION

Most commercially useful fermentations may be classified as either solid-state or submerged cultures. In solid-state fermentations, the micro-organisms grow on a moist solid with little or no 'free' water, although capillary water may be present. Examples of this type of fermentation are seen in mushroom cultivation, bread-making and the processing of cocoa, and in the manufacture of some traditional foods, e.g. miso (soya paste), sake, soyasauce, tempeh (soyabean cake) and gari (cassava), which are now produced in large industrial operations. Submerged fermentations may use a dissolved substrate, e.g. sugar solution, or a solid substrate, suspended in a large amount of water to form a slurry. Submerged fermentations are used for pickling vegetables, producing yoghurt, brewing beer and producing wine and soya sauce.

Solid-state Fermentation

Solid-state (substrate) fermentation (SSF) has been defined as the fermentation process occurring in the absence or near-absence of free water. SSF processes generally employ a natural raw material as carbon and energy source. SSF can also employ an inert material as solid matrix, which requires supplementing a nutrient solution containing necessary nutrients as well as a carbon source. Solid substrate (matrix), however, must contain enough moisture. Depending upon the nature of the substrate, the amount of water absorbed could be one or several times more than its dry weight, which leads relatively high water activity (a_w) on the solid/gas interface in order to allow higher rate of biochemical process. Low diffusion of nutrients and metabolites takes place in lower water activity conditions whereas compaction of substrate occurs at higher water activity. Hence, maintenance of adequate moisture level in the solid matrix along with suitable water activity are essential elements for SSF processes. Solid substrates should have generally large surface area per unit volume (say in the range of 10^3–10^6 m^2/cm^3 for the ready growth on the solid/gas interface). Smaller substrate particles provide larger surface area for microbial attack but pose difficulty in aeration/respiration due to limitation in inter-particle space availability. Larger particles provide better aeration/respiration opportunities but provide lesser surface area. In bioprocess optimisation, sometimes it may be necessary to use a compromised size of particles (usually a mixed range) for the reason of cost effectiveness. For example, wheat bran, which is the most commonly used substrate in SSF, is obtained in two forms, fine and coarse. Former contains particles of smaller size (mostly smaller than 500–600 μ) and the latter mostly larger than these. Most of SSF processes use a mix of these two forms at different ratios for optimal production.

Solid substrates generally provide a good dwelling environment to the microbial flora comprising bacteria, yeast and fungi. Among these, filamentous fungi are the best studied for SSF due to their hyphal growth, which have the capability to not only grow on the surface of the substrate particles but also penetrate through them. Several agro crops such as cassava, barley, etc. and agro-industrial residues such as wheat bran, rice bran, sugarcane bagasse, cassava bagasse, various oil cakes (e.g. coconut oil cake, palm kernel cake, soyabean cake, ground nut oil cake, etc.), fruit pulps (e.g. apple pomace), corn cobs, saw dust, seeds (e.g. tamarind, jack fruit), coffee husk and coffee pulp, tea waste, spent brewing grains, etc. are the most often and commonly used substrates for SSF processes. During the growth on such substrates hydrolytic exo-enzymes are synthesised by the micro-organisms and excreted outside the cells, which create and help in accessing simple products (carbon source and nutrients) by the cells. This in turn promotes biosynthesis and microbial activities. Apart from these, there are several other important factors, which must be considered for development of SSF processes. These include physico-chemical and biological factors such as pH of the medium, temperature and period of incubation, age, size and type of inoculum, nature of substrate, type of micro-organism employed, etc.

Significance of SSF

SSF has been considered superior in several aspects to submerged fermentation (SmF) due to various advantages it renders. It is cost effective due to the use of simple growth and production media comprising agro-industrial residues, uses little amount of water, which consequently releases negligible or considerably less quantity of effluent, thus reducing pollution concerns. SSF processes are simple, use low volume equipment (lower cost), and are yet effective by providing high product titres (concentrated products). Further, aeration process (availability of atmospheric oxygen to the substrate) is easier since oxygen limitation does not occur as there is a increased diffusion rate of oxygen into moistened solid substrate, supporting the growth of aerial mycelium. These could be effectively used at smaller levels also, which makes them suitable for rural areas also.

General aspects of SSF

There are several important aspects, which should be considered in general for the development of any bioprocess in SSF. These include selection of suitable micro-organism and substrate, optimisation of process parameters and isolation and purification of the product. Going by theoretical classification based on water activity, only fungi and yeast were termed as suitable micro-organisms for SSF. It was thought that due to high water activity requirement, bacterial cultures might not be suitable for SSF. However, experience has shown that bacterial cultures can be well managed and manipulated for SSF processes. It has been generally claimed that product yields are mostly higher in SSF in comparison to SmF. However, so far there is not any established scale or method to compare product yields in SSF and SmF in true terms. The exact reasoning for higher product titres in SSF is not well known currently. The logical reasoning given is that in SSF microbial cultures are closer to their natural habitat and probably hence their activity is increased.

Selection of a proper substrate is another key aspect of SSF. In SSF, solid material is non-soluble that acts both as physical support and source of nutrients. Solid material could be a naturally occurring solid substrate such as agricultural crops, agro-industrial residues or inert support. However, it is not necessary to combine the role of support and substrate but rather reproduce the conditions of low water activity and high oxygen transference by using a nutritionally inert material soaked with a nutrient solution. In relation to selection of substrate, there could be two major considerations, one that there is a specific substrate, which requires suitable value-addition and/or disposal. The second could be related with the goal of producing a specific product from a suitable substrate. In the latter case, it would be necessary to screen various substrates and select the most suitable one. Similarly it would be important to screen suitable micro-organisms and select the most suitable one. If inert materials such as polyurethane foam are used, product isolation could be relatively simpler and cheaper than using naturally occurring raw materials such as wheat bran because while extracting the product after fermentation, along with the product, several other water-soluble components from the substrate also leach out and may pose difficulties in purification process. Inert materials have been often used for studying modelling or other fundamental aspects of SSF.

Other relevant issues here could be the selection of process parameters and their optimisation. These include physico-chemical and biochemical parameters such as particle size, initial moisture, pH and pre-treatment of the substrate, relative humidity, temperature of incubation, agitation and aeration, age and size of the inoculum, supplementation of nutrients such as N, P and trace elements, supplementation of additional carbon source and inducers, extraction of product and its purification, etc. Depending upon the kind, level and application of experimentation, single and/or multiple variable parameters optimisation method could be used for these.

Solid-state fermenters

Solid-state fermentation devices vary in technical sophistication, from very primitive banana-leaf wrappings, bamboo baskets and substrate heaps to the highly automated machines used mainly in Japan. Some 'less sophisticated' fermentation systems, e.g. the fermentation of cocoa beans in heaps, are quite effective at large-scale processing. Also, some of the continuous, highly mechanised processes for the fermentation of soya sauce, that are successful in Japan, are not suitable for less highly developed locations in Asia. Thus, fermentation practice must be tailored to local conditions.

The use of pressure vessels is not the norm for solid-state fermentation. The commonly used devices are:

- Tray fermenter
- Static-bed fermenter
- Tunnel fermenter
- Rotary disc fermenter
- Rotary drum fermenter
- Agitated-tank fermenter
- Continuous screw fermenter

These are described below:

Large concrete or brick fermentation chambers, or koji rooms, may be lined with steel, typically Type 304 stainless steel and are corrosion-resistant.

Tray fermenter: This is a simple type of fermenter, widely used in small- and medium-scale koji operations (Fig. 4.1). The trays are made of wood, metal or plastic, and often have a perforated or wire-mesh base to achieve improved aeration. The substrate is fermented in shallow 0.15 m deep) layers. The trays may be covered with cheese cloth to reduce contamination, but processing is non-sterile. Single or stacked trays may be located in chambers in which the temperature and humidity are controlled, or simply in ventilated areas. Inoculation and occasional mixing are done manually, although the handling, filling, emptying and washing of trays may be automated.

Despite some automation, tray fermenters are labour-intensive, and require a large production area. Hence the potential for scaling up production is limited.

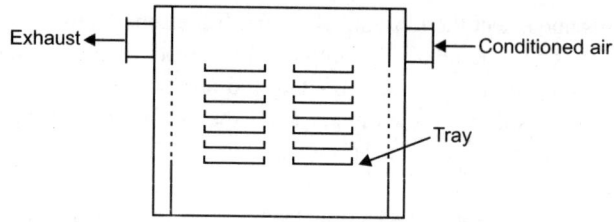

Fig. 4.1: Tray fermenter.

Static-bed fermenter: This is an adaptation of the tray fermenter (Fig. 4.2). It employs a single, larger and deeper, static bed of substrate located in an insulated chamber. O_2 is supplied by forced aeration through the bed of substrate.

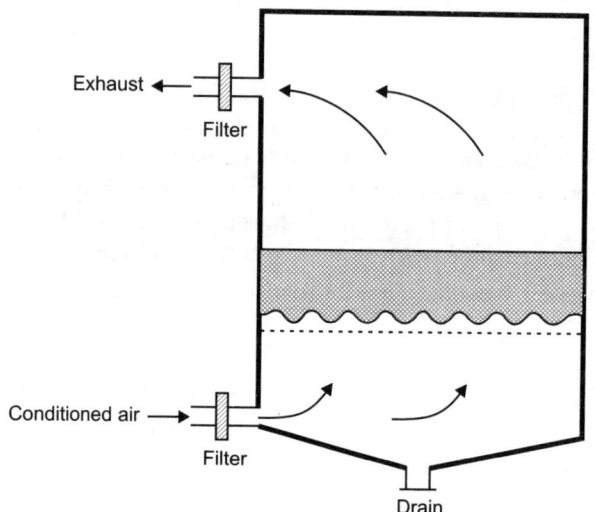

Fig. 4.2: Static-bed fermenter.

Tunnel fermenter: This is an adaptation of the static-bed device (Fig. 4.3). Typically, the bed of solids is quite long but no deeper than 0.5 m. Fermentation using this equipment may be highly automated, by way of mechanisms for mixing, inoculation, continuous feeding and harvest of the substrate.

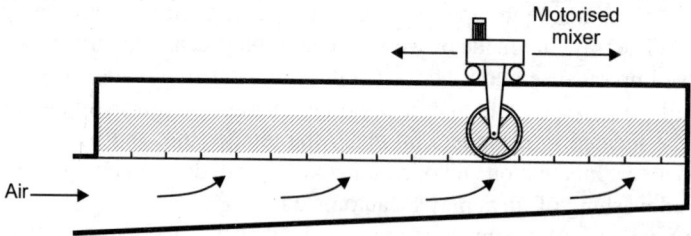

Fig. 4.3: Tunnel fermenter.

Rotary disc fermenter: The rotary disc fermenter consists of upper and lower chambers, each with a circular perforated disc to support the bed of substrate A common central shaft rotates the discs. Inoculated substrate is introduced into the upper chamber, and slowly moved to the transfer screw. The upper screw transfers the partly fermented solids through a mixer to the lower chamber, where further fermentation occurs. The fermented substrate is harvested using the lower transfer screw. Both chambers are aerated with humidified, temperature-controlled air. Rotary disc fermenters are used in large-scale koji production in Japan.

Rotary drum fermenter: The cylindrical drum of the rotary drum fermenter is supported on rollers, and rotated at 1–5 r.p.m. around the long axis (Fig. 4.4). Rotation may be intermittent, and the speed may vary with the fermentation stage. Straight or curved baffles inside the drum aid in the tumbling of the substrate, hence improving aeration and temperature control. Sometimes the drum can be inclined, causing the substrate to move from the higher inlet end to the lower outlet during rotation. Aeration occurs through coaxial inlet and exhaust nozzles.

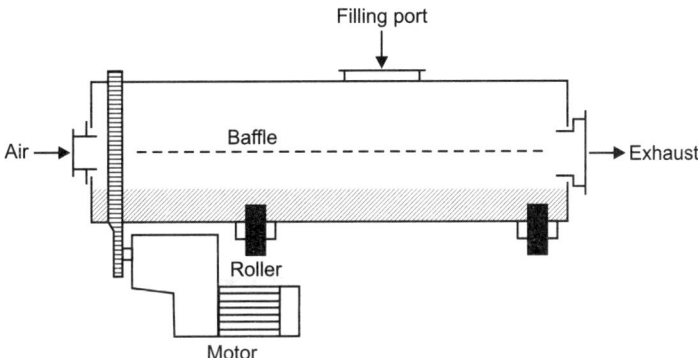

Fig. 4.4: Rotary drum fermenter.

Agitated-tank fermenter: In this type of fermenter, either one or more helical-screw agitators are mounted in cylindrical or rectangular tanks, to agitate the fermented substrate (Fig. 4.5). Sometimes, the screws extend into the tank from mobile trolleys, that ride on horizontal rails located above the tank. Another stirred-tank configuration is the paddle fermenter. This is similar to the rotary drum device, but the drum is stationary and periodic mixing is achieved by motor-driven paddles supported on a concentric shaft.

Continuous screw fermenter: In this type of fermenter, sterilised, cooled and inoculated substrate is fed in through the inlet of the non-aerated chamber (Fig. 4.6). The solids are moved towards the harvest port by the screw, and the speed of rotation and the length of the screw control the fermentation time. This type of fermenter is suitable for continuous anaerobic or micro-aerophilic fermentations.

Submerged Fermentations

Fermentation systems

Industrial fermentations may be carried out either batchwise, as fedbatch operations, or as continuous cultures (Fig. 4.7). Batch and fedbatch operations are quite common, continuous fermentations being relatively rare. For example, continuous brewing is used commercially, but most beer breweries use batch processes.

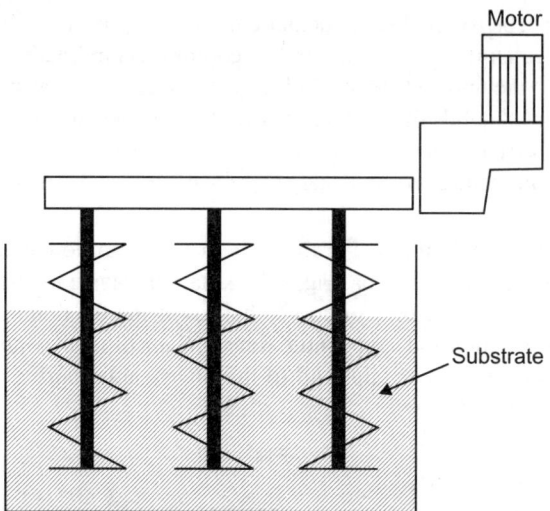

Fig. 4.5: Agitated-tank fermenter.

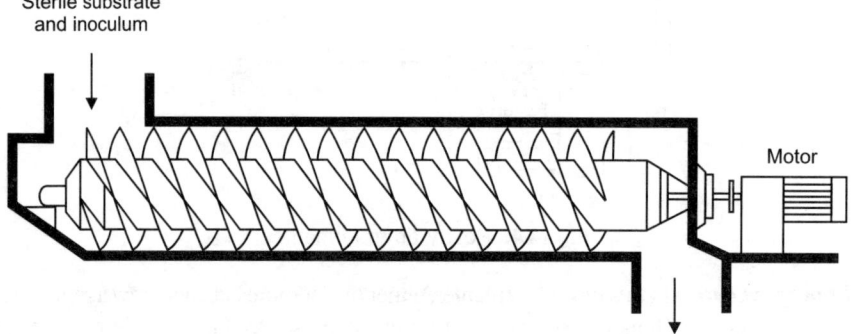

Fig. 4.6: Continuous screw fermenter.

In batch processing, a batch of culture medium in a fermenter is inoculated with a micro-organism (the 'starter culture'). The fermentation proceeds for a certain duration (the 'fermentation time' or 'batch time'), and the product is harvested. Batch fermentations typically extend over 4–5 days, but some traditional food fermentations may last months. In fedbatch fermentations, sterile culture medium is added either continuously or periodically to the inoculated fermentation batch. The volume of the fermenting broth increases with each addition of the medium, and the fermenter is harvested after the batch time.

In continuous fermentations, sterile medium is fed continuously into a fermenter and the fermented product is continuously withdrawn, so the fermentation volume remains unchanged. Typically, continuous fermentations are started as batch cultures and feeding begins after the microbial population has reached a certain concentration. In some continuous fermentations, a small part of the harvested culture may be recycled, to continuously inoculate the sterile feed medium entering the fermenter (Fig. 4.7d). Whether continuous inoculation is necessary depends on the type of mixing in the fermenter. 'Plug flow' fermentation devices (Fig. 4.7d), such as long tubes that do not allow back mixing, must be inoculated continuously. Elements of fluid moving along in a plug flow device behave like tiny batch fermenters. Hence, true batch fermentation processes are relatively easily transformed into continuous operations in plug flow fermenters, especially if pH control and aeration are not required. Continuous cultures are particularly susceptible to microbial contamination, but in some cases the fermentation conditions may be selected (e.g. low pH, high alcohol or salt content) to favour the desired micro-organisms compared to potential contaminants.

In a 'well-mixed' continuous fermenter (Fig. 4.7c), the feed rate of the medium should be such that the dilution rate, i.e. the ratio of the volumetric feed rate to the constant culture volume, remains less than the maximum specific growth rate of the micro-organism in the particular medium and at the particular fermentation conditions. If the dilution rate exceeds the maximum specific growth rate, the micro-organism will be washed out of the fermenter.

Industrial fermentations are mostly batch operations. Typically, a pure starter culture (or seed), maintained under carefully controlled conditions, is used to inoculate sterile Petri dishes or liquid medium in the shake flasks. After sufficient growth, the pre-culture is used to inoculate the 'seed' fermenter. Because industrial fermentations tend to be large (typically 150–250 m^3), the inoculum is built up through several successively larger stages, to 5–10% of the working volume of the production fermenter. A culture in rapid exponential growth is normally used for inoculation. Slower-growing micro-organisms require larger inocula, to reduce the total duration of the fermentation. An excessively long fermentation time (or batch time) reduces productivity (amount of product produced per unit time per unit volume of fermenter), and increases costs. Sometimes inoculation spores, produced as seeds, are blown directly into large fermentation vessels with the in going air.

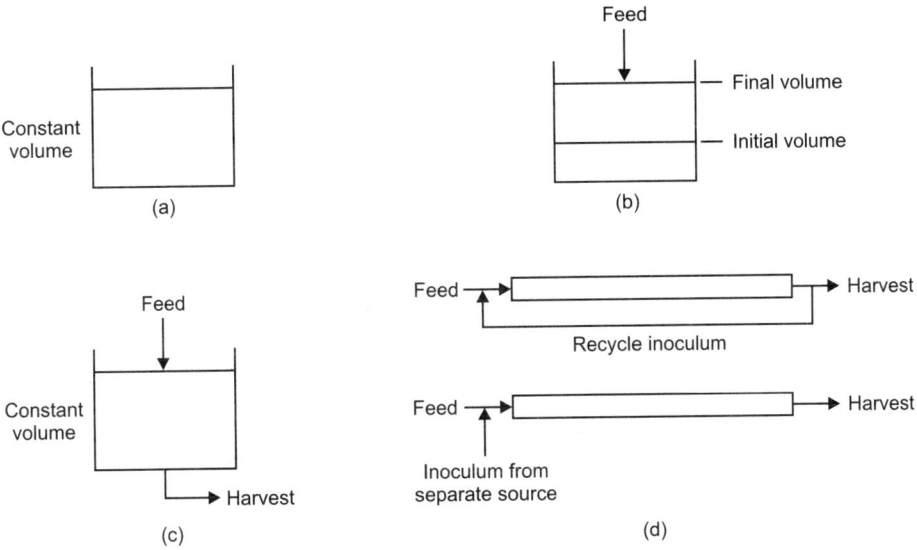

Fig. 4.7: Fermentation methodologies. (a) Batch fermentation. (b) Fedbatch culture. (c) Continuous-flow well-mixed fermentation. (d) Continuous plug flow fermentation, with and without recycling of inoculum.

Submerged-culture fermenters

The major types of submerged-culture bio-reactor are:

- Stirred-tank fermenter.
- Bubble column.
- Airlift fermenter.
- Fluidised-bed fermenter.
- Trickle-bed fermenter.

These are shown in Fig. 4.8.

Stirred-tank fermenter (see Fig. 4.8a.) This is a cylindrical vessel with a working height-to-diameter ratio (aspect ratio) of 3–4. A central shaft supports three to four impellers, placed about 1 impeller-diameter apart. Various types of impeller, that direct the flow axially (parallel to the shaft) or radially (outwards from the shaft) may be used. Sometimes axial- and radial-flow impellers are used on the same shaft. The vessel is provided with four equally spaced vertical baffles, that extend from near the walls into the vessel. Typically, the baffle width is 8–10% of the vessel diameter.

Bubble column (see Fig. 4.8b) This is a cylindrical vessel with a working aspect ratio of 4–6. It is sparged at the bottom, and the compressed gas provides agitation. Although simple, it is not widely used because of its poor performance relative to other systems. It is not suitable for very viscous broths or those containing large amounts of solids.

Airlift fermenters (see Figs 4.8c and 4.8d) These come in internal-loop and external-loop designs. In the internal-loop design, the aerated riser and the unaerated downcomer are contained in the same shell. In the external-loop configuration, the riser and the downcomer are separate tubes that are linked near the top and the bottom. Liquid circulates between the riser (upward flow) and the downcomer (downward flow). Generally, these are very capable fermenters, except for handling the most viscous broths. Their ability to suspend solids and transfer O_2 and heat is good. The hydrodynamic shear is low. The external-loop design is relatively little-used in industry.

Fluidised-bed fermenter (see Fig. 4.8e) These are similar to bubble columns with an expanded cross section near the top. Fresh or recirculated liquid is continuously pumped into the bottom of the vessel, at a velocity that is sufficient to fluidise the solids or maintain them in suspension. These fermenters need an external pump. The expanded top section slows the local velocity of the upward flow, such that the solids are not washed out of the bioreactor.

Trickle-bed fermenter (see Fig. 4.8F.) These consist of a cylindrical vessel packed with support material (e.g. woodchips, rocks, plastic structures). The support has large open spaces, for the flow of liquid and gas and the growth of micro-organisms on the solid support. A liquid nutrient broth is sprayed onto the top of the support material, and trickles down the bed. Air may flow up the bed, countercurrent to the liquid flow. These fermenters are used in vinegar production, as well as in other processes. They are suitable for liquids with low viscosity and few suspended solids.

Safe Fermentation Practice

The micro-organisms used in certain industrial fermentations are potentially harmful. Certain strains have caused fatal infections in immuno compromised individuals, and rare cases of fatal disease in previously healthy adults have also been reported. Microbial spores and fermentation products, as well as microbes, have been implicated in occupational diseases. Most physiologically active fermentation

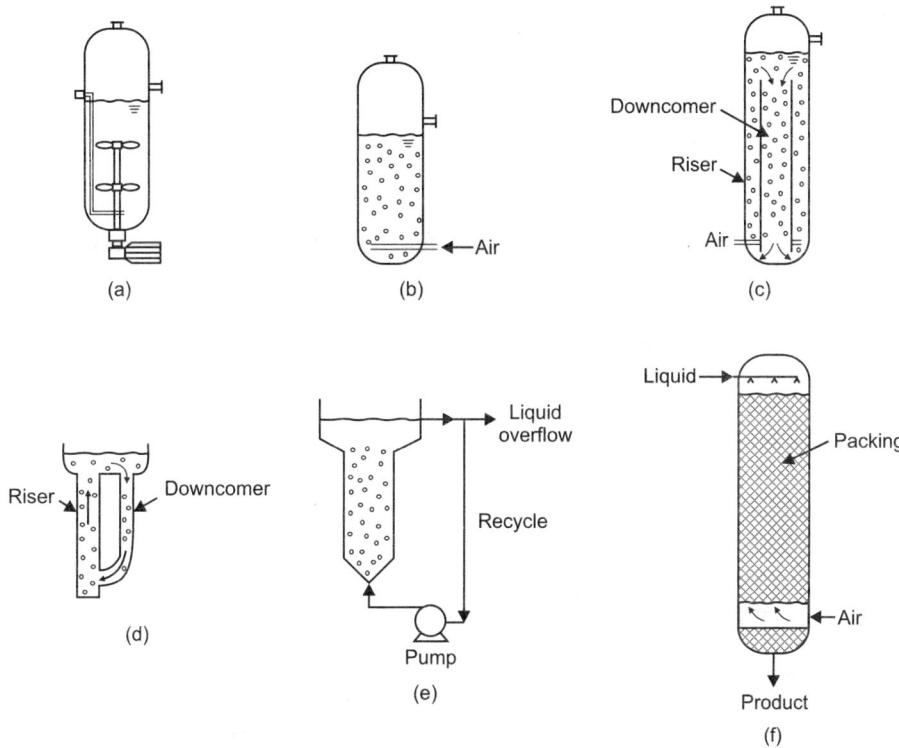

Fig. 4.8: Types of submerged-culture fermenter. (a) Stirred-tank fermenter. (b) Bubble column. c) Internal-loop airlift fermenter. (d) External-loop airlift fermenter. (e) Fluidised-bed fermenter. (f) Trickle-bed fermenter.

products are potentially disruptive to health, and certain products are highly toxic. The product spectrum of a given micro-organism often depends on the fermentation conditions. Under certain environmental conditions, some organisms, e.g. *Aspergillus flavus* and *A. oryzae,* are known to produce lethal toxins, and specific strains of the blue-veined cheese mold *Penicillium roqueforti* also produce mycotoxins under narrowly defined environmental conditions. Poor operational practice and failings in process and plant design can increase the risks.

RECOVERY AND PURIFICATION OF PRODUCTS

Downstream processing is usually considered a specialised field in biochemical engineering, itself a specialisation within chemical engineering, though many of the key technologies were developed by chemists and biologists for laboratory-scale separation of biological products.

Downstream processing and analytical bioseparation both refer to the separation or purification of biological products, but at different scales of operation and for different purposes. Downstream processing implies manufacture of a purified product fit for a specific use, generally in marketable quantities, while analytical bioseparation refers to purification for the sole purpose of measuring a component or components of a mixture, and may deal with sample sizes as small as a single cell.

Stages in Downstream Processing

A widely recognised heuristic for categorising downstream processing operations divides them into four groups which are applied in order to bring a product from its natural state as a component of a tissue, cell or fermentation broth through progressive improvements in purity and concentration.

Removal of insolubles: Removal of insolubles is the first step and involves the capture of the product as a solute in a particulate-free liquid, for example the separation of cells, cell debris or other particulate matter from fermentation broth containing an antibiotic. Typical operations to achieve this are filtration, centrifugation, sedimentation, flocculation, electro-precipitation, and gravity settling. Additional operations such as grinding, homogenisation, or leaching, required to recover products from solid sources such as plant and animal tissues, are usually included in this group.

Product isolation: Product isolation is the removal of those components whose properties vary markedly from that of the desired product. For most products, water is the chief impurity and isolation steps are designed to remove most of it, reducing the volume of material to be handled and concentrating the product. Solvent extraction, adsorption, ultrafiltration, and precipitation are some of the unit operations involved.

Product purification: Product purification is done to separate those contaminants that resemble the product very closely in physical and chemical properties. Consequently steps in this stage are expensive to carry out and require sensitive and sophisticated equipment. This stage contributes a significant fraction of the entire downstream processing expenditure. Examples of operations include affinity, size exclusion, reversed phase chromatography, crystallisation and fractional precipitation.

Product polishing: Product polishing describes the final processing steps which end with packaging of the product in a form that is stable, easily transportable and convenient. Crystallisation, desiccation, lyophilisation and spray drying are typical unit operations. Depending on the product and its intended use, polishing may also include operations to sterilise the product and remove or deactivate trace contaminants which might compromise product safety. Such operations might include the removal of viruses or depyrogenation.

A few product recovery methods may be considered to combine two or more stages. For example, expanded bed adsorption accomplishes removal of insolubles and product isolation in a single step. Affinity chromatography often isolates and purifies in a single step.

Removal of insoluble components

Filtration: Filtration is a mechanical or physical operation which is used for the separation of solids from fluids (liquids or gases) by interposing a medium through which only the fluid can pass. Oversize solids in the fluid are retained, but the separation is not complete, solids will be contaminated with some fluid and filtrate will contain fine particles (depending on the pore size and filter thickness).

Liquid filtration: There are many different methods of filtration, all aim to attain the separation of substances. Separation is achieved by some form of interaction between the substance or objects to be removed and the filter. The substance that is to pass through the filter must be a fluid, i.e. a liquid or gas. Methods vary depending on the location of the targeted material, i.e. whether it is in the fluid phase or not.

Centrifugation

Centrifugation is a process that involves the use of the centrifugal force for the separation of mixtures, used in industry and in laboratory settings. More-dense components of the mixture migrate away from the axis of the centrifuge, while less-dense components of the mixture migrate towards the axis. Chemists and biologists may increase the effective gravitational force on a test tube so as to more rapidly and completely cause the precipitate ('pellet') to gather on the bottom of the tube. The remaining solution is properly called the 'supernate' or 'supernatant liquid'. The supernatant liquid is then either quickly decanted from the tube without disturbing the precipitate, or withdrawn with a Pasteur pipette.

Microcentrifuges and superspeed centrifuges: In microcentrifugation, centrifuges are run in batch to isolate small volumes of biological molecules or cells (prokaryotic and eukaryotic). Nuclei is also often purified via microcentrifugation. Microcentrifuge tubes generally hold 1.5–2 ml of liquid, and are spun at maximum angular speeds of 12000–13000 rpms. Microcentrifuges are small and have rotors that can quickly change speeds. Superspeed centrifuges work similarly to microcentrifuges, but are conducted via larger scale processes. Superspeed centrifuges are also used for purifying cells and nuclei, but in larger quantities. These centrifuges are used to purify 25–30 ml of solution within a tube. Additionally, larger centrifuges also reach higher angular velocities (around 30000 rpm) and also use a larger rotor.

Ultracentrifugation: Ultracentrifugation makes use of high centrifugal force for studying properties of biological particles. While microcentrifugation and superspeed centrifugation are used strictly to purify cells and nuclei, ultracentrifugation can isolate much smaller particles, including ribosomes, proteins, and viruses. Ultracentrifuges can also be used in the study of membrane fractionation. This occurs because ultracentrifuges can reach maximum angular velocites in excess of 70000 rpm. Additionally, while microcentrifuges and supercentrifuges separate particles in batch, ultracentrifuges can separate molecules in batch and continuous flow systems.

In addition to purification, analytical ultracentrifugation (AUC) can be used for determination of macromolecular properties, including the amino acid composition of a protein, the protein's current conformation, or properties of that conformation. In analytical ultracentrifuges, concentration of solute is measured using optical calibrations. For low concentrations, the Beer-Lambert law can be used to measure the concentration. Analytical ultracentrifuges can be used to simulate physiological conditions (correct pH and temperature). In analytical ultracentrifuges, molecular properties can be modelled through sedimentation velocity analysis or sedimentation equilibrium analysis. In sedimentation velocity analysis, concentrations and solute properties are modelled continuously over time. Sedimentation velocity analysis can be used to determine the macromolecules shape, mass, composition, and conformational properties. During sedimentation equilibrium analysis, centrifugation has stopped and particle movement is based on diffusion. This allows for modelling of the mass of the particle as well as the chemical equilibrium properties of interacting solutes.

Sedimentation

Sedimentation is the tendency for particles in suspension or molecules in solution to settle out of the fluid in which they are entrained, and come to rest against a wall. This is due to their motion through the fluid in response to the forces acting on them: these forces can be due to gravity, centrifugal acceleration or electromagnetism.

Flocculation

Flocculation is, in the field of chemistry, a process where colloids come out of suspension in the form of floc or flakes. The action differs from precipitation in that, prior to flocculation, colloids are merely suspended in a liquid and not actually dissolved in a solution.

Product Isolation

Adsorption

Adsorption is the accumulation of atoms or molecules on the surface of a material. This process creates a film of the adsorbate (the molecules or atoms being accumulated) on the adsorbent's surface. It is different from absorption, in which a substance diffuses into a liquid or solid to form a solution. The term sorption encompasses both processes, while desorption is the reverse process of 'adsorption'.

Precipitation

Precipitation is the formation of a solid in a solution during a chemical reaction. When the reaction occurs, the solid formed is called the precipitate, and the liquid remaining above the solid is called the supernate. Powders derived from precipitation have also historically been known as flowers. An important stage of the precipitation process is the onset of nucleation. The creation of a hypothetical solid particle includes the formation of an interface, which requires some energy based on the relative surface energy of the solid and the solution. If this energy is not available, and no suitable nucleation surface is available, supersaturation occurs.

Chromatography

Chromatography is the collective term for a set of laboratory techniques for the separation of mixtures. It involves passing a mixture dissolved in a 'mobile phase' through a stationary phase, which separates the analyte to be measured from other molecules in the mixture based on differential partitioning between the mobile and stationary phases. Subtle differences in compounds partition coefficient results in differential retention on the stationary phase and thus separation.

Chromatography may be preparative or analytical. The purpose of preparative chromatography is to separate the components of a mixture for further use (and is thus a form of purification). Analytical chromatography is done normally with smaller amounts of material and is for measuring the relative proportions of analytes in a mixture. The two are not mutually exclusive.

Column chromatography: Column chromatography is a separation technique in which the stationary bed is within a tube. The particles of the solid stationary phase or the support coated with a liquid stationary phase may fill the whole inside volume of the tube (packed column) or be concentrated on or along the inside tube wall leaving an open, unrestricted path for the mobile phase in the middle part of the tube (open tubular column). Differences in rates of movement through the medium are calculated to different retention times of the sample.

Planar chromatography: Planar chromatography is a separation technique in which the stationary phase is present as or on a plane. The plane can be a paper, serving as such or impregnated by a substance as the stationary bed (paper chromatography) or a layer of solid particles spread on a support such as a glass plate (thin layer chromatography). Different compounds in the sample mixture travel different distances according to how strongly they interact with the stationary phase as compared to the mobile phase. The specific retardation factor (R_f) of each chemical can be used to aid in the identification of an unknown substance.

Paper chromatography: Paper chromatography is a technique that involves placing a small dot or line of sample solution onto a strip of chromatography paper. The paper is placed in a jar containing a shallow layer of solvent and sealed. As the solvent rises through the paper, it meets the sample mixture which starts to travel up the paper with the solvent. This paper is made of cellulose, a polar substance, and the compounds within the mixture travel farther if they are non-polar. More polar substances bond with the cellulose paper more quickly, and therefore do not travel as far.

Thin layer chromatography

Thin layer chromatography (TLC) is a widely-employed laboratory technique and is similar to paper chromatography. However, instead of using a stationary phase of paper, it involves a stationary phase of a thin layer of adsorbent like silica gel, alumina, or cellulose on a flat, inert substrate. Compared to paper, it has the advantage of faster runs, better separations, and the choice between different adsorbents. For even better resolution and to allow for quantitation, high-performance TLC can be used.

Gas chromatography: Gas chromatography, also sometimes known as gas-liquid chromatography, (GLC), is a separation technique in which the mobile phase is a gas. Gas chromatography is always carried out in a column, which is typically 'packed' or 'capillary'.

Gas chromatography (GC) is based on a partition equilibrium of analyte between a solid stationary phase (often a liquid silicone-based material) and a mobile gas (most often helium). The stationary phase is adhered to the inside of a small-diameter glass tube (a capillary column) or a solid matrix inside a larger metal tube (a packed column). It is widely used in analytical chemistry, though the high temperatures used in GC make it unsuitable for high molecular weight biopolymers or proteins (heat will denature them), frequently encountered in biochemistry, it is well suited for use in the petrochemical, environmental monitoring, and industrial chemical fields. It is also used extensively in chemistry research.

Liquid chromatography: Liquid chromatography (LC) is a separation technique in which the mobile phase is a liquid. Liquid chromatography can be carried out either in a column or a plane. Present day liquid chromatography that generally utilises very small packing particles and a relatively high pressure is referred to as high performance liquid chromatography (HPLC).

Affinity chromatography: Affinity chromatography is based on selective non-covalent interaction between an analyte and specific molecules. It is very specific, but not very robust. It is often used in biochemistry in the purification of proteins bound to tags. These fusion proteins are labelled with compounds such as His-tags, biotin or antigens, which bind to the stationary phase specifically. After purification, some of these tags are usually removed and the pure protein is obtained.

Supercritical fluid chromatography: Supercritical fluid chromatography is a separation technique in which the mobile phase is a fluid above and relatively close to its critical temperature and pressure.

Ion exchange chromatography: Ion exchange chromatography uses ion exchange mechanism to separate analytes. It is usually performed in columns but can also be useful in planar mode. Ion exchange chromatography uses a charged stationary phase to separate charged compounds including amino acids, peptides, and proteins. In conventional methods the stationary phase is an ion exchange resin that carries charged functional groups which interact with oppositely charged groups of the compound to be retained. Ion exchange chromatography is commonly used to purify proteins using fast protein liquid chromatography (FPLC).

Size exclusion chromatography: Size exclusion chromatography (SEC) is also known as gel permeation chromatography (GPC) or gel filtration chromatography and separates molecules according to their size (or more accurately according to their hydrodynamic diameter or hydrodynamic volume). Smaller molecules are able to enter the pores of the media and, therefore, take longer to elute, whereas larger molecules are excluded from the pores and elute faster. It is generally a low resolution chromatography technique and thus it is often reserved for the final, 'polishing' step of a purification. It is also useful for determining the tertiary structure and quaternary structure of purified proteins, especially since it can be carried out under native solution conditions.

Crystallisation

Crystallisation is the (natural or artificial) process of formation of solid crystals precipitating from a solution, melt or more rarely deposited directly from a gas. Crystallisation is also a chemical solid-liquid separation technique, in which mass transfer of a solute from the liquid solution to a pure solid crystalline phase occurs.

The crystallisation process consists of two major events, nucleation and crystal growth. Nucleation is the step where the solute molecules dispersed in the solvent start to gather into clusters, on the nanometer scale (elevating solute concentration in a small region), that becomes stable under the current operating conditions. These stable clusters constitute the nuclei.

Many compounds have the ability to crystallise with different crystal structures, a phenomenon called polymorphism. Each polymorph is in fact a different thermodynamic solid state and crystal polymorphs of the same compound exhibit different physical properties, such as dissolution rate, shape (angles between facets and facet growth rates), melting point, etc. For this reason, polymorphism is of major importance in industrial manufacture of crystalline products.

For crystallisation to occur from a solution it must be supersaturated. This means that the solution has to contain more solute entities (molecules or ions) dissolved than it would contain under the equilibrium (saturated solution). This can be achieved by various methods, with: (i) solution cooling, (ii) addition of a second solvent to reduce the solubility of the solute (technique known as antisolvent or drown-out), (iii) chemical reaction, and (iv) change in pH being the most common methods used in industrial practice. Other methods, such as solvent evaporation, can also be used. The spherical crystallisation has some advantages (flowability, bioavailability) for the formulation of pharmaceutical drugs.

FACTORS AFFECTING FERMENTATION ECONOMICS

This section discusses the economics of fermentation processes. Relatively high efficiency, coupled with availability and cost of agricultural raw materials, allows use of biological systems to produce needed chemical products. Prior to commercialisation of a fermentation process, extensive economic evaluation is necessary. Process design follows both conventional engineering needs as well as those derived from unique biochemical inputs from laboratory and pilot plant data. The capital and gross operating cost of a conventional fermentation plant is roughly independent of the product produced. The fermentation operating costs per unit volume and per unit time will vary somewhat but will hold generally within a reasonably narrow range.

That is, barring use of an exotic raw material, the cost of labour, utilities, and materials would not vary greatly if put on a unit volume and unit cycle time basis. Also, processing or extraction can vary in complexity but, by and large, they too will hold to a reasonably narrow cost range. The capital and operating costs of all auxiliary operations for fermentation and extraction *per se* also exert a fly wheel effect relative to total installation cost and cost of operation. Thus, the unit cost of the bulk product is much a function of fermentation yield and fermentation cycle time.

Market Potential

- A process is considered commercially successful when it produces high yield of any fermentation product.

- The selling price of the product should be such that it recover the production cost with satisfactory profit.
- On the other hand, the demand or market potential of the product decide to what extent product should be produce.
- Placement of any new fermentation product in the market is a time consuming and costly affair as it require approval by the government agencies like Food and Drug Administration or other before they introduced in the market.
- There are two ways to equate demand and supply ration. First, whether the market for the product may already exist or not and the same product has previously been sold by others or presently under selling. Second, what is the potential of establishment of market for a newly discovered fermentation product which has not been sold previously in the market, i.e. it is required to establish a market for that product.
- Sometimes, due to fewer uses and less demand of certain fermentation products, it not easy to place it in the market.
- The market of some products already exists due to consumers acceptability and long term persistence of these products in the market.
- Market potential, production cost, market demand and sale and competitions amongst existing products are important aspects to be consider at the time of launching a new product in the market.
- Many fermentation products are not sale directly in the market but used internally for an industrial concern. For example, a company that manufactures and sells product A might produce B by fermentation of C, followed by a chemical conversion of the B to A.
- On the other hand, a fermentation product can be sold directly to a different industrial used which chemically transforms the fermentation product before sale in the market. Doing so, it is possible to reduce the extra costs of advertising, packaging, and distribution to retail outlets which apply to fermentation products marketed directly to public. Addition to this, generally it do not require any trademark, proprietary name and generic name.
- Foreign sales present an added market potential for many fermentation products. For example, exporting product and selling in the foreign marker and on the other hand, manufacturing the same product in foreign country and selling in their market makes a wide difference in the profit.

FERMENTATION AND PRODUCT RECOVERY COSTS

For any fermentation process, the economic position of product is decided by its production and distribution.

Medium Constitute

- Usually, the costs of different constituents of the production medium decides the competitive position and potential profits of a fermentation product.
- For example, inoculum media are usually less expensive because they are designed to promote fast growth of the producer instead of converting a large amount of carbon substrate in the product. At the same time they are require in less volume.
- In contrast, production media which are aimed to produce the product required in more volume and may time high-cost of a single medium can affect the selling price the fermentation product.

- As any component of the medium to reduce the medium cost one should try to make provision for the alternate and substitute low-cost replacement of certain medium constituents while formulating the production media.
- As a matter of fact every components are directly or indirectly influence the economics of the fermentation process.
- For example, because of the world political situation, if the availability and cost of cane black-strap molasses becomes too high, one cannot afford to use this as a carbon source. Under such situations there should be provision for an alternative carbon source which is cost effective and made available easily.
- One should remember that sometime the use of an alternate medium requires the use of a different fermentation microorganism or strain of the organism.
- Several media need decontamination to remove some contaminated chemical species as they influence the product accumulation and many time make the recovery process difficult. For example, certain metal ions can be removed by ion-exchange resins.
- Many media requires pretreatment before they employed. For example, starch must be pre-treated with amylase to release fermentable sugars and proteins must be degraded by proteolytic enzymes to release amino acid for yeasts growth.
- Acid or alkali are required to adjust the pH of the medium during production process. As these reagents are not so costly, but considerable reagent may be required for adjusting the pH value of a large volume of medium which many time lead to a considerable high cost.
- Same way, media rich in protein components produce foam and increase the chances of contamination. Under such condition controlling of foam either by chemical antifoam agents with mechanical device also add the cost of production.
- Recovery and purification of fermentation product is very important step of down step process and markedly increase the overall product cost. Recovery and purification should be fast, easy and should occupy less step with application of low cost chemicals.

Fermentation Incubation Period

- Fermentation process with short incubation period are less costly compared to processes with prolonged incubation with reference to the inoculum build up and to production.
- One can harvest more batch in case of short incubation period fermentation with a larger turnover of fermentation equipment.
- Fermentation requiring long incubation periods require: (i) additional requirement, (ii) extra labour cost and (iii) have more possibility of getting contamination.

Contamination and Sterilisation

Many fermentation processes are more prone to contamination compared to protected fermentation processes. For example:

- Fermentations processes in which foaming is the common problem.
- Prolonged incubation requiring processes.
- Processes in which the product producing organisms poorly compete with contaminants for fermentation nutrient.

- Processes in which contaminants degrades or chemically change the fermentation products.
- Certain fermentation processes sensitive to lytic phage (for example certain bacterial and actino-mycetes fermentations).

Those fermentation processes in which media sterilisation cost is not affordable certain alternative methods to control the growth of contaminants are adapted. These are:

- Adjusting low pH of the medium.
- Selecting a substrate poorly amenable to attack by contaminants.
- Partial sterilisation of the medium.
- Selecting certain chemicals which retard the growth of contaminants.

If the fermentation medium is severely contaminated, then it must be discarded and this adds more cost to the production. Even though the medium is not gravely contaminated and may not be serious enough to discard although also affect the overall yield of the product. Many time mutation in genetically unstable production strain results in a population of low-yielding cells

Yields and Product Recovery

- High yield and adequate recovery of any fermentation product is of prime importance in any fermentation process. For any fermentation product, high yield with proper recovery and purity affect its position on the open market.
- For any fermentation product, to maintain a competitive market position it requires to have continuous research and product development programme to improve and increase its yields and a better methods of product recovery. The cost of these important downstream processes are very high but extremely important too.

Product Purity

- The purity of fermentation product decides its future stand and long term market value. The costs of product is directly associated to its purity.
- For example, some antibiotic preparations useful for human applications must be sterile and free from pyrogens. On the other hand, some products which are used with other antibiotic preparations, are sold in crude form for mixture with animal feeds.
- Some fermentation products can also be marketed at different level of purity and at a more than one concentration For example, lactic acid is sold at strengths ranging from approximately 20 to 85 per cent and at purity levels ranging from the relatively crude to high purity edible and U.S.P grades and each of these grades of lactic acid has a place on the market.
- Purification steps like solvent extraction and followed by crystallisations for a fermentation product significantly contribute to the overall costs.

Waste Disposal

The overall production cost of any fermentation product is influenced by outflow of capital used for waste treatment and disposal and it depends on following two aspects:

1. Acceptance of fermentation waste with or without pretreatment by municipal waste-treatment facilities or any other local service providers.
2. Maintenance of waste water treatment plant by the fermentation.

As per state and federal regulations, it is not possible to dispose fermentation wastes into rivers, streams or any other stagnant water bodies for long time. In certain case, for example, waste generated in the fermentation processes which exploit plant or animal pathogens as the fermentation organisms requires an additional expense of sterilisation before discarding to avoid biohazards.

There are many sources through which waste is generated. These are:
- Waste generated from the actual fermentation.
- Wastes from recovery processes.
- Waste form cleaning water.
- Waste from cooling waters.
- Waste from various staged of product recovery and purification, etc.

Many time it is feasible to use waste water for some other purposes, for example, cooling waters can be utilised in media makeup for similar or other fermentations. Many time, it is possible to recover by-products from the fermentation wastes. For example, riboflavin from the acetone-butanol fermentation. It is imperative to state that recovery of economically valuable by-products from a fermentation process recover the overall cost and improve profit picture for many fermentation product.

Labour Cost

The cost which incur to pay for non-technically and technically trained personal working in any industry at all level of is called labour costs. This include labour related to:
- Cultures handling
- Inoculum
- Production
- Product recovery and purification
- Maintenance of product sterility
- Packaging of the product
- Steam production
- Equipment maintenance and cleaning
- Administration, etc.

This labour costs depends on type of fermentation process (batch of continuous) level of containment to be employed, type of organisms under use (genetically engineered or non-genetically engineered), volume of the product, etc.

Research Costs

- To maintain a competitive commercial stand, the cost picture of a fermentation process which include expenses incurred in the research and discoveries for the development of the process should be clear.
- Many time high cost research without positive outcome in terms of innovations and novelty is of no use for industries facing financial crisis.
- Investing in research is a long term venture. It give incredible financial success provided it is done in a right direction.

Capital Expenditure

- Commercial production of a newly developed fermentation is expensive and may require outlay of capital.
- Establishment of expensive fermentation equipment which are in continuous use for the production along need the maximum requirement for capital expenditure with other facilities.
- The newly developed fermentation process may have need of the setting up of newly designed fermentation equipment. Installation of this equipment to a great extent adds additional cost to the fermentation costs

Patent Position

- The profit picture of any fermentation process or product is remarkably influenced by sound patent position.
- Patent position of process or product helps in reducing the extensive commercial competition which remarkably influences the profit picture.
- A promising patent position provides greater potential for cost recovery and an adequate profit.
- The costs for getting a patent are relatively small. However, in case any infringement proceedings are instigated, the patents become a costly affair as.
- Patents, can yield revenue to the holder through its utilisation in the production and marketing of the fermentation product.
- For a return of royalties, it is possible to licensed then patent to other fermentation companies and even competing companies. Even after doing so, the patent holder can still produce the fermentation product.
- Many time, the patent holder may not have the capital or facilities to produce and market the product. In such cases, it may be to the advantage of the patent holder to license the patent without producing the fermentation product, or to sell the patent outright

Overhead Expenses

- General expense incurred for the running of a business is called overhead expenses which include taxes, rent, insurance, heat, light, audit and accounting, depreciation and other routine office expenses.
- These all expenses contribute the overall cost of the products before they introduce in the market for sale.

PROCESS ASSESSMENT

The prior deliberations must be evaluated keeping in mind the present and future market conditions to assess the economic potential for any fermentation process development. It is equally important to re-evaluate the process during commercial production. Following aspects should be considered during overall evaluation of any fermentation process:

- Estimation of present and future availability and price of fermentation substrates.
- Costs of labour.
- Overhead expense.

- Public demand for the product.
- Competition in the market.
- Potential for bettering yields.
- Product recoveries.
- Capability and facilities to meet market demands for the product.
- Consideration of all present and future costs.
- Selling price and desired profit for the product.

All of these dimensions must be studied thoroughly to decide before the production of the fermentation products.

SECTION II

Genetically Modified Crops

Chapter 5

Feeding the World and Eradicating Hunger

INTRODUCTION

Hundreds of millions of people around the world cannot grow or obtain food in sufficient quantity or quality to sustain healthy life. The Food and Agriculture Organisation of the United Nations (FAO) finds that the number of undernourished people worldwide and in developing nations has risen over the past four years for which data is available. Currently, 985 million people worldwide are believed to be undernourished, nearly three times the population of the United States, and 923 million (as on 2015) of these live in developing nations. Recently, the effects of drought on food production have put additional millions of people in sub-Saharan Africa at risk of not just malnutrition but starvation.

The topic of global hunger has become a prominent backdrop for the worldwide debate over genetically modified (GM) food crops. The possible use of biotechnology to boost food production and quality in developing countries has become a focal point for biotechnology advocates and critics alike.

In some respects, the debate about the appropriateness of GM crops for developing countries is not all that different from the debate occurring in the industrialised world. Proponents of the technology point to the potential benefits of the technology to increase food production, reduce crop losses from diseases, insects, and drought, and improve the nutritional content of traditional foods. Critics point to possible human health risks from GM foods, such as new allergens or toxins, or potential environmental and economic concerns, such as the spread of a transgenic trait through wild ecosystems or conventional crops that could threaten biodiversity or trade with nations that reject GM crops.

While this 'benefit versus risk' argument raises issues similar to those in the debate about GM foods in the developed world, the social and economic context of the debate regarding the developing world differs considerably. It is one thing to discuss the impact of a new agricultural technology on a society that already produces abundant, safe, diverse, and affordable foods, where farmers have experience with technology and access to capital for technology investments, where public and private research and development meet evolving agricultural needs, and where a well developed regulatory system is in place. It is quite another question to understand the net impact of the technology in a society that fails to produce enough food to feed its people, where one or a few foods dominate diets, where farmers lack the basic infrastructure to transport, store, and sell the food they do grow, where farmers lack the income and access to credit necessary for investments in technology, where little public or private investment exists for developing appropriate technologies, or where there is little, if any, capacity to manage possible risks associated with the technology. These broader social and economic factors may have as much to do with the potential impact of biotechnology as the narrower issues of specific benefits and risks. As a result, one central question is whether biotechnology addresses the underlying causes of hunger in developing nations. Some observers argue that the cause of hunger is not inadequate food production, but inequitable food distribution, and that biotechnology—indeed, any agricultural technology—does not address that inequity. Critics also argue that the private sector has little financial incentive to develop GM crops to benefit developing country subsistence farmers. Moreover, if such crops were to be developed, critics are concerned that large multinational seed companies could exercise undue market power over small farmers by controlling seed supplies.

On the other hand, others contend that any technology that can increase yields or improve nutrition in places where there is inadequate food wields great promise and should be welcomed. Many would assert that public investment in agricultural research on how to better utilise technology to solve the problems of the poor in developing countries is essential because it would not only impact hunger but could alleviate some of the concerns about private control of the technology and access in these regions. Toward that goal, some advocate increased involvement of the international donor community to enable more public sector investment in GM crops in developing countries.

This chapter discusses some of the key issues pertaining to the appropriate role of biotechnology in addressing hunger and food production in the developing world. It does not attempt to resolve the debate, but rather aims to give the reader a basic understanding of the contours of the debate. Its focus is on hunger reduction, as opposed to poverty reduction. Therefore, the discussion is oriented toward the adoption of GM crops primarily by small holder and subsistence farmers to increase production of food crops and enhance their nutritional characteristics, as opposed to generating income through commercial scale farm operations.

GLOBAL FOOD PRODUCTION AND HUNGER

Green Revolution

The latter half of the twentieth century saw a dramatic worldwide increase in food production generated by coupling higher yielding plant varieties with such increasingly intensive technologies as irrigation and chemical fertilisers and pesticides. This dramatic increase in food production was dubbed the Green Revolution and is credited with staving off the worst predictions of global food shortages that some feared would accompany the world's burgeoning population. The population growth experienced over the last thirty years was partly offset by an 20 per cent overall increase in food production (measured as calories per person), combined with growing cereal exports to developing countries.

Hunger Remains

The encouraging global statistics cited above, however, mask the persistence of hunger in a number of countries. Increases in food production have not been evenly distributed. Consequently, while the world currently produces enough food on a caloric basis to nourish the global population, food production and access to food vary widely among countries. The Food and Agriculture Organisation of the United Nations reports that: After 50 years of modernisation, world agricultural production today is more than sufficient to feed six billion human beings adequately. Cereal production alone could to a large extent cover the energy needs of the whole population if it were well distributed. However, cereal availability varies greatly from one country to another. Moreover, within each country, access to food or the means to produce food is very uneven among households. World food security, therefore, is not an essentially technical, environmental or demographic issue in the short-term: it is first and foremost a matter of grossly inadequate means of production of the world's poorest peasant farmers who cannot meet their food needs.

Countries that already were the most productive realised the greatest gains, although the Green Revolution did result in significant gains in India, the Philippines, and other targeted nations. Despite recent increases, the overall number of undernourished people in the developing world was reduced during the last 30 years from 956 million to 798 million. This decline was realised while per capita food production simultaneously fell in the majority of developing countries as populations grew. Global hunger and malnutrition persist as a continually growing population demands more food and as gains in agricultural productivity begin to slow. A report jointly prepared by seven international academies of science titled Transgenic Plants and World Agriculture states that. The global increase in food production, therefore, reflects an increasing disparity between the most productive and least productive nations. This is particularly apparent in sub-Saharan Africa, where the number of chronically undernourished people has more than doubled in the past 30 years. During this time, the rate of food production increases also slowed, from 3% per year in the 1970s to 1% per year in the 1990s.

Constraints to Further Increases in Food Production in Developing Nations

Constraints on food production in less developed countries are both environmental and economic. High quality agricultural land is concentrated in relatively few countries and, while the equitable distribution of land is the prevailing problem in some developing countries, many less developed countries struggle due to lower quality lands and inhospitable climates. Although the technologies of the Green Revolution helped compensate for these deficiencies, significant additional production gains using conventional crops may not be possible without prohibitively expensive inputs (pesticides, herbicides, fertilisers, machinery, fuel, and water). In addition, the Green Revolution's irrigation- and chemical-intensive farming practices can have adverse environmental impacts. Pinstrup-Andersen and Schipler explain in their book, *Seeds of Contention*, that. One reason for the success of the Green Revolution was that it was a package deal. Fertiliser enabled high-yielding plants to more fully exploit their yield potential, and agrochemicals restricted losses due to weeds, pests, and diseases. But in many places—again with Africa and parts of Asia as notable exceptions— the consumption of agricultural inputs has reached dangerously high levels. There is a logical limit to how much extra expenditure can be justified in return for only marginal gains in yield—quite apart from the strain on the environment.

Even where existing technologies might increase production on a sustainable basis, there are practical and economic barriers to adoption of those technologies. Farmers cannot effectively apply technology without sufficient infrastructure for transfer of knowledge and continuing education. In addition, input-intensive farming requires initial capital, access to inputs, and access to markets where prices are high enough to yield a profit as costs of production rise. In various areas of the world, limited transportation infrastructures impede the movement of seeds, chemicals, machinery, and fuel to farmers. Civil strife also disrupts production, markets, transportation systems, and the flow of any potential capital. All of these factors create barriers to the agricultural development necessary for increasing food security. As a result, many now are concerned that conventional agricultural technologies promise only a limited increase in food production. A statement from the Director-General of the United Nations Food and Agriculture Organisation states.

We can no longer depend on bringing significant new areas of virgin lands into the food production chain and further expansion of food production must come from increased yields on the lands already farmed by the poorest of small farmers and the larger farms alike. This raises the twin challenges of raising productivity on the more fertile lands farmed by the better-off farmers together with an improvement in the output and range of food crops that can be grown on the less well-endowed fragile marginal lands. It is now widely recognised that we are at a post-Green Revolution standstill and that yield ceilings of the main food crops have already been reached in conventional breeding programmes.

Constraints on Equitable Distribution

While insufficient production levels continue to contribute to hunger among subsistence farming communities in developing countries, many point out that hunger could be greatly minimised with improved mechanisms for distributing the food that is produced. Many of the constraints on food production cited above also serve as barriers to equitable food distribution at regional, national and international levels. Consequently, even if populations in developing countries had sufficient money to purchase food, problems with infrastructure and social stability would continue to impede distribution and access to food commodities.

Over eighty developing countries lack sufficient food to feed their populations and the money to import food supplies. In response, international development focuses both on poverty reduction and hunger reduction. Individual country agencies such as the US Agency for International Development and international institutions like the World Bank provide funds to both improve agricultural productivity and support business development through education and training and access to capital and technology.

The international community attempts to alleviate global hunger through short-term relief efforts, as well as longer-term agricultural and economic development programmes, but the solutions to hunger are as elusive as its causes are complex. The United Nations recently estimated that it will take 130 years, i.e. by 2140 to eliminate global hunger at the current pace of progress.

Barriers to Development

Nations who seek to develop self-sufficient agricultural systems face many obstacles including poverty, poor soils, environmental degradation, drought, plant diseases, limited crop diversity, civil strife, epidemic illness, and poor agricultural and transportation infrastructures. Each of these obstacles contributes to hunger and malnutrition. Some barriers to development are more easily overcome than others. Such discrete tasks as building irrigation systems, starting small businesses, establishing community facilities, and creating lending programmes may be manageable. Other development challenges, such as inadequate transportation systems, insufficient regulatory and legal systems, limited access to national markets, and a lack of trained professionals, are more formidable. Finally, the ability to produce sufficient food to sustain individual families or local and regional populations is severely limited where key agricultural inputs (land, water, fertiliser, pesticides, and seed) are unaffordable or unobtainable, or where combinations of weather and disease take a frequent and substantial toll.

BREEDING AND GENETICALLY MODIFIED CROPS

Conventional Breeding

In the late nineteenth century, plant breeders began using hybridisation and cross breeding more aggressively as a means of improving agricultural crops. Since then, the application of a growing body of knowledge about plant genetics, combined with the application of statistical methods to plant breeding and the development of effective field trial protocols, has led to the development of hardier and more productive varieties of corn, wheat, and other staple crops. The use of these techniques to enhance desirable traits (the physical characteristics of a plant generated by its genetic makeup) dominated plant breeding efforts for much of the twentieth century, famously marked by the awarding of the 1970 Nobel Peace Prize to Norman Borlaug for his development of high-yield wheat varieties using these methods. The application of advanced conventional breeding technologies has made much of the production gains of the Green Revolution possible.

Development of Modern Biotechnology

As these advances in conventional plant breeding led to the development of hardier and more productive varieties of food crops, scientists began using recombinant DNA (rDNA) techniques as a new tool for developing crops with beneficial traits. Recombinant DNA techniques allow scientists to isolate certain desired gene sequences from various organisms and introduce (recombine) them into other organisms (or insert multiple or reverse copies of an organisms genes to alter various metabolic functions).

Using this new technique, scientists have modified such traditional food crops as corn and soyabeans to incorporate new traits that protect them from harmful insects and create resistance to specific herbicides, which has the potential to both increase crop yields and decrease the labour required of farmers. *Bt* corn and cotton, for example, are GM varieties of those crops into which scientists inserted a gene sequence from the bacterium *Bacillus thuringiensis (Bt)* which produces an insecticide in the tissue of the plant to kill certain species of insect pests. However, some see rDNA techniques as fundamentally different from conventional plant breeding, which is generally based on sexual reproduction, because rDNA techniques can introduce genes into a crop plant from sexually incompatible plants and even from other living organisms as is the case with *Bt* crops. This ability to introduce novel genes into food crops has led to concerns about the possibility of introducing unrecognised toxins or allergens into the food supply, or disrupting the environment by spreading novel genes to wild relatives of the modified plants, potentially threatening biodiversity.

Status of GM Crops Worldwide

GM crops currently approved for commercialisation in the United States possess at least one of six general traits—insect resistance, herbicide resistance, virus resistance, delayed ripening, altered oil content, and/or pollen control.

The vast majority—over 99 per cent of the acreage of commercialised GM crops are corn, soyabeans, cotton, and canola that incorporate genes for resistance to insects and herbicides. These GM varieties account for almost one-fifth of the total global acreage of these crops. Although most of the acreage of GM crops is in just four countries, over three-quarters of the 5.5 million farmers planting GM crops in 2002 were cotton farmers in China and South Africa. Adoption of GM crops has been rapid. Additional GM crops that have been planted (although several only at an experimental level) include rice, wheat, beet, potato, tomato, peanut, rapeseed, and sweet pepper in China, and rice, rapeseed, potato, eggplant, and cauliflower in India.

Scientists have also developed genetically modified avocadoes, pineapples, and mangos. In the US, GM crops at the field test stage include apples, broccoli, carrots, grapes, lettuce, pears, peppers, plums, and strawberries. However, some of the most economically significant crops, such as wheat, still have not been commercialised.

POTENTIAL BENEFITS OF BIOTECHNOLOGY

Agricultural biotechnology is intended to address some of the challenges faced by farmers. These challenges are similar for farmers in developed and developing countries and include threats to crop yield from disease, pests, weeds, and weather, and lack of critical growing inputs such as nutrients and water. Typically, farmers in developing countries need crops that offer disease resistance (to viruses, fungi, and bacteria), insect resistance, environmental stress resistance (to heat, drought, salinity, flooding, soil pH), quality improvement (nutritional improvement, increased yield), and reduced post-harvest losses. In addition, ease of use is important for farmers who may have limited access to educational or extension services. Crop improvements and other technologies that reduce manual labour may be of particular value in developing countries where labour savings could be used for other activities including water and fuel collection, childcare, and education. While conventional technology has resulted in many hardier crops, biotechnology proponents believe that the new technology offers significant additional opportunities to meet the need for improved crop varieties. Scientists can modify both agronomic traits (traits that determine how well a plant is suited for its environment, such as drought tolerance) and quality traits (such as the oil content of a soyabean) of crops.

From a technological standpoint, one advantage of agricultural biotechnology is that the technological input (in the case of biotechnology, the new trait) is contained in the seed itself. Relative to other technologies such as intensive irrigation or chemical pesticides, seeds are relatively easy to transport and require no new technical expertise to plant. However, as discussed in more detail later, the proper management of certain kinds of GM crops may require new expertise. Nonetheless, if certain traits were made available at an affordable cost, farmers could adopt GM crops without disrupting traditional agricultural practices. In theory, a farmer could reduce intensive irrigation by using a crop variety modified for drought resistance. This could lower the cost of production in areas with high irrigation costs or enable production in areas that lack sufficient water altogether.

While this discussion has focused on the potential of biotechnology to increase yields or improve nutrition, many also argue that associated environmental benefits could be realised. Higher yields from land already under cultivation also may make it possible to avoid the conversion of biologically diverse natural habitats such as rainforests, wetlands, or grasslands to farmland. Drought tolerant crops, for example, could reduce the use of irrigation in areas with a limited water supply or reduce production risks in areas without irrigation. Insect-resistant crops also could reduce the use of more persistent or toxic pesticides. Higher yields may obviate the needs to cultivate steep hillsides or wetlands critical for watershed protection, flood control, and soil retention. On the other hand, some argue that the increased ability to use marginal lands that would otherwise be unsuitable for cultivation acts as a disincentive to critical conservation efforts.

Agronomic Traints

Improved agronomic traits can help increase crop yields in numerous ways: (i) increasing the actual amount of food produced per plant, (ii) reducing crop loss due to pests, disease, or weeds, (iii) compensating for inhospitable environments that limit or prohibit planting, (iv) extending growing and harvest seasons, and (v) reducing production risks. The potential benefits of improved agronomic traits are lower costs of production, greater consistency of production, higher yields, and the ability to use lower quality lands that would otherwise be inhospitable for agricultural production. Increased productivity may result in surplus crops for sale at local markets and reduced labour requirements which can create time for other income-generating activities, as well as adult and child education.

Pest resistant plants

Although consistent data on global crop losses is difficult to find, some estimates indicate that pests destroy over half of global crop production, and that pre- and post-harvest crop losses due to disease and pests represent an annual cost of approximately $100 billion. Regional losses compared to actual production show that losses from pathogens, insects, and weeds may equal the quantity of successfully harvested crops in regions such as Africa and be nearly 50 percent of the amount successfully harvested in regions such as North America. For rice alone, one study estimates that 50 million metric tons are lost each year from fungal diseases, million tons from insects, and another 10 million tons from viral diseases.

GM crops clearly hold significant potential to boost crop production through insect loss prevention. One study estimates that the adoption of existing and possible future *Bt* crops (cotton, rice, corn, fruits, and other vegetables) could save over $2.6 billion of the over $8 billion spent annually on insecticides. This study also suggests that widespread adoption of *Bt* fruit and vegetable crops could dramatically reduce crop losses to pests and thus benefit numerous developing countries, including nearly all of Africa, Asia, and Latin America, and that the introduction of *Bt* corn and rice could likewise benefit China, India, Bangladesh, Philippines, Thailand, Indonesia, Vietnam, Colombia, Paraguay, Peru, Bolivia, Mexico, Argentina, Costa Rica, Ghana, Cameroon, Zimbabwe, Nigeria, Tanzania, and Ethiopia. A number of studies have been conducted to estimate the benefits GM crops have afforded various countries and regions in comparison with traditional varieties. GM cotton was found to produce 5–80 per cent more than non-GM cotton.

In addition to potential gains in crop yields, adoption of pest-resistant plants could confer environmental and human health benefits to the extent that such adoption would reduce the use of chemical pesticides. Reports on pesticide reductions due to adoption of GM crops vary, but the National Center for Food and Agricultural Policy recently reported that the adoption of insect resistant corn and cotton, herbicide tolerant canola, corn, cotton, and soyabeans, and virus resistant papaya and squash in the United States led to a total reduction in pesticide use in 2001 of 46 million pounds. One estimate indicates that the use of *Bt* cotton in China has reduced the amount of insecticide applied by 20 per cent or as much as 78000 T. While recent studies have suggested that herbicide use may be on the rise in some parts of the world employing herbicide-tolerant GM crops, other studies have noted that these crops promote the use of conservation tillage practices and of herbicides that are less persistent.

Many competing factors may be responsible for the variance observed in the magnitude of yields obtained and pesticide used on GM crops compared with non-GM varieties. These include differences in insect pressures, horticultural practices, land quality, access to inputs (e.g. irrigation), and annual weather occurrences in various regions. In addition to creating new varieties of insect- and herbicide-resistant crops, biotechnology can develop new varieties of virus-resistant crops. Virus-resistant varieties of squash and papaya, for example, are among the GM foods approved in the United States. In Africa, work is ongoing to develop a virus resistant sweet potato, and scientists are performing similar research around the world on such crops as banana and cassava. Because certain viruses can wipe out a large portion of a seasons crop, enhanced virus resistance could significantly improve the food resources and security of a community or country that is highly dependent on a staple crop such as the sweet potato.

Hardier plants

Scientists are also developing crop varieties with improved environmental hardiness with the potential to significantly increase crop yields and open up or restore agricultural lands. In addition, GM plants that mature more quickly could help raise production in areas with shorter growing seasons. Among the many traits in various stages of research and development are drought tolerance (beans, groundnuts), cold tolerance (sorghum), and aluminum tolerance. In areas that have been subject to intensive irrigation, for example, salts that have accumulated in the soil can render it useless. In an attempt to address this agricultural challenge, a number of researchers have investigated genes that may confer resistance to salt stress in various plant species. These kinds of GM crops could prove beneficial in countries where these hostile conditions limit agricultural production.

Quality Traits

Biotechnology also has the capability to develop new crop varieties with improved qualities beyond those developed by conventional methods. Improved quality traits may benefit both the farmer and the consumer. Two crops genetically engineered to include quality traits have received marketing approval in the United States (altered oil content in oilseeds and delayed ripening in tomatoes), and a variety of others are in the research and development stage (such as delayed ripening in a variety of fruits and improved nutrition in crops such as carrots and potatoes). Value-added traits could help farmers in less developed countries realise a higher return on their marketed crops and lower prices for consumers—a significant factor for poor consumers who traditionally spend 50–80 per cent of their income on food. Traits that enhance nutritional content and improve handling characteristics also offer important opportunities to mitigate malnutrition and hunger.

Nutrition

The use of biotechnology to modify the nutritional make-up of crops has also been cited as a means to reducing malnutrition in developing countries. Researchers in India recently announced the creation of a genetically modified potato in which protein content is increased by a third (including the essential amino acids lysine and methionine). Proponents of the GM potato, which has been submitted for regulatory approval, say the enriched potato could improve nutritional deficiencies in the diets of the country's poorest populations, particularly for children for whom these amino acids are essential for proper development. Other advances in nutritional enhancement include the continuing development of so-called 'golden rice,' rice genetically modified to produce beta-carotene, which serves as a source for the vitamin A lacking in the diets of populations for whom rice is a staple crop. Researchers are also working to develop 'golden mustard' that would yield cooking oil high in beta-carotene. Other traits in the research and development stage that could enhance quality include: elevated levels of iron (rice), modified starch content (cassava), reduced alkaloid content (potatoes), and reduced phytic acid, which interferes with the body's absorption of iron (corn). Although currently in the research stage, modifications such as these could significantly improve the diets of hundreds of millions of people who live primarily on one or a few types of crops. The International Food Policy Research Institute explains how work toward increased nutritional value is complimentary that toward increased food production: In the case of trace minerals (iron and zinc, in particular), the objectives of breeding for higher yield and better human nutrition do largely coincide. That is, mineral-dense crops offer various agronomic advantages, such as greater resistance to infection (which reduces dependence on fungicides), greater drought resistance, and greater seedling vigour, which in turn, is associated with higher plant yield.

Handling

In areas with slow transportation or inadequate storage facilities, the ability to preserve a harvested crop on its way to market could provide considerable economic value. Research regarding ways to delay ripening or provide post-harvest pest resistance through genetic modification could mitigate current barriers to food distribution in developing countries. For example, altering a fruit's production of the ripening agent ethylene could reduce rot by extending shelf life and increasing the amount of time the fruit could spend in transportation. Researchers have accomplished this in tomatoes, development is underway for delayed ripening raspberries, strawberries, bananas, and pineapples.

Safety

As previously discussed, because pest-resistant GM crops may decrease the need for pesticides, the occupational exposure of farmers and their family members to these chemicals could be reduced. In addition to this safety advantage, agricultural biotechnology may provide enhanced food safety to consumers. Enhanced food safety may be realised if scientists can modify conventional crops to eliminate or reduce allergens or toxins. Crops for which allergen removal research is ongoing include soyabeans and peanuts, and researchers are also exploring ways to remove the ricin toxin from castor plants to make the waste products from castor processing safer. Cassava, which requires proper preparation because it contains potentially harmful levels of cyanide, is an example of a staple crop where genetic modification could improve safety.

POTENTIAL RISKS OF BIOTECHNOLOGY

As with benefits, some of the concerns expressed about biotechnology in developing nations do not differ from the concerns in developed nations. The concerns about the possible risks from GM food to human health or the possible risks from GM plants to the environment are fundamentally similar. However, the impacts realised from those risks may differ considerably in developing and developed countries.

ENVIRONMENTAL RISKS

Some of the environmental concerns raised in response to the Green Revolution focus on sustainability. In a March 2003 speech, Gordon Conway stated that:

The Green Revolution had real failings. The new rice and wheat varieties were designed for irrigated land. They had short stems and required only shallow root systems. They were bred to put all of their growth energy into seed production — that is, wheat and rice grain. To thrive, they needed more water and fertiliser than traditional varieties, a lot more. Increased use of pesticides created pesticide resistance, while killing off some beneficial insects. Fertilisers allowed farmers to avoid costly and labour-intensive work on agricultural preparation and maintenance, such as soil aeration, crop rotation, and working organic matter into the soil. But these steps, we understand today, are key to long-term sustainability. The results of omitting them were soil erosion, nutrient depletion, falling water tables and salinisation, even in some of the world's most fertile regions like India's vast Punjab. US and European intensive agricultural systems have similar drawbacks.

Indeed, critics have voiced similar concerns about the sustainability of biotechnological modifications to agricultural systems. Specifically, some believe our growing dependence on 'technology based' approaches for global food production is not sustainable and that the economic value in the trade-off between cost of labour and cost of mechanisation is questionable. The history of agricultural technology is one of continuing innovation both in methods of production and in the crops produced. To date, society has adopted increased mechanisation of production because of the significant gains in productivity. However, critics have also argued that 'modern agriculture is intrinsically destructive of the environment.'

The application of biotechnology to crop production raises new questions about the ongoing challenges of balancing environmental sustainability with the inherent ecological interventions of farming—the manipulation of the natural environment to produce food and fibre.

Beyond broad sustainability questions, genetically modified crops raise some specific environmental concerns. One such frequently cited environmental risk is the potential for movement of novel genes from genetically modified crops to other plants ('gene flow'). With respect to gene flow, there is concern that a GM crop might breed with a wild relative ('outcross') to produce a 'super weed'—a hardier plant that could displace native plants or become an agricultural pest that interferes with crop cultivation. Some have expressed concern that the movement of new genes from GM crops into indigenous relatives may reduce the presence of native strains, which might be of particular concern in areas aiming to preserve biological diversity. For example, researchers recently reported evidence of a gene sequence from *Bt* corn in native maize landraces in Mexico. While this study has been strongly criticised, a general agreement exists that gene flow from transgenic corn into conventional corn and native maize varieties is likely unless steps are taken to prevent it. In fact, conventionally bred crops have long been known to outcross with wild relatives, it is highly unlikely that transgenic crops would behave differently without specifically designed control measures.

The ramifications of these environmental concerns are not limited to ecology. Both ecological and economic consequences of outcrossing between GM crops and conventional or wild relatives exist, the significance of which depend on the specific genes introduced, traits expressed, and the environment which such varieties are grown. For example, if a genetically modified variety contaminated a conventionally bred food crop via outcrossing, it is possible that all of that commodity would become ineligible for export to countries that had not approved the specific genetic modification. Such trade ramifications could induce significant economic harm for countries involved in food export. Similarly, niche markets such as those served by organic farmers, which can often command higher prices for their crops, may face adverse impacts from outcrossing with GM varieties, this also raises the possibility that GM crops may negatively impact markets.

As a result, many continue to advocate for the development and implementation of technical, regulatory, and enforcement approaches to minimise gene flow with either non-GM commercial crops or wild relatives of GM crops. Some research has focused on limiting the possibility of gene flow biologically. Just as scientists can alter the genome of plants using rDNA techniques to incorporate new agronomic or nutritional traits, some have also investigated methods of altering the ability of a plant to reproduce. Insertion of genes that confer infertility, frequently referred to as 'terminator genes,' would serve multiple purposes. This technology would protect the developers of genetically modified crops by limiting illicit use of their seeds as well as limiting the ability for outcrossing to produce viable new undesirable varieties. Efforts to limit outcrossing between GM and other crop and wild plant varieties by inserting terminator genes were greeted by many with hostility. Limiting the reproductive viability of GM varieties was viewed by some as an attempt by large international corporations to control markets and undermine the independence of small resource-poor farmers in the very countries suffering from the worst hunger.

The risk of gene flow from any particular GM plant and methods for managing this risk will depend on a large variety of factors and will require assessment on a case-by-case basis. As developers incorporate a wider variety of new traits into more GM crops that are then introduced into new areas of the world, the potential for gene flow must be addressed at each stage of research, development, testing, and regulation.

A second set of environmental concerns frequently cited by critics highlight the interactions of pest resistant crops with both non-target organisms (organisms that are not the intended target of the pesticide) and on the pests themselves. Some suggest that plants engineered to produce pesticides could harm populations of non-target species. Concern has also been expressed that such crops could accelerate the development of pesticide resistance in pest populations.

Finally, the effect this technology may have on patterns of land use is also somewhat uncertain. While some GM crops may increase yields on existing agricultural land and reduce the need to press additional environmentally sensitive land into production, GM crops with agronomic traits that expand the range of environments in which crops can be grown also could increase pressures to farm on marginal lands with potentially harmful impacts on the ecosystem. The development of hardier GM crops, better suited for previously inhospitable environments, raises the possibility that fragile lands, ill suited for intensive agriculture, will be degraded by new or increased production (through increased erosion, for example). Expanding or intensifying agricultural areas could lead further to indirect negative effects on the ecology of the surrounding ecosystem (by altering forage resources for native animals, for example).

To some extent, the environmental issues discussed here are not exclusive to agricultural biotechnology. Conventional breeding techniques have been used to develop new varieties of crops that, like transgenic crops, can also mate with native and wild relatives and potentially threaten regional biodiversity. The use of conventional chemical pesticides, commonly applied to conventionally bred plants, presents significant risks to non-target organisms and may accelerate the development of pesticide resistance. Lastly, conventionally bred plants have also expanded land use to areas otherwise not used for agricultural development. Although these parallels suggest that risks associated with agricultural biotechnology could be evaluated in the context of traditional practices, genetic modification allows for the incorporation of an otherwise unachievable variety of genes and related traits. The unique potential offered to crop development via agricultural biotechnology may also require a unique approach when considering the risks associated with GM crops.

Human Health Risks

GM crops have been consumed widely and no harm to human health has yet been verified. Despite this, it is still appropriate to monitor GM crops for potential health effects, especially those that may be a result of long-term exposure. Companies seeking to bring new GM foods to market routinely assess for a variety of food safety and quality factors, however, an effective regulatory system might require such an analysis.

The human health risk most commonly discussed with respect to GM crops is the potential for these crops to produce allergens and toxins which in turn could enter the food supply. A protein produced by a gene in a GM crop that was previously not in that crop, or not in the traditional food supply at all, could be an allergen. Likewise, a new gene introduced into a GM crop conceivably could produce a new toxin or alter the levels of a naturally occurring toxin. Another potential health risk arises from the possibility that the genetic modification of a GM crop might negatively affect its nutritional makeup (by lowering the level of an important nutrient, for example).

It is also theoretically possible that the health risks to consumers from GM crops might vary among populations. Consumers whose diet is dominated by a limited number of staple crops, for example, may be more vulnerable to risks associated with those crops. An allergenic protein in a staple food of a certain population might have an effect that would not occur in a population that consumes the same food as a much smaller part of a more diverse diet. Similarly, nutritional deficiencies in a GM staple crop could be more serious for populations that rely heavily on that crop as a predominant part of the diet. Food modifications could potentially affect a population suffering from chronic malnutrition or diseases that compromise the immune system in ways that would not take place in a healthy population.

CONCERNS AND CHALLENGES: DELIVERING BENEFITS, MANAGING RISKS

Most of the potential benefits of, and possible concerns about, agricultural biotechnology for small-scale developing nation farmers do not differ remarkably from those that exist for developed nations. The context of those impacts is what makes the debate so different. Given the underlying institutional, environmental, and financial constraints that make it so difficult for conventional agriculture to meet the needs of the local population, biotechnology poses both opportunities and unique challenges and concerns for developing nation small farmers.

Development, Deployment and Capacity

The extent to which biotechnology will be an effective tool in the fight against global hunger depends, in part, on whether new GM crop varieties will meet the specific needs of small farmers in developing nations such as drought tolerance in a regional staple crop. A number of significant barriers exist to the development and deployment of GM technology in these countries. First, it is unlikely that the potential market for such products is large enough to entice the private sector to invest in the needed research and development. Even if incentives are developed to encourage private sector participation or partnerships, the public sector would likely be the primary source for such research and development funding. Second, deployment of the technology to the farmers requires an institutional infrastructure for distribution and education. Finally, ensuring food and environmental safety requires a regulatory capacity. In order to overcome each of these barriers, the public sector must commit scarce resources, which is a difficult proposition for many developing nations already hard-pressed to deliver basic services. In addition, governments in developing nations face many immediate needs and therefore find it difficult to make long-term investments in agricultural research, thus leading to a continuing cycle of lost opportunities.

As previously mentioned, the private sector may have little market incentive to invest in the development and marketing of crops raised by subsistence farmers in the developing world. It is also unclear whether subsistence farmers could generate the income needed to buy commercial GM seed varieties, particularly if developers do not permit farmers to continue the practice of saving seed for future plantings. Similarly, while only a small number of GM crop traits have been commercialised to date, others that hold promise for less developed countries are either at earlier stages of research or are only theoretical. Thus, the possibility of biotechnology providing immediate and dramatic benefits to alleviate hunger or poverty seems relatively low, except perhaps in specific areas with a limited number of crops.

Biotechnology companies have much of the essential research expertise and have led in investments in the research and development of GM crops. The private sector finances approximately 50 per cent of biotechnology research in industrialised countries, but only about 10 per cent in developing countries. As much as 77 per cent of GM crop field trials worldwide are attributed to private industry. Most private sector investment aims to develop GM varieties of commercially significant commodity crops, such as corn, wheat, rice, and soyabeans. Due to the limited commercial promise in developing countries, making GM varieties available to the small developing nation farmer will probably require development by universities, governments, and international public research centers and distribution at subsidised prices. The Assistant Director General of the Agriculture Department of the Food and Agriculture Organisation of the United Nations (FAO), Louise O. Fresco, recently characterised the gap between public and private funding as a critical barrier to realising the full promises of biotechnology for food security in the developing countries. She stated that: It is no exaggeration to say that we are witnessing a molecular divide. The gap is widening between developed and developing countries, between rich and poor farmers, between research priorities and needs, and above all between technology development and actual technology transfer. Today 85% of all plantings of transgenic crops globally are herbicide-resistant soyabean, insectresistant maize and genetically improved cotton varieties, designed to reduce input and labour costs in large scale production systems, not to feed the developing world or increase food quality. There are no serious investments in any of the five most important crops in the semi-arid tropics—sorghum, pearl millet, pigeon pea, chickpea and groundnut. This is largely because 70% of the agricultural biotechnology investments are by multinational private sector research, mostly in developed or advanced developing countries. These investments concentrate on GMOs and biotic stresses. Barring a few initiatives here and there, there are no major public sector programmes to tackle more critical problems of the poor and the environment or targeting crops such as cassava or small ruminants.

The widening molecular divide which generates a gap between promise and reality of the impact of biotechnology is a cause for concern. In some cases where market incentives have not warranted commercial investment, the private sector has indicated a willingness to donate expertise and intellectual property to support public research undertakings. For example, Monsanto announced two years ago that it would provide royalty-free licenses to all of its technologies that could help the development of golden rice and other pro-vitamin A-enhanced rice varieties. At the same time, Monsanto opened up its rice genome sequence database to the International Rice Genome Sequencing Project (IRGSP).

More recently, the Rockefeller Foundation announced the formation of the African Agricultural Technology Foundation, designed to secure royalty-free access to patented processes and materials and provide technical assistance for the deployment of new GM crops in Africa. Biotechnology companies, including Monsanto, DuPont, Syngenta, and Dow AgroSciences have announced their support of the Foundation's efforts. The technological capacity of developing countries to perform their own research and development for GM crops varies widely. A few countries such as India, China, Brazil, and South Africa have well established agricultural research systems, while many others have much more limited research capacity. Of course, some existing research and development may also be of use to developing countries, as is the case with *Bt* cotton (which numerous farmers plant in China and Africa).

Furthermore, many countries have no regulatory system for governing the import, development, testing, and use of GM crops, as well as the intellectual property rights involved. Even where such systems exist, the scientific and legal capacity to implement and enforce regulations may be very limited.

International agreements affecting biotechnology trade and transfer are additional factors that developing countries must consider as they build legal and technical capacity and seek access to biotechnology. The Agreement on Trade-Related Aspects of Intellectual Property Rights (TRIPS) of the World Trade Organisation establishes baseline principles of patentability and protection that member countries must provide. TRIPS includes a specific reference ensuring the patentability of plant varieties, although TRIPS would recognise the laws of most developing nations that preserve rights for both breeders and farmers to make extensive use of protected varieties. The Cartagena Protocol on Biosafety to the Convention on Biological Diversity establishes various agreements and mechanisms for making regulatory risk assessment decisions, sharing information, and informing countries of when genetically modified organisms are transported across national borders.

A number of international institutions provide resources aimed at developing and distributing GM crops in developing countries and helping these countries build scientific and legal capacity to properly manage crop development, risk assessment, and intellectual property management. The Consultative Group on International Agricultural Research (CGIAR), an association of public and private donor agencies that funds sixteen international research centers, conducts research to support agricultural productivity, including biotechnology, in developing countries. The International Service for the Acquisition of Agri-Biotech Applications (ISAAA), whose activities are supported by a number of private companies, non-profit organisations, and government agencies, focuses on the identification, assessment, and adoption of new crop biotechnology applications in Africa and Asia. The World Bank, the United Nations, and a number of other public institutions also provide various forms of assistance for biotechnology transfer to and among developing countries.

Donor country agencies and private institutions also provide both direct funding for research and development of biotechnology transfer and management capacity. In the United States, for example, the Agency for International Development's Collaborative Agricultural Biotechnology Initiative encompasses a number of programmes that fund biotechnology research and adoption. Many of the projects funded by USAID are executed by international collaborators such as a programme intended to develop vitamin A enriched sweet white corn for African nations. The International Maize and Wheat Improvement Center in Mexico, the International Institute of Tropical Agriculture in Nigeria, Wageningen University of the Netherlands, the University of Illinois, Iowa State Univeristy, and Monsanto are all participants in this USAID sponsored endeavor. Likewise, the Canadian International Development Agency supports agricultural development in a number of developing countries. This support includes strengthening the human and institutional resources required to integrate and react to advances such as biotechnology in agricultural systems. As a private institution, the Rockefeller Foundation has spent over $100 million on plant biotechnology research and has trained hundreds of scientists from Africa, Asia, and Latin America. Since 1993, the McKnight Foundation has committed over $50 million to fund crop improvement research, some of it using GM methods, in partnerships led by developing public sector scientists linked with advanced labs around the world.

To effectively deploy GM crops throughout many developing countries, it may be necessary to build appropriate scientific and regulatory expertise in those countries, and to transfer genetic resources, technology, and intellectual property rights to those institutions and companies that can develop GM crops appropriate for targeted countries. In addition, education and communications programmes will have to be established to ensure that farmers are informed users of the technology and that consumers can make informed purchasing decisions.

Risk Management

Just as the potential benefits of GM crops will be realised only through a combination of technology, resources, and infrastructure, the potential risks of GM crops can be identified and managed properly only when requisite scientific expertise is applied under an appropriate regulatory system. The risks and benefits of GM crops are highly dependent upon the specific traits, specific crops, and the context in which they are grown. Meaningful risk assessment must be conducted on a case-by-case basis. Whether *Bt* corn presents a risk of outcrossing, for example, depends on the specific conditions and region in which it will be grown, and risk management will have to be tailored to those conditions. The Human Development Report of 2001 asks: Could the genes flowing from genetically modified organisms into non-target organisms endanger non-target populations? It depends on how genetically modified organisms interact with their environment. Whether or not these harms could possibly occur is a matter of science— but if the possibilities are real, the extent to which they become risks depends on how the technologies are put to use. Debates today, however, sometimes proceed as if risks about specific products can be isolated from the context in which they occur. Therefore, concerns have been raised about the capacity of less developed countries to properly manage GM crops. These concerns may apply to both the governmental and individual levels, since GM crop management requires not only competent regulation, but also farmers educated in risk management practices. Although it was pointed out earlier that planting a GM seed theoretically requires little technical support or infrastructure, management of the GM crop itself may entail new practical requirements (such as planting refuges of conventional corn to mitigate the development of insect resistance to *Bt* corn).

Furthermore, the context of risk management matters significantly. Less developed countries may weigh the benefits of a particular GM crop more heavily, and the risks less heavily, than those not directly affected by food shortages or oppressive poverty. The balancing of benefits and risks is a value judgment that those societies must make. As the Human Development Report of 2001 explains: Even when societies and communities consider all sides, they may come to different decisions because of the variety of risk and benefits they face and their capacity to handle them. European consumers who do not face food shortages or nutritional deficiencies see few benefits of genetically modified foods, they are more concerned about possible health effects. Undernourished farming communities in developing countries, however, are more likely to focus on the potential benefits of higher yields with greater nutritional value, the risks of no change may outweigh any concerns over health effects. Choices may differ even between two developing countries that need the nutritional benefits of genetically modified crops, as one may be better able to handle the risk. Also, effective regulatory systems facilitate the apportioning of responsibility and ensure accountability. Some fear that biotechnology companies will have too free a hand in less regulated countries without an appropriate balance set by government policy to protect the interests of farmers, consumers, and the environment. Some also raise concerns about corrupt governments in a number of developing nations, calling into question decisions about use of the technology and enforcement of regulatory protections.

Socio-economic Issues

The socio-economic impact of GM crops is also of concern. Some see negative aspects of globalisation in the rise of the biotechnology industry and use biotechnology examples to make broader points about global capitalism. Some suggest that large multi-national companies will gain inappropriate control over indigenous farmers through control of the technology. Similarly, some fear that biotechnology companies will acquire property rights in genes obtained from indigenous plants without appropriately compensating host countries or indigenous societies.

Others believe the adoption of GM crops will simply exacerbate what they see as existing problems in modern agriculture such as the use of technology-intensive practices and monoculture farming that may not be compatible with sustainable land management, as well as rising technology and input costs that demand farmland consolidation and accelerate divisions between relatively wealthy and relatively poor farmers. These critics raise concerns about directing public sector research and development funds into transgenic technologies rather than into what they see as sustainable agriculture and agroecology.

Some developing nations also fear being flooded with exports of GM crops from the United States and Europe, undermining local markets and local food production. They see local food production as essential to providing the income and jobs necessary for alleviating in part the poverty that is at the heart of hunger. The transformation of traditional subsistence farm economies into global market economies unquestionably involves change that can pose challenges to important traditional social and cultural values and structures. Some view such change as emblematic of traditional values succumbing to global capitalism and technology for technologys sake. To the extent that biotechnology is seen as a tool for economic change, it may invoke these fears. On the other hand, the intensive farming practices of the developed world have been adopted around the world because they are effective at increasing food production to meet the needs of growing human populations.

To sum up, poverty and the uneven distribution of food are fundamental sources of global hunger. While world hunger indeed could be significantly alleviated if current food production or global income were more equitably shared, distribution is only one piece of a much more complex solution. Given the various complex barriers to global food or income redistribution, this is not a promising short-term solution. In addition, the argument that world hunger should be solved solely through food redistribution (through aid and assistance programmes) seems partly at odds with the goals of economic development and self-sustainability that are advanced on behalf of developing countries. Local food production will continue to be a primary way of addressing hunger. There are, however, major systemic barriers to increasing agricultural production in many developing nations. Civil strife, weak governmental institutions, lack of public funds and private capital, lack of access to agricultural inputs, and inadequate agricultural and transportation infrastructure are all barriers to adequate production. Agricultural biotechnology as a whole does not offer solutions to these broad systemic problems.

Agricultural biotechnology may, however, provide the means for developing crop varieties tailored for particular regions that could play an important role in addressing hunger. Traits such as disease, pest, and drought resistance could help to increase food production and thereby help meet local food needs. Increased food production and reduced pest control and labour costs could also help to address rural hunger through increased income. Surplus production and reduced post-harvest losses may also help deal with hunger in urban areas. In addition to helping address hunger through increasing production and availability of food, biotechnology may help address critical nutritional deficiencies by enhancing the nutritional value of staple crops in the developing world.

While biotechnology offers the prospect for hardier crops, it also raises environmental and human health considerations, risks that must be considered during the development and deployment of new genetically modified crops. Such an assessment will frequently require a case-by-case examination of specific crops in specific environments that also considers the potential environmental benefits of specific applications. If significant risks are identified, parallel concerns will arise about the capacity of small farmers and regulators to manage those risks without training and resources.

To capture the improvements in crop yield and food nutrition that biotechnology promises while managing risks may require a commitment of public resources from nations that have few, if any, public resources to spare. Each nation is likely to face those choices with a different perception of how to weigh the benefits against the risks, and how to deal with the changes that new technologies often bring.

Finally, it must be realised that these decisions are being made in the context of a broader political and philosophical dialogue about the impacts technology and globalisation have on our world. Where some see biotechnology as a means of assuring food security for impoverished populations and argue that there are high risks to not taking advantage of it, others see this technology and the complications associated with access to it as a potential vehicle for capitalist exploitation. Many of these differences are rooted in perspectives that are far broader than the biotechnology debate itself and cannot be resolved solely within the confines of that debate.

Chapter 6

Genetically Modified Crops

INTRODUCTION

There is a scientific consensus that currently available food derived from GM crops poses no greater risk to human health than conventional food, but that each GM food needs to be tested on a case-by-case basis before introduction. Nonetheless, members of the public are much less likely than scientists to perceive GM foods as safe. The legal and regulatory status of GM foods varies by country, with some nations banning or restricting them, and others permitting them with widely differing degrees of regulation.

Biotechnology offers a variety of potential benefits and risks. It has enhanced food production by making plants less vulnerable to drought, frost, insects, and viruses and by enabling plants to compete more effectively against weeds for soil nutrients. In a few cases, it has also improved the quality and nutrition of foods by altering their composition. However, the use of biotechnology has also raised concerns about its potential risks to the environment and people. For example, some people fear that common plant pests could develop resistance to the introduced pesticides in GM crops that were supposed to combat them. Genetic engineering provides a means to introduce genes into plants via mechanisms that are different in some respects from classical breeding. A number of commercialised, genetically engineered (GE) varieties, most notably canola, cotton, maize and soyabean, were created using this technology, and at present the traits introduced are herbicide and/or pest tolerance. Gene technology enables the increase of production in plants, as well as the rise of resistance to pests, viruses, frost, etc. Gene transfer is used to modify the physical and chemical composition and nutritional value of food. Gene transfer in animals will play a part in boundless possibilities of improving qualitative and quantitative traits. The yield, carcass composition and meat characteristics the use of nutritive substances – not sure what is being said here? and resistance to diseases can be improved. On the other hand, negative effects of gene technology on animals, human, and environment should be considered.

Genetically modified organisms (GMOs) are defined as organisms (except for human beings) in which the genetic material has been altered in a way that does not occur naturally by mating and/or natural recombination. GMOs have widespread applications as they are used in biological and medical research, production of pharmaceutical drugs, experimental medicine, and agriculture. The use of gene technology in food production has become interesting due to increased needs of food as well as its improved quality. With the application of gene technology to plants and animals, goals can be achieved more quickly than by traditional selection. Consequently, ethical dilemmas are opened concerning the eventual negative effects of production of genetically modified food. It seems that supplementation of nutraceuticals and wild foods as well as wild lifestyle may be protective, whereas western diet and lifestyle may enhance the expression of genes related to chronic diseases. Our human health. genes or pathways are most likely regulated by microRNA. The prevalence and mortality due to multifactorial polygenic diseases, hypertension, coronary artery disease (CAD), diabetes and cancer vary depending upon genetic susceptibility and environmental precursors because they have identifiable Mendelian subsets. Rapid changes in diet and lifestyle may influence heritability of the variant phenotypes that are dependent on the nutraceutical or functional food supplementation for their expression. It is possible to recognise the interaction of specific nutraceuticals, with the genetic code possessed by all nucleated cells. There is evidence that South Asians have an increased susceptibility to CAD, diabetes mellitus, central obesity and insulin resistance at younger age, which may be due to interaction of gene and nutraceutical environment. The negative consequences can affect the health, environment, etiology, society and ethics.

APPLICATIONS OF TRANSGENIC PLANTS IN HUMAN NUTRITION

Genetically modified foods are classified into three categories according to their usage and legal regulations.

1. Food is genetically modified (potato, tomato, soya, maize, sunflowers, rice, pumpkins, melons, rape, etc.).
2. Food contains components of genetically modified plants (starch, oil, sugar, aminoacids, vitamins, etc.).
3. Food contains genetically modified organisms (yoghurt contains transgenic micro-organisms).

Gene technology enables higher yields in plants, resistance to pests and frost, as well as mechanical properties of fruits, etc. We can also modify physical and chemical composition in order to improve nutritional and physiological value of foods. Transgenic plants also enable production of more healthy food (more unsaturated fatty acids, transfer of proteins from legumes into wheat, increased content of essential amino acids, transfer of proteins from sunflowers into maize, etc.). Thus, dangers of heart diseases, allergies are diminished and malignancy prevented.

GENETICALLY MODIFIED CROPS

Bt Cotton

Cotton is an important fibre crop of India being cultivated over an area of about 9.5 million hectares (mha) representing approximately one quarter of the global area of 35 million hectares under this crop. After China, India is the largest producer and consumer of cotton. Much of this success owes itself to the introduction of *Bt* cotton in 2002 prior to which cotton production suffered huge losses due to its susceptibility to insect pests. Among the insects, cotton bollworms are the most serious pests of cotton in India causing annual losses of at least US$300 million. Insecticides valued at US$660 million are used annually on all crops in India, of which about half are used on cotton alone. *Bt* or *Bacillus thuringiensis* is a ubiquitous soil bacterium first discovered in 1901 by Ishiwata, a Japanese microbiologist. Later it was found that some *Bt* strains (*Cry+*) were highly toxic to larvae of certain insect species which are also plant pests. *Bt* was first sold as a spray formulation in 1938 in France for the management of

European corn borer. Subsequent research has revealed that *Bt* carries proteinaceous crystals that cause mortality in those insects which carry receptor proteins in gut membranes that bind to *Bt* proteins. Other organisms that do not contain receptors to *Bt* proteins are not affected by the toxin. The advent of genetic transformation technology made it possible to incorporate cry genes and thus the ability to produce *Bt* proteins in plant cells so that target insect larvae infesting the crop plants are effectively killed. The first *Bt* crops viz *Bt* cotton, *Bt* corn and *Bt* potato were commercialised in USA in 1996. *Bt* crops are currently cultivated in 23 countries over an area of 46 mha. The advent of genetic transformation technology made it possible to incorporate cry genes and thus the ability to produce *Bt* proteins in plant cells so that target insect larvae infesting the crop plants are effectively killed. The first *Bt* crops viz *Bt* cotton, *Bt* corn and *Bt* potato were commercialised in USA in 1996. *Bt* crops are currently cultivated in 23 countries over an area of 46 mha. It is also recognised that GM technology may entail rare unintended risks and hazards to environment, and human and animal health. These risks include toxicity and allergenicity, emergence of new viruses, development of antibiotic resistance in micro-organisms, adverse effects on non-target organisms, erosion of crop diversity, and development of new weeds. *Bt* cotton is in many ways an ideal candidate for introduction as a transgenic commercial crop. It is basically grown as a fibre crop, while cotton seed oil used for consumption is free of proteins, including *Bt* protein. The safety of *Bt* toxins in terms of toxicity and allergenicity towards mammals and other non-target organisms is well documented.

Lack of receptors that bind to *Bt* toxins and their instant degradation in human digestive system makes them innocuous to human beings. Community exposure to *Bt* spray formulations over a period of last six decades has not resulted in any adverse effects. Lack of homology to any allergenic protein/epitope sequences makes *Bt* toxins nonallergenic. The safety of *Bt* crop-derived foods has also been well established. In recent years, the effects of *Bt* crop cultivation on non-target organisms including insect predators, parasitoids and pathogens have been investigated quite extensively.

Golden Rice

The bright orange colour of carrots comes from betacarotene, which forms vitamin A in our bodies. Yet 250 million people suffer from vitamin A deficiency. Each year a half million children become blind from lack of vitamin A and over half of these die within months. Ideally, everyone would have a varied diet with lots of produce that supplied ample vitamin A and other nutrients. Better nutrition could prevent up to two million deaths in children under the age of four each year. But that requires more prosperity for much of the world – something that's a long way off. Nearly half the world's population survives on a daily bowl of white rice, which contains no vitamin A. Making rice more nutritious, could improve people's lives tremendously.

A team of researchers decided to try creating rice that contains beta-carotene (the compound we convert to vitamin A). They were inspired by the bright yellow daffodil. How did it produce beta-carotene? They found that several daffodil enzymes manufacture beta-carotene from other molecules. Rice has those other molecules, but it doesn't produce the enzymes to rearrange them into beta-carotene in its kernel. Could they give rice the genes for those enzymes and get them to work together? Previous researchers had inserted several genes that worked individually to make separate products. No one had successfully inserted a group of genes that had to work in sync to make one product. They tried putting the genes in a gene gun and shooting them into rice cells. That didn't work, so they put two genes in one *Agrobacterium* and another gene in another *Agrobacterium*. Both bacteria 'infected' the rice cells, inserted the new genes, and soon the lab grew rice plants carrying all three genes. It was easy to see that the genes worked because of the kernels' golden glow. A bowl of this 'golden rice' provides enough vitamin A to keep a person healthy. Meanwhile, researchers are working on a related nutritional problem. White rice also contains very little useable iron, and without iron, children don't grow or learn well. Iron deficiency causes 40 million mothers to have premature and low weight babies. Many of these mothers and babies die of anemia. The solution also involves several genes from several sources: a fungus, another kind of rice, and a green bean. These genes produce proteins in the rice kernel that help the human body absorb and store iron. Again, they are using *Agrobacterium* to get the genes into rice. Someday, researchers may crossbreed the rice plant that makes betacarotene with one that makes iron to produce a hybrid that makes both essential nutrients. The research team worked ten years on golden rice. They are working out legal issues so they can donate this rice to farmers in developing countries.

Potatoes

Many poor countries can't afford vaccines or can't get them to remote villages. Clinics often can't refrigerate the vaccines or sterilise needles. These problems make safeguarding millions of children extremely difficult. In addition, most vaccines are made from the infectious organism that causes the disease. Every once in a while such vaccine can cause harmful side effects, even the disease they are supposed to prevent. In 1991 the World Health Organisation challenged scientists to create a simpler, safer, cheaper way to vaccinate children. Some scientists began to brainstorm about plants. Since plants naturally make a number of different compounds, they could be reprogrammed to make edible vaccines.

Researchers tried making a cholera vaccine using plants. Cholera is a bacterial disease that causes deadly diarrhea. It spreads rapidly where people don't have clean water and it kills two to three million children each year. Researchers pinpointed part of the cholera bacterium that the human immune system can recognise, so it could be used as a vaccine. Scientists found the genes that make that bacterial part. After some trial and error, they put those genes into potatoes to turn potatoes into a handy vaccine. Potatoes grow in many areas of great health need, and they can withstand long shipping and storage. But there is a snag. People don't eat raw potatoes. So scientists cooked them and found that some of the vaccine still survives. When people ate these cooked potatoes, their bodies made some of the antibodies that can protect them from cholera. Imagine getting your vaccines and boosters from potatoes or some other food instead of painful shots! But that's still a ways off. With the cholera vaccine, researchers need to adjust the dose in each bite and find ways to package them. Of course, people will get their vaccine bits from nurses and clinics, not from the supermarket. Ideally, edible vaccines wouldn't spoil, which would cut the cost and difficulty of delivering them in the developing world. They'd be more pleasant, too. In industrialised countries, most people don't suffer from too little food. They suffer from too much. Obesity is a major health problem even for children. We all know that we should avoid greasy French fries and sugary sodas, but it is hard! If we can't take the junk food away from people, maybe we can take the 'junk' out of food – but keep the taste in. Again, scientists are looking at the potato. When it's fried, oil replaces the water in the potato. But the starchier the potato, the less oil it soaks up. Restaurants pay a premium price for high starch potatoes because they make crisper, less greasy fries. Scientists are trying to develop potatoes with even more starch so they will soak up even less oil. Another way to make a healthier fry is to make healthier oil. Scientists have already modified plants like soyabean and canola to produce a less saturated, healthier fat. Future plants may make even healthier oils that actually strip away fatty deposits from your arteries. What about that soda with your fries? Scientists are working on that, too. They are modifying the sugar beet to produce an enzyme that changes sugar (sucrose) to fructan. Fructan tastes like sugar, but we don't digest fructan so it adds no calories. They have also cloned the gene for a protein in an African plant that tastes a thousand times sweeter than sugar! We could get the same sweetness with a thousand times less sweetener.

Bt Brinjal

The Genetic Engineering Approval Committee's approval of *Bt* brinjal, the first genetically modified crop for human consumption in India, has sparked off protests across the country. On October 15, 2009, the Genetic Engineering Approval Committee (GEAC) of the ministry of environment, the regulatory body for approving genetically modified crops (GM crops) in India, approved *Bt* brinjal, the first GM crop for human consumption in India, for commercial use. The approval came following the review of reports submitted by the Maharashstra Hybrid Seeds Company Limited (Mahyco), the Indian subsidiary of the US-based company Monsanto, that uses biotechnology to produce high yielding, pest resistant crops. *Bt* Brinjal is a genetically modified plant in which a gene from the soil bacterium *Bacillus thuringensis* is inserted into the genome of the brinjal, which can then produce a protein, *Cry*1Ac. This protein behaves as a toxin against the shoot and fruit borer (SFB), a pest that commonly affects brinjal. The gene modification also includes the addition of two antibiotic resistance marker genes.

GENETICALLY MODIFIED ANIMALS AND HUMAN NUTRITION

Important advancement in production and processing of transgenic plants has encouraged studies in animals. Like in plants, microinjection and similar techniques are used to inject foreign gene (DNA) into the nucleus of fertilised egg-cell in animals. When egg is developed to blastula it is transferred to the uterus of an animal where transgenic organism develops. Genetic linkage maps for cattle, pigs and sheep elucidating chromosomal regions for economically important traits will considerably contribute to better quality and amounts of meat. Gene technology is prosperous in farm animal production and in improvement of quality and quantity traits. Gene technology stimulates the yields, higher nutrient consumption, and animal welfare. These traits can be improved directly by gene transfer or using growth hormones, vaccines, antibodies, immunity stimulants and anti-allergy DNA produced by genetic engineering. Gene transfer is expected to improve those production traits in animals that are poorly inherited (low heritability rate, h2), for example number of weaned piglets per sow reported that transgenic plants that produced vaccines, which animals consumed with forage, were produced. The gene for resistance enables breeding of animals resistant to diseases. Vaccine for immune castration of animals, which is painless in male animals and diminishes aggressiveness while female animals are free of negative effects of oestrus, positively affects the economically important trait carcass composition. The possibilities of biotechnological interventions are numerous but the application depends on economic, social and cultural conditions. Transgenic technique can improve the carcass traits and meat quality. The percentage of meat in carcass increases, taste and water binding improve, diminish the percentage of fat and improve the fatty acid composition of meat (more nonsaturated fatty acids. Milk has been modified with transgenes and in most cases without any harm to transgenic animals. Proteins that are used in pharmaceutical industry were obtained from milk of transgenic animals, like human antitrypsin in sheep, plasminogene activator in goat and human protein C in pig.

Transgenic milk can be used as: (i) Food for wide use, (ii) raw materials for milk products, (iii) food for infants, and (iv) source of biologically active substances for pharmaceutical industry.

Even non-protein compounds of human milk, like oligosaccharides, are highly appreciated in milk of transgenic animals. Mammary gland produces milk proteins and lactose under the influence of hormones during late pregnancy and lactation period. Caseins and lactoglobulines are synthesised only during lactation period. Genes from mentioned compounds are used for transgenic milk production that is used for cheese production and for substitute to human milk for infant nutrition reported on wide use of bovine growth hormone (somatotropin) in cattle to increase production of milk and meat. The bovine growth hormone gene had implied as the prediction of the possibilities of production of ideal pork with ultra low fat content and favorable fatty acids composition with transgenic pigs took place.

HEALTH RISK OF GENETICALLY MODIFIED ORGANISMS

'Several animal studies indicate serious health risks associated with GM food,' including infertility, immune problems, accelerated ageing, insulin regulation, and changes in major organs and the gastro-intestinal system.

GMOs ARE INHERENTLY UNSAFE

There are several reasons why GM plants present unique dangers. The first is that the process of genetic engineering itself creates unpredicted alterations, irrespective of which gene is transferred. This creates mutations in and around the insertion site and elsewhere. The biotech industry confidently asserted that gene transfer from GM foods was not possible, the only human feeding study on GM foods later proved that it does take place. The genetic material in soyabeans that make them herbicide tolerant transferred into the DNA of human gut bacteria and continued to function. That means that long after we stop eating a GM crop, its foreign GM proteins may be produced inside our intestines.

GM DIET SHOWS TOXIC REACTIONS IN THE DIGESTIVE TRACT

The very first crop submitted to the FDA's (Food & Drug Administration) voluntary consultation process, the FlavrSavr tomato, showed evidence of toxins. Out of 20 female rats fed the GM tomato, 7 developed stomach lesions. The type of stomach lesions linked to tomatoes could lead to life-endangering hemorrhage, particularly in the elderly who use aspirin to prevent blood clots. Dr. Pusztai believes that the digestive tract, which is the first and largest point of contact with foods, can reveal various reactions to toxins and should be the first target of GM food risk assessment. Mice fed potatoes engineered to produce the *Bt*-toxin developed abnormal and damaged cells, as well as proliferative cell growth in the lower part of their small intestine (ileum). Rats fed potatoes engineered to produce a different type of insecticide (GNA lectin from the snowdrop plant) also showed proliferative cell growth in both stomach and intestinal walls.

GM DIETS CAUSE LIVER DAMAGE

Rats fed the GNA lectin potatoes had smaller and partially atrophied livers Rats fed Monsanto's Mon 863 corn, engineered to produce *Bt*-toxin, had liver lesions and other indications of toxicity. Rabbits fed GM soy showed altered enzyme production in their livers as well as higher metabolic activity. Rats fed Roundup Ready soyabeans also showed structural changes in their liver.

GM FEED ANIMALS HAD HIGHER DEATH RATES AND ORGAN DAMAGE

The cells in the pancreas of mice fed Roundup Ready soy had profound changes and produced significantly less digestive enzymes, in rats fed a GM potato, the pancreas was enlarged. In various analysis of kidneys, GM fed animals showed lesions, toxicity, altered enzymes production or inflammation. Enzyme production in the hearts of mice was altered by GM soy, and GM potatoes caused slower growth in the brain of rats.

REPRODUCTIVE FAILURE AND INFANT MOTALITY

The testicles of both mice and rats fed roundup ready soyabeans showed dramatic changes. In rats, the organs were dark blue instead of pink. In mice, young sperm cells were altered. Embryos of GM soy-fed mice also showed temporary changes in their DNA function, compared to those whose parents were fed non-GM soy.

GM CROPS TRIGGER IMMUNE REACTIONS AND MAY CAUSE ALLERGIES

Allergic reactions occur when the immune system interprets something as foreign, different and offensive and reacts accordingly. All GM foods, by definition have something foreign and different. And several studies show that they provoke reactions. GM potatoes caused the immune system of rats to responded more slowly. And GM peas provoked an inflammatory response in mice, suggesting that it might cause deadly allergic reactions in people. In addition to the herbicide tolerant protein, GM soyabeans contain a unique, unexpected protein, which likely came about from the changes incurred during the genetic engineering process. Scientists found that this new protein was able to bind with IgE antibodies, suggesting that it may provoke dangerous allergic reactions. Organic farmers and others have sprayed crops with solutions containing natural *Bt* bacteria as a method of insect control. The toxin creates holes in their stomach and kills them. Genetic engineers take the gene that produces the toxin in bacteria and insert it into the DNA of crops so that the plant does the work, not the farmer. The fact that we consume that toxic pesticide in every bite of *Bt* corn hardly appetising. Studies verify, however that natural *Bt*-toxin is not fully destroyed during digestion and does react with mammals. The *Bt*—toxin produced in GM crops is vastly different from the bacterial (*Bt*-toxins) used in organic and traditional farming and forestry. The plant produced version is designed to be more toxic than natural varieties. Just like the GM soy protein, the *Bt* protein in GM corn varieties has a section of its amino acid sequence identical to a known allergen (egg yolk), the protein is too resistant to break down during digestion and heat. If *Bt*—toxin causes allergies, then gene transfer carries serious ramifications. If *Bt* genes relocate to human gut bacteria, our intestinal flora may be converted into living pesticide factories, possibly producing *Bt*-toxin inside of us year after year.

SAFETY ASPECTS OF GMO FOOD

It has been well discussed whether the consumption of DNA in approved novel foods and novel foods ingredients can be regarded as safe as consumption of DNA in existing form. All DNA, including DNA from GMOs are composed of the same 4 nucleotides. Genetic modification results in the re-assortment of sequences of nucleotides leaving their chemical structures unchanged. Therefore, DNA from GMOs is chemically equivalent to any other DNA. The only uniqueness is restricted to differences in the DNA sequence, which occurs also in natural variations. The present use of recombinant techniques in the food chain does not introduce changes in the chemical characteristics of the DNA. There is no difference in the susceptibility of recombinant DNA and other DNA to degradation by chemical or enzymatic hydrolysis. There are no indications that ingested DNA has allergenic or other immunogenic properties that would be of relevance for consumption of food derived from GMOs. Uptake integration and expression of any residual extracellular DNA fragments from foods by micro-organisms of the gastrointestinal trait can not be excluded. Each of these circumstances is a rare event and would have happen sequentially. *In vivo* uptake of DNA fragments by mammalian cells after oral administration has been observed. There are effective mechanisms to avoid genomic insertion of foreign DNA. There is no evidence that DNA from dietary sources has ever been incorporated into the mammalian genome studied the animal nutrition with GMOs. Their conclusions are similar as they from. They didn't find differences in physiological and nutritive values in food of animal's products when the animals are feed with GM plants. Adverse health effects need to be screened for, because health effects are dependent upon the modifications made. Most feeding trials have observed no toxic effects and saw that GM foods were equivalent in nutrition to unmodified foods, although a few reports attribute physiological changes to GM food. However, some scientists and advocacy groups such as Greenpeace and World Wildlife Fund consider that the available data do not prove that GM food does not pose risks to health, and call for additional and more rigorous testing before marketing genetically engineered food. A 2008 review published by the Royal Society of Medicine noted that GM foods have been eaten by millions of people worldwide for over 15 years, with no reports of ill effects.

However, a 2009 review in *Nutrition Reviews* found that although most studies concluded that GM foods do not differ in nutrition or cause any detectable toxic effects in animals, some studies did report adverse changes at a cellular level caused by some GM foods, concluding that 'More scientific effort and investigation is needed to ensure that consumption of GM foods is not likely to provoke any form of health problem'. A study published in 2009 found clear negative impact on liver and kidney function in rats consuming GM maize varieties for 90 days.

However, if the product has no natural equivalent, or shows significant differences from the unmodified food, then further safety testing is carried out. Worldwide, reports of allergies to all kinds of foods, particularly nuts, fish and shellfish, seem to be increasing, but it is not known if this reflects a genuine change in the risk of allergy, or an increased awareness of food allergies by the public. A 2005 review in the journal *Allergy* of the results from allergen testing of current GM foods stated that 'no biotech proteins in foods have been documented to cause allergic reactions'

To sum up, the latest development of biotechnology, particularly molecular biology, genetic engineering and transgenic technology has a very large number of potential applications in food production, including micro-organisms, plants and animals. Transgenesis is much more difficult to apply to farm animals than to plants or micro-organisms. Genetic modification has increased production in some crops. But the technology has too few challenges in few crops. Genetic modification is not a good in itself but it is a tool where public & private science can balance each other. Genetically modified foods have various advantages like high yield, salinity tolerant, insect resistance, etc. GM foods have a lot of health effects on living beings. GM foods have both positive and negative effects. These may be either direct effects, on organisms that feed on or interact with the crops, or wider effects on food chains produced by increases or decreases in the numbers of other organisms. As an example of benefits, insect-resistant *Bt*-expressing crops will reduce the number of pest insects feeding on these plants, but as there are fewer pests, farmers do not have to apply as much insecticide, which in turn tends to increase the number of non-pest insects in these fields. Other possible effects might come from the spread of genes from modified plants to unmodified relatives, which might produce species of weeds resistant to herbicides.

SECTION III

Biotechnology of Probiotics and Functional Foods

Chapter 7

Probiotics in Food

INTRODUCTION

Probiotics are live micro-organisms thought to be beneficial to the host organism. Probiotics are usually introduced to food, condiments and beverages as a component of fermentation process at appropriate stage. Lactic acid bacteria (LAB) and *bifidobacteria* are the most common types of microbes used as probiotics, but certain yeasts and *bacilli* may also be helpful. Probiotics are commonly consumed as part of fermented foods with specially added active live cultures, such as in yogurt, soya yogurt or as dietary supplements. Due to their long time survival and multipurpose capacity. There are different rout of administered mechanisms based on age class interval. It can be taken orally in the form of capsule or probiotic food. In order to crate and supply health effect, probiotic cells are constantly viable in the food carries and adapt extreme harsh environment of Gastro intestinal transit (GIT). In another hand, it must be stable during Gastro intestinal transit and fulfill at least cell count become 10^6 CFU g^{-1}. Despite the fact that dairy-based items are proposed to be the principle bearers for the conveyance of probiotics, other nondairy based items. The term probiotics includes countless bacteria and other different micro-organisms, for example, yeast, and every probiotic strain acts particularly. In this manner, when a specific medical advantage has been built up for one strain, this does not imply that all probiotics can present this medical advantage. Each strain must be tried for a specific impact independently.

Numerous microbial species have been used as probiotics. They can be yeast, bacteria or molds. But most commonly, bacterial species are predominant.

According to the currently adopted definition, by WHO/Food and Agriculture Organisation probiotics are an oral supplement or a food product that contains a sufficient number of viable micro-organisms to alter the micro flora of the host and has the potential for beneficial health effects on host, which when it is administered in adequate amounts.

The first recorded probiotic was fermented milk for human consumption. After that, probiotics became popular with animal nutrition. The role of fermented milk in human diet was known even in Vedic times. But, the scientific interest in this area boosted after the publication of the book entitled the prolongation of life by Ellie Metchinkoff in 1908. He suggested that people should consume fermented milk containing *lactobacilli* to prolong their lives. Accelerated ageing is because of autointoxication (chronic toxemia), which is due to the toxins produced by gut microflora. Bulgarian peasants who were subjected to the experiments on longevity had consumed large quantities of sour milk. The pathological reaction might be removed and life expectancy could be enhanced by implanting lactic acid bacteria from Bulgarian yogurt. Since then, researchers started investigations relating to the role of lactic acid bacteria in human and animal health. Probiotics have been used as growth promoters, for lactose intolerance, antitumour and anticholestrolaemic effects. Probiotics have been extensively studied under *in vitro* and *in vivo* conditions. The main fields of research with respect to probiotics are heart diseases, allergic reaction, cancer, diarrhoea, etc. The use of probiotics resulting in alleviation of lactose intolerance due to increased concentration of β-galactosidase in the small intestine, relief from constipation by increased bowel function, antitumour activities due to inhibition of tumour cells, destruction of carcinogens, etc. have been well documented. Intestinal infections caused by *Escherichia coli, Campylobacter fetus* subsp. *jejuni, Clostridium perfringens* and *C. botulinum* were reduced in the presence of *Lactobacillus* supplements. The *Lactobacillus* has shown promising results and *Bifidobacterium longum* has been successfully used to reduce the after-effects of antibiotic therapy.

FEATURES AND COMPOSITION OF PROBIOTICS

Features of Probiotics

A good probiotic agent needs to be non-pathogenic, nontoxic, resistant to gastric acid, adhere to gut epithelial tissue and produce antibacterial substances. It should persist, albeit for short periods in the gastro-intestinal tract influencing metabolic activities like cholesterol assimilation, lactose activity and vitamin production. The survival of probiotic organisms in the gut depends on the colonisation factors that they possess, organelles which enable them to resist the antibacterial mechanisms that operate in the gut. In addition to the antibacterial mechanisms, they need to avoid the effects of peristalsis, which tend to flush out bacteria with food. This can be achieved either by immobilising themselves or by growing at a much faster rate than the rate of removal by peristalsis. The probiotic strain needs to be resistant to the bile acid, e.g. *Bifidobacteria* strains proved significantly less acid-resistant than the *Lactobacillus* strains, when exposed to human gastric juice.

Composition of Probiotics

Probiotics can be bacteria, molds, yeast. But most probiotics are bacteria. Among bacteria, lactic acid bacteria are more popular. *Lactobacillus acidophilus, L. casei, L. lactis, L. helviticus, L. salivarius, L. plantrum, L. bulgaricus, L. rhamnosus, L. johnsonii, L. reuteri, L. fermentum, L. delbrueckii, Streptococcus thermophilus, Enterococcus faecium, E. faecalis, Bifidobacterium bifidum, B. breve, B. longum* and *Saccharomyces boulardii* are commonly used bacterial probiotics. A probiotic may be made out of a single bacterial strain or it may be a consortium as well. Probiotics can be in powder form, liquid form, gel, paste, granules or available in the form of capsules, sachets, etc.

CHARACTERISTICS OF A GOOD PROBIOTICS

In different research studies, probiotics have unique potential properties. To be considered for use and selection as probiotic its safety, technological, and functional characteristics must be sought. Moreover, the following criteria need to be fulfilled:

1. Probiotics should be able to create a beneficial effect on host animal by increasing resistance to diseases.

2. Probiotics must be from human origin.

3. Probiotics needed to have excessive cell viability.

4. Probiotics should be non-pathogenic and non-toxic.

5. It should be able enough to interact or send signals to immune modulator activity.

6. It must have ability to influence local metabolic activity.

7. It ought to be fit for surviving and processing in the gut condition like resistance to low pH and organic acids.

8. Probiotics must be stable, safe, effective and equipped for staying viable for periods under storage and field conditions.

9. It must have power of restore and replace the intestinal micro flora.

10. It should have anti-carcinogenic and anti-mutagenic activity, cholesterol lowering effects, can maintain mucosal integrity and can enhance bowel motility.

11. It should be able to speed up, facilitate and colonise/maintain the digestive tract.

12. They must have the ability to resist gastric juices and the exposure to bile acid which seems to be crucial for oral administration.

13. Adhesion to mucosal and epithelial surfaces, an important property for successful immune modulation, competitive exclusion of pathogens, as well as prevention of pathogen adhesion and colonisation.

14. Antimicrobial activity against pathogenic bacteria.

15. Bile salt hydrolase activity.

16. Antibiotic resistance may help them to survive in the presence of administered drugs and other antimicrobial compounds.

17. Fast multiplication, with either permanent or temporary colonisation of the gastrointestinal tract.

18. Stabilisation of the intestinal micro flora and nonpathogenicity.

19. Survival on passing through gastrointestinal tract at low pH and in contact with bile.

The minimum requirements needed for probiotic status include:

1. Assessment of strain identity (genus, species and strain level).

2. *In vitro* tests to screen potential probiotics: such as resistance to gastric acidity, bile acid, and digestive enzymes as well as antimicrobial activity against potentially pathogenic bacteria.

3. Safety assessment: requirements for proof that a probiotic strain is safe and without contamination in its delivery form.

4. *In vivo* studies for substantiation of the health effects in the target host.

Mechanism of Probiotics

Probiotics have various mechanisms of action currently, three major ways of action of probiotics have been revealed. The first one is a competition for nutrients and for ecological niche at this time the indigenous anaerobic flora limits the concentration of potentially pathogenic flora in the digestive tract. Probiotics can have a direct effect on other micro-organisms through inhibition of pathogen adhesion this kind of major defense mechanism is used to maintain internal health condition. *Lactobacilli* and *bifidobacteria* have been shown to inhibit a broad range of pathogens by performing colonisation of pathogenic bacteria and finally by doing antagonistic activity against gastrointestinal pathogens. This principle in many cases is crucial for the prevention and treatment of infections and restoration of the microbial equilibrium in the gut. The second mechanism is involved in the production of anti-micro-organism substances, bacteriocins, toxins, organic acids, short chain fatty acid production, lowering of gut pH. These substances are responsible for inhibit the growth of other harmful microbes such as foodborne pathogens and spoilage organisms in GIT environment then lead to the death of the pathogen by creating antagonistic condition, and such action may result in the inactivation of toxins. Probiotic mode of effects are carried out based on microbial products which is determine a specific probiotic action and its effective application for the prevention or treatment of a certain disease by destruction of target cell.

The third mechanism is the stimulation/modulation of specific and nonspecific immune response by T-cell activation, to cytokine production/throughout immunomodulation by inducing phagocytosis and IgA secretion, modifying T-cell responses, enhancing Th1 responses, and attenuating Th2 responses. This mode of action is most likely important in the prevention and therapy of infectious diseases. Probiotic bacteria can exert an immunomodulatory effect. These bacteria have the ability to interact with epithelial and dendritic cells (DCs) and with monocytes/macrophages and lymphocytes. In various strategies they are interact and modulate the immune system in a good manner. The immunological advantages of probiotics can be because of activation of local macrophages and modulation of IgA production locally and systemically, to changes in pro/anti-inflammatory cytokine profiles, or to the modulation of response towards food antigens.

The probiotics have a three step action mechanism:

1. It stimulates and modulates immune response.
2. It normalise intestinal microflora by ensures colonisation resistance and controls irritable bowel syndrome and other inflammatory bowel diseases.
3. The final mechanism is also have the metabolic effects like-bile salt deconjugation and secretion, lactose hydrolysis, reduction in toxigenic and mutagenic reactions in gut, Supply of nutrients to colon epithelium.

FACTORS AFFECTING VIABILITY IN FOODS

Some factors, both intrinsic and extrinsic, may influence the survival of probiotics in food, and so have to be considered in all stages of probiotic food manufacturing.

1. Physiological state of the added probiotic in the food.
2. Physico-chemical conditions of food processing.
3. Physical conditions of product storage, like temperature.
4. Chemical composition of the product, such as content of nutrients, oxygen or pH.
5. Interactions with other product components, that can be inhibitory or protective.

Physiological State

The physiological state of the bacteria when prepared, and the physiological state of bacteria in the product itself is an important factor for the survival of the probiotics. The dryness affect to the state of the probiotic, a dry food product will keep the probiotic in a quiescent state during storage, while a wet product will keep the bacteria in a vegetative state with potentially active metabolism. The temperature could modify the shelf-life of the bacteria. With low temperature, long-term survival will be reached.

Bacteria can respond to stressful environments by the induction of various stress tolerance mechanisms. One of them is the induction of stress proteins by exposure of the cells to sublethal stresses so they can condition probiotics to better tolerate environmental stresses in food production, storage, and gastro-intestinal transit. Different probiotic strains have their own intrinsic tolerances to environmental conditions, tolerance can also be influenced by how the culture is prepared, and some cross-protection can be observed, providing protection against other stresses by the exposure to only one stress. Stress responses can be explored to make probiotic strains more resilient and likely to survive in food matrices, with high industrial importance.

Temperature

The temperature at which probiotic organisms grow is an important factor in food applications where fermentation is required, is also a critical factor influencing probiotic survival during manufacture and storage. As discussed above, the lower the temperature the more stable probiotic viability in the food product will be. During processing, temperatures over 45–50°C will be detrimental to probiotic survival, this means that the higher the temperature, the shorter the time period of exposure required to severely decrease the numbers of viable bacteria, ranging from hours or minutes at 45–55°C to seconds at higher temperatures.

Therefore it is obvious that probiotics should be added downstream of heating/cooking/pasteurisation processes in food manufacture to avoid the high temperatures. Elevated temperature also has a detrimental effect on stability during the product process of shipping and storage. Again, the cooler a product can be maintained, the better probiotic survival will be, like in vegetative probiotic cells in liquid products, where refrigerated storage is usually essential. If the product is dried, the bacteria will be in a quiescent state, so acceptable probiotic viability can be maintained in dry products stored at ambient temperatures for 12 months or more. Producing and maintaining low water activities in the foods is the key to maintaining probiotic viability during nonrefrigerated storage because there is a remarkable interaction between temperature and water activity.

pH

Some bacteria like *Lactobacilli* and *bifidobacteria* can tolerate lower pH levels because produce organic acid and products from carbohydrate metabolism. Indeed, numerous *in vitro* and *in vivo* studies have demonstrated that in gastric transit where the cells are exposed to low pH values and with a time of exposure relatively short, some probiotic organisms can survive. In fermented milks and yogurts with pH values between 3.7 and 4.3 *lactobacilli* are able to grow and survive, while *Bifidobacteria* tend to be less acid tolerant, with most species surviving poorly in fermented products at pH levels below 4.6. *B. animales* subsp. lactis is most commonly used in acidic foods because it is more acid tolerant than human intestinal species, and *B. thermoacidophilum*, is even more tolerant to low pH (and heat), but has not yet been characterised thoroughly for probiotic traits and is not used commercially.

Regarding to fruit juices (pH 3.5–4.5) commercially successful products have been produced, such as Gefilus (Valio Ltd., Finland), which contains *Lactobacillus rhamnosus* GG. The viability at low pH can be improved with carriers such as dietary fibers. Survival of *lactobacilli* in low pHs has also been enhanced in the presence of metabolisable sugars, that allow cell membrane proton pumps to operate and prevent lowering of intracellular pH. This can improve survival during gastric transit, but may not be applicable to improving probiotic survival over the time stages of shelf-storage.

Water Activity

For quiescent probiotic bacteria, water activity is a crucial determinant of survival in food products during storage. The higher moisture levels and water activity, the lower survival of probiotics. There is a substantial interaction between water activity and temperature with respect to their impact on the survival of quiescent probiotics. As the storage temperature is increased, the detrimental impact of moisture is magnified. Here, the osmotic stresses appear to play a role, with the presence of smaller molecules resulting in poorer bacterial survival, although the exact cell death mechanisms have not been elucidated yet. There may be technological limitations to reducing water activity to low levels for improving survival. These include the energy costs of drying, adverse impacts on the taste of foods and difficulties in wetting and dispersing powders. Moisture barrier packaging may be applied to prevent the development of moisture from the environment during storage. Maintaining probiotic viability in moderate water activity foods (0.4–0.7) is a great challenge and solutions such as microencapsulation or incorporation of probiotics into fat phases of products can provide improved survival.

Oxygen

Both *bifidobacteria* and *lactobacilli* are considered strict anaerobes and oxygen can be detrimental to its growth and survival. However, the degree of oxygen sensitivity varies considerably between different species and strains, for example, *lactobacilli*, which are mostly microaerophilic, are more tolerant of oxygen than *bifidobacteria*, to the point where oxygen levels are not an important consideration in maintaining the survival of *lactobacilli*. Most probiotic *bifidobacteria* do not grow well in the presence of oxygen, although, many *bifidobacteria* have enzymatic mechanisms to limit the oxygen toxicity. For oxygen sensitive strains, some strategies can be used to prevent oxygen toxicity in food products. Antioxidant ingredients have been shown to improve probiotic survival, as well as the use of oxygen barrier or modified atmosphere packaging. Therefore, it is advisable to minimise processes that are highly aerating, particularly when using *bifidobacteria*.

Toxicity of ingredients

Interactions between probiotics and other ingredients could happen and those interactions can be protective, neutral or detrimental to probiotic stability. Obviously, the inclusion of antimicrobial preservatives can inhibit probiotic survival and elevated levels of ingredients such as salt, organic acids, and nitrates can inhibit probiotics during storage, while starter cultures can sometimes inhibit the growth of probiotics during fermentation through the production of specific bacteriocins.

Growth factors, protective, and synergistic ingredients

Probiotic *lactobacilli* and, in particular, *bifidobacteria* are only weakly proteolytic and grow relatively slowly or poorly in milk. The growth of *bifidobacteria* can be improved by the presence of suitable companion cultures, which can aid in protein hydrolysis and through the production of growth factors. Some growth substrates such as carbon sources, nitrogen sources, and growth factors or antioxidants, minerals, and vitamins can be added to improve growth. Finally, the food matrix itself can be protective like in the cheese, where the anaerobic environment, high fat content and buffering capacity of the matrix helps to protect the probiotic cells both in the product and during intestinal transit.

Freeze-thawing

The damages made to cell membranes freezing probiotics is detrimental to survival, and also can make the cells more vulnerable to environmental stresses. To prevent or at least mitigate cell injury, protectants are usually added to cultures to be frozen or dried. Once frozen, probiotics can survive well over long shelf-lives in products such as frozen yogurts and ice-cream. Using alternative methods of freezing, such as slow-cooling rates or pre-freezing stress, can significantly improve cell survival. Repeated freeze-thawing cycles are highly detrimental to cell survival and should be avoided.

Sheer forces

Probiotic *lactobacilli* and *bifidobacteria* are gram-positive bacteria with thick cell walls that are able to tolerate the sheer forces generated in most standard food production processes such as high-speed blending or homogenisation, that may result in cell disruption and losses in viability.

Side-effects

In some situations, such as where the person consuming probiotics is critically ill, probiotics could be harmful. In a therapeutic clinical trial conducted by the Dutch Pancreatitis Study Group, the consumption of a mixture of six probiotic bacteria increased the death rate of patients with predicted severe acute pancreatitis. In a clinical trial conducted at the University of Western Australia, aimed at showing the effectiveness of probiotics in reducing childhood allergies, researchers gave 178 children either a probiotic or a placebo for the first six months of their life. Those given the good bacteria were more likely to develop a sensitivity to allergens.

Some hospitals have reported treating *lactobacillus septicaemia*, which is a potentially fatal disease caused by the consumption of probiotics by people with lowered immune systems or who are already very ill. There is no published evidence that probiotic supplements are able to replace the body's natural flora when these have been killed off, indeed bacterial levels in feces disappear within days when supplementation ceases. Probiotics taken orally can be destroyed by the acidic conditions of the stomach. A number of micro-encapsulation techniques are being developed to address this problem. Recent studies indicate that probiotic products such as yogurts could be a cause for obesity trends. However, this is contested as the link to obesity, and other health related issues with yogurt may link to its dairy and calorie attributes. Some experts are skeptical on the efficacy of many strains and believe not all subjects will benefit from the use of probiotics.

PROBIOTIC PRODUCT SPECIFICATIONS, QUALITY ASSURANCE AND REGULATORY ISSUES

Regulatory Issues

Government regulations differ among countries, however the status of probiotics as a component in food is currently not established on an international basis. For the most part, probiotics come under food and dietary supplements because most are delivered by mouth as foods. These are differentiated from drugs in a number of ways, especially with respect to claims. Drugs are allowed to claim effectiveness in the treatment, mitigation or cure of a disease, whereas foods, feed additives and dietary supplements can only make general health claims. In order to understand where probiotic products currently fall in terms of regulatory agencies, and the claims that can be made with their use.

- A 'health claim' is defined as 'a statement, which characterises the relationship of any substance to a disease or health-related condition, and these should be based upon well-established, generally accepted knowledge from evidence in the scientific literature and/or recommendations from national or international public health bodies. Examples include 'protects against cancer'.

- A structure/function claim is defined as 'a statement of nutritional support that describes the role of a nutrient or dietary ingredient to affect the structure or functioning of the human body, or characterises the documented mechanism by which a nutrient or dietary ingredient acts to maintain such structure or function. Examples include 'supports the immune system'. Claims that substances can treat, diagnose, cure or prevent a disease are not structure/function claims.

The new paradigm of risk analysis is making its way into regulatory food safety and focuses on a functional separation of the science-based risk assessment and risk management. However, the issue of communication is now also considered an important integrated part of risk analysis. Communication includes exchange between assessors and managers and two-way interaction with other interested parties. Within this concept, the transparency of the decision making process for food safety regulatory action is emphasised, as well as the importance of providing a vehicle for consumers and others to participate in the development process. Therefore communication efforts relative to the use of probiotics should be considered as an integrated part of the development of regulatory initiatives.

Appropriate Labelling

To clarify the identity of a probiotic present in the food, it is recommended that the microbial species be stated on the label. If a selection process has been undertaken at the strain level, the identity of the strain should also be included, since the probiotic effect seems to be strain specific. There is a need to accurately enumerate the probiotic bacteria in food products in order to include them on the label. The label should state the viable concentration of each probiotic present at the end of shelf life.

Manufacturing and Handling Procedures

To ensure that any given culture maintains the beneficial properties, the stock culture should be maintained under appropriate conditions and be checked periodically for strain identity and probiotic properties. Furthermore, viability and probiotic activity must be maintained throughout processing, handling and storage of the food product containing the probiotic, and verified at the end of shelf-life.

Adequate quality assurance programmes should be in place. Good manufacturing practices should be followed in the manufacture of probiotic foods. It is recommended that the Codex General Principles of Food Hygiene and Guidelines for Application of HACCP be followed.

Powdered milk products

The purpose of this description is to address the health and nutritional properties of milk powder with live lactic acid bacteria, it was considered necessary to further address the issue in this section. Methods of production of dried probiotic powders should be such that adequate numbers of viable probiotic bacteria are maintained in the dried powder following manufacture, and also retention/stability of probiotic properties should be ensured throughout shelf-life.

Careful consideration should be given to factors such as the following, with respect to viability of the probiotic:

- Drying method.
- Type of packaging.
- Size of packaging.
- Storage conditions (temperature, humidity, etc.).
- Powder milk quality (Standard reference).

- Rehydration procedure.
- Handling of rehydrated product.

Prebiotics

Prebiotics as an area is distinct from probiotics and therefore, will not be covered in detail in this section. Smith and others recognised both the potential benefits of prebiotics with respect to probiotics, in addition to their ability to stimulate indigenous beneficial bacteria in the host.

Prebiotics are generally defined as 'nondigestible food ingredients that beneficially affect the host by selectively stimulating the growth and/or activity of one or a limited number of bacterial species already established in the colon, and thus in effect improve host health'.

The concept of prebiotics essentially has the same aim as probiotics, which is to improve host health via modulation of the intestinal flora, although by a different mechanism. However, there are some cases in which prebiotics may be beneficial for the probiotic, especially with regard to *bifidobacteria*, that is the synbiotic concept. Synbiotics are defined as 'mixtures of probiotics and prebiotics that beneficially affect the host by improving the survival and implantation of live microbial dietary supplements in the gastrointestinal tract of the host. If a synbiotic relationship is intended, then it should be verified scientifically.

Prebiotics vs Probiotics

While prebiotics and probiotics sound similar, these supplements are very different and have different roles in the digestive system (or gut).

Prebiotic fibre is a non-digestible part of foods like bananas, onions and garlic, Jerusalem artichoke, the skin of apples, chicory root, beans, and many others. Prebiotic fibre goes through the small intestine undigested and is fermented when it reaches the large colon.

This fermentation process feeds beneficial bacteria colonies (including probiotic bacteria) and helps to increase the number of desirable bacteria in our digestive systems (also called the gut) that are associated with better health and reduced disease risk.

Probiotics are live beneficial bacteria that are naturally created by the process of fermentation in foods like yogurt, sauerkraut, miso soup, kimchi, and others. Probiotics are also available in pill form and as an added ingredient in products like yogurt and health drinks.

Kefir, a milk drink that has been fermented using kefir grains, is an especially potent source of probiotics. According to Jeannette Hyde, a nutritional therapist, and high-profile advocate for kefir, 'it contains *lactobacilli* and *bifidobacteria* in high doses, and also helps diversity too – more than 50 different types of bacteria can be found in kefir. When you drink kefir (it has the consistency of a drinking yoghurt), these bacteria travel through the digestive tract to colonise the colon.'

While many types of bacteria are classified as probiotics, most come from two groups:

- *Lactobacillus* – the most common probiotic found in yogurt and other fermented foods. Can help with diarrhea and may help with people who can't digest milk sugar (lactose).

- *Bifidobacterium* – also found in some dairy products. May ease symptoms of irritable bowel syndrome (IBS) and related conditions. Naturally present in the large intestine, bifidobacteria fight harmful bacteria in the intestines, prevent constipation and give the immune system a boost. Furthermore, evidence indicates that bifidobacteria help reduce intestinal concentrations of certain carcinogenic enzymes.

A helpful metaphor to understand the difference between a prebiotic and a probiotic may be a garden. You can add seeds—the probiotic bacteria—while the prebiotic fibre is the water and fertiliser that helps the seeds to grow and flourish.

Benefits of probiotics

The beneficial effects of probiotics have been widely demonstrated. Health professionals often recommend probiotics in supplement form to patients on antibiotics in an attempt to repopulate the colon with desirable bacteria after the course of antibiotics has wiped out both beneficial and undesirable bacteria. Some find taking probiotics can combat gastrointestinal side effects of the medication and reduce the bacterial growth leading to yeast infections. Since each body is different, it is necessary to determine which probiotics will be helpful to one's own system. In addition, it is important to make sure the bacteria in probiotic supplements are alive. Probiotic bacteria are fragile and can easily be killed by stomach acid, time, and heat.

Benefits of prebiotics

Researchers have found that prebiotics are helpful in increasing the helpful bacteria already in the gut that reduce disease risk and improve general well being. Prebiotic fibre is not as fragile as probiotic bacteria because it is not affected by heat, stomach acid, or time. Nor does the fermentation process differ depending on the individual.

Foods rich in prebiotic fibre:

Chicory root: About 65% of the chicory root is fibre by weight and is an extraordinarily rich source of prebiotic fibre.

Onions and garlic: 2 grams of fibre per ½ cup – about 17% is prebiotic fibre.

Oatmeal: 2 grams of fibre per ½ cup—very high in prebiotic fibre content.

Wheat bread with wheat bran: About 1 gram of fibre per slice, nearly 70% of the total fibre in wheat bran is prebiotic fibre.

Asparagus: 2–3 grams of prebiotic fibre per 100 gram serving (about ½ cup).

Dandelion greens: 4 grams of fibre per 100 gram serving (about ½ cup) – most of this fibre is prebiotic.

Jerusalem artichoke: 2 grams of fibre per 100 gram serving (about ½ cup) – 76% comes from inulin prebiotic fibre.

Barley: 3–8 grams of prebiotic fibre per 100 gram serving (about ½ cup).

Apple with skin: 2 grams of fibre per ½ apple (mainly in the skin) Pectin, which has prebiotic benefits, makes up about 50% of the total fibre in the apple.

Table 7.1 shows compares between prebiotics and probiotics.

Table 7.1: Compares between prebiotics and probiotics.

Prebiotics	*Probiotics*
Prebiotics are a special form of dietary fibre that acts as a fertiliser for the good bacteria in your gut	Probiotics are live bacteria that can be found in yogurt and other fermented foods. There are hundreds of probiotic species available. Which of these species are best for the average healthy person is still unknown

(Cont'd...)

Prebiotics	*Probiotics*
Prebiotic powders are not affected by heat, cold, acid, or time	Probiotic bacteria must be kept alive to be active. They may be killed by heat, stomach acid, or simply die with time.
Prebiotics nourish the good bacteria that everyone already has in their gut	Probiotics must compete with the over 1000 bacteria species already in the gut
Research has determined that supplementing with an oligofructose enriched inulin-based (OEI) Prebiotic fibre can be helpful with a wide range of conditions and disorders, including digestive disorders, obesity, and bone loss	Certain probiotic species have been shown to be helpful for childhood diarrhea, irritable bowel disease, and for recurrence of certain bowel infections such as C. difficile.

RECOMMENDATIONS

1. Potential probiotic strains must be identified by methods including internationally accepted molecular techniques and named according to the International Code of Nomenclature, and strains should preferably be deposited in a reputable internationally recognised culture collection.

2. In order to be termed a probiotic, the probiotic micro-organism must be able to confer defined health benefits on the host.

3. There is a need for refinement of *in vitro* and *in vivo* tests to better predict the ability of probiotic micro-organisms to function in humans.

4. There is a need for more statistically significant efficacy data in humans.

5. Good manufacturing practices must be applied with quality assurance, and shelflife conditions established, and labelling made clear to include minimum dosage and verifiable health claims.

6. The regulatory status of probiotics as a component in food has to be established on an international level.

7. The regulatory framework be established to better address issues related to probiotics including efficacy, safety, labellling, fraud and claims.

8. Probiotic products shown to confer defined health benefits on the host should be permitted to describe these specific health benefits.

9. Surveillance systems, including trace-back and post marketing surveillance, should be put in place to record and analyse any adverse events associated with probiotics in food. Such systems could also be used to monitor the long-term health benefits of probiotic strains.

10. Efforts should be made to make probiotic products more widely available, especially for relief work and populations at high risk of morbidity and mortality.

11. Further work is needed to address criteria and methodologies for probiotics.

GUIDELINES FOR THE EVALUATION OF PROBIOTICS IN FOOD

In order to claim that a food has a probiotic effect, the following guidelines should be followed.

Genus/Species/Strain

It was recognised that it is necessary to know the genus and species of the probiotic strain. The current state of evidence suggests that probiotic effects are strain specific. Strain identity is important to link a strain to a specific health effect as well as to enable accurate surveillance and epidemiological studies.

A possible exception is the ability in general of *S. themophilus* and *L. delbrueckii* ssp. *bulgaricus* to enhance lactose digestion in lactose intolerant individuals. In this case, or in other cases where there is suitable scientific substantiation of health benefits that are not strain specific, individual strain identity is not critical. Speciation of the bacteria must be established using the most current, valid methodology. It is recommended that a combination of phenotypic and genetic tests be used.

Nomenclature of the bacteria must conform to the current, scientifically recognised names. Protracted use of older or misleading nomenclature is not acceptable on product labels. The use of incorrect names does not properly identify the probiotic bacterium in the product and forces consumers and regulatory agencies to make assumptions about the identity of the real bacterium being sold.

DNA-DNA hybridisation is the reference method to specify that a strain belongs to a species, however, as it is time consuming and beyond the resources of many laboratories, requiring a large collection of reference strains, the use of DNA sequences encoding 16S rRNA is suggested as a suitable substitute. In this case, it is recommended that this genotypic technique be combined with phenotypic tests for confirmation.

Patterns generated from the fermentation of a range of sugars and final fermentation products obtained from glucose utilisation are key phenotypes that should be investigated for identification purposes.

Strain typing has to be performed with a reproducible genetic method or using a unique phenotypic trait. Pulsed Field Gel Electrophoresis (PFGE) is the gold standard. Randomly Amplified Polymorphic DNA (RAPD) can also be used, but is less reproducible. Determination of the presence of extrachromosomal genetic elements, such as plasmids can contribute to strain typing and characterisation. It is recommended that all strains be deposited in an internationally recognised culture collection.

In vitro Tests to Screen Potential Probiotics

In vitro tests are critical to assess the safety of probiotic microbes. In addition, *in vitro* tests are useful to gain knowledge of strains and the mechanism of the probiotic effect. However, it was noted that the currently available tests are not fully adequate to predict the functionality of probiotic micro-organisms in the human body. It was also noted that *in vitro* data available for particular strains are not sufficient for describing them as probiotic.

Probiotics for human use will require substantiation of efficacy with human trials. Appropriate target-specific *in vitro* tests that correlate with *in vivo* results are recommended. For example, *in vitro* bile salts resistance was shown to correlate with gastric survival *in vivo*. A list of the main currently *in vitro* tests used for the study of probiotic strains is shown in Table 7.2. All of these tests require validation, however, with *in vivo* performance.

Table 7.2: Main currently used *in vitro* tests for the study of probiotic strains.

Resistance to gastric acidity

Bile acid resistance

Adherence to mucus and/or human epithelial cells and cell lines

Antimicrobial activity against potentially pathogenic bacteria

Ability to reduce pathogen adhesion to surfaces

Bile salt hydrolase activity

Resistance to spermicides (applicable to probiotics for vaginal use)

SAFETY CONSIDERATIONS

Antimicrobial Resistance Profiles of Probiotics

As with any bacteria, antibiotic resistance exists among some lactic acid bacteria, including probiotic micro-organisms. This resistance may be related to chromosomal, transposon or plasmid located genes. However, insufficient information is available on situations in which these genetic elements could be mobilised and it is not known if situations could arise where this would become a clinical problem.

There is concern over the use in foods of probiotic bacteria that contain specific drug resistance genes. Bacteria, which contain transmissible drug resistance genes, should not be used in foods. Currently, no standardised phenotypic methods are available which are internationally recognised for *lactobacilli* and *bifidobacteria* (non-pathogens). Thus, it recognises the need for the development of standardised assays for the determination of drug insensitivity or resistance profiles in *lactobacilli* and *bifidobacteria*. Also the plasmids exist in *lactobacilli* and *bifidobacteria*, especially in strains isolated from the intestine, which have genes encoding antibiotic resistance. Due to the relevance of this problem, it is suggested that further research be done relating to the antibiotic resistance of *lactobacilli* and *bifidobacteria*. When dealing with selection of probiotic strains, it is recommended that probiotic bacteria should not harbour transmissible drug resistance genes encoding resistance to clinically used drugs. Research is required relating to the antibiotic resistance of *lactobacilli* and *bifidobacteria* and the potential for transmission of genetic elements to other intestinal and/or foodborne micro-organisms.

Safety of Probiotics in Humans

In terms of safety of probiotics, it is believed that a set of general principles and practical criteria need to be generated to provide guidelines as to how any given potential probiotic micro-organism can be tested and proven to have a low risk of inducing or being associated with the etiology of disease, versus conferring a significant health benefit when administered to humans. These guidelines should recognise that some species may require more vigorous assessment than others. In this respect, the evaluation of safety will require at least some studies to be performed in humans, and should address aspects of the proposed end use of the probiotic strain.

Information acquired to date shows that *lactobacilli* have a long history of use as probiotics without established risk to humans, and this remains the best proof of their safety. Also, no pathogenic or virulence properties have been found for *lactobacilli*, *bifidobacteria* or *lactococci*. Thus, under certain conditions, some *lactobacilli* strains have been associated with adverse effects, such as rare cases of bacteremia. However, a recent epidemiological study of systematically collected *lactobacilli* bacteremia case reports in one country has shown that there is no increased incidence or frequency of bacteremia with increased usage of probiotic *lactobacilli*.

It is also acknowledged that some members of lactic acid bacteria, such as *enterococci* may possess virulence characteristics. For this and other reasons, recommended that *Enterococcus* not be referred to as a probiotic for human use. The rationale is based upon:

1. Strains can display a high level of resistance to vancomycin, or can acquire such resistance. If this resistance is present, transfer to other micro-organisms may occur and this could enhance the pathogenesis of such recipients.

2. Certain strains of vancomycin resistant *enterococci* are commonly associated with nosocomial infections in hospitals.

Smith and others recognised some strains of *Enterococcus* display probiotic properties, and may not at the point of inclusion in a product display vancomycin resistance. However, the onus is on the producer to prove that any given strain cannot acquire or transfer vancomycin resistance or be virulent and induce infection.

Safety aspects and harmful side effects of probiotics

Probiotics may be responsible for four types of side effects in susceptible individuals: systemic infections, deleterious metabolic activities, excessive immune stimulation, and gene transfer. When the dose of intake is very high extends to causes of infections in humans not only in all age groups but also in immuno-compromised individuals. Three approaches can be used to assess the safety of a probiotic strain: studies on the intrinsic properties of the strain, studies on the pharmacokinetics of the strain (survival, action in the digestive tract, dose–reaction connections, fecal and mucosal recuperation) and studies hunting down for interaction between the strain and the host. Symptoms of side effects are accepted to come about because of bacteria-host interactions in which the probiotic supplement might be contrary with the present living space of the user's microbiota, eventually setting off a response. Cases of normal reactions from probiotics include: abnormal bowel movements, bloating, flatulence, gurgling, and stomach aches. It is happened in rare case. May be producing an active infection, although this risk is quite low but, allow to stimulate the situation in immunosuppressed patients. Administration during pregnancy and early infancy is considered safe. Available data indicate that not those much harmful effects have been observed in controlled clinical studies with *lactobacilli* and *bifidobacteria*. Thus, the scientists should conduct research in the area about its negative impact in human all over status profoundly.

The capacities of probiotics to survive and be metabolically active in the GIT and to associate with the gastrointestinal mucosa and gastrointestinal microflora have prompted four zones of worry about safety:

1. Potential for bacteria to translocate/transmigrate, crossing the gastrointestinal tract boundary and bringing about intrusive infection. Translocation by intestinal bacteria is encouraged by various elements including intestinal mucosal damage, immunodeficiency, gut prematurity and abnormal bacterial flora and adherence of the bacteria to the mucosal surface.

2. The likelihood for some probiotic life forms is to harbour protection from anti-infection agents (antibiotics), prompting a potential for antibiotic resistance that is to be exchanged from probiotic bacteria to other possibly pathogenic bacteria. With the goal that such organisms may harbour genes that may add to opportunistic infections on the grounds that the antibiotic resistance gene can be exchanged by conjugation, transduction or transformation way.

3. Metabolic activity and immunologic effects of probiotics leading to possible deleterious metabolic effects and excessive immune stimulation.

4. Last but not the least all the concerns that we have is all about product quality, since products that does not contain the probiotic on the label, or that contain contaminants may likewise put the consumer in danger.

Due to potential impact of the use of probiotics on gastrointestinal physiology, Gastrointestinal toxicity studies should be studied as one part of safety concern, as there may be production of metabolites that are undesirable, chance that there can be generation of metabolites that are unfortunate, chance that the probiotic bacteria might lead to crate, encourage or increment the danger of various physiological and anatomical issues.

Future Perspectives of Probiotics

Now-a-day, technological innovations contribute a mechanism to solve the problem of probiotic stability and viability. Pure and active viability of cells is very necessary in food processing and gastro intestinal transit to reach the intended site of action in sufficient numbers. Most of the time probiotics are loss their function and useful property/viability. Due to passage through the low pH environment of the stomach and high bile salt conditions in the intestine. The only way of to overcome the challenge is introducing to sublethal stress, applying encapsulation and using in food matrix/ carriers. Encapsulation is a mechanical or physicochemical process that traps a potentially sensitive material and provides a protective barrier between it and the external conditions. The new microencapsulation technologies/methods have been developed to protect the bacteria from damage caused by external environment through a protective outer coating. Microencapsulation of probiotics enables storage of viable bacteria at room temperature and may allow incorporation of probiotics into a wide range of food products. The spray-drying, emulsion and extrusion techniques are well known encapsulation methods for the production of microcapsules containing probiotics. The future attitude regarding to improve overall characteristics of the strain and to get power full desired trait is apply genetic engineering on the area.

Thus, the consumption of probiotics helps to lead a healthy life. Currently, this is globally a well-accepted concept and guarantee for the next generation. Probiotics are widely used in order to solve and simplify particular diseases. In the future highly emphasise further *in vitro* and *in vivo* experiments should be designed and conducted to identify true probiotics and to select the most suitable ones for the prevention/treatment of diseases. Lastly recommend further practical studies need confirmation about its effect in human health with in high quality research and well-designed clinical trials.

Chapter 8

Health Benefits of Probiotics

INTRODUCTION

The major origin of probiotics are fermented non-digestible carbohydrate compounds, food supplements, dairy based compounds, non-dairy fermented food and non-intestinal sources. Probiotic micro-organisms can isolated, screened, identified and characterised from numerous natural substrates. The sources of the power full strain in recent years, wide and up to now still growing. In addition to this, peoples are explored live cell containing food because it is enhance nutritive quality, bioavailability of the micro nutrients and possess anti oxidative property. Anti-oxidative property helps to fight oxidative stress, strengthens host anti oxidative defense mechanism and delays ageing. Bioavailability of the micro nutrients, possess anti oxidative property also. Anti-oxidative property helps to fight oxidative stress, strengthens host anti oxidative defense mechanism and delays ageing. Therefore, many probiotic foods can effective in full fill the interest of the people at all age.

The beneficial effects of probiotic foods on human health and nutrition are increasingly recognised by health professionals. Recent scientific work on the properties and functionality of living micro-organisms in food have suggested that probiotics play an important role in immunological, digestive and respiratory functions, and that they could have a significant effect on the alleviation of infectious diseases in children and other high-risk groups. In parallel, the number and type of probiotic foods and drinks that are available to consumers, and marketed as having health benefits, has increased considerably.

The beneficial effects of food with added live microbes (probiotics) on human health, and in particular of milk products on children and other high-risk populations, are being increasingly promoted by health professionals. It has been reported that these probiotics can play an important role in immunological, digestive and respiratory functions and could have a significant effect in alleviating infectious disease in children.

GUIDELINES FOR THE ASSESSMENT OF PROBIOTIC MICRO-ORGANISMS

In order to assess the properties of probiotics, the following guidelines should be used. For use in foods, probiotic micro-organisms should not only be capable of surviving passage through the digestive tract but also have the capability to proliferate in the gut. This means they must be resistant to gastric juices and be able to grow in the presence of bile under conditions in the intestines, or be consumed in a food vehicle that allows them to survive passage through the stomach and exposure to bile. They are Gram positive bacteria and are included primarily in two genera, *Lactobacillus* and *Bifidobacterium*.

Selection of Probiotic Strains for Human Use

Probiotics must be able to exert their benefits on the host through growth and/or activity in the human body. However, it is the specificity of the action, not the source of the micro-organism that is important. Indeed, it is very difficult to confirm the source of a micro-organism. Infants are born with none of these bacteria in the intestine, and the origin of the intestinal microflora has not been fully elucidated. It is the ability to remain viable at the target site and to be effective that should be verified for each potentially probiotic strain. There is a need for refinement of *in vitro* tests to predict the ability of probiotics to function in humans. The currently available tests are not adequate to predict the functionality of probiotic micro-organisms in the intestine.

Classification and Identification of Individual Strains

Classification is the arranging of organisms into taxonomic groups (taxa) on the basis of similarities or relationships. Nomenclature is the assignment of names to the taxonomic groups according to rules. Identification is the process of determining that a new isolate belongs to one of the established, named taxa. It is recommended that probiotics be named according to the International Code of Nomenclature to ensure understanding on an international basis. But for the sake of full disclosure, probiotic strains be deposited in an internationally recognised culture collection. Since probiotic properties are strain related, it is suggested that strain identification (genetic typing) be performed, with methodology such as pulse field gel electrophoresis (PFGE). It is recommended that phenotypic tests be done first, followed by genetic identification, using such methods as DNA/DNA hybridisation, 16S RNA sequencing or other internationally recognised methods. For the latter, the RDP (ribosomal data base project) should be used to confirm identity.

HEALTH BENEFITS OF PROBIOTICS

Dairy strains of lactic acid bacteria (LAB) have a long history of utilisation. LAB, including diverse types of *Lactobacillus* and *Enterococcus* species, that has been consumed daily since humans started to use fermented milk as food. Probiotic impacts are strain particular the impacts depicted for one strain can't be specifically applied to others and every individual. Probiotic bacterial strain has its own health benefits. The major beneficial effects are correlated against various disease conditions. Probiotics have a colossal criticalness and application in controlling different kinds of microbial infections. Probiotics are applicable in human health improvement, infection control, diseases treatment and management.

Defining and Measuring the Health Benefits of Probiotics

A number of health effects are associated with usage of probiotics. There are differing degrees of evidence supporting the verification of such effects. Also there are reports showing no clinical effects of certain probiotic strains in specific situations. The use of probiotic micro-organisms to confer health benefits on the host must indicate the dosage regimens and duration of use as recommended by the manufacturer of each individual strain or product based upon scientific evidence, and as approved in the country of sale. While this practice is not currently in place, and it is strongly recommended that each product should indicate the minimum daily amount required for it to confer specific health benefit(s). Such evidence should, where possible result from *in vitro*, animal (where appropriate) and human studies. Examples have been cited below to illustrate studies on specific strains and clinical outcomes. In doing so, the emphasis should not be on one particular strain being termed as superior to another, rather that the benefit conferred and the methods used to obtain and measure said benefits are of most importance.

Prevention of diarrhea caused by certain pathogenic bacteria and viruses

Infectious diarrhea is a major world health problem, responsible for several million deaths each year. While the majority of deaths occur amongst children in developing countries, it is estimated that up to 30% of the population even in developed countries are affected by foodborne diarrhea each year. Probiotics can potentially provide an important means to reduce these problems. It should be noted that some of the studies referenced below utilise probiotics administered in a non-food form.

The strongest evidence of a beneficial effect of defined strains of probiotics has been established using *Lactobacillus rhamnosus* GG and *Bifidobacterium lactis* BB-12 for prevention and treatment of acute diarrhea mainly caused by rotaviruses in children. In addition to rotavirus infections, many bacterial species cause death and morbidity in humans. There is good *in vitro* evidence that certain probiotic strains can inhibit the growth and adhesion of a range of enteropathogens, and animal studies have indicated beneficial effects against pathogens such as *Salmonella*. There is evidence from studies on travellers' diarrhea, where some of the causative pathogens have been presumed to be bacterial in nature, that benefits can accrue with probiotic administration.

It is important to note that probiotic therapy of acute diarrhea should be combined with rehydration if available. Current WHO recommendations state that clinical management of acute diarrhea should include replacement of fluid and electrolytes losses along with nutritional support. Oral rehydration salts (ORS) have been widely used in such disease management, and it is within this context that the combination therapy with probiotics is hereby advocated. Effects such as probiotic restoration of the non-pathogen dominated intestinal microflora secondary to infection, maintaining mucosal integrity and improving electrolyte balance could have a significant impact on programmes of treatment and prevention of acute diarrhea in developing countries.

A major problem associated with antibiotic treatment is the appearance of diarrhea, often caused by *Clostridium difficile*. This organism is not uncommon in a healthy intestinal tract, but the disruption of the indigenous microflora by antibiotics leads to an abnormal elevation of their numbers, and subsequent symptoms related to toxin production. The rationale therefore to use probiotics is that in such patients, administration of exogenous commensal micro-organisms (that is probiotics) is required to restore the microflora to one that more closely reflects the normal flora prior to antibiotic therapy. Some open ended studies have indeed shown that this approach can alleviate the signs and symptoms of *C. difficile* infection. With respect to antibiotic-associated diarrhea, probiotics have proved useful as a prophylactic regimen, and potentially they can also be used to alleviate the signs and symptoms once antibiotic induced diarrhea has occurred. It must be recognised that evidence for therapeutic effects against *C. difficile,* and other disorders has been obtained using certain probiotic strains, such as *L. rhamnosus* GG. It is important to note that such effects may also be conferred by other strains, but scientific evidence may not yet be available or the micro-organisms involved may not be included.

Helicobacter pylori infection and complications

A new development for probiotic applications is activity against *Helicobacter pylori,* a Gram negative pathogen responsible for type B gastritis, peptic ulcers and gastric cancer. *In vitro* and animal data indicate that lactic acid bacteria can inhibit the growth of the pathogen and decrease urease enzyme activity necessary for the pathogen to remain in the acidic environment of the stomach. Human data is limited, but there is some evidence of an effect induced by *L. johnsonii* La1. In terms of measuring probiotic effects, feasible end points include the suppression of the infection (which may be reversible upon cessation of treatment), combination treatment with antibiotics leading to fewer side effects such as acid reflux, and lower risk of recurrent infection. Placebo-controlled trials are needed before specific claims can be made for probiotic anti-*Helicobacter pylori* benefits in humans with respect to prevention and treatment. Such studies are warranted given the preliminary evidence to support these effects.

Probiotics and inflammatory bowel disease

Incorporation of probiotic bacteria has an ability to become stable the immunological barrier in the gut mucosa by declining the generation of local pro-inflammatory cytokines. Probiotics is used for treatment of the inflammatory bowel disease, such as ulcerative colitis, Crohn's disease and Pouchitis. Potential mechanisms include suppression of growth or epithelial binding and invasion by pathogenic bacteria, production of antimicrobial substances, improved epithelial barrier function, and immunoregulation. The effects of probiotic are probably both strain-dependent and dose dependent.

Inflammatory bowel diseases, such as pouchitis and Crohn's disease, as well as irritable bowel syndrome, may be caused or aggravated by alterations in the gut flora including infection. These are new avenues of investigation, although it is premature to state a firm action of probiotics in these conditions. Some studies support the potential role of probiotics in therapy and prophylaxis and illustrate that combinations of strains may have a role to play in remediation.

The intestinal microflora likely plays a critical role in inflammatory conditions in the gut, and potentially probiotics could remediate such conditions through modulation of the microflora. Clinical and mechanistic studies are urgently required to better understand the interface between the microbes, host cells, mucus and immune defenses, and to create efficacious interventions. Such studies should include molecular examination of the intestinal (not only fecal) flora and long-term (5–10 years) effects of probiotic micro-organisms.

Cancer

There is some preliminary evidence that probiotic micro-organisms can prevent or delay the onset of certain cancers. This stems from the knowledge that members of the gut microflora can produce carcinogens such as nitrosamines. Therefore, administration of *lactobacilli* and *bifidobacteria* could theoretically modify the flora leading to decreased b-glucuronidase and carcinogen levels.

Furthermore, there is some evidence that cancer recurrences at other sites, such as the urinary bladder can be reduced by intestinal instillation of probiotics including *L. casei* Shirota. *In vitro* studies with *L. rhamnosus* GG and *bifidobacteria* and an *in vivo* study using *L. rhamnosus* strains GG and LC-705 as well as *Propionibacterium* sp. showed a decrease in availability of carcinogenic aflatoxin in the lumen. However, it is too early to make definitive clinical conclusions regarding the efficacy of probiotics in cancer prevention.

Thus, there is sufficient proof of a correlation between probiotics and specific anti-cancer effects, and urged that extensive studies are required. Such studies must utilise internationally recognised markers for cancer, or risk of cancer, and evaluate such markers and presence of carcinogenic lesions or tumours over a suitably long period of time for prevention of primary cancer, and reduction of the incidence of recurrences.

Constipation

The ability of probiotic therapy to alleviate constipation (difficulty in passing stool, excessive hardness of stool, slow transit through the bowel) is debatable, but may be a feature of selected strains. Randomised placebo controlled efficacy studies aimed at exploring these effects are strongly recommended.

Immunologic enhancement/immunity stimulation

Probiotics have biological effect in Immunological functionality. The immunological benefits of probiotics can be due to activation of local macrophages and modulation of IgA production locally and systemically, to changes in pro/anti-inflammatory cytokine profiles, or to the modulation of response towards food antigens. The intrinsic properties of *lactobacilli* to modulate the immune system make them appealing for wellbeing applications. The proposed systems engaged with reinforcing of nonspecific and antigen-specific defense against infection and tumours, adjuvant impact in antigen-particular immune responses, Regulating/affecting Th1/Th2 cells, production of anti-inflammatory cytokines, improving phagocytic action of granulocytes, cytokine discharge in lymphocytes, and increases immunoglobulin-emitting cells in blood in order to scale up antibody production. This is ordinary reactions of probiotics, which are all demonstrative of changes in the immune system. An inflammatory immune response delivered cytokine-actuated monocytes and macrophages, causing the arrival of cytotoxic particles fit for lysing tumour cells and pathogens in the body.

Mucosal immunity: The innate and adaptive immune systems are the two compartments traditionally described as important for the immune response. Macrophages, neutrophils, natural killer (NK) cells and serum complement represent the main components of the innate system, in charge of the first line of defence against many micro-organisms. However, there are many agents that this system is unable to recognise. The adaptive system (B and T cells) provide additional means of defence, while cells of the innate system modulate the beginning and subsequent direction of adaptive immune responses. Natural killer cells, including gamma/delta T cells, regulate the development of allergic airway disease, suggesting that the interleukins play an important role. Intravenous, intraperitoneal and intrapleural injection of *L. casei* Shirota into mice significantly increased NK activity of mesenteric node cells but not of Peyer's patch cells or of spleen cells, supporting the concept that some probiotic strains can enhance the innate immune response.

A number of studies have been performed *in vitro* and in animals which clearly show that probiotic strains can modify immune parameters. Correlating these findings with events taking place in the human body is still somewhat unclear, but evidence is mounting that such effects occur. In a series of randomised, double blind, placebo controlled clinical trials, it was demonstrated that dietary consumption of *B. lactis* HN019 and *L. rhamnosus* HN001 resulted in measurable enhancement of immune parameters in the elderly. Probiotic modulation of host immunity is a very promising area for research. Supportive data is emerging, such as those carried out in humans showing that probiotic micro-organisms can enhance NK cell activity in the elderly and nonspecific host defenses can be modulated.

There is a need to specify whether the activities being advocated are designed to operate in otherwise healthy people or subjects with known diseases. Some of the critical factors involved in the host's defenses have been identified and include the induction of mucus production or macrophage activation by *lactobacilli* signalling, stimulation of sIgA and neutrophils at the site of probiotic action (for example the gut), and lack of release of inflammatory cytokines or stimulation of elevated peripheral immunoglobulins. It is also recognised that in some situations, stimulation of factors such as inflammatory cytokines may confer health benefits on the host.

Future studies should focus on the effect in humans, and elucidate the mechanisms of action within systems which simulate the *in vivo* situation, and link this to bacterial and human genomics.

Probiotics and allergy

Allergies are misguided reactions of the immune system in response to (what should be harmless) particles. Probiotics treat allergies by healing your damaged digestive system, which decreases inflammation, stabilises your immune system, and strengthens your gut lining. An allergy is a hypersensitivity reaction initiated by immunological mechanisms. Probiotics modify the structure of antigens, reduce their immunogenicity, intestinal permeability and the generation of pro-inflammatory cytokines that are eminent in patients with a diversity of allergic disorders. *Lactobacillus GG* and *L. rhamnosus GG* is alleviating the symptoms of food allergies at the same time have significant role in reduction of risk for developing allergic disease. Already known strategies to solve allergic disorder by prevention of antigen translocation into blood stream, improve mucosal barrier function and prevent excessive immunologic responses to increased amount of antigen stimulation of the gut.

In a double-blind, randomised, placebo-controlled trial, *L. rhamnosus* GG was given to pregnant women for four weeks prior to delivery, then to newborns at high risk of allergy for six months with the result that there was a significant reduction in early atopic disease. This study illustrates the potential for probiotic micro-organisms to modulate the immune response and prevent onset of allergic diseases. In other clinical studies with infants allergic to cow's milk, atopic dermatitis was alleviated by ingestion of probiotic strains *L. rhamnosus* GG and *B. lactis* BB-12. The precise mechanisms have not been elucidated, but the premise is based upon the ability of *lactobacilli* to reverse increased intestinal permeability, enhance gut-specific IgA responses, promote gut barrier function through restoration of normal microbes, and enhance transforming growth factor beta and interleukin 10 production as well as cytokines that promote production of IgE antibodies. Whether T-helper-1 (TH1) is enhanced and/or T-helper-2 (TH2) dominance is reduced remains to be determined, as do the time-points of these types of events. Certain micro-organisms can contribute to the generation of counter-regulatory T-helper cell immune responses, indicating that use of specific probiotic micro-organisms could redirect the polarised immunological memory to a healthy one.

Cardiovascular disease

There is preliminary evidence that use of probiotic *lactobacilli* and metabolic by-products potentially confer benefits to the heart, including prevention and therapy of various ischemic heart syndromes and lowering serum cholesterol. Thus, these findings are important, more research and particularly human studies are required before it can be ascertained that probiotics confer health benefits to the cardiovascular system.

Urogenital tract disorders

Excluding sexually transmitted diseases, almost all infections of the vagina and bladder are caused by micro-organisms that originate in the bowel. There is a strong correlation between presence of commensals, particularly *lactobacilli* in the vagina with health, and an absence of these micro-organisms in patients with urogenital infections. Disruption of the normal vaginal flora is caused by broad-spectrum antibiotics, spermicides, hormones, dietary substances and factors not, as yet, fully understood. There is some evidence that probiotic micro-organisms delivered as foods and topical preparations have a role in preventing urogenital tract disorders. The criteria for selection of effective probiotic strains have been proposed and should include verification of safety, colonisation ability in the vagina and ability to reduce the pathogen count through competitive exclusion of adherence and inhibition of pathogen growth.

Probiotics and urogenital infections (Bacterial vaginitis)

Several hundred million women are affected by urinary tract infection (UTI) annually. Uropathogenic *Escherichia coli* originating in the bowel is the responsible agent in up to 85% of cases. Asymptomatic bacteruria is also a common finding in women, and sometimes it is followed by symptomatic UTI. There is evidence, including randomised controlled data to suggest that once weekly vaginal capsules of freeze dried *Lactobacillus* strains GR-1 and B-54 prepared with addition of skim milk, and once daily oral capsule use of *Lactobacillus* strains GR-1 and RC-14, can result in the restoration of a *lactobacilli* dominated vaginal flora and lower risk of UTI recurrences. By creating a *lactobacilli* barrier in the vagina, it is believed that fewer pathogens can ascend into the bladder, thereby blocking the infectious process.

Bacterial vaginosis is an abnormal vaginal condition that is characterised by vaginal discharges and results from an overgrowth of atypical bacteria in the vagina. A urinary tract infection is an infection involving the kidneys, ureters, bladder, or urethra. These are the structures that urine passes through before being eliminated from the body. Urogenital infection occurs due to change in vaginal environmental in which *lactobacilli* decrease in concentrations or absent. *Lactobacillus spp*, are the prominent microbial factors that governs the presence, growth, colonisation and persistence of nonendogenous micro-organisms in vagina. As the *Lactobacillus spp*. count decreases, the protection provided by them against uropathogens also decreases. It is also proposed that *lactobacilli* produce biofilms, which cover the urogenital cells. *Lactobacilli* use in bacterial vaginosis is supported by positive results obtained in clinical trials. Probiotic capsules for example *Lactobacillus rhamnosus*, *Lactobacillus crispatus*, *Lactobacillus gasseri*, *Lactobacillus vaginalis*, *Lactobacillus acidophilus*, *Lactobacillus reuteri* and *Streptococcus thermophilus* are effectiveness for recurrent bacterial vaginosis prevention. The principal mechanisms by which *lactobacilli* exert their protective functions in urogenital health care are:

1. Stimulation of the immune system.
2. Competition with other micro-organisms for nutrients and for adherence to the vaginal epithelium, urinary and vaginal tract cells.
3. Reduction of the vaginal pH by the production of organic acids, especially lactic acid.
4. Production of antimicrobial substances and competitive exclusion Inhibitor production, such as bacteriocins, and hydrogen peroxide.

Bacterial vaginosis

Bacterial vaginosis (BV) is a disease of unknown etiology resulting from the overgrowth of various anaerobic bacterial species and associated with the disappearance of *lactobacilli*, which dominate the normal vagina. Many women with BV are asymptomatic yet are at risk of more serious complications such as endometriosis, pelvic inflammatory disease and complications of pregnancy including pre-term labour. There is some clinical evidence to suggest that oral and vaginal administration of *lactobacilli* can eradicate asymptomatic and symptomatic BV. Oral administration of *Lactobacillus acidophilus* and yogurt has been used in the prevention and therapy of candidal vaginitis, although no efficacy data have yet been generated. The necessity for the *lactobacilli* to produce hydrogen peroxide has been proposed, but given that these micro-organisms are more prone to being killed by spermicides, the combination of two or more strains, one of which produces hydrogen peroxide and others which resists spermicidal killing, may prove to be more therapeutic.

Yeast vaginitis

Yeast vaginitis is a very common ailment, often precipitated by antibiotic use, exposure to spermicides or hormonal changes as yet not fully understood. Unlike BV and urinary tract infection, yeast vaginitis is not necessarily due to loss of *lactobacilli*. Few *Lactobacillus* strains are able to inhibit the growth and adhesion of *Candida albicans* or other *Candida* species, and there is no solid evidence to indicate that intravaginal administration of *lactobacilli* can eradicate yeast infection. However, there is some evidence to suggest that *lactobacilli* ingestion and vaginal use can reduce the risk of recurrences and further studies are warranted since this disease is widespread and debilitating.

Probiotics and blood pressure

It has also been demonstrated that probiotics and their products can improve Blood pressure through mechanisms including improving total cholesterol and low-density lipoprotein cholesterol levels. Reducing blood glucose level and insulin resistance, regulating the renin–angiotensin system and significant reduction takes place in blood or serum cholesterol when cholesterol is elevated. Interestingly, probiotic supplementation might positively help in reducing Blood pressure in the hypertensive conditions. *Lactobacillus helveticus, Saccharomyces cerevisiae, Lactobacillus rhamnosus GG, Lactobacillus casei, Lactobacillus acidophilus, Lactobacillus rhamnosus, Lactobacillus bulgaricus, Bifidobacterium breve, Bifidobacterium longum, Streptococcus thermophiles, Lactobacillus delbrueckii* ssp. *Bulgaricus, Lactobacillus kefiri* are the common one used for anti-hypertension.

Probiotics and liver diseases

Micro flora resident in intestinal lumen plays a significant role in hepatocytes function. Alterations to the type and amount of micro-organisms that live in the intestinal tract can result in serious and harmful liver dysfunctions such as cirrhosis, nonalcoholic fatty liver disease, alcoholic liver disease, and hepatic encephalopathy. Probiotic is used as a novel treatment strategy against liver disease in a mechanism of regulation, restoration and alteration of gut micro flora and immune function. Probiotics are useful in the treatment of chronic liver diseases as they block entry of micro-organisms to blood flow and ultimately to liver by increasing the strength of intestinal barrier.

Probiotics and cholesterol assimilation

Probiotic strains, particularly lactic acid microscopic organisms (bacteria) have a noteworthy part to play in the cholesterol by bringing down the mechanism. The cholesterol levels can be cut down direct or indirect by using probiotics. Direct mechanism involves inhibition of *de novo* synthesis or decrease in the intestinal absorption of dietary cholesterol. The decrease in dietary cholesterol retention can be diminished by three ways -assimilation, binding or by degradation. Probiotic strains absorb the cholesterol for their own particular digestion. Probiotic strains can attach to the cholesterol particle, and they are capable for debasing cholesterol to its catabolic products.

The cholesterol level can be decreased in an indirect way by deconjugating the cholesterol to bile acids, in this way lessening the aggregate body pool. Reduction of total cholesterol to be done in *B. animalis subsp. lactis MB 202/DSMZ 23733, B. bifidum, B. breve*. Hypercholesterolemia (elevated blood cholesterol level) is considered a major risk factor for the development of coronary heart disease. Therefore, lowering the serum cholesterol level is important to prevent the disease. The cholesterol removing ability of LAB isolates was assessed *in vitro* and *in vivo* mechanisms. *Lactobacillus pentosus LP05, L. brevis LB32, L. reuteri* and *L. plantarum* are powerful.

Probiotics and dental caries

Dental caries is a multifactorial disease of bacterial origin that is described by corrosive demineralisation of the tooth enamel. It seems following changes in the homeostasis of the oral environment prompting multiplication of the bacterial biofilm, composed notably of *streptococci* from the mutans group. To have a helpful impact in restricting or averting dental caries, a probiotic must have the ability to stick to dental surfaces and coordinate into the bacterial groups making up the dental biofilm. It must also compete with and antagonise the cariogenic bacteria and thus prevent their proliferation. Finally, metabolism of food-grade sugars by the probiotic should result in low acid production. The advantage of incorporating probiotics into dairy products lies in their capacity to neutralise acidic conditions. For instance, it has just been accounted for that cheese prevents demineralisation of the enamel and advances its remineralisation.

Probiotics and orthodontic treatment

White spot lesions are caused by *streptococcus* mutans and they are the basic scars found amid and after orthodontic treatment. The wellbeing advancing microbes can address the lopsidedness in the oral biofilm by intensely hindering the pathogens and moving the oral mileau to a higher pH thereby, turning around the demineralisation. Fixed orthodontic appliances are considered to endanger dental wellbeing because of gathering of micro-organisms that may cause enamel demineralisation, clinically visible as white spot lesions. Besides, the intricate plan of orthodontic bands and brackets may make a biological environment that encourages the foundation and development of cariogenic mutans *streptococci* strains. White spot lesion formation can be viewed as imbalance between mineral loss and mineral gain and the latest orderly audits have examined methods to prevent this side effect of orthodontic treatment. Studies are required to clear up if utilisation of probiotics can be powerful as an alternative method for the prevention of demineralisation and white spots. *Lactobacilli brevis, Bifidobacterium animalis subsp. Lactis BB-12 and Bifidobacterium lactis* derived probiotic through a lozenge tablet could reduce the levels of *S. mutans* in plaque around orthodontic brackets.

Probiotics and oral health

A standout amongst the most imperative advantages of probiotics in the oral cavity is lessening of inflammation. Probiotics can help to destroy the harmful microbes in the oral cavity by fighting against them and helps in maintaining healthy gums and teeth. Since probiotics is an all-natural treatment it should not have any side effects. Both *lactobacillus acidophilus* and *bifidobacterium lactis* have well known antifungal property.

Probiotics and voice prosthesis

Probiotics emphatically diminish the occurrence of pathogenic bacteria in voice prosthetic biofilms. Effectively disposes of biofilm development on indwelling voice prostheses, possibly related to the presence of *Streptococcus thermophiles* and *Lactobacillus bulgaricus*.

Probiotics and periodontal diseases

Studies have demonstrated that the pervasiveness of *lactobacilli*, especially *Lactobacillus gasseri* and *L. fermentum*, in the oral cavity was more prominent among healthy participants than among patients with chronic periodontitis. Different studies have detailed the limit of *lactobacilli* to repress the development of periodonto pathogens, including *P. gingivalis, Prevotella intermedia* and *A. actinomycetemcomitans*.

Together, these perceptions recommend that *lactobacilli* living in the oral cavity could play a role in the oral ecological balance. *L. brevis*, *L. casei*, *L. salivarius*, reuteri strains, *Bacillus subtilis*, *L. reuteri* and *L. brevis* the involvement cared out in anti-inflammatory activity decreasing the number of pathogens in periodontal tissues.

Probiotics and HIV

Probiotics appear to support maintenance of a strong gut epithelia layer, improve gut barrier function and stimulation of innate immunity which act as the first layer of defense against translocation of viral particles and bacterial pathogens. When immune system is well developed, able to prevent HIV replication and slow down the progression of AIDS in host. Daily consumption of probiotics over a prolonged period of time can improve CD4 count in people living with HIV. A screening of saliva taken from several volunteers demonstrated that some Lactobacillus strains created proteins that are fit for binding a specific type of sugar, called mannose, found on HIV envelope. The binding of the sugar empowers the microscopic organisms (bacteria) to adhere to the mucosal coating of the mouth and gastric tract and colonise them. One of the strain indicated copious mannose-binding protein particles into its surroundings which binded to the sugar coating henceforth neutralising HIV. It is also observed that the trapped immune cell by *lactobacilli* leads to formation of clumps leading immobilisation of any immune cells harbouring HIV and preventing them to infect other cells.

Probiotics and Halitosis

Halitosis or bad breath is the condition when the breath has unpleasant odour. It has many causes, for example, utilisation of specific foods, metabolic disorders, respiratory tract infections and related with an irregularity of the commensal microflora of the oral cavity. Essentially, it is started from the activity of anaerobic bacteria that corrupt salivary and food proteins to create amino acids, which are thusly changed into volatile sulphur compounds, including hydrogen sulphide and methane thiol. *Streptococcus salivarius* act as a commensal probiotic of the oral cavity this strain screened and recognised typically from people groups without halitosis. *S. salivarius* is known to create bacteriocins, which could add to lessening the quantity of microscopic organisms that produce volatile sulphur compounds. The utilisation of gum or capsules containing *S. salivarius K12* (BLIS Technologies Ltd., Dunedin, New Zealand) diminished levels of volatile sulphur compounds among patients diagnosed to have halitosis. Take a probiotic supplement regularly. There is good evidence it helps to regulate the growth of harmful bacteria. *S. salivariu*, *L. salivarius*, *L. reuteri*, *L.casei* and *W. Cibaria* was supplied for management option.

Use of probiotics in otherwise healthy people

Many probiotic products are used by consumers who regard themselves as being otherwise healthy. They do so on the assumption that probiotics can retain their health and well being, and potentially reduce their long-term risk of diseases of the bowel, kidney, respiratory tract and heart. Several points need to be made on this assumption and its implications. Thus, it was recognised that the use of probiotics should not replace a healthy lifestyle and balanced diet in otherwise healthy people. Firstly, there is no precise measure of 'health' and subjects may actually have underlying and undetectable diseases at any given time. Secondly, no studies have yet been undertaken which analyse whether or not probiotic intake on a regular basis helps retain life-long 'health' over and above dietary, exercise and other lifestyle measures. One study of day care centres in Finland showed that probiotic use reduced the incidence of respiratory infections and days absent due to ill health. Thus, these studies should be done to give credibility to the perception that probiotics should be taken on a regular basis by healthy men, women and children. Such studies should be multi-centred and require randomisation on the basis of age, gender, race, nutritional intake, education, socio-economic status and other parameters.

It is currently unclear as to the impact of regular probiotic intake on the intestinal microflora. For example, does it lead to the depletion or loss of commensal micro-organisms which otherwise have beneficial effects on the host? While there is no indication of such effects, the issue needs to be considered. Furthermore, the concept of restoring a normal balance assumes that we know what the normal situation in any given intestinal tract comprises. Thus, it was deemed important to further study the various contributions of gut micro-organisms on health and disease. Another point worthy of note is that, to date, the ingestion of probiotic strains has not led to measurable long-term colonisation and survival in the host. Invariably, the micro-organisms are retained for days or weeks, but no longer. Thus, use of probiotics likely confers more transient than long-term effects, and so continued intake appears to be required.

In newborn children, where a commensal flora has not yet been established, it is feasible that probiotic micro-organisms could become primary colonisers that remain long-term, perhaps even for life. While such probiotic usage can prevent death and serious morbidity in premature, low birth weight infants, the alteration of flora in healthy babies is a more complex situation. Just so, an implication of the Human Genome Project is that selected probiotics may be used at birth to create a flora that improves life-long health. These issues are very important for the future, and will require full discussion including human ethical considerations.

TESTING METHODS FOR ESTABLISHING HEALTH BENEFITS CONFERRED BY PROBIOTIC MICRO-ORGANISMS

Proper *in vitro* studies should establish the potential health benefits of probiotics prior to undertaking *in vivo* trials. Tests such as acid and bile tolerance, antimicrobial production and adherence ability to human intestinal cells should be performed depending on the proposed health benefit.

In order to ascertain that a given probiotic can prevent or treat a specific pathogen infection, a clinical study must be designed to verify exposure to the said pathogen (preventive study), or that the infecting micro-organism is that specific pathogen (treatment study). If the goal is to apply probiotics in general to prevent or treat a number of infectious gastroenteritis or urogenital conditions, the study design must define the clinical presentation, symptoms and signs of infection, and include appropriate controls.

For *in vivo* testing, randomised double blind, placebo controlled human trials should be undertaken to establish the efficacy of the probiotic product. Thus, it was recognised that there is a need for human studies in which adequate numbers of subjects are enrolled to achieve statistical significance. It would be preferable to have such findings corroborated by more than one independent center. For some foods, it may be difficult to separate a probiotic effect from an effect related to the general product characteristics of the food. Therefore, it is essential that proper controls be included in these human trials. Furthermore, data obtained with one specific probiotic food cannot be extrapolated to other foods containing that particular probiotic strain or to other probiotic micro-organisms.

With respect to measuring the health benefits in human studies, consideration should be given to clinically relevant outcomes in the population being studied. For diarrheal studies, this might be preventing death in some countries, while in others it might be prevention of a defined and statistically significant weight loss, decreased duration of watery/liquid stools, and faster recovery to normal health, as measured by restoration of normal bowel function and stool consistency. Although it is known that certain probiotics can elicit beneficial effects, little is known about the molecular mechanisms of the benefits reported. The mechanisms may vary from one probiotic to another (for the same benefit via different means) and the mechanism may be a combination of events, thus making this a very difficult and complex area. It could involve the production of a specific enzymes or metabolites that act directly on the micro-organisms, or the probiotic could also cause the body to produce the beneficial action.

Examples of possible probiotic mechanisms of action, in the control of intestinal pathogens include:

- Antimicrobial substance production.
- Competitive exclusion of pathogen binding.
- Competition for nutrients.
- Modulation of the immune system.

Thus, it is suggested that clear experiments (*in vitro* and/or *in vivo*) should be designed at the molecular level to elucidate the mechanisms of probiotic beneficial effects. Appropriate experiments including genetic analysis to elucidate the mechanism of actions should be performed. Probiotic bacteria containing β-galactosidase can be added to food to improve lactose maldigestion. However, a similar health effect is also observed for lactose fermenting starter bacteria such as *L. delbrueckii.* ssp. *bulgaricus* and *S. thermophilus* in fermented milk products like yogurt. These traditional starters are not considered probiotics since they lack the ability to proliferate in the intestine.

Safety Considerations: Requirements for Proof that a Probiotic Strain is Safe and without Contamination in its Delivery Form

Historically, *lactobacilli* and *bifidobacteria* associated with food have been considered to be safe. Their occurrence as normal commensals of the mammalian flora and their established safe use in a diversity of foods and supplement products worldwide supports this conclusion. However, probiotics may theoretically be responsible for four types of side-effects:

1. Systemic infections.
2. Deleterious metabolic activities.
3. Excessive immune stimulation in susceptible individuals.
4. Gene transfer.

Documented correlations between systemic infections and probiotic consumption are few and all occurred in patients with underlying medical conditions. The following is a list (including some microbes used in non-food applications) of infections reported to be associated (although not necessarily proven) with the consumption of commercial products:

- Two cases of *L. rhamnosus* traced to possible probiotic consumption.
- Thirteen cases of *Saccharomyces* fungemia due to vascular catheter contamination.
- *Bacillus* infections linked to probiotic consumption include three reports detailing seven cases of *B. subtilis* bacteremia, septicemia and cholangitis, all in patients with underlying disease.

No cases of infections from *Bifidobacterium* have been reported. *Enterococcus* is emerging as an important cause of nosocomial infections and isolates are increasingly vancomycin resistant. Smith and others recognised that some strains of *Enterococcus* display probiotic properties, and may not at the point of inclusion in a product display vancomycin resistance. However, the onus is on the producer to prove that any given probiotic strain is not a significant risk with regard to transferable antibiotic resistance or other opportunistic virulence properties.

In recognition of the importance of assuring safety, even among a group of bacteria that is Generally Recognised as Safe (GRAS), Smith and others recommends that probiotic strains be characterised at a minimum with the following tests:

1. Determination of antibiotic resistance patterns.
2. Assessment of certain metabolic activities (e.g. D-lactate production, bile salt deconjugation).
3. Assessment of side-effects during human studies.
4. Epidemiological surveillance of adverse incidents in consumers (post-market).
5. If the strain under evaluation belongs to a species that is a known mammalian toxin producer, it must be tested for toxin production. One possible scheme for testing toxin production has been recommended by the EU Scientific Committee on Animal Nutrition.
6. If the strain under evaluation belongs to a species with known hemolytic potential, determination of hemolytic activity is required.

Assessment of lack of infectivity by a probiotic strain in immunocompromised animals would add a measure of confidence in the safety of the probiotic.

In vivo Studies using Animals and Humans

In some cases, animal models exist to provide substantiation of *in vitro* effects and determination of probiotic mechanism. Where appropriate, the Working Group encourages use of these prior to human trials. The principal outcome of efficacy studies on probiotics should be proven benefits in human trials, such as statistically and biologically significant improvement in condition, symptoms, signs, well-being or quality of life, reduced risk of disease or longer time to next occurrence, or faster recovery from illness. Each should have a proven correlation with the probiotic tested.

Probiotics have been tested for an impact on a variety of clinical conditions. Standard methods for clinical evaluations are comprised of Phase 1 (safety), Phase 2 (efficacy), Phase 3 (effectiveness) and Phase 4 (surveillance). Phase 1 studies focused on safety are discussed in Section 3.3 above. Phase 2 studies, generally in the form of randomised, double blind, placebo-controlled (DBPC) design, measure efficacy compared with placebo. In addition, phase 2 studies measure adverse effects. A general recommendation for the testing of probiotic foods is that the placebo would be comprised of the food carrier devoid of the test probiotic. Sample size needs to be calculated for specific endpoints. Statistically significant differences must apply to biologically relevant outcomes.

Health Claims and Labelling

Currently in most countries, only general health claims are allowed on foods containing probiotics. Smith and others recommends that specific health claims on foods be allowed relating to the use of probiotics, where sufficient scientific evidence is available, as per the guidelines set forth in this section. Such specific health claims should be permitted on the label and promotional material. For example, a specific claim that states that a probiotic 'reduces the incidence and severity of rotavirus diarrhea in infants' would be more informative to the consumer than a general claim that states 'improves gut health'. This would better comply with Codex General Guidelines on Claims to avoid misleading information.

It is recommended that it be the responsibility of the product manufacturer that an independent third party review by scientific experts in the field be conducted to establish that health claims are truthful and not misleading.

Smith and others recommended that the following information be described on the label:

- Genus, species and strain designation. Strain designation should not mislead consumers about the functionality of the strain.
- Minimum viable numbers of each probiotic strain at the end of the shelf-life.
- The suggested serving size must deliver the effective dose of probiotics related to the health claim.
- Health claims.
- Proper storage conditions.
- Corporate contact details for consumer information.

RECOMMENDATIONS

1. Adoption of the definition of probiotics as 'Live micro-organisms which when administered in adequate amounts confer a health benefit on the host'.
2. Use and adoption of the guidelines in this section should be a prerequisite for calling a bacterial strain 'probiotic'.
3. Regulatory framework to allow specific health claims on probiotic food labels, in cases where scientific evidence exists, as per the guidelines set forth in this section.
4. Promotion of these guidelines at an international level.
5. Good manufacturing practices (GMP) must be applied in the manufacture of probiotic foods with quality assurance, and shelf-life conditions established.
6. Further development of methods (*in vitro* and *in vivo*) to evaluate the functionality and safety of probiotics.

Chapter 9

Functional Foods

INTRODUCTION

Functional foods are foods or dietary components that claim to provide health benefits aside from basic nutrition. These foods contain biologically active substances such as antioxidants that may lower the risks from certain diseases associated with ageing. Examples of functional foods include fruits and vegetables, whole grains, soya, milk, enhanced foods and beverages and some dietary supplements.

Diet and health are closely related. Thus, crops are now being enhanced through biotechnology to increase levels of important biologically active substances for improved nutrition, to increase body's resistance to illnesses, and to remove undesirable food components. Let your food be your medicine. Today, food has indeed become central for the cultural pleasure of feeding a family and greeting friends in a social setting.

Are foods also medicine? Products intended to cure diseases are medicinal products and not foods. But on the other hand, a healthy diet consisting of foods with functional properties can help promote well-being and even reduce the risk of developing certain diseases. WHO stresses the importance of a healthy diet in preventing non-communicable diseases. However, a diet can only be healthy if the combination of individual foods is good. Moreover, a healthy diet is not just about limiting certain components of concern such as saturated or trans fatty acids or simply delivering nutrient intake, it is also about including those elements that may provide an extra benefit.

Above all, eating should be a pleasurable experience and is often celebrated as a social event, helping communication and maintaining relations with family members and friends. Why not combine pleasure and functionality then? To keep alert when you have to work late at night, why not enjoy some food or drink that at the same time helps keeping you alert? To promote a healthy bowel function, why not integrate the goodness of dietary fibre into your daily food? To help prevent cardiovascular disease, why not naturally incorporate the preventive elements into your daily diet? We all eat many different foods every day, lets make sure we eat what suits us best for any particular need we have at that time, or what will promote our overall health in the long run.

It is easy to say: 'eat healthy', but do we really know what is the best food for a particular need? What type of evidence do we need to have? It is certainly not enough to study certain parameters simply because we are able to measure them. Rather we need to ensure that the parameters we use to measure health are really valid and relevant biomarkers for the target functions we are interested in. This is where functional food science can help by providing the tools to answer some of these questions.

LIFESTYLE, DIET AND PHYSICAL ACTIVITY

Human societies have become less rural and more urban and economic activity has become proportionally less agricultural and more industrially based. These changes have brought benefits in terms of improvements in many aspects of quality of life and health and they have also brought changes to the way individuals interact, the amount of leisure time they have, their access to food and their levels of physical activity. In developed countries improvements in the quality of life, not the least of which is a safer, more varied diet, are associated with increased and increasing longevity. In general, individuals have more leisure time, they have greater access to a wider variety of foods and their daily routine requires less physical activity. These changes, however, have made it increasingly difficult to balance energy intake and expenditure, resulting in increased frequency of overweight and obesity worldwide.

As these changes have taken place, there has been an increase in the prevalence of chronic, non-communicable diseases such as increased blood pressure, cardiovascular disease and type 2 diabetes. It is clear that aspects of lifestyle, diet and physical inactivity all play a role in the increasing incidence of these diseases and the success of measures to reduce them will depend on striking the right balance between these three factors. This has led to a realisation that diet can contribute to long-term health in ways that have not been recognised previously and this is leading to an appreciation of ways in which foods can bring positive influences to health and well-being beyond simply providing basic nutrition.

History and the Concept of a Balanced Diet

Through much of its history, nutrition has concerned itself with the observation that deficiencies in the diet lead to disease states and that deficiency diseases can be avoided by ensuring an adequate intake of the relevant dietary components. For the first half of the twentieth century the focus of nutrition science was on establishing the minimum requirements for essential nutrients that ensure the avoidance of deficiency diseases. This led to the establishment of reference values, such as dietary reference intakes (DRIs), population reference intakes (PRIs), and dietary reference values (DRVs), for vitamins, minerals, proteins, fats, carbohydrates and energy. These guide strategies for nutrition policy and practice directed to ensuring that intakes are adequate to meet the needs of the average, healthy consumer for normal growth and development, body maintenance and physical activity. The concept of a balanced diet evolved, in which the ideal diet consisted of a sufficient variety of food groups (fruit, vegetables, cereals, meat, fish, dairy products, etc.) to meet these target intake values for the essential nutrients.

The concept of a balanced diet has further evolved to embrace the need for dietary intakes to accommodate changes in lifestyles with correspondingly reduced energy expenditure and caloric needs, but with a continuing unaltered level of need for other key nutrients. There is a need to strike a balance in selecting a sufficiently varied choice of foods to ensure an adequate intake of essential nutrients, while avoiding excessive intakes of energy and the associated diseases. Increasing longevity increases the risk of developing non-communicable diseases if the balance of lifestyle, diet and physical activity is disturbed. The selection of a balanced diet, which provides adequate intakes of nutrients while keeping energy intake in balance with basic metabolism and physical activity, remains the cornerstone of sound nutrition. Thus, there is a need to ensure the availability of diets and foods with appropriately increased amounts of nutrients relative to energy (that is, foods with an appropriate nutrient density). This need can be met by selection of natural products, but the lifestyle and the expectations of modern living do not make this easy and, as a result, opportunities have arisen for manufactured foods that meet these requirements. Concomitantly, an improved appreciation of the potential beneficial or adverse effects of nutrients and other components in the diet has led to the realisation that it is possible to create food items with specific characteristics that are capable of influencing body function over and above meeting the basic nutrition needs. These foods have come to be known as 'functional'. Although of course at some level all foods are functional, these foods would additionally have the potential to promote long-term health, improving both physical and mental health and well-being.

From a practical point of view, a functional food can be:

• A natural, unmodified food.
• A food in which one of the components has been enhanced through special growing conditions, breeding or biotechnological means.
• A food to which a component has been added to provide benefits.
• A food from which a component has been removed by technological or biotechnological means so that the food provides benefits not otherwise available.
• A food in which a component has been replaced by an alternative component with favourable properties.
• A food in which a component has been modified by enzymatic, chemical or technological means to provide a benefit.
• A food in which the bioavailability of a component has been modified.
• A combination of any of the above.

However the functional food may be constituted (whether modified or not), it has to comply with the general requirement that it must be safe. In any discussion of food functionality, in either a regulatory or a scientific context, there is no consideration of a trade off between health benefit and health risk. Whether a food is considered to be functional or not, it must always be safe for its intended use

ASSESSMENT OF FOOD FUNCTIONALITY

A direct measurement of the effect a food has on health and well-being and/or reduction of disease risk is often not possible. This may be because the endpoint (the state of health and well-being) does not always lend itself to quantifiable measurement. Also, in the case of a disease like cancer, the time frame for development of the disease is very long or it would be unethical to monitor its development under the conditions of a controlled study.

Instead, functional food science works from knowledge of the key processes in the attainment of optimal health or in the development of a disease to identify markers that can be used to monitor how those key processes are influenced by foods or food components. Provided that the role of those key processes in the attainment of optimal health or disease development is well established and the markers are chosen accurately to reflect the process, it is possible to study the effect of consuming the food on the final endpoint (the improved state of health or reduction of disease risk) by measurement of the markers. The markers could be chosen to reflect either some key biological function (markers of a target function) or a key stage in development that is unequivocally linked to the endpoint under study, in which case they serve as markers of an intermediate endpoint. Measurements made in the short term on carefully chosen markers of intermediate endpoints can be used to make inferences about effects on final endpoints that would only otherwise be accessible through long-term study. Where the underlying target functions or intermediate endpoints are unequivocally linked to the risk of disease, the markers are also referred to as risk factors for the disease.

ROLE OF FUNCTIONAL FOODS

Some of the functional foods are discussed below:

- Early development and growth.
- Regulation of energy balance and body weight.
- Cardiovascular function.
- Defence against oxidative stress.
- Intestinal function – the gut microflora.
- Mental state and performance.
- Physical performance and fitness.

A short illustrative explanation of each topic is given, followed by a summary of some potential functional food components that have been, or might be, developed to improve health in each area. This list is not exhaustive and other areas of physiology also have potential for the development of functional foods. Also, examples of food components are given to illustrate the general nature of possible candidates. In many cases, further research is needed to confirm unequivocally whether or not they are able to fulfil their potential.

Early development and Growth

The feeding of mothers during pregnancy and lactation and of their infants and young children is of great biological importance. Nutritional factors during early development may have not only short-term effects on growth, body composition and body functions but may also exert longer-term effects. The development of high blood pressure and heart disease, for example, can be affected by early nutrition – a phenomenon called 'metabolic programming'. The interaction of nutrients and gene expression may form the basis for many of these programming effects and offers exciting potential for functional food development. The course of pregnancy and childbirth, as well as the composition of breast milk and the short- and long-term development of the child, are influenced by the intake of nutrients, particularly polyunsaturated fatty acids (PUFA), folic acid, iron, zinc and iodine, as well as by total energy intake.

The evaluation of dietary effects on child growth may require epidemiological studies, as well as the evaluation of specific cell and tissue growth. Growth factors and conditionally essential nutrients (e.g. amino acids and polyunsaturated fatty acids – PUFAs) may be useful as ingredients in functional foods. Intestinal growth, maturation and adaptation, as well as longer term function, may be influenced by food ingredients such as probiotics and oligosaccharides that function as prebiotics. Pregnancy and the first postnatal months are critical periods for the growth and development of the human nervous system, processes for which adequate nutritional supplies are essential. Early diet may have long-term effects on some sensory and cognitive abilities, as well as behaviour. Possible long-term effects of early exposure to tastes and flavours on later food choice preferences could have an impact on public health.

Exciting possibilities have been suggested by the beneficial effects of some functional foods on the developing immune response, for example the effects of antioxidant vitamins, trace elements, fatty acids, L-arginine, nucleotides, pro-, pre- and synbiotics, and altered allergenic components of infant foods.

Peak bone mass at the end of adolescence can be increased by dietary means. This is expected to be of importance for reducing the risk of osteoporosis in later life. The combined effects of calcium and prebiotic fructans, together with other constituents of growing bone, such as proteins, phosphorus, magnesium and zinc, as well as vitamins D and K and fluoride, offer many possibilities for the development of functional foods, although many still need to be confirmed by research. Some examples of opportunities for modulation of target functions related to growth, development and differentiation, and the food components that might be used to achieve them.

Regulation of Energy Balance and Body Weight

Dietary intakes and balance influence all metabolic and physiological processes. An optimally balanced diet is usually expressed in terms of its energy and content of macronutrients (carbohydrates, fats and proteins). Within these broad classes of macronutrients, there are sub-classes, which have differing nutritional impacts. Amongst the carbohydrates, a most important functional and metabolic distinction is between those that are digested and absorbed in the small intestine, for example, glucose, sucrose and available starch, and those that are not, for example, dietary fibres, resistant starch, sugar alcohols (polyols) and certain oligosaccharides. Amongst fatty acids, such distinctions depend on the length and saturation (that is the number of carbon atoms and whether adjacent carbon atoms are linked by double or single bonds) of their carbon chain. Thus, major functional components of lipids are saturated fatty acids (SFA) on the one hand, and monounsaturated fatty acids (MUFA) and polyunsaturated fatty acids (PUFA) on the other.

Energy balance and obesity

Obesity is defined as an excessive accumulation of body fat. Its prevalence varies between 5% and 50% in different populations, depending on the definition applied. The epidemic of obesity with its accompanying health risks is now recognised to be one of the major health challenges in both the developed and developing world. People with central obesity are most at risk. Central obesity appears to be a reflection of increased amounts of internal (as opposed to subcutaneous) fat.

Obesity is associated with an increased risk of heart disease, type 2 diabetes, high blood pressure and some forms of cancer. The interaction of genetic predisposition and environmental factors, such as a sedentary lifestyle and a high-energy intake, is the most commonly accepted model for the cause of human obesity.

Diabetes

Diabetes mellitus is a disease characterised by inappropriately increased plasma glucose concentrations. Insulin is the main hormone that controls blood glucose levels, and diabetes results from impaired insulin secretion or reduced insulin action at its target sites (insulin resistance).

Two main forms of diabetes mellitus are defined by clinical manifestations and causes. Type 1 or insulin-dependent diabetes usually develops in young, lean individuals and is the result of an almost complete destruction of the pancreatic beta cells, usually as a consequence of an autoimmune process. Because it is the beta cells that produce insulin, type 1 diabetes is characterised by plasma insulin levels that are very low. Type 2 or non-insulin-dependent diabetes usually develops in overweight and/or older individuals. It has a slow onset (the subject may be without clinical symptoms for several years) and is characterised by insulin resistance (reduced sensitivity of body tissues to insulin), resulting in chronically elevated plasma insulin and glucose levels. Insulin resistance syndrome.

Apart from being associated with higher than normal levels of insulin and glucose, insulin resistance is also associated with characteristic changes in lipid metabolism. Lipids, which are generally water-insoluble, are transported in the blood in the form of lipoprotein particles composed of specific proteins and lipids (triacylglycerol (TAG), cholesterol and phospholipids). Low-density lipoproteins (LDL) and very low-density lipoproteins (VLDL) contain high concentrations of triacylglycerol (TAG) and cholesterol, and are termed 'low density' on the basis of comparison with the density of water. Elevated levels of LDL and VLDL are recognised risk factors for coronary and other cardiovascular diseases. High-density lipoproteins (HDL) contain lower concentrations of cholesterol and are believed to be beneficial. Insulin resistance syndrome is characterised by increased concentrations of TAG, decreased concentrations of HDL cholesterol, and high blood pressure.

Seen in association with central obesity as characterised, for example, by waist circumference and waist-to-hip ratio, this pattern of risk factors is also referred to as 'metabolic syndrome'.

Functional foods for optimising metabolism

This area offers many opportunities for the development of functional foods. The approach to controlling glucose levels is based on choosing foods that cause a slower absorption of glucose into the blood stream, so that blood glucose fluctuations are less pronounced and, consequently, insulin requirements are lowered. The rate of glucose uptake is influenced by the structural properties of foods, such as the presence of intact cells or starch granules. It is also influenced by the type of carbohydrate constituents, for example by certain types of oligosaccharides and starch, and by the content of soluble, viscous types of dietary fibre (pectin, gums, oat β-glucan and psyllium seed husk). Organic acids and other components are known also to influence the rate of glucose uptake.

The descriptor 'low glycaemic index' is reserved for foods with carbohydrates that are absorbed in the gut but which cause only a slow and small rise in blood glucose levels. Examples of such foods are bread with whole grains and/or sour dough, legumes, whole grain pasta and products enriched in soluble viscous types of dietary fibre. Although the role of 'low glycaemic foods' remains to be fully established, an increasing body of knowledge is becoming available in this area on which development of functional foods with optimised release of carbohydrates can be based.

Examples of opportunities for modulation of target functions related to the regulation of metabolic processes, and the food components which might be used to achieve them.

Cardiovascular Function

Cardiovascular diseases (CVD) are a group of degenerative diseases of the heart and blood circulatory system and include coronary heart disease (CHD), peripheral artery disease and stroke.

CHD is a major health problem in most industrialised countries and a rapidly emerging problem in developing countries and countries in transition. The predominant clinical pictures are angina (chest pain), myocardial infarction (heart attack), heart failure and sudden cardiac death. The arteries supplying blood to the heart are narrowed by atherosclerosis and may become suddenly blocked when an atherosclerotic plaque on the artery wall ruptures and is impacted by a blood clot. This leads to lack of oxygen to the heart muscle, damage to the heart muscle tissue and life-threatening loss of heart function. To fully appreciate the potential role functional foods can play in the prevention of CVD, it is necessary to understand the diversity of risk factors associated with its development. These include high blood pressure, inflammation, inappropriate blood lipoprotein levels, insulin resistance (see the discussion under 'Regulation of energy balance and body weight') and control of blood clot formation. The interdependence of these factors has not been fully characterised. Because only 50% of the incidence of CVD can be explained by these known risk factors, other contributory and interactive factors are certainly involved. For example, genetic predispositions, smoking, and levels of physical activity play a role. Some specific examples of markers known to be risk factors for CVD follow.

High blood pressure

CVD is directly related to high blood pressure and any measures taken to reduce high blood pressure should lower the risk of coronary disease. High blood pressure increases the risk of arterial injury. Genetic predisposition and obesity are involved in the aetiology of high blood pressure, but diet and lifestyle have a substantial impact, with overweight, physical inactivity, high sodium intake and low potassium intake being among the main contributors.

Integrity of artery lining

Damage to the endothelial cells that line the arteries, as well as more general structural damage at susceptible points in the arteries (such as at 'forks'), increases the risk of CVD. Oxidation is now believed to be a major contributor to atherosclerosis, because it converts LDL into an oxidised form. Oxidised LDL has been found in damaged arterial walls and has been shown to have several actions that could contribute to the initiation and progression of arterial damage. The extent of LDL oxidation is related to the extent of atherosclerosis.

Elevated blood lipids

A raised plasma concentration of LDL is a strong risk factor for CVD. High levels of other lipoproteins, high TAG concentrations and low levels of HDL are also risk factors. Raised levels of lipids, especially TAG, after a meal appear to be a stronger risk factor than fasting levels.

High homocysteine levels

Epidemiological data suggest that high plasma levels of homocysteine, an amino acid, are associated with increased risk of CVD. Several mechanisms for the effects of homocysteine on atherosclerosis and thrombosis have been suggested, but none has been confirmed.

Increased blood clot formation

The control of blood clotting is likely to be an important element in the reduction of the risk of CVD. Risk factors include those that increase the clumping of platelets and those that increase the activity of the clotting factors. These are counterbalanced by factors that promote the breakdown of the clot.

Functional foods to promote optimal heart health

Balance of dietary lipids: The levels of blood lipids can be influenced by dietary fatty acids, an influence usually related to their molecular size and shape and the degree of saturation of their hydrocarbon chains. Fatty acids with a hydrocarbon chain that contains no double bonds are saturated fatty acids (SFAs). SFAs with chain lengths of 12–16 carbon atoms increase plasma LDL cholesterol concentrations more than they increase plasma HDL concentrations. In their favour, they cannot become oxidised.

Unsaturated fatty acids are those in which the hydrocarbon chain contains at least one double bond. Monounsaturated fatty acids (MUFA) contain one double bond, PUFAs contain two or more. Most naturally occurring unsaturated fatty acids are *cis*-fatty acids, in which the two hydrogen atoms around the double bonds (one at each end) are positioned on the same side of the fatty acid chain. This causes a bend in the hydrocarbon chain at that point. In contrast, *trans*-fatty acids have the hydrogen atoms at each of the double bonds on opposite sides of the fatty acid chain and, as a result, are straight and more like SFAs. They are formed during some manufacturing processes and are therefore consumed in products such as hard margarines and baked goods. They are also formed in the rumen of animals such as cows and, consequently, a portion of the *trans*-fatty acids in the diet (it has been estimated at around 20%) comes from the consumption of dairy products and meat. Dietary *trans*-unsaturated fatty acids can increase plasma LDL and reduce HDL cholesterol concentrations. Diets low in SFAs and *trans* fatty acids could therefore reduce the risk of CVD. The *cis*-unsaturated fatty acids with 18 carbon atoms – oleic (mono-unsaturated), linoleic and alpha-linolenic acids (polyunsaturated) – reduce plasma concentrations of LDL cholesterol and some may also raise plasma concentrations of HDL cholesterol. Functional foods enriched in these unsaturated fatty acids could also be used to reduce the risk of CVD. Like alpha-linolenic acid, the long-chain, highly unsaturated PUFAs found in fish oils belong to the *n*-3 family. They can promote improvements in endothelial and arterial integrity as well as counteract blood clotting and reduce blood pressure. They also reduce plasma TAG levels and may have suppressive effects on the cellular immune system. One of the focus areas of functional food development concerns the incorporation of *n*-3 fatty acids into foods.

Other food components with functionality for optimal heart health

Soluble viscous types of dietary fibre can reduce LDL cholesterol concentrations, particularly in people with high lipoprotein levels. Diets rich in antioxidants, including plant flavonoids, can inhibit LDL oxidation and inhibit the formation of cell-to-cell adhesion factors, which are implicated in damage to the arterial endothelium and in the formation of blood clots. However, the importance of this for CVD remains to be established. Evidence suggests the possibility of protecting vascular integrity through beneficial modulation of risk indicators such as high plasma homocysteine concentrations and high blood pressure. Folate has the potential to reduce cardiovascular risk by lowering the plasma level of homocysteine. Evidence supports its ability to reduce homocysteine levels but, to date, its effectiveness in reducing CHD has not been confirmed in clinical trials. An increase in the intake of potassium and calcium and a reduction in sodium can help to reduce blood pressure. Consumption of certain fatty acids and peptides derived from milk proteins has also been reported to be beneficial.

Two significant areas of functional food development are the use of plant sterol and stanol esters and soya protein to reduce levels of LDL cholesterol. Plant sterols are natural constituents of plants, including trees and a number of common crops, such as soya and maize. They play a similar role to that of cholesterol in animals as a metabolic precursor and as a structural molecule. It has been known for 50 years that plant sterols can interact with cholesterol in the intestinal tract to bring about a reduction of cholesterol absorption and a subsequent reduction in blood cholesterol. More recently, a number of studies have confirmed the ability of plant sterols and stanols to reduce LDL cholesterol under a variety of conditions. They are present naturally in the diet at levels below those shown by the studies to be necessary for effect. However, recent technological advances providing for the extraction and esterification of plant sterols, either as the sterols themselves or as stanols, allow them to be solubilised in the matrix of food fat and so to be incorporated easily into food products at effective levels.

Defence Against Oxidative Stress

Oxygen is essential to human life. Without it, we cannot survive. Paradoxically, oxygen is also involved in toxic reactions and, therefore, is a constant threat to the wellbeing of the human body.

Most of the potentially harmful effects of oxygen are believed to be the result of the formation and activity of reactive oxygen species (ROS). These act as oxidants and are believed to be major contributors to ageing and to many of the diseases associated with ageing, including cardiovascular diseases (CVD), cancer, cataracts, agerelated decline in the immune system, and degenerative diseases of the nervous system, such as Parkinson's and Alzheimer's diseases.

The human body has several mechanisms for defence against ROS. The various defences are complementary to one another because they act on different oxidants or in different cellular compartments. One important line of defence is a system of antioxidant enzymes. Nutrition plays a key role in maintaining these enzymatic defences. Several essential minerals and trace elements, including selenium, copper, manganese and zinc, are involved in the structure or catalytic activity of these enzymes. If the supply of these is inadequate, enzymatic defences might be impaired. Another line of defence is the group of small-molecular weight compounds that act as antioxidants, examples are glutathione, and some vitamins (e.g. vitamins C and E), that regenerate the buffer capacity of the body's antioxidant systems. If exposure to external sources of oxidants is high, for example from tobacco smoke or atmospheric pollution, the body's antioxidant defences may be put under pressure to cope. The result is a condition called oxidative stress, an imbalance between pro-oxidants and antioxidants. In the normal situation, pro-oxidant factors are adequately counterbalanced by antioxidant defences. An increase either in the production of oxidants or in a deficiency in the defence system could disturb this balance, causing oxidative stress.

Functional foods to promote optimal defence against oxidative stress

The body's own defences could be supported by a wide variety of small-molecular-weight antioxidants found in the diet, and these give ample scope for functional food components. The best known are vitamin E, vitamin C, carotenoids and polyphenols, including flavonoids. Many of the antioxidant compounds in the diet are of plant origin. Plant leaves are exposed to visible and ultraviolet light and other radiation and are especially susceptible to damage by activated forms of oxygen. Hence, they contain numerous natural antioxidant constituents that can either counter reactive oxygen species (ROS) directly or boost the regeneration system to restore antioxidant capacity. Trials to date with single antioxidants have not provided evidence of effect. So, if an advantage is to be obtained, it might best come from consuming a battery of antioxidants such as occur naturally in foods.

Examples of opportunities for modulation of target functions related to defence against oxidative stress, and the food components that might be used to achieve them.

Intestinal Function – the Gut Microflora

The human large intestine (colon) is a highly metabolically active organ. The colon contains an extremely complex microbial ecosystem. In fact, bacterial cells account for around 90% of the total cells in the body. The majority of these bacteria are anaerobic (they die in the presence of oxygen). The most common species in the adult human colon are those of the bacteroides, bifidobacterium and eubacterium.

The gut microflora provides the basis for a barrier that prevents harmful bacteria from invading the GI tract. Moreover, it plays a major role in eliciting, at an early age, an immune system which has a measured response to foreign proteins as potential antigens, balanced with an effective resistance to infection. The intestinal microflora, together with the gut's own immune system, allows the resident bacteria to perform a protective function, especially against the proliferation of pathogens. Both large bowel integrity and colonic microflora are important in determining stool characteristics, such as weight, consistency, frequency and total intestinal transit time – perhaps the most reliable markers of general colonic function. The composition and the metabolic and enzymatic activities of the faecal microflora, which may themselves be either beneficial or risk factors, can be good markers of the status of the resident gut microflora. The infant's colon is sterile at birth, and the organisms that make up the microflora are acquired during delivery and in subsequent days from the mother and the environment. The initial colonising species create a habitat that favours the growth of strict anaerobes. Thereafter, differences in the composition of the microflora are believed to depend largely on the nature of the diet as well as the host. The microflora of breast-fed infants is dominated by *bifidobacteria*. In contrast, formula-fed infants have a more complex gut microflora, which includes *bifidobacteria*, bacteroides, *clostridia* and *streptococci*. After weaning, a new adult-like equilibrium is set up, depending again on the host and diet. Each individual has his or her own specific gut microflora, which, in its uniqueness, is comparable to a fingerprint.

Bacterial numbers and composition vary considerably along the GI tract, but the large intestine is by far the most intensively populated part of the gut microbial ecosystem, with several hundred species accounting for a total of between 10^{11} and 10^{12} bacteria per gram of contents. Quantitatively, the most important genera of intestinal bacteria in humans are the bacteroides and the *bifidobacteria*, which can account for 35% and 25%, respectively, of the known species. Traditional gut microbiological methodologies are based on morphological and biochemical properties of the organisms. Because these methods rely on an ability to culture the organisms taken by sampling and not all the organisms present can be cultured, it is likely that there are many more species of bacteria present in the gut than have yet been identified. However, recent advances in molecular genetics for quantitative and qualitative monitoring of the nucleic acids from human gut microflora have revolutionised their characterisation and identification.

The different components of the colonic microflora exist in a delicate balance. Some bacteria are considered actively beneficial (e.g. *bifidobacteria* and *lactobacilli*), and others benign (e.g. certain eubacterium). Both types of bacteria are believed to suppress the growth of a third group of bacteria that are potentially harmful to human health. These harmful bacteria include the proteolytic bacteroides and *clostridia*, sulphate-reducing bacteria and the pathogenic species of the *Enterobacteriaceae*. Thus a symbiosis has evolved between the host and its gut microflora. Increasing evidence suggests that gastro-intestinal (GI) well-being and function may be compromised by modern lifestyles (e.g. eating habits, antibiotic use, or stress) and that this is related to disturbances in gut microflora composition and function.

The main substrates for bacterial fermentation are dietary carbohydrates that have escaped digestion (i.e. dietary fibre) and endogenous carbohydrates (e.g. mucus). Dietary fibre includes starch that enters the colon (resistant starch), as well as non-starch polysaccharides (e.g. celluloses, hemicelluloses, pectins and gums), nondigestible carbohydrates (such as fructo- and galactooligosaccharides added as prebiotics, and polydextrose). In addition, proteins and amino acids can be used as growth substrates by colonic bacteria.

The main end products of fermentation in the colon are short-chain fatty acids (SCFA), such as acetic, propionic and butyric acids, which play various important metabolic roles. In particular, butyric acid is of importance for colonocyte health. In addition, the process of fermentation induces a number of changes in the metabolic environment of the gut lumen that are believed to be beneficial to health. These include a lowering of the pH (increased acidity), increase in faecal water, a decrease in the toxicity of the luminal contents and, sometimes, laxative properties including softening of faeces. A stimulation of colonic mineral absorption (magnesium, calcium) also has been reported. The gut microflora is a complex interactive community of organisms, and its functions are a consequence of the combined activities of all the microbial components. It is commonly agreed that gut microflora can play an important role in GI infections, constipation, irritable bowel syndrome, inflammatory bowel diseases and, perhaps, colorectal cancer.

Functional foods to promote gut health

The gut is an obvious target for the development of functional foods, because it acts as an interface between the diet and all other body functions. One of the most promising areas for the development of functional food components lies in the use of probiotics and prebiotics to modify the composition and the metabolic activity of the gut microflora.

A probiotic is defined as a live microbial food ingredient that confers a health benefit on the consumer. Various species of *lactobacilli* and *bifidobacteria* combined (or not) with *Streptococcus thermophilus* are the main bacteria used as probiotics in yoghurts or fermented dairy products. Their major health benefits, demonstrated in humans, are alleviation of lactose intolerance and reduction of the incidence or severity of GI infections. They have also been shown to reduce the incidence of precancerous lesions in carcinogen-treated animals, but clinical trials are needed to confirm the importance of this observation for humans. Because probiotic bacteria are only transient in the intestinal tract and do not become part of the host's gut microflora, regular consumption is necessary for the maintenance of favourable effects.

One mechanism by which probiotic bacteria may promote health in the digestive tract is by altering the local immune response. Survival of bacteria during intestinal transit and adhesion to intestinal cells seem to be important for modifying the host's immune reactivity. A favourable modification of immune responses is the likely explanation for the reduction seen in the risk of atopic disease in children after the ingestion of probiotics. A prebiotic is a non-digestible food ingredient that beneficially affects the host by selectively stimulating the growth and/or modifying the metabolic activity of one or a limited number of bacterial species already established in the colon that have the potential to improve host health. Key criteria for a food ingredient to be classified as a prebiotic are (i) that it must not be hydrolysed or absorbed in the upper part of the GI tract so that it reaches the colon in significant amounts, and (ii) it must be a selective substrate for one or more beneficial bacteria that are stimulated to grow. It may also induce local (in the colon) or systemic effects through bacterial fermentation products that are beneficial to host health.

Apart from their potential to modify the gut microflora and its metabolic activities in a beneficial manner, many other helpful effects of prebiotics are being investigated. These include their ability to modulate gut function and transit time, to activate the immune system, to increase the production of butyric and other short-chain fatty acids, to increase the absorption of minerals such as calcium and magnesium and to inhibit lesions that are precursors of adenomas and carcinomas. Thus, they could have the potential to help reduce some of the risk factors involved in the causes of colorectal diseases. Strategies for developing prebiotic products as functional foods aim to provide specific fermentable substrates for bacteria such as *bifidobacteria* and *lactobacilli*. These may provide beneficial amounts and proportions of fermentation products, especially in the lower part of the colon where the effects are believed to be most favourable. Mixtures of probiotics and prebiotics, which favourably modify the gut flora and its metabolism by increasing the survival of health-promoting bacteria, are described as synbiotics.

The major applications for probiotics are in dairy foods. Prebiotics are added to dairy products, table spreads, baked goods and breads, breakfast cereals and bars, salad dressings, meat products and some confectionery items. Examples of opportunities for modulation of target functions related to intestinal physiology, and the food components that might be used to achieve them.

MENTAL STATE AND PERFORMANCE

Some effects of foods or food components are not directly related to disease or health in the traditional sense but they, nevertheless, fulfil an important function in changing mood or mental state. Such changes may affect appetite or the sensation of satiety, cognitive performance, mood and vitality, and an individual's reaction to stress, with consequent changes in behaviour.

Behaviour is probably the most varied and complex of all human responses. This is because it is the cumulative outcome of two distinct influences: (i) biological factors, including genetics, gender, age, body-mass, etc. and (ii) socio-cultural aspects, including tradition, education, religion, economic status, etc. As a result, perceptions about the effects of food components on behaviour and mental performance are characterised by a high degree of subjectivity, with large differences in response among people. Age, weight and sex are among a number of crucial parameters to take into account when evaluating the power of food components to alter behaviour. In addition, the effects seen immediately after the first time a food component is ingested may be different from the longer-term effects of the same food component as a part of the habitual diet. In some cases, adaptation may occur so that with repeated ingestion the effects may be diminished or lost.

Key aspects of behaviour that may be affected by foods include performance in mental functions (such as vigilance, memory, attention and reaction time) and aspects of eating behaviour (such as eating frequency, food preferences and dietary selection). In general, two types of effects can be discriminated: (i) the immediate effects, such as those on reaction time, attention focus, appetite and satiety, and short-term effects on memory, and (ii) longer-term effects, such as changes in memory and mental processes in ageing. Adaptation effects may be a crucial consideration in the case of agents that are used to modify appetite (palatability enhancers, artificial flavours, colours, etc.) and satiety (e.g. fibre, carbohydrate and protein content) over long periods where the aim is to assist in weight control.

Functional foods to promote optimal mental performance

Some functional foods – such as the ideal lunch food that will not induce, or might even prevent, a dip in vigilance in the post-lunch period – are desirable to everyone. Other foods could be functional for students who want to face exams with the maximum intellectual readiness, for those people at an emotional low point who expect to obtain a lift by ingesting foods such as chocolate, sugars or alcohol, or for the elderly and others who may have failing memory. In considering functional foods to promote optimal mental performance, the specific needs of the target consumer are of key importance.

Glucose has been reported to exert general beneficial influences on mental performance, including improvements in working memory and decision time, faster information processing and better word recall. Caffeine also can lead to an improvement in most measures of cognitive performance (reaction time, vigilance, memory and psychomotor performance), especially in the morning hours.

Meals high in carbohydrate help to produce feelings of sleepiness and calmness. In addition, the amino acid tryptophan reduces sleep latency. Tyrosine and tryptophan may help in recovery from jet lag, but only a limited amount of scientific evidence supports this effect. Sweet foods, such as sucrose, may relieve distress in young infants and may reduce pain perception in members of the general population.

Meals high in protein reduce hunger and increase satiety, which may help in body weight control. Intake of alcohol is both traditional and widespread in Europe. It is one of the few substances to affect all major areas of psychological and behavioural functions (appetite, cognitive performance, mood and stress), and the effects are conspicuously dependent on the dose.

Examples of opportunities for modulation of target functions related to behaviour and mental performance, and the food components that might be used to achieve them.

Physical Performance and Fitness

During physical stress such as exercise, demands are high for food components (the substrates) that act as the starting material for reactions that release energy. Diet can play a crucial role in improving the level of performance. Training and competition will increase the daily energy expenditure by between 500 and 1000 kilocalories per hour of exercise, depending on intensity. Large sweat losses may pose a risk to health by inducing severe dehydration, impaired blood circulation and heat transfer. This may ultimately lead to heat exhaustion and collapse. Insufficient replacement of carbohydrates may lead to low blood glucose levels, fatigue and exhaustion. An ever-increasing amount of daily, high-intensity training leads to high stress on the metabolic machinery – the musculoskeletal and hormonal systems. A growing body of evidence supports observations that the supply of food ingredients or food-derived substances may interact with the biochemical and physiological systems involved with physical and mental performance. The results may impact recovery from intensive training and, hence, affect the physical well-being and health of the athlete.

Functional foods to promote optimal physical performance and recovery

Requirements for specific nutrients and water depend on the type, intensity and duration of physical effort. Specific nutritional measures and dietary interventions can be devised to be appropriate for the distinct phases of preparation, competition and recuperation.

Oral rehydration products for athletes were one of the first categories of functional foods and drinks for which scientific evidence was obtained on all levels of functionality. Among these functions are rapid gastric emptying, fast intestinal absorption, improved water retention, improved thermal regulation, improved physical performance and delayed fatigue.

Liquid food formulae, established to deliver fluid and available carbohydrates and electrolytes in a convenient and easily digestible form, have been shown to be of benefit to athletes. Exercise-induced losses of nitrogen, minerals, vitamins and trace elements should be replenished by ingesting larger amounts of high quality, micronutrientdense foods at meal times. However, this may be difficult in those circumstances in which low-energy diets are combined with intense training or in the case of multipleday events, such as cycling competitions.

The use of special meals or food products and micronutrient supplementation can help ensure adequate intakes under these conditions. Specific types of carbohydrates with moderate-to-high glycaemic index in combination with each other and with protein have been shown to influence physical performance and enhance recovery of athletes, and this offers potential for the development of functional foods.

Some Further Considerations

Significance of risk factors

In the preceding discussion, risk factors for various diseases have been used to identify where functional foods might be used to beneficially affect health and, specifically, to reduce the risk of disease. While any particular risk factor may present a suitable target for change by a functional food, it has to be borne in mind that diseases are the end result of complex biological processes. Many factors may contribute to the establishment of a disease, either acting together or independently through multiple pathways. Although it may be possible significantly to influence any one risk factor by the consumption of a functional food, that alone may not reduce the overall risk that the disease will eventually become established. In order to achieve an actual reduction in the incidence of the disease, it may be necessary to influence simultaneously several risk factors in a positive way.

Similarly an optimal state of health and well-being is the outcome of a balance of complex biological processes. Its maintenance will most likely be achieved by seeking to influence several target functions in a positive way in concert. The assessment of food functionality can be simplified by considering the impact of food components on single risk factors and target functions. However, in reality, the interaction between risk factors and between the underlying processes, represented by target functions, provides multiple pathways for achieving a reduced risk of disease and an improved state of health and well-being in which a spectrum of functional foods is eaten as part of a broad and balanced diet.

Functional foods and drugs

Functional foods are not medicines. Although they are intended to modify physiological functions within the body in a positive way, their mode of action is to restore, reinforce or maintain normal body processes in ways consistent with normal physiology. They may restore or enhance body functions within normal ranges in order to optimise health and well-being or they may reduce factors known to be associated with the risk of contracting diseases. Medicines on the other hand function by intervening in disturbed physiological processes or by amplifying physiological processes beyond normal extremes in order to achieve an effect. Their function is to treat or prevent diseases, or to heighten physiological performance outside the normal range. However, there is no absolute boundary between foods and drugs in terms of their functionality. The distinction has to be made case by case, taking into account the type of product (food, supplement or pill) and its effect.

As a general rule, functional foods are intended to be consumed as part of a normal diet and they take the form of foods. Medicines are intended to be taken as part of a controlled regimen, often under the supervision of a medical practitioner, and they usually take the form of tablets, pills, capsules or syrups, which can be administered in precise doses. Ultimately, the manufacture and marketing of medicines is subject to different regulatory controls than those that apply to the manufacture and marketing of foods. Functional foods fall within the control of the food legislation.

Food supplements, like medicines, often take the form of pills or capsules but they cannot legally be presented as treating or preventing diseases, and they are controlled by food law. Although from a legal perspective they are foods, they do not usually have the form of foods and are not consumed in the same way as typical foods as part of the normal diet. Consequently they do not fall within the concept of functional foods as presented and discussed in this section.

DELIVERING THE BENEFITS

Role of Biotechnology

Biotechnology, in the form of food processing, is an established part of the food chain. It serves to convert raw materials into edible, safe and nutritious food with the taste and texture, shelf life and convenience to suit everyday needs. Biotechnology also provides the means to extract components with functionality from foods and raw materials and to optimise their form and chemical structure to make them suitable for inclusion in new food products. The extraction of phytosterols from plant sources and their esterification, either as sterols or in hydrogenated form as stanols, to enable them to be incorporated into products for use in reducing serum LDL-cholesterol (described above) provides an example of this.

As more food components with the desired functionality are identified, technology has the potential to maximise their accessibility and availability so that they become available on an everyday basis in a form that suits consumers' needs and preferences.

Biotechnology can help to achieve this goal in three ways:

1. By creating new functional food components in traditional materials, in new raw materials or by synthesising.
2. By maximising the presence of functional food components already existing in foods and raw materials by improving their preservation, modifying their function or increasing their bioavailability.
3. By providing the means to monitor the amount and effectiveness of functional components in foods and raw materials to ensure that they are retained to the maximum degree at all stages in the food chain.

COMMUNICATING THE BENEFITS

How do Consumers Learn about the Health Benefits of Functional Foods?

Functional foods and food components may be available in the market place but in order for consumers to access the benefits of functional foods, they must be informed about them. The simplest and most direct way for consumers to learn about functional foods is through food labelling but the information provided by labelling must be understandable and it must be reliable.

General food labelling laws already in place almost everywhere require that information provided on food labels and through advertising is factually truthful and not misleading. For the most part, these general laws are sufficient to ensure that statements made on labels and in advertisements properly and adequately convey information to consumers. However, in the case of claims made about the health benefits of foods there is extra scope for confusion and uncertainty. Firstly, this scope for confusion arises because of the potential overlap between foods conveying health benefits and medicines. Secondly, a full understanding of the meaning and underlying bases of health claims usually requires expert knowledge.

The first source of possible confusion has been resolved historically by legislation to prevent health claims being made about foods, such claims may only be made for medicines and are controlled by legislation concerning medicines. Recently this situation has changed (and continues to evolve) with categories of health claims being permitted for foods in various countries. These are subject to a clear distinction being made between health benefits relating to support or enhancement for normal body functions and reduced risk of disease (in the case of foods), and the treatment or prevention of disease (in the case of medicines).

Legislators have sought to resolve the second source of possible confusion, the need for expert knowledge, by requiring health claims for foods to be evaluated and agreed by experts before they can be used. The acceptance of the principle that health claims for foods should be allowed, is still relatively recent and legislation is developing differently in different countries. Nevertheless, the types of claims that are envisaged have been categorised by authorities and expert groups into (more or less) two classes:

- Health claims other than those referring to reduction of disease risk.
- Claims referring to reduction of disease risk.

Some authorities (e.g. Codex Alimentarius) have subdivided the first class into two relating to 'nutrient function claims' and 'other function claims', respectively. In the European Union, claims referring to children's development and health are considered as a special subcategory of this class which have to undergo a more rigorous appraisal by the authorities. Essentially all of these claims describe the physiological role of a nutrient in growth, development and normal functions of the body, or they refer to specific beneficial effects of foods or food components beyond their generally accepted nutritional effects, other than the reduction of disease risk. For example, a claim might say that calcium aids in the development of strong bones and teeth and then draw attention to the calcium content of the food for which the claim is made. Or, it might say that consumption of a particular food helps to maintain a high state of mental alertness or physical performance.

Claims referring to reduction of disease risk are claims that consumption of a food or one of its components reduces a risk factor in the development of a human disease.

In some countries, such as the USA, an evaluation process for health claims is mandatory but, once approved, the claims can be used for all foods fulfilling the stipulated compositional criteria, that is, the claims are generic. In other countries, like Sweden, the Netherlands and the UK, voluntary codes of practice have been established by the food sector in co-operation with the authorities and provide procedures for evaluating claims. The evaluation process has not been part of the legislation and claims have been allowed, subject to the general food labelling rules. But, in enforcing the labelling rules, the authorities have taken into account whether claims have been evaluated and endorsed by suitably qualified experts. There is an expectation that health claims will undergo a process of substantiation before they appear on food labels. New legislation on nutrition and health claims has recently been adopted in the European Union and is currently in the process of being implemented. The rules apply in all Member States of the EU (replacing the codes of practice previously in place in Sweden, the Netherlands and the UK) and will, when they are fully implemented, require health claims to be substantiated by experts before they can be used.

Since food labelling laws, in most countries where they exist, regard advertising and promotional information provided in relation to specific food products to be equivalent to labelling, claims made in these ways must comply with the same rules.

How are Claims to be Substantiated?

For a claim to be truthful, it must accurately reflect its underlying basis. Substantiation of the claim should be based on a systematic review of the evidence relevant to the claim and an assessment of whether the wording of the claim is fully consistent with scientific evidence.

Informing Other Stakeholders – The Broader Picture

Claims made on food labels, and the associated advertising and promotional literature for specific food products, provide the most direct way of communicating the benefits of functional foods to consumers. Incorporation of the concept of food functionality into awareness about nutrition through general education will ensure that consumers are equipped to understand the messages conveyed by claims, both so that they are not mislead and also so that they can take full advantages of the benefits on offer.

However, consumers are not the only stakeholders where diet and health are concerned. A growing body of scientific evidence points to the fact that functional foods have the potential to contribute positively to long-term health and well-being and disease risk reduction. Those potential benefits can best be delivered if all stakeholders (commerce, industry, agriculturists, educators, government and consumers) are sufficiently well-informed to motivate the introduction of functional food products into the marketplace. Food manufacturers, retailers, primary food producers, caterers, health care workers and governments all have roles to play and benefits to gain in ensuring that functional food science becomes part of general knowledge concerning nutrition, in the same way that knowledge about the role of vitamins and minerals in basic nutrition has become part of general knowledge in the past.

FUTURE PERSPECTIVES

Functional food science is still at an early stage in its development. As knowledge about the functional effects of foods increases and the functionality of particular foods and food components is more extensively recognised, technology will have a continuing role to play in making those foods and food components more widely available and accessible. Basic education in nutrition will also have a continuing role to play in ensuring that the benefits of functional foods are understood by all stakeholders in order to ensure that the benefits are enjoyed to the full. These aspects of future development are a continuation of activities already underway.

Role of 'Omics' in the Future Development of Functional Foods

Rather more exciting for the future of functional foods is the potential for the use of knowledge being gained in the fields of genomics, proteomics and metabolomics. Genetic variation between populations and between individuals has long been recognised as a source of the variation that is evident in the outward appearance of individuals and in many aspects of their susceptibility to disease.

There are well-established genetic links to conditions such as haemophilia, sickle cell anaemia and familial hypercholesterolaemia. There is also evidence that obesity is influenced by genetic factors. There is growing evidence that genetic factors influence the relationship between diet and the protective and risk factors for disease, and the ways in which different protective and risk factors can lead to the actual incidence of disease.

Until now, study of these interactions has been hampered by lack of a complete knowledge of the way an individual's genetic makeup determines their physical and physiological profiles. This is, in part, because of the large number of genes involved, it has been estimated that the number of human genes must be counted in the tens of thousands. However, recent technologies provide the possibility of measuring large numbers of biological datum points simultaneously. It is now possible to visualise differences between the genetic profiles of individuals at the molecular level and to begin to understand how they relate to differences between individuals' responses to physiological factors at the level of the whole organism. Similar techniques can be applied to gain an understanding of the way differences between other aspects of an individual's molecular biology affect their response to physiological factors. These techniques are known collectively as 'omics'.

The study of the total genetic makeup (the genome) is known as genomics, and the study of the total metabolic profile (the metabolome) is known as metabolomics. Transcriptomics and proteomics are the terms that describe the study of the total gene expression and the total protein complement of an organism, respectively. A greater understanding of the relationships between physiological responses at the level of the individual and their underlying bases at the molecular level has the potential to lead to the development of new markers of intermediate endpoints of relevance for improved health and reduced risk of disease. The accessibility of such markers to measurement by the application of 'omics'- based methods may provide additional means to study the influence of dietary factors on human health and disease, leading to the identification of new routes to food functionality. The availability of markers responsive to short-term interventions and accessible to rapid, noninvasive measurement holds the promise of providing greater means to substantiate health claims, as well as to identify groups of individuals who might respond well or not so well to dietary interventions. As greater insight into the basis for the variation between individuals' physiological responses is gained from the application of 'omics', it may be possible for individuals to know whether they are at risk from, say, heart disease and to choose foods and diets that have the functionality to suit their particular needs. A synergy between developments in functional food science and 'omics' may, in the future, result in a situation where it is possible for individuals to make truly informed choices about which foods provide the best opportunities for health, well-being and reduced risk of disease.

Opportunities and Challenges for Developing Countries

Functional foods through biotechnology can provide developing countries food sources with increased nutritional value. Staple starchy crops such as cassava and yams have been modified to lower the amylopectin content of starch, which has been associated with diet-related conditions such as type 2 diabetes. In areas of drought and poor soil quality, where high quality proteins are scarce, genetic modification has been undertaken on some legumes and in soyabean to increase the levels of high quality proteins.

Currently, commercialisation of genetically-modified nutritionally-enhanced crop is very limited due to many factors that include the cost of introducing a new product to the market and the lack of suitable regulatory controls. In addition, the development and marketing of functional foods require significant research efforts because most markets require scientific evidence and proof of functionality.

Thus, functional foods sprung from the desire to prevent the onset of diseases associated with an ageing population. Developing countries, especially China and countries in Latin America, face increasing health problems related to life style: diabetes and cardiovascular diseases among others. In these cases foods with improved nutritional qualities and added function would be useful, hence, developing countries need to increase the investment on rigorous scientific research on potential functional foods.

Chapter 10

Health Benefits of Functional Ingredients from Marine Bioresources

INTRODUCTION

The marine ecosystem represents a vast and dynamic array of bioresources attributed with its huge diversity and considered as potential untapped reservoirs for the development of functional foods for future health markets. Basically, marine micro-organisms, sponges, algae, invertebrates such as crustaceans and mollusks along with marine fish species can be considered as marine bioresources, which can be utilised to obtain different health benefits for humans, directly or after processing. Most of the biomolecular components, such as lipids and proteins from these marine bioresources, which can be extracted in large scale using the modern and advanced biotechnological approaches, are suitable drug candidates for the pharmaceutical industry as well as functional food ingredients for the food industry. Moreover, the furtherance of high throughput molecular biological techniques has already been incorporated with identification, mining and extraction of molecular components from marine bioresources. This chapter discusses potential marine bioresources with respect to their extractable biomolecules in details, while explaining the present and prospective methods of identification and extraction, which are integrated with advanced techniques in modern biotechnology. In addition, this provides an overview of future trends in marine biotechnology.

The life of the earth has been originated from the ocean. The oldest fossil evidences showed that rise of environment oxygen by oxygen producing eukaryotes and cyanobacteria was appeared over 2.45–2.32 billion years ago. Afterwards, many of organisms or evolutionary forms have been evolved from that primitive age to contemporary state by having extraordinary structural variations. On the other hand, abiotic factors such as light intensity, temperature, nutrients and salinity levels make much influence for their biological functions. Furthermore, extreme fluctuations of climate according to seasons have been given a major outbreak through the variations and to be survived in the competitive environmental conditions. However, the evolution of the diverging forms with respect to their biochemical and morphological traits are still being taking place with the adaptive establishments in the ocean. In that case, marine bacteria, fungi, sponges, algae (macro or micro), crustaceans, mollusks, fish and small vertebrates have been identified as the major classes that comprise the diversity of marine biomass. As the consequences of this diversity, biochemically and ecologically significant differences have brought with the vast marine biomasses and have been associated with a broad spectrum of secondary metabolites. However, the significant gains are continually being improved due to the exploration of novel secondary metabolites from the ocean. Therefore, it is logical to consider that mining of natural products and functional ingredients from the marine biomasses would be a productive commencement in the view point of pharmacology.

There has been existed long tradition of consuming seafood as a delicacy along with the human diet for many centuries. Ancient people in many parts of the world would have believed that the steady enhancement of life expectancy and long term health effects, quality and care could be rendered by the marine functional ingredients. Consumption of seafood is considered as a preventive strategy against lifestyle diseases and fruitful solution for the prospective health challenges. Moreover, substantial scientific evidences have been shown that direct consumption of seafood or food supplements may contribute for the health promoting effect. A sufficient intake of seafood nutrients has been linked for preventing chronic heart disease and relieving from many more health problems. In addition, largely derived human clinical studies have encompassed the vast array of health benefits from marine foods. Furthermore, seafood commodities including fish, shellfish and seaweeds are commonly used for delicacies in many parts of the world and highly demanded natural comestibles in the global food market. This can be well understood by the tremendous diversity of the seafood and their nutritional characteristics including poly unsaturated fatty acids (PUFA) such as eicosapentaenoic acids (EPA) and docosahexaenoic acids (DHA) as well as proteins, bioactive peptides, free amino acids, enzymes, vitamins, minerals and other functional ingredients which are proven to play significant beneficial roles in human health.

The exploration of marine bioresources is an indefinite challenge for biomining researchers. However, the immersing diversity of marine environment has revealed that the untapped resources are having broad probe to meet the curiosity of the future scientific community. Therefore, this endeavor leads to overview the recent biotechnological aspects on valuable health products, mainly marine proteins and lipids from marine bioresources through the future prospects, with respect to their increasing sustainability. Hence, this biotechnological and pharmacological scenario will be emphasised due to increasing demands on the marine bioresources, in order to strengthen the development of marine functional foods, nutraceuticals and associated medicinal benefits.

POSSIBLE APPLICATIONS ON EXTRACTIONS AND ISOLATION OF FUNCTIONAL INGREDIENTS

Marine functional materials can elicit positive health benefits with long term efficacy. Therefore, these materials are believed to be implemented a novel strategy for the functional foods in the marine biotechnology. Over the many years, a vast array of structural and biochemical diversity in terms of the various marine organisms might have targeted for the screening projects. Effective ways of screening, biomining and advancing of bioactive natural products using appropriate techniques, in order to commercialise the respective products are highly concerned matters in the modern field of marine biotechnology. Furthermore, the development of functional materials for the functional foods and identification of possible extraction methods are performed with the respective parameters such as heat resistance, solubility, molecular weights. In this regard, two possible extraction methods are described preliminary, designated as enzymatic hydrolysation and organic solvent extraction. Even though numerous extraction methods are currently employed, only few of them are suitable for industrial scale. In addition, solid-liquid extraction (SLE), liquid-liquid extraction (LLE) and soxhlet extraction (SE) techniques are characterised as less economical strategies. Furthermore, high volumes of solvents should be used with long time proceedings in those procedures, which may in turn lead to remain toxic residues with the final extracts. Therefore, sophisticated green methodologies are described to increase the yield, purity and reproducibility of the target compounds. However, enzymatic hydrolysation can be preferred to isolate proteins, lipids and their bioactive constituents, with the safe, time and cost effective way from the heterogeneous sources. Despite the fact, chemical extractions (using organic solvents) are also applicable as a popular method to isolate biofunctional materials in the traditional way. For example, recovering of PUFA as free fatty acids from the marine oil followed by saponification, isolation, purification and crystallisation techniques were performed depending on the solubility of the particular chemical and available distillation methods. However, purification of proteins, mainly enzymes, peptides and free amino acids can be purified with the ultramembrane filtration (UF), gel or size exclusive column chromatography and couple with the high performance liquid chromatography (HPLC). The mass determination and structure elucidations may further subjected using liquid chromatographymass spectrophotometer (LC-MS) or mass-mass spectrophotometer (MS-MS) along with the nuclear magnetic resonance spectrophotometer (NMR). However, the commercialisation of proteins and lipids would become an interesting view for the functional and medicinal industries as new chemicalentities for lead-compounds or innovative drugs from marine bioresources.

Therefore, in recent years, the pharmaceutical firms have a look towards the investigation and utilisation of new isolation methodologies under optimised conditions, which would become efficient with respect to the time consumption and cost in industrial scale, rather than the laboratory scale. The technological advances in industrial scale (i.e. combining conventional membrane filtration with electrophoresis on separating highly charged bioactive peptides) might render fast separation rather than chromatography. In addition, the highest degree of technologies such as supercritical fluid extraction (SCFE), pressurised liquid extraction (PLE), pressurised hot water extraction (PHWE), accelerated solvent extraction (ASE) as well as ultrasoundassisted (UAE) and microwaveassisted (MAE) extraction have been used in recent marine bioresource extractions and functional food formulations. Moreover, bioactivity guided fractionations lead to determine the fastest and accurate way to reach the last step of separation procedures. However, the sequential chromatographic steps involve further purification and characterisation of innovative secondary metabolites from the marine natural sources. Moreover, the purified functional materials can be used to scale up the activities, testing the efficacy of *in vitro* assays, analytical techniques to quantify minor components and animal or human clinical studies may further needed to establish the health prospects before the development and commercialisation (Fig. 10.1).

PHARMACOLOGICAL AND NUTRITIONAL VIEW OF MARINE PROTEINS AND LIPIDS

The very long evolutional period of marine life than the terrestrial life has generated a massive biodiversity throughout the species at genomic level. Thus marine bioresources are showing a broad spectrum of biomedical constituents. Therefore, it is believed that the vast number of reports have encountered the significances of medicinal effects from marine bioresources. Furthermore, various interests over the pharmacological and nutraceutical effects of the proteins including enzymes, peptides and free amino acids, and also lipids including saturated and unsaturated fatty acids from marine organisms will be discussed in the next part of this section.

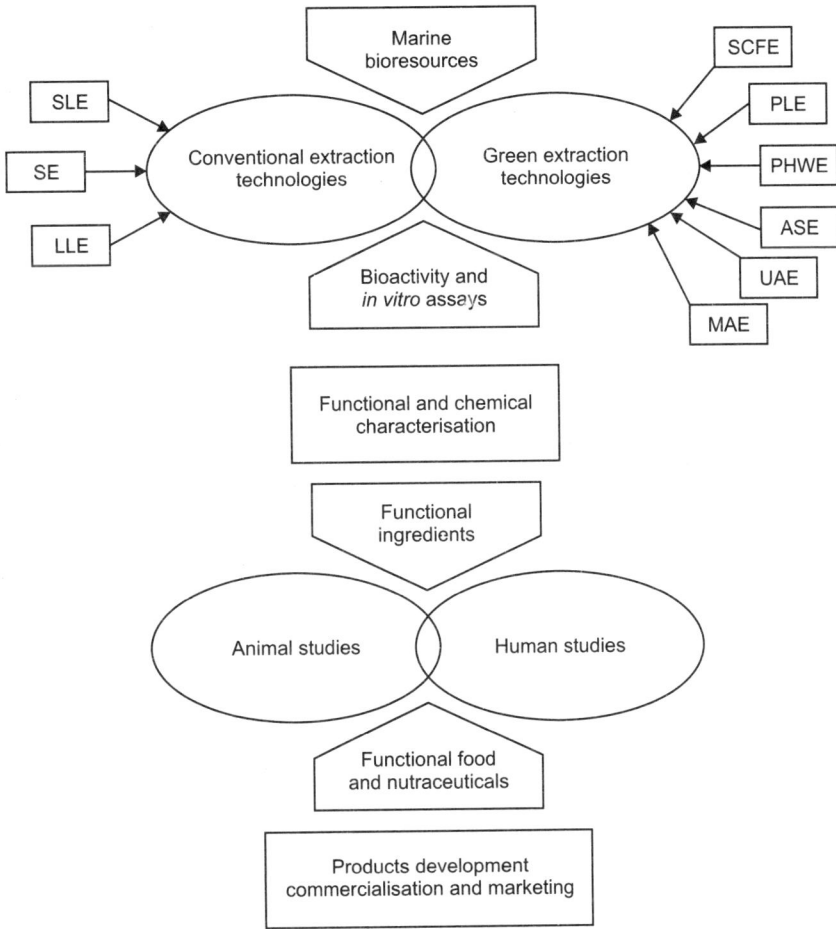

Fig. 10.1: Flow chart of the possible exploration and development of the functional ingredients from marine bioresources through the conventional and green extraction technologies for future pharmanutra applications.

Marine Proteins and Protein Derived Constituents

Proteins, bioactive peptides and free amino acids derived from marine sources have gained a much attention due to promising health benefits. Basically most of the marine bioresources are comprised of high content of proteins than the other functional materials. Interestingly, marine microalgae, muscle tissues of fish, shellfish and other invertebrates showed the range of 15–65% (w/w) proteins. In addition, marine fish and shellfish waste along with processing by-products are also considered to be consisted with high quality and quantities of proteins [10–23% (w/w)]. These sources are targeted as a potential source for mining of active components. In fact, extractions of these proteins or protein derived constituents from different sources have been extensively studied. Usual extraction protocols of those proteins from organisms are initiated by the disruption of cells of the source and releasing proteins. However, the bioactivity of the proteins may not directly correlate with the intact form of parent proteins.

Free forms of amino acids are also used in high levels of production of food supplements, energy formulations and nutraceuticals in the infant and adult food industries. Taurine and creatine are two main amino acids used commonly in those industries. In terms of the biofunctionalities of the taurine, it showed very high activity and found in most of the organisms as dominant free form.

Furthermore Smith and others have reported that the both amine and sulphonic subunits of taurine can undergo ionisation and involve in biological and physiological activities by their dissociation components. Moreover, they were known to be involved in antioxidative reactions immune stimulations homeostasis cell membrane stabilisations anti-inflammatory developmental regulation of renal functions endocrine and metabolic effects in humans. In fact, raw flesh fish, raw clams and raw mussels are also known to contain high taurine levels, among the marine organisms. On the other hand, creatine is also being considered as an important free or phosphorylated form of amino acid that furnishes positive physiological effects for humans. Moreover, skeletal muscle regeneration, contraction by performing on cardial muscles and fat free mass promotion has also been encountered by creatine level in human body. Importantly, bioactive peptides showed much significant roles in the pharmanutraceutical applications. Marine food sources have exerted many of bioactive peptides and the peptide sequences may short and consisted of 3–20 amino acid residues. Therefore, isolation and purification of bioactive peptides could be described by facilitating upon the proteolytic enzyme hydrolysation and follow the activity guided fractionations. Moreover, the composition and number of amino acid residues of the isolated peptides exert their bioactivity. Hence, the identified bioactivities have given a wide range of physiological applications.

Most of these bioactive peptides have been obtained from enzymatic hydrolysation of marine organisms including algae, fish, shellfish fermented marine food sources processed by micro-organisms and marine processing by-products. These peptides demonstrated the overall effect of the human health claims and the knowledge of the diet make the awareness of functional foods and nutraceuticals. Interestingly, bioactive peptides and amino acids may act as alternative molecules to small molecular drugs. Moreover, those have shown a great advantage over the conventional drugs with high bioavailability and biospecificity to the targets. The desirable properties of these molecules, such as low toxicity, structural diversity and least or no accumulations in the body tissues have rendered the interest of many scientists to use for the therapeutic purposes. Furthermore, during the gastrointestinal digestion, bioactive peptides may not be broken down and can provoke the positive health effects that have confirmed by *in vivo* studies. Anti-ACE and anti-obesity studies have linked with the active peptides and mediated through stimulation of hormones and regulatory satiety. In addition, Quantitative Structure-Activity Relationship (QSAR), that is modelling information, can be used to predict the peptide structures with the ability to cross membrane barriers in the specific target sites. Furthermore, the improvement of the structures, stabilities and binding capacities of the bioactive peptides with the respective binding sites of human model systems incorporation with synthetic peptides can be strengthen the knowledge of peptides.

Enzymes based biotechnologies have been emphasised and released a broad range of functional metabolites from marine bioresources over the last few years. With the growing interest of industrial applications and development of new extractions protocols, the exploration of new enzymes or enzyme sources have given a new era for the future biotechnology. In particular, a high specificity and a very high reactivity even at low concentrations have been facilitated in different enzymes to use in multiple purposes. In this regard, marine organisms have targeted to isolate stable novel enzymes at hyperthermal and extremely acidic conditions even from the coldest habitats of the marine organisms which are being extensively studied. Especially enzymatic applications in food processing, diagnostics, molecular biology research, pharmacological and health care as well as environment monitoring applications have been developed.

Marine Lipids

Marine bioresources are characteristically comprised of a significant lipid composition which takes a greater attraction for many lipids, including monogalactosyl diacylglycerols (MGDG), digalactosyl diacylglycerols (DGDG), and phosphatidylglycerols (PG). In addition, Bruno and others have shown that these biomolecules are attributed with health promotion effects. Over the many years, the extensive studies have documented that the lipid fractions of marine food sources are primarily consisted of poly unsaturated fatty acids (PUFA). This represents omega-3 and omega-6 fatty acids, including eicosapentaenoic (EPA) or docosahexaenoic acids (DHA).

Consumption of seafood is proven to be preventing the life style diseases and giving a long term health claims. It has been noticed and widely accepted that the long chain omega-3 fatty acid molecules help to prevent cardiovascular diseases. In addition, abundant evidences have shown that fish oil, shellfish and seaweed PUFA can reduce the risk of certain cardiovascular diseases. Furthermore, consumption of fish for one or two times per week may enhance the protective effect against coronary heart diseases, especially reducing the death risk by 36% and 17% from total mortality due to high content of DHA and EPA in the diet. A noteworthy finding is that intake of EPA and DHA at 250 mg or range from 850 ~ 4000 mg per day can expected to provide promising health protective effects by primary and secondary prevention levels. In addition, the intake of a sufficient amount of these functional ingredients also has been associated to relieving effects of some symptoms or medical conditions such as inflammatory diseases, physiological conditions, diabetes and cancers. Basically, Riediger and others have reported that the development of protective effect was mechanistically being proven by epidemiological studies due to pleiotropic effects including inflammation, platelet aggregation, hypertension and hyperlipidemia.

Very recently marine algae have been demanded as materials for production of oil due to the content of lipids for the potential of biodiesel production. As reported, the content of lipids exists within the range of (1–5%) and (15–50%) in the dry weight of seaweed and microalgae, respectively. For example, a marine diatom, *Phaeodactylum tricornutum* showed 30–45% of PUFA, accounted for EPA up to 20–40% of the total fatty acids. However, the same significant value has been taken on the fish waste and processing by-products with the prolific lipid sources. For example, fish by-product commodities such as sardine, mackerel, shark, and cod contain more than 30% of PUFA including essential omega-3 and 6 lipids for commercialisation. In particular, salmon head consider as a good source of PUFA as major part of the by-products. However, most of these by-products have marketed by recovering their essential lipids as nutraceutical or food supplements recently. These evidences suggest that the future biotechnology on functional foods and health products may directly depend upon the marine waste and processing by-products. Thus the understanding of the sustainability of the marine bioresources might be performed much significant role in future.

Impact of Biotechnological Advances on Marine Bioresources

The idea of an implementation of the green chemistry and biotechnology for the isolation of marine bioactive components is to sustain, design and develop in environmental friendly way with focusing on new protocols, methods, procedures and systems. The above mentioned technologies have come up with new trends in the future biotechnology and pharmacology applications, in order to improve health benefits without damage to the bioresources and marine environment as well. Progress of the biotechnology is currently being served for the exploration of new functional materials from marine life. Thus, this can cure health disorders and consider as a challenge to the upcoming diseases as well. However, recent biotechnological aspects are expected to hold the promising effects against the numerous disciplines, where it can be approached to focus on research tools and strategic importance, such as algae biomass culturing, aquaculture, harnessing biomaterials, bioremediation, bioinformatics, marine genomics, genetic engineering and research infrastructures. These evidences suggest that the importance of the utilisation of sustainability of marine bioresources and protection of marine environment along with its biodiversity for the future benefits of mankind.

Molecular Biological Approaches on Investigation and Development of Marine Bioresources

Molecular biology is a remarkable field of biology, which is basically dedicated to study on structures and functions of macro molecules like nucleic acids and proteins, with respect to their roles in life processes. However, applications of this promising discipline on marine bioresources such as micro-organisms, algae, invertebrates, vertebrates including economically important fish species have become precious recently. In fact, tremendous collaboration with other related disciplines including marine biology, chemistry and microbiology lead to the sustainable exploration of marine life, mainly for human health and welfare. As described in previous sections, marine ecosystem consisted of a huge diversity of organisms and that is a prominent source of wide array of therapeutic agents which can be developed as pharmaceuticals. Novel propagations of molecular biology, including genomics, metagenomics, transcriptomics, proteomics and bioinformatics, along with development of its technologies, such as genetic engineering and bioactivity screening offers a unique opportunity to establish marine natural products as interesting drug candidates (Fig. 10.2).

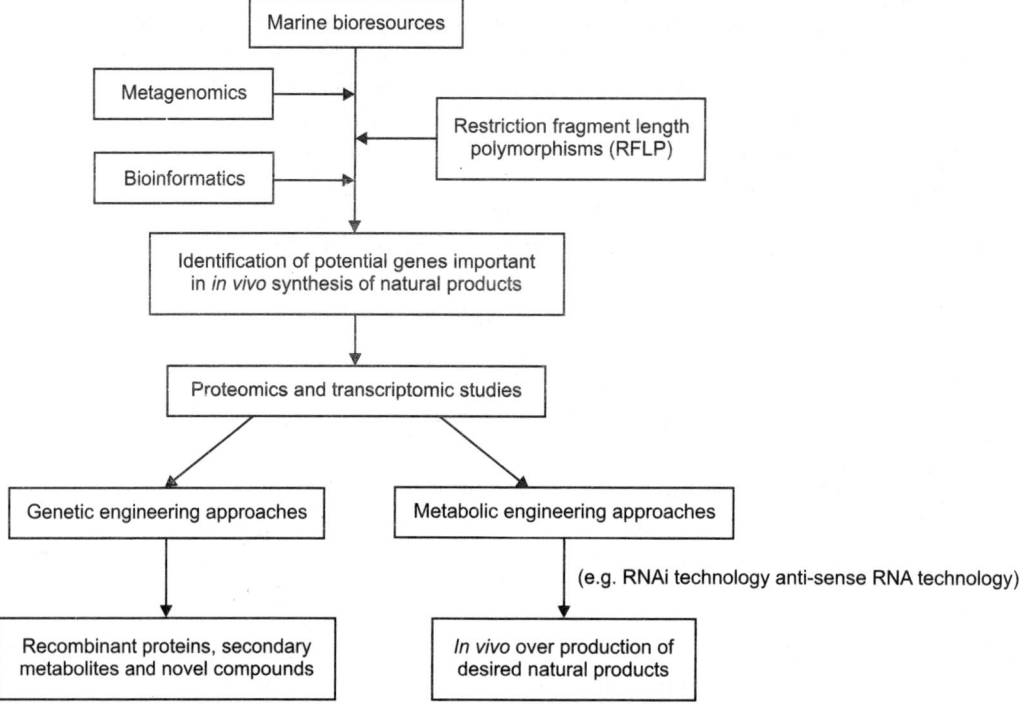

Fig. 10.2: Flow chart of the investigation and development of functional ingredients via Molecular Biology studies for future biotechnology.

Most of the micro-organisms present in the marine environments are not readily cultivable. Therefore culture independent technique is needed for the yielding of natural products from those organisms. This bottleneck can be overcome by metagenomic approaches, which can further maximise the diversity of libraries of marine natural products, through direct accessibility of DNA from marine organisms. Basically in these genetic strategies, genomic DNA from the interested micro-organisms is isolated and digested into large fragments using restriction enzymes. Subsequently, the fragments are cloned into artificial vector systems and sequenced through high throughput sequencing techniques to screen the putative clones for the potent genes or gene clusters encoding the protein components involved in biosynthetic pathways of the corresponding natural products. After detecting particular genes or gene clusters, gene sequence can be analysed using bioinformatics tools to reveal chemical structure of the desired genes. This approach may in turn leads to investigate the regulatory mechanisms of the gene expression with respect to the existing biosynthetic pathways of natural products or to combine them with genes in other pathways, to synthesise novel compounds.

Single cell genomics is an another extension of molecular biology, which can be used as an promising tool to study the entire biochemical potential of single uncultured microbes from complex microbial communities, using a strategy known as multiple displacement amplification. Several explorations have been made based on the genomic and metagenomic approaches, including the identification of putative bryostatin polyketide synthase gene cluster from the marine bacterium *Candidatus Endobugulase ertula*, which encodes crucial constituents in biosynthesis of pharmaceutically potent bioactive compounds, known as bryostatins and characterisation of polyketide biosynthesis with respect to the corresponding genes, such as polyketide synthase (PKs) and nonribosomal peptide synthetase (NRPS), which is responsible for the production of antitumour molecules, from an uncultivated bacterial symbiont of the marine sponge, *Theonellaswinhoe*.

In addition, polyketides coding genes were identified in *Streptomyces* species, convincing the antitumour and antibiotic secretary possibility of marine bacteria as their secondary metabolites. Particularly, PKs and NRPS can be used in screening approaches to preselect new isolates with promising bioactivities for natural product analysis.

Genome sequence tags (GST) are identified genes, involved in natural product biosynthesis, which can be employed as probes to screen clonal libraries, in order to select putative clones for further screening for novel natural product gene clusters in marine organisms. Restriction fragment length polymorphism (RFLP) is a basic molecular biological technique, which is now applied in screening of marine microbial genomic libraries and especially in identification of microbes exhibiting bioactive potential, using PCR amplified fragments of 16S and 18S rRNA coding genes in marine microbial drug discovery programmes. Nevertheless, more efforts are in progress to identify and analyse a set of genes present in the genome of marine organisms that encodes a chain of enzymes, further synthesising bioactive compounds to disclose the hidden potential biosynthetic pathways of marine organisms.

Discovering novel antimicrobial peptides (AMP) can be originated from marine invertebrates. For instance penaedin, crustin and defensing, etc. have been facilitated by molecular iological approaches, especially through isolation of cDNA transcripts or express sequence tag (EST) clones which show homology with already characterised AMP sequences, furnishing ease and successful investigations as desirable features. For example, Ls-stylclin was able to be identified in shrimp *Litopenaeus stylirostris* using sequence information of its transcripts.

Using this strategy, around 20 peptide sequences were found to be potentially antimicrobial peptides, including iPAP-A and ci-MAM-A which exhibit broad spectrum of antimicrobial activity. Large scale extractions and supply of the complex natural chemicals from marine bioresources are environmentally unsound and economically unviable. In this regard, genetic engineering provides a better solution through production of recombinant purified compounds. Purity, safety and effective positive impact on human health care are some of the prevailing advantages of recombinant products against natural extracts, with respect to the pharmaceutical compounds obtained from marine biomasses. Intriguingly, a successful attempt was made by researchers from Australian Institute of Marine Science and London School of Pharmacy, on shotgun cloning and heterologous expression of the patellamide gene cluster, to produce antineoplastic metabolites, patellamides.

Metabolic engineering is a remarkable extension of genetic engineering, which is commonly experienced in terrestrial plants and micro-organisms which stands for modification of cellular metabolic pathways to enhance the metabolic composition. Complying with the basic objective, improvement of the metabolic composition can be accomplished through producing new compounds, enhancing the production of existing compounds and eliminating undesirable products of the biosynthetic pathways through introducing novel genes or total pathway and over expression of endogenous pathways. These modifications can be achieved by several possible genetic manipulations, such as upregulation of transcriptional factors, which in turn enhance the transcription of genes involve in biosynthetic pathway, down regulation of endogenous genes, using antisense RNA technology or RNA interference (RNAi) technique to eliminate the obstructive regulation of the synthesis of desired product and increasing the copy number of the gene or genes cluster of interest or driving the expression of the desired genes under strong promoters to over express the genes.

In most of the above approaches, transgenic technology is commonly used to introduce the exogenous gene into the desired organism. Among the marine organisms, especially marine micro-organisms, algae and sponge like invertebrates can also be considered as prospective candidates to be subjected to the genetic improvements, as described above, to improve them as sustainable resources of bioactive compounds. Eventually, the overall insights into the impact of molecular biology on the field of investigation and development of marine bioresources convince us a flourishing future, especially with respect to the drug discovery programmes.

Future Challenges and Prospective of Marine Biotechnology

Marine organism cultivation has been practiced long time ago, however, marine micro and macro algae (seaweeds) are much significant and mainly cultivated by two systems such as open culture and closed (photobioreactors) or artificial culture. Particularly, each one of the culturing systems has either desirable or undesirable, since the photosynthetic organisms and their mass production may directly correlate with the source of energy. Open culture systems can be established for seaweed or some of the microalgae farming and can be achieved by low cost, high capacity and free access solar energy for high photosynthetic efficiencies with long term durability.

However, impossibility to control temperature, invasion of parasitic algae or other algae strains may result in negative drawbacks. Despite these situations, only selected few microalgae strains can be used in controlled conditions for the cultivation in particular environments like, high alkaline environment for *Spirulina splatensis* and high salinity condition for *Dunaliella salina*. However, the production of β-carotene from *D. salina* is interested in closed culturing system over the open system. Therefore the closed culture system is preferable for culturing particular microalgae. The controlled optimum condition with the operational inputs such as salt, dissolved CO_2, water, nutrients, pH and O_2 provide a great opportunity for the steady environment without being contaminated in the closed cultures. However, a certain type of conditions may lead to the high cost of the closed system than the open system. In fact, these cell factories have gained many opportunistic advantages and still remain a competitive cultivation options for the production of marine algae. In addition, production of marine functional ingredients from marine algae sources is one of the leading industries that correlate with mass culture techniques and uses for the applications in food, pharmaceuticals and cosmetics. For example, marine pigments, mainly carotenoids including β-carotene, astaxanthine, lutein, zeaxanthin as well as phycocyanin and chlorophyll have been documented along the production and extraction parameters recently. In addition, there is a growing awareness on the single cell oils (SCO) due to the richness of PUFA in marine microalgae. Therefore, optimising photobioreactors for cultures and scaling up approaches have been shown a great deal against different microalgae while opening up a wide market place for them.

Over the many years, there is a new trend behind the cultivation of micro-organisms or micro-organism associated to invertebrate symbionts by marine scientists. Besides the knowledge of the both open seabased aquaculture and alternative growing methods are likely to be an important with requiring the detailed information about the biology and life cycle of the culturing organisms and the different symbiotic associations. It is not a surprised fact that the marine organisms and symbiotic associations are important for new functional materials. In fact, the way of applying biotechnology upon the marine eukaryotes and prokaryotes, along the cultivation is believed as a new research for feeding sources on desired organism and improving the growth rate as well as disease prevention. On the other hand, bioremediation is also an important discipline associated with the marine micro-organisms. It is basically degradation of toxic pollutants into nontoxic products through marine microbial community.

Marine pollutants including, heavy metals, oil spills, industrial wastes and other nuclear contaminants have become more problematic issues and exerts the bad effects always on marine enviornments. However, marine biotechnology and marine bioresources can be provided sustainable solutions along the bioremediation. This is an effective and efficient treatment for cleaning up hazardous materials than the conventional methods. Microbial communities and biological remediation will be not damage to the fragile marine ecosystems and described as natural phenomenal approaches. Hence, identification and isolation of specific marine microbes which could accumulate toxic contaminants for bioremediation would be the key successor. In fact, characterising the culture conditions for surviving in affected areas and for the habitat protection would be expected.

Despite the fact, the aquaculture industry has been developed fast enough to produce more than 16 million tons of fish and shellfish annually for the world food market. Moreover, in terms of the utilisation for human consumption, nearly 15.7% of world population accounted for fish proteins and fish production has been exponentially increased in recent past years. On the other hand, the total marine functional ingredients have been growing faster than the other food ingredients. Importantly, marine functional ingredients have attributed in the utilisation at pharmaceutical, nutraceutical, health care, food and feed markets. Therefore, a wide range of social, economical, institutional and biotechnological considerations are implemented for the development of fish, shellfish and marine microbes cultures. This has been further flourished due to the new vaccine therapies, hormone therapies for gaining resistance and transgenic techniques in advancing on the establishment of aquaculture.

However, this would spread more by introducing new species in extensive cultivation, raising the diversity of marine foods or biomedical products and also developing the sustainable practices through the physiological health benefits.

As a consequence, the rapid growth of marine bioresources and a significant expansion of marine biotechnological aspects are expected within the next few decades, In fact, marine genomics research may involve generating new tools, functional molecular markers, bioinformatics, and new knowledge about statistics as well as inheritance phenomena that could increase the efficiency and precision of marine biotechnological applications.

In addition, in order to meet the requirements of better understanding with the infrastructures of scientific communities, integrations are needed to fulfill through the collaborative researches. Collectively, capitalising the knowledge and experience of the marine biotechnologists and strengthening the awareness of interdisciplinary expertise can secure the sustainability of marine bioresources.

SECTION IV

Biotechnology of Dairy and Milk Products

Chapter 11

Milk and Dairy Products

INTRODUCTION

All dairy products start with receiving of raw milk from the farm. The raw milk generally is transported by way of tanker trucks and is typically already refrigerated to 7°C. When the raw milk is unloaded into the processing facility, it is sometimes also sent through a centrifuge to remove particulates, a process known as clarification and cooled to 4°C via a heat exchanger on its way to a refrigerated storage tank. Stored raw milk is kept at a 4°C prior to processing, usually by way of a jacketed storage tank and agitation.

The first step of all dairy processes is standardisation, the object of which is to ensure the proper fat content and Solids Non Fat (SNF) content for the desired finished product. Ensuring the proper fat content can be done one of two ways. Both processes use a centrifuge to separate the very low fat content and dense skim portion from the high fat content and less dense cream portion.

One process involves analysing the raw milk's fat content prior to processing and calculating the proportion of fat to remove during centrifugation. The other process involves completely separating raw milk as it is unloaded from the tanker truck and individually storing the two phases. These two streams are then recombined in the proportions required by the specific product as the first step of processing. The latter method is used primarily by larger operations with diverse products, giving them the flexibility to quickly switch the product being produced without having to retest the milk and recalculate the degree of separation. For some dairy products, such as yogurt and cheese, some water must be removed to achieve the solids nonfat content required for later process steps (such as fermentation) and desired finished product attributes. This water removal can be done via a single stage evaporator or ultrafiltration, both of which are described below. Manufacturers of these products may also add solids in the form of cream or powdered skim milk (the latter of which is quite energy intensive to produce).

Dairy milk is an opaque white liquid produced by the mammary glands of mammals (including monotremes). It provides the primary source of nutrition for newborn mammals before they are able to digest other types of food. The early lactation milk is known as colostrum, and carries the mothers antibodies to the baby. It can reduce the risk of many diseases in the baby. The exact components of raw milk varies by species, but it contains significant amounts of saturated fat, protein and calcium as well as vitamin C. Cow's milk has a pH ranging from 6.4 to 6.8, making it slightly acidic.

The specific gravity of milk ranges from 1.029 to 1.039. The specific gravity of milk decreases with the increasing fat content and increases with increasing amounts of proteins, sugar and salts. The freezing point of milk is almost a constant value at 0.53–0.55°C and is suitable indicator for detection of dairy cheese and yoghurt are +0.05 and –0.15 V respectively.

PHYSICAL AND CHEMICAL STRUCTURE OF MILK

Milk is an emulsion or colloid of butterfat globules within a water-based fluid. Each fat globule is surrounded by a membrane consisting of phospholipids and proteins; these emulsifiers keep the individual globules from joining together into noticeable grains of butterfat and also protect the globules from the fat-digesting activity of enzymes found in the fluid portion of the milk. In unhomogenised cow milk, the fat globules average about four micrometers across. The fat-soluble vitamins A, D, E and K are found within the milkfat portion of the milk.

The largest structures in the fluid portion of the milk are casein protein micelles: aggregates of several thousand protein molecules, bonded with the help of nanometer-scale particles of calcium phosphate. Each micelle is roughly spherical and about a tenth of a micrometer across. There are four different types of casein proteins, and collectively they make up around 80 per cent of the protein in milk, by weight. Most of the casein proteins are bound into the micelles. There are several competing theories regarding the precise structure of the micelles, but they share one important feature: the outermost layer consists of strands of one type of protein, kappa-casein, reaching out from the body of the micelle into the surrounding fluid. These kappa-casein molecules all have a negative electrical charge and therefore repel each other, keeping the micelles separated under normal conditions and in a stable colloidal suspension in the water-based surrounding fluid.

Both the fat globules and the smaller casein micelles, which are just large enough to deflect light, contribute to the opaque white colour of milk. Skimmed milk, however, appears slightly blue because casein micelles scatter the shorter wavelengths (blue compared to red).

The fat globules contain some yellow-orange carotene, enough in some breeds—Guernsey and Jersey cows, for instance to impart a golden or 'creamy' hue to a glass of milk. The riboflavin in the whey portion of milk has a greenish colour, which can sometimes be discerned in skim milk or whey products. Fat-free skim milk has only the casein micelles to scatter light, and they tend to scatter shorter wavelength blue light more than they do red, giving skim milk a bluish tint.

Milk contains dozens of other types of proteins besides the caseins. They are more water-soluble than the caseins and do not form larger structures. Because these proteins remain suspended in the whey left behind when the caseins coagulate into curds, they are collectively known as whey proteins. Whey proteins make up around twenty per cent of the protein in milk, by weight. Lactoglobulin is the most common whey protein by a large margin. The carbohydrate lactose gives milk its sweet taste and contributes about 40 per cent of whole cow milk's calories. Lactose is a composite of two simple sugars, glucose and galactose. In nature, lactose is found only in milk and a small number of plants. Other components found in raw cow milk are living white blood cells. Mammary-gland cells, various bacteria and a large number of active enzymes are some other components in milk.

NATURAL COMPONENTS IN MILK

Lactoferrin, lactoperoxidase and xanthine oxidase are naturally present in milk and have some specific properties that have been found to be beneficial for the shelf life and quality of dairy products. These compounds are nonimmune antimicrobial proteins that have been investigated by several researchers.

Lactoferrin

Lactoferrin is an iron-binding glycoprotein in milk that inhibits the growth of pathogenic bacteria by its high affinity with iron. The antimicrobial mechanism of lactoferrin is more complex than simple binding of iron. Lactoferrin also disrupts bacterial cell membranes by binding bacterial lipopolysaccharide and modifies membrane permeability by binding porin molecules in the outer membrane. This interaction with the cell membranes facilitates the bactericidal properties of lactoferrin. When the antimicrobial effects of lactoferrin on *Escherichia coli*, *Salmonella typhimurium*, *Shigella dysenteriae* and *Listera monocytogenes* were investigated, lactoferrin was shown to inhibit the growth of these microbes.

Lactoperoxidase

Lactoperoxidase is an antibacterial enzyme naturally present in colostrum and milk. This enzyme catalyses the oxidation of thiocyanate (SCN^-) in the presence of hydrogen peroxide (H_2O_2), producing hypothiocyanite ($OSCN^-$), which is a toxic intermediary oxidation product. This product inhibits bacterial metabolism by oxidation of essential sulphydryl groups in proteins. This reaction produces a severe change in the cytoplasmic membrane of spoilage bacteria. The use of lactoperoxidase is not approved in the United States because its activation requires a thiocyanide compound, which is unsafe for children. The activation of the lactoperoxidase system in refrigerated raw milk retarded the growth of psychrotrophic bacteria for several days. The lactoperoxidase system has been used to extend the shelf life of raw, pasteurised and ultra high temperature (UHT)-treated milk and to preserve cream, cottage cheese, mozzarella cheese and yogurt.

Xanthine Oxidase

Xanthine oxidase is a complex metallo-flavo enzyme present in the fat globule membrane. The enzyme catalyses the reaction to produce bactericide superoxide radicals and hydrogen peroxide in the presence of oxygen. Hydrogen peroxide can also be used to activate the lactoperoxidase system. Xanthine oxidase was studied for its activity in dairy products such as raw, evaporated and powdered milks; ice cream; yogurt; cheese; and butter and creams. Nielsen reported that xanthine oxidase, an oxido-reductase enzyme, catalyses the oxidation of purine bases and reduces nitrate to nitrite. Nitrate inhibits the germination of spore butyric acid bacteria in cheese.

MILK PROCESSING OPERATIONS

The steps in milk processing operations carried out in a dairy plant are: (i) blending, (ii) clarification and cream separation, (iii) heat treatment, primarily pasteurisation, (iv) homogenisation, and (v) bottling. The process steps are shown in the Fig. 11.1.

Blending of different batches of milk in cold condition (~5°C) is carried out to obtain a specified fat content. In the second step, the blended milk is clarified by a centrifugal clarifier to remove any sediments, body cells from the cow's udder and some-bacteria. A high speed centrifuge called bactofuge is used for almost complete removal of bacteria. Cream separation is often achieved simultaneously in this step. This operation is carried out at 40°C at 5000–6000 rpm.

The clarified milk is then given appropriate heat treatment depending on the requirement. Pasteurisation is carried out to destroy lipase activity and other milk enzymes and also to destroy pathogenic organisms in batch or harding method by heating the milk at 62–65°C for at least 30 minutes or by HTST method (high temperature short time method) at 71–74°C for 15 seconds or in a short time process at 85°C for 2 seconds. Pasteurised milk is not sterile and hence it is quickly cooled to prevent multiplication of

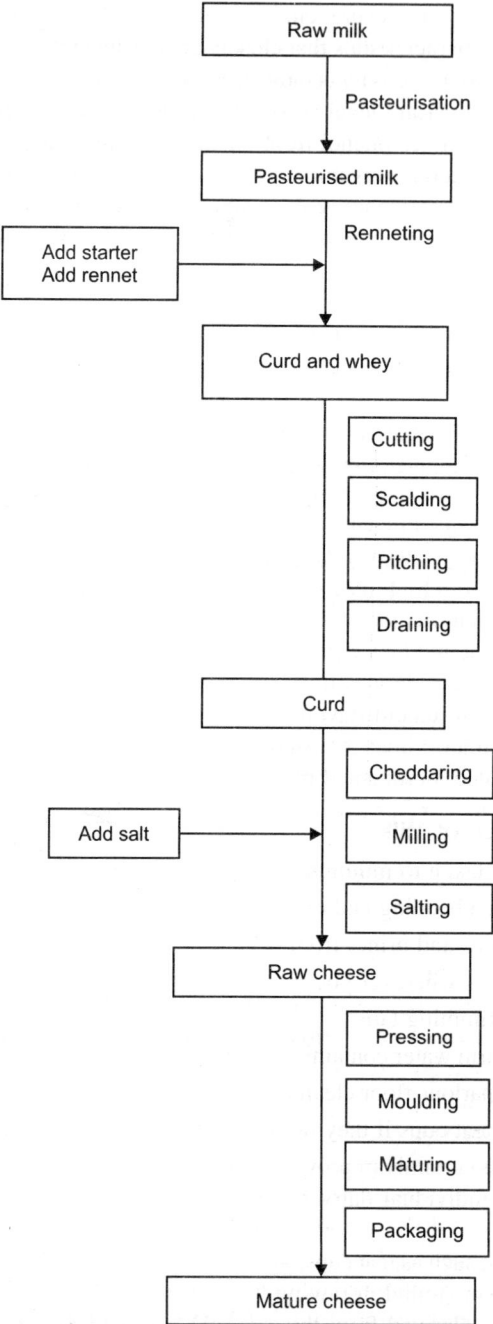

Fig. 11.1: Production of milk, curd and cheese.

souring bacteria. Raw milk contains several enzymes of which alkaline phosphatase is important. This enzyme has heat destruction characteristics that closely approximate the time-temperature exposures of proper pasteurisation and hence its activity is indicative of the effectiveness of pasteurisation. If alkaline phosphatase activity beyond a certain level is found in pasteurised milk, it is indicative of inadequate pasteurisation. The enzyme liberates phenol from phenol phosphoric acid compounds and free phenol gives a deep blue colour with certain organic compounds, which forms the basis for phosphatase test. Ultra high temperature (UHT) treatment involves indirect heating by coils or plates at 135–140°C for 6–10 seconds or by direct heating by live stream injection at 140–150°C for 2–4 seconds followed by aseptic packaging. Milk may also be sterilised in retail packages by heating in autoclaves at 110–120°C for 15–20 minutes.

Homogenisation may be carried out before or after pasteurisation. This process makes a stable emulsion of milk fat and milk serum by mechanical treatment. Homogenisation is achieved by passing milk or cream through a small aperture under high pressure and velocity. The fat globules of milk with varying sizes in the range of 0.1 to 20 μm in diameter have a tendency to gather into clumps and rise due to their lower density than milk, thereby separating into a cream layer. Homogenisation breaks the fat globules to uniform size of about 2 μm or less which are covered by an adsorbed layer of plasma proteins including casein micelles. This stabilises the milk emulsion. Homogenised milk has creamy structure, bland flavour and a whiter appearance. It has greater whitening power when added to coffee and tea compared to skim milk. A soft curd is formed from homogenised milk, which is more easily digested than curd obtained from unhomogenised milk. In the manufacture of evaporated milk and ice cream, homogenisation reduces the chance of separation of fat and hence provides smoother texture.

Homogenisation, however, accelerates lipase activity, particularly if homogenisation is carried out before pasteurisation leading to rancid flavour. High pressure homogenisers, low pressure rotary type homogenisers and sonic vibrators may be used for homogenisation of milk. The homogenised and pasteurised milk is then cooled, bottled and marketed.

Minimising Contamination of Milk

Following measures can be taken to minimise milk contamination:

1. Providing enough clean bedding and replacing it as necessary.
2. Removing slurry (faeces and urine) from concrete areas at least twice daily.
3. Preventing muddy areas wherever possible.
4. Shaving udders and trimming tails.
5. Washing teats with warm water containing disinfectant and drying individually with paper towels.
6. Keeping the milking parlour floor clean during milking.
7. Thoroughly cleaning teat cups if they fall off during milking and discarding foremilk.

Although such procedures certainly improve the microbiological quality of milk, economic constraints such as increasing size of individual dairy herds and decreased manning levels in milking parlours encouraged their neglect.

Milk-handling equipment such as teat cups, pipework, milk holders and storage tanks, is the principal source of the micro-organisms found in raw milk. As the overall quality of the milk decreases so the proportion of the microflora derived from this source increases. Milk is a nutritious medium and, if equipment is poorly cleaned, milk residues on surfaces that are frequently left wet will act as a focus for microbial growth which can contaminate subsequent batches of milk. Occasional neglect of cleaning and sanitising procedures is usually less serious since, although it may contribute large numbers of micro-organisms to the product, these tend to be fast growing bacteria that are heat sensitive and will be killed by pasteurisation. They are also sensitive to sanitising practices used and will be eliminated once effective cleaning is resumed. If cleaning is persistently neglected though, the hydrophobic, mineral-rich deposit known as milks tone can build up on surfaces, particularly heated ones. This will protect organisms from sanitisers and allow slower growing organisms to develop such as micrococci and enterococci. Many of these are thermoduric and may not be removed by pasteurisation.

In most developed countries milk is chilled almost immediately after it issues from the cow and is held at a low temperature thereafter. It is stored in refrigerated holding tanks before being transported by refrigerated lorry to the dairy where it is kept in chill storage tanks until use. Throughout this time, its temperature remains below 7°C and the only organisms capable of growing will be psychrotrophs. There are many psychrotrophic species, but those most commonly found in raw milk include gram-positive rods of the genera *Pseudomonas, Acinetobacter, Alcaligenes, Flavobacterium*, psychrotrophic coliforms, predominantly, *Aerobacter* spp., and gram-positive *Bacillus* spp.

Types of Milk and Milk Products

A variety of milk and milk products are produced in the dairy depending on consumer requirements. These include the following categories of products.

Vitamin D milk

Vitamin D content in cow's milk depends on the cattle feed and exposure of the cow to sunlight. The diet of most children is deficient in Vitamin D and hence Vitamin D enriched milk is marketed. Vitamin D activity in milk is enhanced by irradiating milk with UV light which converts the milk sterol, 7-dehydrocholesterol into Vitamin D_3. Alternatively, Vitamin D concentrates may be added to milk to bring the potency to about 400 units per quart.

Low sodium milk

Low sodium milk is required by people with high blood pressure or edema. It is prepared by passing milk through an ion-exchange resin that replaces sodium with potassium. Low sodium milk contains 3–10 mg of sodium per 100 ml compared to about 50 mg in ordinary milk.

Concentrated milk

Evaporated milk is whole milk from which most of the water has been evaporated. Raw whole milk is clarified and concentrated in the vacuum pans at a temperature of 74–77°C. It is then fortified with the Vitamin D, homogenised, and filled into can and sterilised at 118°C for 15 minutes and cooled. This heat treatment gives evaporated milk a light brown colour due to sugar-protein interaction and also its characteristic flavour. Khoa is a semi solid obtained by evaporating milk in open pans and is used in preparing Indian sweetmeats. Malai is made by simmering milk to a thick layer of milk fat and coagulated proteins consumed with or without sugar.

Sweetened condensed milk

This is made from pasteurised milk by concentration and sucrose is added to sweeten the concentrate to the extent of 65 per cent in the final product. Sweetened condensed milk is not sterile but microbial growth is prevented by the added sugar. Kheer is concentrated milk, obtained by evaporating milk to nearly half its original volume, resembling sweetened condensed milk.

Dry milk

Whole milk is dehydrated to the extent of 97 per cent by spray drying or vacuum drying. Skim milk and low-fat milk are also dehydrated to manufacture milk powder. Vitamins A and D are added to enhance the nutritive value. The low-cost dry milk has a long shelf life and can be reconstituted to fluid milk by mixing with the required amount of water. Dried milk is stored in dry air tight containers.

Skim milk

Whole milk from which fat has been removed by centrifugation is called skim milk. It contains all other constituents of milk except fat and fat-soluble Vitamins A and D. These vitamins can be added to skim milk. By varying the amount of fat removed from whole milk, low fat milk (containing 0.5–2 per cent fat) is prepared. Condensed skim milk finds extensive use in baking and confectionery industries.

Cream

Concentrated milk fat is called cream. Cream is formed as a layer of fat globules, which rise to the top in unhomogenised milk. Cream is separated from milk by centrifugation. Table cream (coffee cream) contains about 18 per cent fat while whipping cream contains about 35 per cent fat. Cream, used for butter making, contains about 40 per cent fat. Sour cream, which is extensively used in bakery items as salad dressings, is prepared by pasteurising cream with about 18 per cent fat at 75°C for 30 minutes to kill all bacteria and then inoculating with a controlled culture of lactic acid bacteria to develop the desired acid tang in the finished product.

Yoghurt

Yoghurt (or yogurt) or dahi in India is a fermented milk product with a fine curdled gel like consistency having a sour and aromatic flavour. It is obtained by fermenting homogenised pasteurised milk with about 3 per cent thermophylic lactic acid bacteria (mixed cultures of Streptococcus thermophilus or Lactobacillus bulgaricus) at 42–45°C for about 3 hr. During fermentation, the curdled gel-like consistency and acid flavour due to formation of carbonyl compounds such as diacetyl and acetaldehyde develop. Sugar and fruit pastes may be added to the final product to achieve distinct fruity flavours.

Sour milk

Sour milk or buttermilk is fermented fluid milk either by spontaneous souring (by *Streptococcus lactis* or *cremoris*) or by fermentation with aroma forming bacteria (*Streptococcus diacetylactis* or *Betacoccus citrovorus*) or by fermentation with pure bacterial cultures to give cultured buttermilk. During fermentation lactose is converted into lactic acid which coagulates casein at pH 4.5–5 to give the final sour tasting curdled buttermilk.

Kefir and kumiss

Kefir and kumiss are sparkling, carbonated alcoholic beverages derived from milk. The microflora such as *Torula* yeast (for alcoholic fermentation) and *Streptococcus lactis* and *Lactobacillus caucasicus* (for lactic acid fermentation) which form clotted milk particles are added to fluid milk to give kefir. Kumiss is made from goat's or mare's milk by fermentation with *Thermobacterium bulgaricus* and yeast Candida.

Taette milk

Milk is fermented with *Streptococcus lactis* var. *hollandicus* to yield a sour viscous thread-like product due to the symbiotic growth of the lactic acid bacteria and yeast. Taette is popular in Sweden, Norway and Finland.

PASTEURISATION OF MILK

Pasteurisation is usually, but not necessarily, the next step in most dairy processes. This process can also be substituted with sterilisation or some other kind of heat treatment, depending on the specific product. The object of pasteurisation is to inactivate pathogenic micro-organisms (in the interest of human health) and a majority of microbes that can spoil the milk. Pasteurisation does not kill spores or some thermophilic micro-organisms, which leads to the relatively short shelf life and refrigeration requirement of most pasteurised dairy products. Sterilisation treats the milk with a more extreme heat treatment to inactivate all micro-organisms and spores, leading to a longer shelf life without refrigeration required. Though sterilised milk requires more heat up front for the sterilisation, it can save significant energy on refrigeration through the rest of the milk's life. Because it requires more intense heating, though, sterilisation significantly changes the flavour of the milk. This type of milk represents a small market share in the US and England, but a large market share in Germany and France. Finally, some dairy products, such as yogurt, require a more intense heat treatment than required for pasteurisation. In addition to pasteurising the milk, the heat treatment functions to alter certain milk properties, such as inactivating specific enzymes, in order to obtain the desired finished product attributes.

Pasteurisation, sterilisation and other heat treatments are occasionally done via a batch process, where a tank of the milk is heated to a specific temperature and held for a specific length of time. However, by far the most common method used is a continuous process. In a continuous process, a gear pump or a flow regulator is used to deliver a constant and accurate flow rate to the pasteurisation process.

The stream is passed through a heat exchanger, which heats the milk to the desired temperature. It is then pumped through a specific length of piping to hold it at this temperature for a specified period of time and then it is cooled back down. Most dairy processors use a process called regeneration to cut down on energy costs. Regeneration cools the outlet stream by using it to heat the incoming stream, recovering approximately 85–90% of the thermal energy. A small amount of steam is used to finish heating the inlet stream and a small amount of cooling is used to finish cooling the outlet stream.

Almost all dairy products are subjected to homogenisation at some point during processing. Milk is composed of a water soluble component and a fat soluble component that will separate if not homogenised, resulting in the phenomenon of 'creaming.' The purpose of homogenisation is to break up the fat globules into smaller sizes and disperse them in the water soluble component, which prevents them from coalescing and forming the separate layer. This is done using a three or five piston pump to create a large pressure drop across a small opening that the milk stream is forced through. Often a two-stage homogeniser is used, where the first stage is primarily used for cavitation to break up the fat globules and the second stage creates turbulence to break up aggregates and disperse the small fat globules.

In several dairy products, a cooking or fermentation step is required to allow the desired biological/chemical changes to occur to the product. Yogurt, sour cream, cottage cheese and cheese are the primary products that require such a step. The standardised milk or cream is filled into a jacketed tank, along with any enzymes, microbial cultures and/or other ingredients required. The mixture is then heated to a specified temperature for a specific period of time, usually several hours, to allow the enzymes and micro-organisms to perform their biological transformations into cheese curds, yogurt, or sour cream.

For yogurt production, an alternative process is sometimes used to create 'set yogurt.' In this process, packaging occurs immediately after all the ingredients are mixed together. The sealed containers then undergo fermentation in controlled temperature air-blast tunnels. Cheese generally needs mechanical work performed on it to achieve its final state. During the cooking step, solid cheese curds are formed, leaving liquid whey as the by-product. After the cooking step, the liquid whey is drained, leaving the cheese curds. The curds are then washed and pressed together. Though there are several ways to press the curds together into blocks, most modern industrial processes use pneumatic conveyance to transport the dried curds to the top of a tall tower, where the weight compresses the curds at the bottom of the tower into a single block, to be cut with a mechanical cutter. Cheeses with a 'stringy' consistency, such as mozzarella, utilise mechanical dough kneaders to stretch the cheese in hot water.

To create butter, the cream phase of milk must be churned. Cream is an oil-in-water emulsion, meaning that small drops of fat soluble components are surrounded by and dispersed in the water soluble component. Churning inverts the emulsion into a water-in-oil emulsion, which means small drops of water soluble component are now surrounded by and dispersed in the fat soluble component. In larger operations, churning is done via a large, mechanical beater. When the water-in-oil emulsion is made, a water phase, known as buttermilk, is released and drained. The curds of butter are then sometimes (but not always) washed with water and allowed to drain. Curds are then pressed together to form a single mass. The single mass is then extruded and cut into the proper shape and size.

Freezing is a process used in the production of ice cream and other frozen desserts. Batch freezers are commonly used by smaller ice cream manufacturers, where the ice cream mixture is cooled in an agitated, jacketed tank until the mixture is partially frozen. Continuous freezers, usually used in large operations, utilise heat exchangers to achieve the partial freezing. The second step of ice cream freezing is a process called hardening. The partially frozen ice cream is packaged and subjected to blast air freezing at temperatures of -20 to $-30°F$ to fully freeze the ice cream.

Evaporation is the process most commonly used to remove water from dairy products. While used to some degree in other products, evaporation is mostly used in the production of evaporated milk, condensed milk, milk powder and whey powder. While there are several types of evaporators, the most common type of evaporator is a falling film evaporator, due to its higher efficiency, lower operating temperature, easy maintenance and ability to have more 'effects' than other types, which is described below.

In falling film evaporation, the liquid falls by gravity down the inside surfaces of tubes arranged in a shell-and-tube heat exchanger configuration, with steam generally used as the heating medium. Evaporators are commonly operated under vacuum to lower the required operating temperature. A common approach to energy efficiency for evaporators is to use the hot vapour that boils out of the liquid in one evaporator as the heating medium in another effect, which is operated at a lower pressure. The liquid moves from one effect to the next, becoming more and more concentrated in each effect. This approach is called 'multi-effect' evaporation. In practice, up to 5 effects can be used in dairy evaporation (up to 7 are possible if using falling film evaporators).

An additional way to improve energy efficiency in multi-effect evaporators is to include a vapour recompression system. The thermal energy of the outlet vapour is increased by injected steam (thermal vapour recompression) or with a turbo compressor (mechanical vapour recompression) before it is used as the heating medium for the next effect. Since evaporation can only remove a certain amount of water, drying is used to create dry milk powder and whey powder. The most common method for drying is spray drying. Spray drying involves atomising the liquid and spraying it into a tower or chamber of flowing hot air that removes the moisture, leaving a dry powder at the outlet. Recently, multistage dryers have appeared, in which the powder is mostly dried in the spray dryer and then transferred to a fluidised bed chamber to finish the drying. The fluidised bed step requires less air to finish the drying compared to the spray dryer, making the drying process more efficient. A less common technique is roller drying. In roller drying, the liquid flows over a rotating, heated drum, typically under vacuum and the dried powder is scraped off of the drum. Although roller dryers tend to be smaller and less costly to operate, the quality of the powder can be affected. Drying processes are quite energy intensive and measures taken to increase the energy efficiency of such processes can have large energy savings benefits.

Membrane concentration can be applied to remove water from milk in lieu of or as a precursor to traditional evaporation methods. In membrane concentration, water can be separated from milk solids or cream using pressure as a driving force across a semi-permeable membrane. Because membrane concentration does not require a phase change (in contrast to traditional evaporation methods), it can offer a more energy-efficient option of dairy product concentration. Common types of membrane concentration used in the dairy industry are microfiltration, ultrafiltration and reverse osmosis, each indicating a different range of pore sizes. The type of filtration recommended for use is highly dependent on the application.

Cold storage is used for most dairy products after production. Refrigerated storage for fluid milk, yogurt, cheese, butter and other products is used to delay the growth of micro-organisms that can spoil the final products. It is used to also prevent undesirable physical and chemical changes to the products, such as drying, oxidation, or melting/deformation of butter and soft cheeses. Frozen storage is used to keep ice cream and other frozen desserts in their desired frozen state, since thawing and refreezing will change the product attributes. Sterilised, concentrated and dry dairy products generally do not need or use cold storage after production.

KEY PRODUCTS IN DAIRY INDUSTRY

This section presents representative process flow diagrams for several key products in the dairy processing industry. While not inclusive of all processes employed and all products manufactured in the dairy industry. Figures 11.2 to 11.3 depict process flow diagrams for the major products manufactured in the dairy processing industry: (i) fluid milk, (ii) yogurt, (iii) butter, (iv) cheese, (v) dry, sweetened condensed and evaporated milk, (vi) dry whey powder and (vii) ice cream.

Fluid milk: Figure 11.2 illustrates the typical process for producing fluid milk. Upon entering the facility, raw milk is (sometimes) clarified and then cooled a few degrees prior to being transferred to cooled storage tanks. To produce pasteurised milk, the most common type of milk in the US, the milk is standardised and pasteurised, with the homogenisation step usually occurring prior to the milk being cooled back down. The cooled milk is then packaged and kept in refrigerated storage until shipment. Ultra high temperature milk (UHT), which is quite common in Europe, replaces the pasteurisation step with a sterilisation step. The milk is then aseptically packed into sterilised packaging. Due to its long shelf life, UHT milk is commonly stored at room temperature.

Yogurt production: Figure 11.3 shows the typical process for yogurt production. As with other milk products, the raw milk is (sometimes) clarified and then cooled prior to processing. Different types of yogurt require a wide range of compositions, which are achieved by standardisation.

In addition, some yogurts may require a solids nonfat level higher than normal milk contains. For these situations, the milk may be concentrated by water removal via ultrafiltration or evaporation. After standardisation, the milk is homogenised at an elevated temperature (typically around 70°C) and then subjected to a heat treatment process that also serves as pasteurisation. The more intense heat treatment denatures and breaks down certain milk proteins to encourage fermentation. The milk is then cooled to

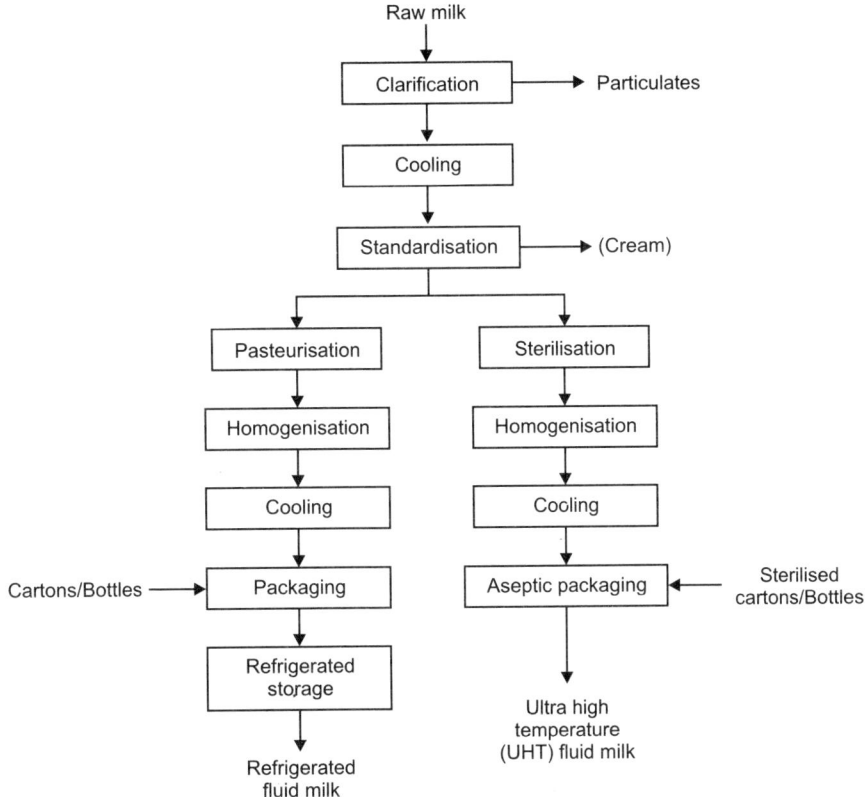

Fig. 11.2: Process diagram for fluid milk production.

fermentation temperature (typically around 42°C) and inoculated with the proper bacterial cultures. 'Tank fermented' yogurt is then held at the fermentation temperature and allowed to ferment in the tank before being cooled and packaged. 'In cup fermented' yogurt is first packaged and then placed in blast-air tunnels which hold at fermentation temperature before cooling. Both types are then held in refrigerated storage until shipment.

Butter manufacturing: A typical butter manufacturing process is depicted in Fig. 11.4. Cream is harvested from the standardisation process of other milk products. After the cream is pasteurised, it is held at high temperature to 'ripen.' This allows for proper crystallisation of the milkfat, which makes churning easier and reduces fat lost in buttermilk. The cream is then churned to air break open the milkfat globule membrane and create the primarily fat phase that becomes butter, releasing the aqueous phase as buttermilk. The butter curds are then washed to remove excess buttermilk. An optional salting step is performed and the butter curds are pressed and extruded into the proper sizes prior to packaging and storage. There are an astonishing number of cheese varieties. The difference between the varieties is dependent upon several factors, including the proportions of milkfat, solids nonfat and water, strains of bacteria used and the parameters of each processing step. Hence, Fig. 11.5 shows a generic process for cheese production. Specific cheese types may have specified resting/ageing times, or repeated steps not depicted in this process flow diagram.

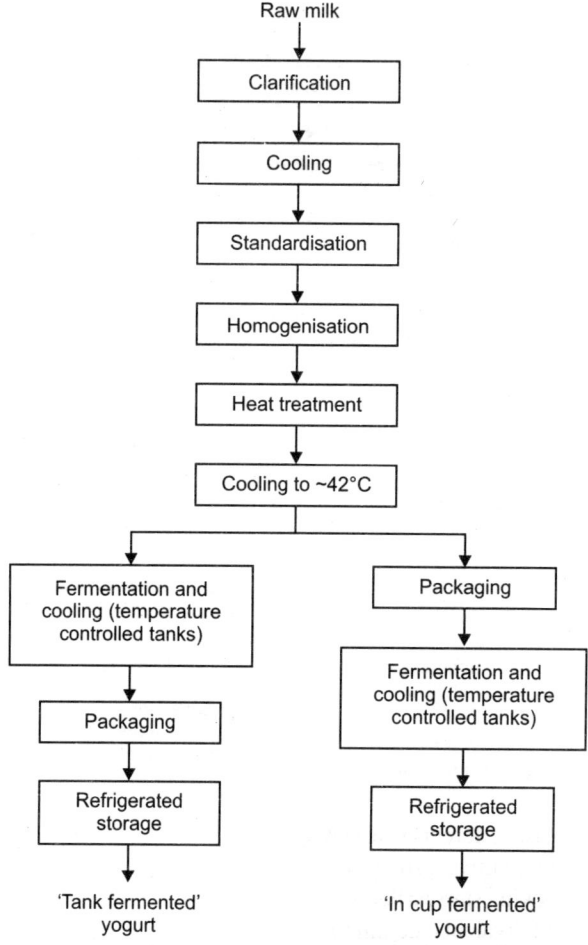

Fig. 11.3: Process diagram for yogurt production.

Cheese manufacturing: Similar to yogurt, raw milk is standardised via centrifuge to achieve the specified level of milkfat. In some cheeses, a higher solids nonfat content is required and ultrafiltration is commonly used as part of the standardisation process to remove water. Other concentrated dairy ingredients may be added such as skim milk powder or cream. Once the desired composition is attained, the milk is pasteurised and filled into a cheese vat. In the cheese vat, rennet, enzymes and/or bacterial cultures are added, depending on the type of cheese to be made. In the United States, cheeses such as cheddar, colby and gouda may have natural colour added. The mixture is then cooked to facilitate the biological processes that create cheese curds. Additional cooking is commonly used to 'age' the cheese to the desired taste and attributes. After the cooking step, the curds are drained from the liquid whey by-product of the cheese making process. After draining, the curds are pressed together to give solid blocks. Some cheeses, such as mozzarella, are also stretched to give a 'stringy' texture. The cheese is then packaged, aged and stored.

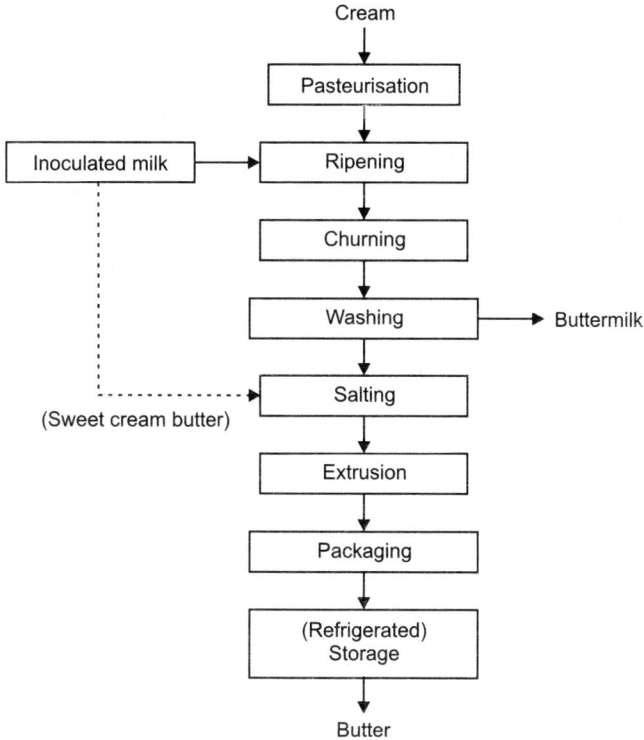

Fig. 11.4: Process diagram for butter production.

Commonly, cheese with visual defects is diverted to be used in the manufacture of processed cheese. The cheese is combined with additives, such as melting/emulsifying salts, into a single vat and cooked to a greater extent than the cooking step earlier in the process. This process promotes protein denaturing—predominantly through loss of divalent calcium linked bridges and other changes—to achieve the desired attributes. During the cooking step, the product is mixed to ensure homogeneity. After cooking, the cheese is cooled, cut into the appropriate size and shape and finally, packaged.

Figure 11.6 illustrates the process flow diagram for dried, condensed and evaporated milk. Evaporated milk is milk that is concentrated and sterilised for a long shelf life. After clarification and cooling upon entering the facility for storage, the milk is standardised and pre-heated to prepare for evaporation. Water is then evaporated off using the evaporation process described earlier and the concentrated milk is homogenised. It is then either packaged into cans and sterilised 'in-package,' or sterilised via heat exchanger, then aseptically filled into sterilised cans.

Sweetened condensed milk is similar, except that sugar is added to the concentrated milk, which · lowers the water activity of the milk to the point that micro-organisms cannot grow, giving it its long shelf life. The milk is first subjected to a heat treatment to kill certain types of micro-organisms and then evaporated similar to evaporated milk. Sugar is then added and the product is subjected to a specific cooling process designed to achieve desired lactose crystallisation. Lactose crystals that are too large are then centrifuged out and the remaining product is packaged.

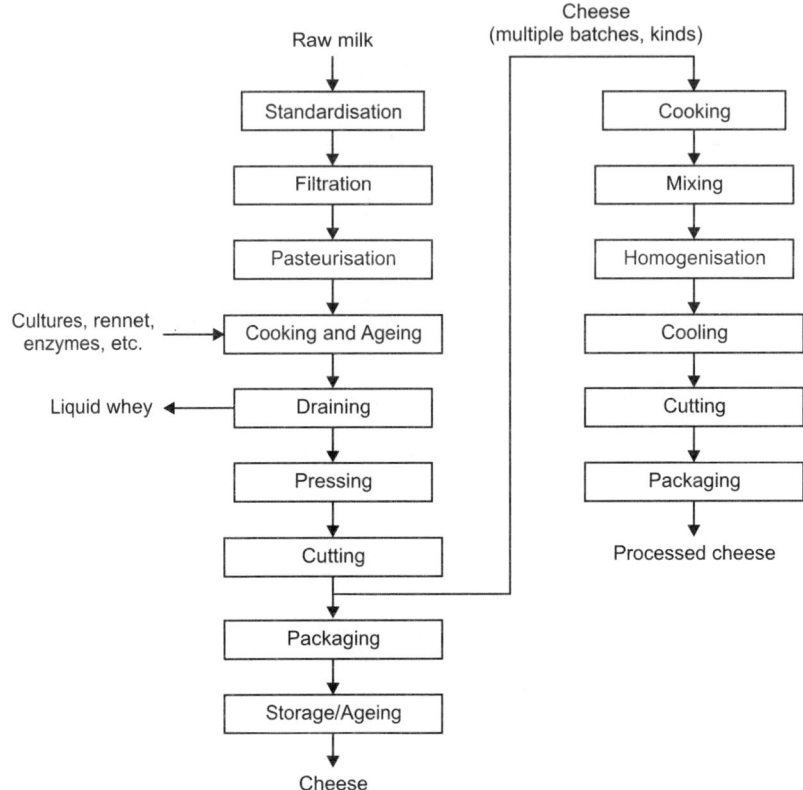

Fig. 11.5: Process diagram for generic cheese production.

Dry milk powder is also subjected to heat treatment to kill micro-organisms. Evaporation, while not required, is often utilised prior to drying, as it is a much less energy intensive way to remove most of the water than by spray drying alone. The product is then homogenised and dried via spray, roller or vacuum drying. The resulting powder is sifted for desired particle size and packaged. A by-product of the cheese making process is liquid whey. Due to its high biochemical oxygen content (BOC), whey is expensive to dispose of in liquid form. Because liquid whey is also perishable, the most common method of handling whey is to dry it into a powder. In powder form, whey can be used for various nutritional supplement applications due to its high protein content. This process, however, requires both high capital investment and high operational energy costs. Figure 11.7 shows the typical processes used to dry liquid whey.

Prior to water removal processes for whey, whey typically will contain small bits of cheese (termed fines) which are removed from the whey typically with screens. Additionally, there is free milkfat in the whey which is removed with the use of a continuous separator. The less dense portion from the whey separator is called whey cream and the more dense portion is termed clarified whey. Both screening and separating require non-trivial amounts of energy.

Originating from the cheese making process, liquid whey is often first subjected to reverse osmosis filtration, microfiltration, ultrafiltration, or some other non-thermal process to remove some of the

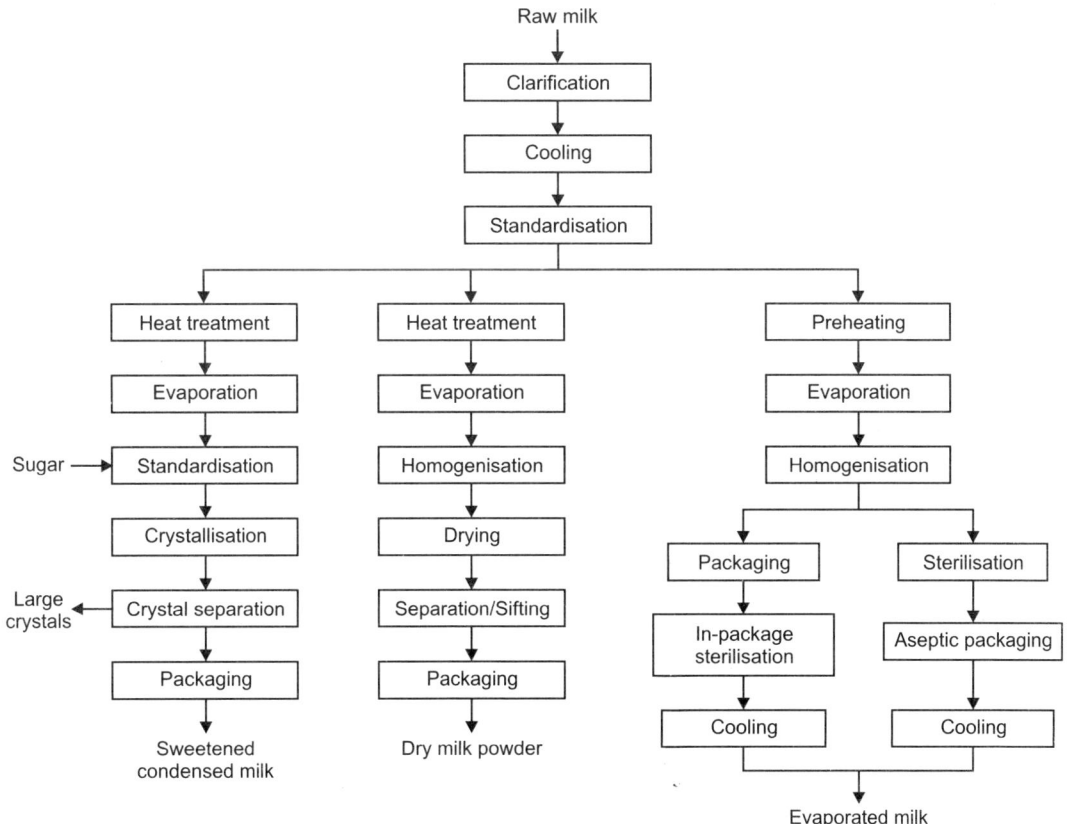

Fig. 11.6: Process diagram for concentrated and dried milk production.

water content (which are both less expensive and more energy efficient). After filtration, the liquid is pasteurised and subjected to evaporation, similar to other dried milk products. At this point, multiple methods can be used to create the final product. The most basic and most energy intensive, involves spray drying the effluent from the evaporation step to the desired moisture content.

An added lactose crystallisation step allows more water to be evaporated prior to spray drying. This step, while decreasing energy use, requires additional process time to perform a cooling/crystallisation step. To reduce operating costs even more, a fluidised bed drying step can be added. This allows the powder to leave the spray drying step at a higher moisture content and finish drying in a less intensive fluidised bed dryer.

Finally, an alternate way to exploit whey crystallisation for increased water removal is to allow the whey to only partially crystallise prior to spray drying. The whey can exit the spray drier at a much higher moisture content and finish crystallisation on a conveyor. The remaining moisture is then dried off in a fluidised bed dryer.

The steps involving a crystallisation step and a fluidised bed step require more equipment and processing time, but use significantly less energy. The leftmost process will use approximately 30% more steam per kilogram of whey than the two processes on the right.

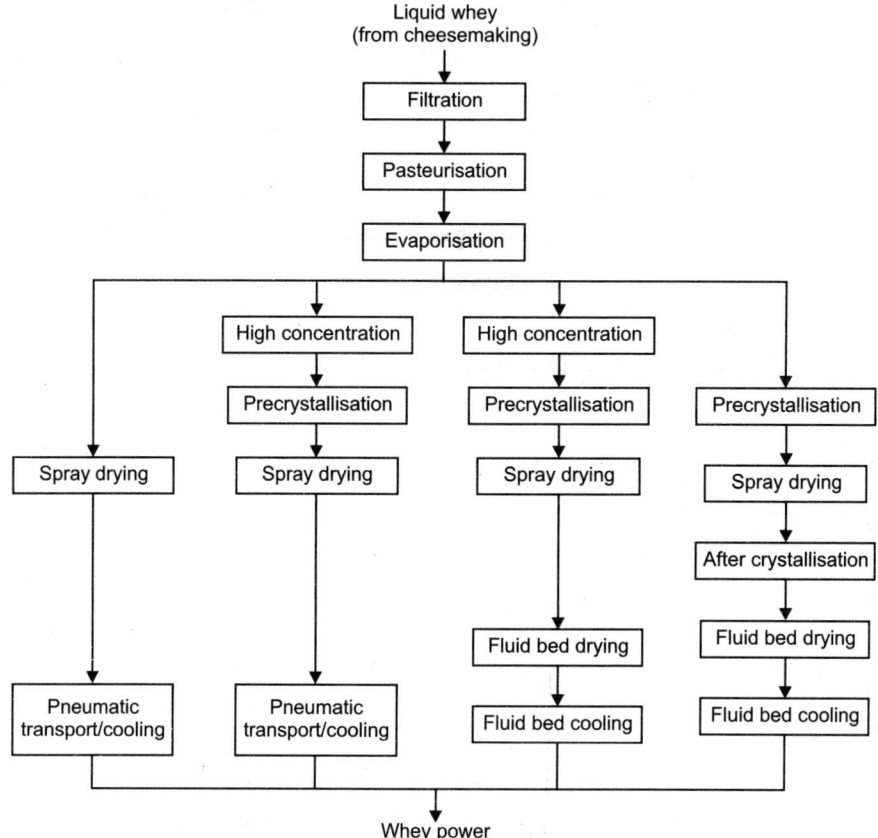

Fig. 11.7: Process diagram for powdered whey production.

Chapter 12

Genetically Modified Cheese

INTRODUCTION

Cheese is a food derived from milk that is produced in a wide range of flavors, textures, and forms by coagulation of the milk protein casein. It comprises proteins and fat from milk, usually the milk of cows, buffalo, goats, or sheep. During production, the milk is usually acidified, and adding the enzyme rennet causes coagulation. The solids are separated and pressed into final form. Some cheeses have molds on the rind or throughout. Most cheeses melt at cooking temperature. Considerable developments have been made to shorten the long ripening period, increase the cheese yield, reduce the bitterness in cheese, produce desired flavour level within short maturation period, etc. The unprecedented accomplishments of biotechnology and genetic engineering since the past decade have added a new margin in designing transgenic cheese.

The cheese technology had been modified time to time in order to surmount the processing challenges and to gift the mankind with novel type of cheese with improved physico-chemical and functional properties. With the triumphal achievements in genetic engineering in the field of food, agriculture and health sector, recent developments in cheese technology have emerged with a new margin as the unprecedented consequence of genetic games alike the conventional and gradual amelioration in cheese making technique as usually expected with the passing generation by the virtue of evolution in order to satisfy the more and more requirements. The controversy concerning the biosafety of Genetically Modified (GM) food, development of Genetically Modified Cheese (GME) as the outcome of good cheese-biotech sorority seems to curtail or minimise a number of processing challenges for cheese manufacture.

GENETIC MODIFICATION

Genes are organised into chromosomes which are found in all living cells. They are a coded form of instructions to make proteins. Most of the proteins manufactured by living cells are enzymes. Enzymes regulate the functional activities. Genetic information mainly is stored in a coded form in deoxyribonucleic acid (DNA) molecules present in genes. So, DNA is a kind of molecular blue print where it stores all the information needed to make a new cell. Therefore, the phrase, 'genetically modified' is commonly used to describe the application of recombinant deoxyribonucleic acid (rDNA) technology for the genetic alteration of micro-organisms, plants and animals. This advanced molecular technology allows for effective and efficient transfer or alter genetic material from one organ to another. An identified single gene responsible for a particular trait can be inserted/transferred among all living organisms. The technology is now well developed and can be applied to any living organism.

GENETICALLY MODIFIED CHEESES

Cheese obtained or manufactured by the application of genetic engineering by adopting the following three approaches can be called as genetically modified cheese. These three approaches may be adopted individually and or in combination to obtain cheese with approved vegetarian sentiment, improved functional attributes, increased nutritional/therapeutic value, accelerated ripening activities and enhanced sensory quality. These approaches are:

- Modification in milk composition.
- Addition of Recombinant coagulating enzymes.
- Application of Modified starter culture.

Among the three approaches, considerable amount of work have been carried out on application of genetically modified micro-organism to obtain improved starter culture and recombinant coagulating enzymes for the cheese production. Work on alteration of milk composition by Genetic manipulation in animals results in the improvement of milk production system which can even produce dairy products with functional characteristics. In this context, main emphasis for genetically modified cheese is now on the processing aspects i.e. on starter culture and coagulating enzymes.

Modification in Milk Composition

It is established that genetic variation in animal causes the different composition in milk. Transgenic cows with altered genetic make up to produce milk with 2 per cent fat with a greater proportion of unsaturated fatty acids in milk fat, higher levels of milk protein, β- and κ-casein and reduced lactose content in milk have been developed. Such milk is suited ideally for people suffering from lactose intolerance and also to produce improved varieties of specialty cheeses with effective cost price efficiency. Genetically modified bovine somatotrophin (BST) also play a role in the regulation of milk yield, growth rate and protein to fat ratio of milk which results in milk composition alteration.

Recombinant Coagulating Enzymes

One of the success stories for the application of genetic engineering is the manufacture of recombinant chymosin and its use as milk coagulating enzyme for commercial cheese production. Concern over the supply of chymosin from traditional source (suckling calves) has led to efforts over the past three decades to develop a recombinant source. Cheese industry has been the major beneficiary of this technological work.

The gene coding for the chymosin enzyme has been cloned in the bacteria *Escherichia coli,* the yeast *Kluyveromyces lactis* and the mold *Aspergilus niger.* The enzymatic properties of the recombinant enzymes are indistinguishable from those of calf chymosin. The cheese making properties of recombinant chymosin produces very satisfactory results and its use in commercial plant have been approved by many countries.

All the three chymosin available are identical to calf chymsoin, considered as vegetarian source and accepted by religion group. More knowledge on genetic engineering combined with better understanding of protein structure might one day give us chymosin with higher activities, lower cost and with flavour enhancing properties of cheese during ripening.

Modified Starter Culture

The application of gene cloning technology to lactic acid bacteria is the potential process in the generation of enhanced starter cultures for the manufacture of cheese and yoghurt. These starter cultures are mainly made of species of *Lactococcus, Lactobacillus* and *Streptococcus.*

Modifications of these micro-organisms were achieved mainly on three directions for cheese making process. These are:

1. Development of phage resistant cheese culture.
2. Organisms with probiotic activity for cheeses.
3. Acceleration of cheese ripening.

Phage resistant cheese culture: The major cause of slow acid production in cheese plants today is bacteriophage (phage). This can significantly upset manufacturing schedules and, in extreme cases, result in complete failure of acid production or 'dead vats'. Phages are viruses that can multiply only within a bacterial cell. They have a head, which contain the DNA and a tail, which is compared of protein.

Morphologically, there are three types of phage for *latococci* as mentioned below:

1. Small isometric headed (spherical headed) phage most common.
2. Prolate—headed (oblong—headed) phage.
3. Large isometric—headed phage.

Phage multiplication occurs in one or two ways, called the lytic and lysogenic cycle. Multiplication of phage is very fast in their hosts. Microbiologists have enhanced cheese culture performance by genetically engineering bacteria increasing the viability of the culture during cheese making. The new strain of bacteria resist phase contamination and are suitable for prolonged use in milk fermentation. Several phase resistance mechanism, including inhibition of phage adsorption, restriction—modification mechanism and abortive infection mechanism are found in LAB. All of these are commonly encoded on plasmids.

A phage resistant starter culture of cheddar cheese, *L. lactis* DPC 5000 have been developed which was shown to embody three effective phage resistant mechanisms. Cheddar cheese manufactured with DPC 5000 compared favourably in term of composition with cheeses manufactured using commercial starter. Two phage resistant thermophilic starter strains DPC 1842 and DPC 5099 have been developed by Cork and others. Ireland which performed well in commercial plants for Mozzarella cheese preparation. The new cultures provide more predictable performance and reduce the chance of vats failure.

Cheese with Increased Yield

Scientists in New Zealand have created the world's first cow clones that produce special milk that can increase the speed and ease of cheese making. The researchers in Hamilton say their herd of nine transgenic cows makes highly elevated levels of milk proteins, i.e. casein-with improved processing properties and heat stability. Cows have previously been engineered to produce proteins for medical purposes, but this is the first time the milk itself has been genetically enhanced. The scientists hope the breakthrough will transform the cheese industry, and if widened, the techniques could also be used to 'tailor' milk for human consumption. But opponents of GM cheese continue to doubt whether such products will be safe.

Probiotic organisms for cheese: Ripened variety of cheeses may offer certain advantages as a carrier of probiotic micro-organisms with specified health benefits. Growing public awareness of diet-related health benefits has fuelled the demand for probiotic foods. Dairy foods including fermented milks and in particular yoghurt are the best accepted food, which can carry probiotic cultures. Among the most important criteria when considering cheddar cheese as a probiotic food is that the micro-organism be able to survive for relatively long ripening time of at least 6 months and/or that they grow in the cheese over this period although the pH is higher in comparison to traditional probiotic dairy foods. The potential health promoting effects achieved by the consumption of dairy foods containing probiotic organisms, such as *Lactobacillus* and *Bifidobacterium* spp., have resulted in intensive research efforts in recent years.

Even though small number of probiotic cheese are currently in the market world wide, a few reports are available concerning cheese as a carrier of probiotic organisms. European manufactures have introduced different varieties of cheese with added *Lactobacillus paracasei* NFBE 338. Dinakar and Mistry incorporated *Bifidobacterium bifidum* into cheddar cheeses as starter adjunct. This strain survived well and retained a viability of 2×10^7 cfu/gm of cheese even after 6 months of ripening, without adversely affecting cheese flavour, texture or appearance.

This suggested that cheddar cheese can provide suitable environment of probiotic organisms. *Bifidobacteria* were used in combination with *L. acidophilus* strain Ki as starter in Gouda cheese manufacture. There was a significant effect on flavour development in the cheese after 9 weeks of ripening. Cheddar cheese was manufactured with either *Lactobacillus salivarius* NFBC 310, NFBC 321 and NFBC 348 or *L. paracasei* NFBC 338 or NFBC 364, isolated from small intestine, as the dairy starter adjunct. Using randomly amplified polymorphic DNA method, it was found that both *L. Paracasel* strains grew and sustained high viability in cheese during ripening, while each of the *L. salivarius* species declined over the ripening period. These data demonstrate that Cheddar Cheese can be an effective vehicle for delivery of some probiotic organisms to the consumer.

Therefore, though the reports on probiotic cheeses are few but there is promising scope to exert probiotic effect on the cheese by enhancing the existing characteristic of probitic organisms with genetic engineering. They may be called as second-generation probiotics. These might include the incorporation of attachment molecules to facilitate colonisation, increased production of antimicrobial compounds directed against common food pathogen, cholesterol metabolism, enhanced immune stimulation and viability which contribute beneficial effect on human health. Attempts are also being made to identify new strains of probiotic culture by identifying the genes involved in such attributes and exploring the advanced molecular technique such as PCR, RFLP, RAPD and DNA finger printing etc. All these modifications in dairy starters can be brought also with the help of genetic engineering with the sole objective of improving their functionality so that cheese industry could benefit from value addition in cheeses/dairy foods through the intervention of these genetically modified organisms.

Another desirable property might be the generation of probiotic genetically modified organism to digest especially prebiotic carbohydrates. The combinations of probiotic and a prebiotic are known a synbiotic and the prebiotic is thought to enhance the survival of the probiotic. Several strains of *Lactobacillus* and *Bifidobacterium* sp. have been engineered to metabolise unusual carbohydrate, but the potential to form enhanced synbiotic has not yet been evaluated.

Acceleration of cheese ripening: The objective of acceleration of cheese ripening is to accelerate the proteolytic process and related events that occur in naturally ripened cheese as closely as possible. The methods used to accelerate ripening fall into six categories:

1. Elevated ripening temperature.
2. Exogenous enzymes.
3. Chemically or physically modified bacterial cells.
4. Genetically modified starters.

5. Adjunct culture.

6. Cheese slurries and enzyme—modified cheeses.

In this context genetically modified starter cultures is of special interest.

Genetically modified starters: Genetically modified starter cultures with enhanced complements of proteinase and/or peptidase, which could be released early and evenly distributed in the curd would be an ideal method of accelerating cheese ripening. Modified/genetically tailored micro-organisms, genetically engineered proteolytic and lipolytic enzymes are now being used for enhancing flavour production in cheese. Enzyme addition is now one of the few preferred methods of accelerated ripening of cheese. Enzyme may be immobilised or encapsulated for long term action on the production for quick action and homogenous distribution in the product. Cheese manufactured with an amino peptidase N-negative clone strain of *Lactococcus* produced bitter off flavour. The possible role of amino peptidase as a debittering agent confirmed by Prost and Chamba after making the Emmetal cheese with *Lb. helveticus* strain L_1 (high amino peptidase activity), L2 or strain L3 (clones selected for lack of amino peptidase activity). They also explained the bitterness in ripened cheese made with *Lb. delbrueckii subsp. lactis* to be due to their very low amino peptidase activity.

The gene for the neutral proteinase (neutrase) of *B. subtilis* has been cloned in *Lc. lactis* UC317. Cheddar cheese manufactured with this engineered culture as the sole starter showed very extensive proteolysis, and the texture became very soft within 2 weeks at 8°C. Cheddar cheese made with *Lc. lactis* subsp. *cremoris* Sk 11 (cloned with proteinase) revealed that starter proteinases are required for the accumulation of small peptides and free amino acids in cheddar cheese. The strain in which the proteinase remained attached to the cell wall appeared to contribute more to proteolysis than the strain that secreted the enzyme. Cheeses made with proteinase positive starter produce more pronounced flavour than those with proteinase negative strain during ripening. The inactivation of genes in a metabolic pathway can be used to alter end products that accumulate from a given pathway. Application of regulated promoters (in genetic engineering) is the controlled expression of lytic genes resulting in autolysis of the starter culture. This would result in rapid release of enzymes (i.e. peptidase) into the cheese matrix and potentially accelerate cheese flavour development.

Therefore, through genetic engineering, specific genes can be implanted/removed to increase or decrease the activity of specific property of the existing strain of LAB. It can be stated from the utilisation of genetically modified starter as: (i) cloning of exogenous proteinases in starter cells leads to enhanced proteolysis, (ii) debittering action of amino peptidase is now well recognised and (iii) starter peptidases and proteinases produce small peptides and amino acids in cheese but may not have a direct impact on flavour.

Chapter 13

Health Benefits of Milk and Functional Dairy Products

INTRODUCTION

Dairy products have so far been in the front line in the development of functional foods. Fermented dairy products have traditionally been considered to have health benefits and thus broadening the product range to other types of health-promoting products is quite natural for the dairy industry. Functional dairy products have recently been increasingly available in the daily-life which has gained increasing popularity in the past few years. Consumer's interest about personal health is reasons in establishing markets for functional dairy products. In the near future we will definitely see more products targeted for special consumer groups. This section discusses milk and its health benefits and functional dairy products. This will benefit to everyone involved with food science and nutrition, research on functional diary product, and food product development.

Milk is considered as a nearly complete food since it is a good source of protein, fat and major minerals. Also, milk and milk products are main constituents of the daily diet, especially for vulnerable groups such as infant's school age children and old age. Several studies have reported the distribution and occurrence of the essential components in various animal milks Milk is one of the most important nutrition food sources besides breast milk for infants and babies. In fact, consumption of dairy products has recently been linked to health benefits that are the direct antitheses of diseases and complexity that related to overweight and obesity. For example, individuals that consume dairy products are more likely to have lower weight, lower blood pressure, and decreased risk of stroke, colon cancer and osteoporosis. There is a wide range of functional foods that were developed recently and many of them are being produced in all over the world including probiotic, prebiotic and synbiotic foods as well as foods enriched with fat-reduced, salt-reduced foods or sugar-reduced foods, antioxidants and phytosterols as shown in Fig. 13.1. Among these foods, probiotic functional food has exerted positive effects on the overall health. We can divide it in both probiotic dairy foods and probiotic non-dairy foods. The market of probiotic dairy foods is increasing annually. An increased demand for dairy probiotic products comes from health promotion effects of probiotic bacteria which are originally initiated from milk products, bioactive

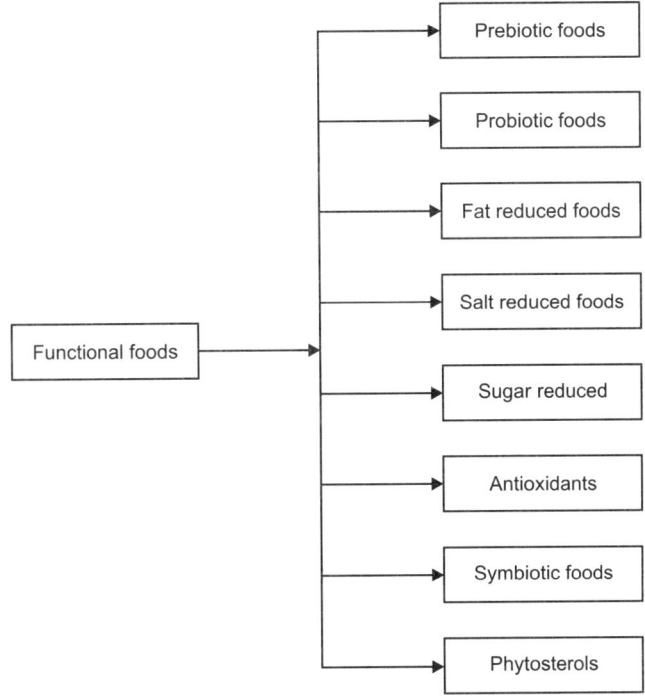

Fig. 13.1: Most of the functional foods that being produced in all over the world.

compounds of fermented dairy products and prevention of lactose intolerance. Therefore, development of these products is a key research priority for food design and a challenge for both industry and science sectors. This mini-review is an attempt to show some advances that have been made connecting milk and functional diary product.

MILK AND ITS HEALTH BENEFITS

Milk is a composite physiological fluid that facilitates postnatal adaptation of baby through digestive maturation simultaneously by providing the bioactive components and nutrients. It supports lymphoid tissues development and establishment of symbiotic micro flora. The importance, potency and the quantity of milk bioactive compounds are possibly more than old consideration. They comprise certain specific organic acids, vitamin A, B12, D, riboflavin calcium, carbohydrates, phosphorous, selenium, magnesium, zinc, proteins, bioactive peptides and oligosaccharides. They mostly emerge during fermentation or digestive processes while in some cases these are components of fresh milk. The possible mechanisms for cholesterol decreasing or removal by probiotic bacteria and fermented dairy products include inhibition of intestinal cholesterol absorption. FAs having medium chain whey proteins and other minerals may add positive result of dairy products on body mass. The dairy proteins play a vital role in food intake regulation, satiety and metabolic distracts relating to obesity. Blood pressure may be affected by lactic acid bacteria, milk proteins, peptides and calcium. Milk fat contains certain components having the functional significance. Antimicrobial effects are exerted by sphingolipids and their active metabolites either directly or upon their digestion. Whey was studied as a medicine as well as an aphrodisiac and skin balm during the Middle Age. Whey proteins, i.e. α-lactalbumin, lactoferrin, lactoperoxidase, serum albumin and β-lactoglobulin acquire important biological and nutritional properties particularly regarding disease prevention. Immunostimulatory, anticarcinogenic and antimicrobial are other whey protein activities that promote health. Milk products and their components take part in regulating the body mass through satiety signals. Therefore, whey proteins include physiological milk components for individuals with metabolic syndrome and obesity. Whey protein in high protein milk products may improve insulin sensitivity and reduce fat deposition. The bioavailability of trace elements and minerals, i.e. manganese, calcium, magnesium, iron, selenium and zinc is also improved by milk proteins and peptides.

The health benefits of milk and dairy products are known to humanity and may be attributed to the biologically active compounds that are existing in milk. Beside the modification of several milk components, proboscis may also act directly as preventive agents, or in therapy of some sever disease. The functional role of fermented dairy products is either directly through interaction with consumed micro-organisms or, indirectly, as a result of action of microbial metabolites like nutrients, generated during the fermentation process. The health promoting mechanisms of probiotic action are mostly based on the positive effect they exert on the immunity response.

Milk facilitates the maturation of digestive tube and cell growth of a baby in gastrointestinal tract (GIT). Donovan reported that milk is a complex combination of nutrients such as particular bioactive saccharides, lipids and proteins content which assist to regulate the development of GIT by representing the important signals. On the other hand, Morrow and Newburg concluded that milk is a source of communication in case of mother and the newborn child that influences the role of mucosal immunity and minimises the risk of infection. Milk has a wide various biologically active substances for instance, enzymes, immunoglobulins, oligosaccharides, antimicrobial peptides, hormones, cytokines, and the growth factors besides the basic saccharides, proteins and lipids as reported by Pouliot. Milk components are particular parts of immune system of newborn and they assist to stimulate and sustain the baby immune homeostasis. Neutrophiles, macrophages and T-lymphocytes as the heterogeneous population of milk cells play an important role in the defense against pathogenic bacteria.

So far, more than 60 different enzymes are recognised in milk and during the heat treatment the most of those enzymes will destroy and become inactive. The heat processing at high level of temperatures causes not only digestion enzymes denaturation (amylases, proteinases, phosphatases, lipases) but also digestion those enzymes having antioxidant and antimicrobial characteristics. These special characteristics are essential in milk stability as well as in the defense against pathogens; catalase, oxidoreductase, xanthine, lysozyme, dismutase, superoxide, lactoperoxidase, ribonuclease and myeloperoxidase. Milk antimicrobial agents have been shown bactericidal and even bacteriostatic behaviour. They are transmitted to progeny where they protect the progeny from highly contagious disorders. Lactoperoxidase, xanthine, oxidoreductase and lysozyme are the other best protecting factors in additions to immunoglobulin. Lactoperoxidase assists in milk storage as well as it inhibits the propagation of psychrotropic bacteria. According to it positively affects the Gram negative catalysed coliform bacteria, pseudomonads, shigella and *salmonella*. Seifu and others reported that lactoperoxidase assembly is utilised as a natural preservation agent in dairy manufacturing in different regions especially tropical areas. The xanthine oxidoreductase has bactericidal effects and reduces the rate and results in cytotoxic nitric oxide production. It results in the production of hydrogen peroxide that acts as substrate for NADPH oxidase, and lactoperoxidase (component of proficient anti microbial systems).

FUNCTIONAL DAIRY PRODUCTS

Dairy products are prominent as natural healthy products that contain the most crucial elements of the balanced diet. In additions to nutritional benefits milk plays a significant role in the control of chronic diseases, example blood pressure was being 'treated' with dairy products. It may not seem obvious to discuss blood pressure in relation to weight management, but the link between dairy components and weight management was initially derived from blood pressure studies. In Europe, dairy products are the major contributors in the functional food market by contributing approximately 60% of the total functional food spellings. They are the second well-liked class of functional foodstuff in the US. The Australian functional foods market is in its early life where probiotic yogurt is being the head in this zone that is growing at 22% and the soya yogurt resides at second. FAO/WHO standards describe the yogurt as 'lactic acid fermentation by the activity of *Lactobacillus delbrueckii* and *Streptococcus thermophilus* (*St. thermophilus*) to produce a coagulated milk.

Food Standards Australia and New Zealand defined low fat yogurt as 'the yogurt synthesised by culturing low fat or skim cow's milk that results in a thickened yogurt and does not have flavouring or fruit. It has 0.3% fat and 6.6% protein on average. Diary probiotic products can be produced by incorporation of probiotic bacteria in both of fermented and unfermented mix as reported by Homayouni and others. The work done on the diary probiotic products is summarised in Table 13.1.

Table 13.1: A summary table for diary probiotic products.

Dairy probiotic foods	Probiotic strains	Characteristics
Probiotic ice cream	*Lactobacillus casei* (Lc01) and *Bifidobacterium lactis* (Bb12)	Highest resistance to simulated acidic, alkaline and ice cream conditions
Petit-suisse cheese	*Bifidobacteria* and *lactobacilli*	The presence of the prebiotics insulin and oligofructose can promote growth rates besides increased lactate and short chain fatty acids production
Conventional yoghurt	*L. acidophilus* and *B. bifidum*	Add extra nutritional and physiological values
Bio-yoghurt	*L. acidophilus* and *B. bifidum*	Have to retain viability and activity in yoghurt as a probiotic at consumption time
Probiotic milk	Lactobacillus acidophilus	Remained viable in sweet acidophilus milk over 28 days at 7°C

Shah quoted that fermented milk is a prepared through mixed starter fermentation by using a culture comprising of *St. thermophilus* and *L. delbrueckii*. In Australia, lactic acid bacteria are allowed to employ as a starter cultures. Consequently, some yoghurt manufacturer use *L. jugurti* and *L. helveticus* for producing of yoghurt. Conversely, the standards in US do not allow any starter culture to be used other than *St. thermophilus* and *Lactobacillus delbrueckii*. The supplementation of different fruit provision in fermented milk products further endorse the healthy image of fermented milk that incorporate the fruits benefits. They provide antioxidants and fibre as described by O Rell. Recently, corn milk soya milk and peanut milk depending on fermented milk products are being synthesised as an alternate of vegetarian bovine milk fermented products that also overcome the allergencity of milk protein. Furthermore, to enhance functionality of fermented milk that by addition of plant extracts such as antioxidative and tea catechin is also significantly considered. Various essential nutrients and different components are provided by fermented milk that regulates various body functions in an optimistic way.

It is confirmed by various scientific evidences that chronic disorders, i.e. coronary heart disease, osteoporosis, hypertension and cancer can be controlled by the ordinary utilisation of probiotic or prebiotic supplemented fermented milk. Therefore the fermented milks meet with the functional food standards.

Thus, in general, dairy products provide a solid nutritional base for losing weight. Diary's dietary minerals may play an important role by influences adipocyte metabolism through calcitrophic hormone, and decrease the energy available from fat in food products by forming undigestable complexes.

The functional dairy components significantly contribute to the prevention of several diseases like hypertension, obesity, cancer, diabetes, and some transmissible diseases. On another hand, there is much kind of applications of these bioactive diary components such as phosphopeptides are currently used as both dietary and pharmaceutical supplements. Many of the components found in milk may have a protective effect against the onset of disease that occurs as a result of overweight, As well as several components found in milk could be sort and use in especial applications for individuals that do not consume dairy or may be lactose intolerant.

SECTION V

Biotechnology of Meat, Fish and Poultry

Chapter 14

Biotechnology of Meat

INTRODUCTION

Meat is one of the most valuable and demanding food products. Worldwide meat consumption is growing and the variety of products available as convenience foods is on the increase. Meat processing in the East European countries, Asia and South America continues on an upward path. These are all factors that contribute to the need for efficient, fully optimised meat processing. The continuously growing focus on automation along the entire meat processing chain is closely linked to the growing use of information technology. Hygiene and traceability are of crucial importance, requiring full monitoring of all processes along with the continuous documentation of origin and production data, in the interests of consumer protection. There are a wide variety of meat products that are attractive to consumers because of their characteristic colour, flavour and texture. The need to improve the processing and quality of these meat products prompted research in the last decades on endogenous enzyme systems that play important roles in these processes, as has been later demonstrated.

It is important to remember that the potential role of a certain enzyme in a specific observed or reported biochemical change can only be established if all the following requirements are met: (i) the enzyme is present in the skeletal muscle or adipose tissue, (ii) the enzyme is able to degrade *in vitro* the natural substance (i.e. a protein in the case of a protease, a tri-acylglycerol in the case of a lipase, etc.), (iii) the enzyme and substrate are located closely enough in the real meat product for an effective interaction and (iv) the enzyme exhibits enough stability during processing for the changes to be developed.

Muscle enzymes: There are a wide variety of enzymes in the muscle. Most of them have an important role in the *in vivo* muscle functions, but they also serve an important role in biochemical changes such as the proteolysis and lipolysis that occur in postmortem meat and during further processing of meat. Some enzymes are located in the lysosomes, while others are free in the cytosol or linked to membranes. The muscle enzymes with most important roles during meat processing are grouped by families and described below.

Muscle proteases: Proteases are characterised by their ability to degrade proteins and they receive different names depending on respective mode of action. They are endoproteases or proteinases when they are able to hydrolyse internal peptide bonds, but they are exopeptidases when they hydrolyse external peptide bonds either at the amino termini or the carboxy termini.

Lysosomal proteases: Lysosomes contain a variety of proteolytic enzymes including cathepsins. The lysosomal cathepsins are present in all mammalian cell types, with the exception of enucleated red blood cells. The concentration of lysosomes, and hence the proteases, varies in different cells and tissues and is particularly high in liver, spleen, kidney and macrophages.

Proteasome complex: The proteasome is a multicatalytic complex with different functions in living muscle. This enzyme has shown degradation of some myofibrillar proteins.

Exoproteases: peptidases: There are several peptidases in the muscle with the ability to release small peptides of importance for taste. Tripeptidyl-peptidases (TPPs) are enzymes capable of hydrolysing different tripeptides from the amino termini of peptides, while dipeptidylpeptidases (DPPs) are able to hydrolyse different dipeptide sequences.

Exoproteases: aminopeptidases: There are five aminopeptidases, known as leucyl, arginyl, alanyl, pyroglutamyl and methionyl aminopeptioases, based on their respective preference or requirement for a specific *N*-terminal amino acid. They are able, however, to hydrolyse other amino acids, although at a lower rate.

Lipolytic enzymes: Lipolytic enzymes are characterised by their ability to degrade lipids and they receive different names depending on their mode of action. Lipases and esterases are located either in the skeletal muscle or in the adipose tissue.

Muscle lipases: Lysosomal acid lipase and acid phospholipase are located in the lysosomes. Both have an optimal acid pH (4.5–5.5) and are responsible for the generation of long-chain free fatty acids.

Glycolysis: Glycolysis consists in the hydrolysis of carbohydrates, mainly glucose, either that remaining in the muscle or that formed from glycogen, to give lactic acid as the end product. As lactic acid accumulates in the muscle, pH falls from neutral values to acid values around 5.3–5.8. The contribution of glycolysis to pH drop is restricted to a few hours postmortem, although it is also important in fermented meats where sugar is added for micro-organism growth.

Lipolysis: Lipolysis makes an important contribution to the quality of meat products by the generation of free fatty acids, some of which have a direct influence on flavour and others that, with polyunsaturations, may be oxidised to volatile aromatic compounds, acting as flavour precursors. In addition, the breakdown of triacylglycerols affects the texture of the adipose tissue and an excess of lipolysis/oxidation may contribute to the development of rancid aromas or yellowish colours in fat.

MEAT AND MEAT PRODUCTS

Meat is animal flesh that is used as food. Most often, this means the skeletal muscle and associated fat, but it may also describe other edible tissues such as organs, livers, skin, brains, bone marrow, kidneys or lungs. The word meat is also used by the meat packing and butchering industry in a more restrictive sense — the flesh of mammalian species (pigs, cattle, etc.) raised and prepared for human consumption to the exclusion of fish, poultry and eggs. Originally meat was a term used to describe any solid food, but has now come to be applied, almost solely to animal flesh. As such, it has played a significant role in the human diet since the days of hunting and gathering, and animals (sheep) were first domesticated at the beginning of the Neolithic revolution around 9000 BC. Though abjured by some on moral or religious grounds, meat eating remains widely popular today. In the main, this is due to its desirable texture and flavour characteristics, although meat protein does also have a high biological value.

STRUCTURE AND COMPOSITION

Edible animal flesh comprises principally the muscular tissues but also includes organs such as the heart, liver and kidneys. Most microbiological studies on meat have been conducted with muscular tissues and it is on these that the information presented here is based. Though in many respects the microbiology will be broadly similar for other tissues, it should be remembered that differences may arise from particular aspects of their composition and microflora.

Structurally muscle is made up of muscle fibres, long, thin, multi-nucleate cells bound together in bundles by connective tissue. Each muscle fibre is surrounded by a cell membrane, the sarcolemma, within which are contained the myofibrils, complexes of the two major muscle proteins, myosin and actin, surrounded by the sarcoplasm.

The approximate chemical composition of typical adult mammalian muscle after *rigor mortis* is presented in Table 14.1. Its high water activity and abundant nutrients make meat an excellent medium to support microbial growth. Though many of the micro-organisms that grow on meat are proteolytic, they grow initially at the expense of the most readily utilised substrates—the water soluble pool of carbohydrates and nonprotein nitrogen. Extensive proteolysis only occurs in the later stages of decomposition when the meat is usually already well spoiled from a sensory point of view.

Table 14.1: Chemical composition of typical adult mammalian muscle after *rigor mortis*.

Chemical composition of mammalian muscle		*% weight*
Water	–	75.0
Protein	–	19.0
Myofibrillar	11.5	–
Sarcoplasmic	5.5	–
Connective	2.0	–
Lipid	–	2.5
Carbohydrate	–	1.2
Lactic acid	0.9	–
Glycogen	0.1	–
Glucose and glycolytic intermediates	0.2	–
Soluble non-protein nitrogen		1.95
Creatine	0.55	
Inosine mono phosphate	0.30	–
NAD/NADP Nucleotides	0.30	–
Nucleotides	0.10	–
Amino acids	0.35	–
Carnosine, anserine	0.35	–
Inorganic	–	0.85
Total soluble phosphorus	0.20	–
Potassium	0.35	–
Sodium	0.05	–
Magnesium	0.02	–
Other metals	0.23	–
Vitamins		

The carbohydrate content of muscle has a particularly important bearing on its microbiology. Glycogen is a polymer of glucose held in the liver and muscles as an energy store for the body. During life, oxygen is supplied to muscle cells in the animal by the circulatory system and glycogen can be broken down to provide energy by the glycolytic and respiratory pathways to yield carbon dioxide and water.

After death the supply of oxygen to the muscles is cut off, the redox potential falls and respiration ceases, but the glycolytic breakdown of glycogen continues leading to an accumulation of lactic acid and a decrease in muscle pH. Provided sufficient glycogen is present, this process will continue until the glycolytic enzymes are inactivated by the low pH developed. In a typical mammalian muscle the pH will drop from an initial value of around 7 to 5.4–5.5 with the accumulation of about 1 per cent lactic acid. Where there is a limited supply of glycogen in the muscle, acidification will continue only until the glycogen runs out and the muscle will have a higher ultimate pH. This can happen if the muscle has been exercised before slaughter but can also result from stress or exposure to cold. When the ultimate pH is above 6.2, it gives rise to dark cutting meat, a condition also known as dry, firm, dark (DFD) condition. Because the pH is relatively high, the meat proteins are above their isoelectric point and will retain much of the moisture present. The fibres will be tightly packed together giving the meat a dry, firm texture and impeding oxygen transfer. This, coupled with the higher residual activity of cytochrome enzymes, will mean that the meat has the dark colour of myoglobin rather than the bright red oxymyoglobin colour. The higher pH will also mean that microbial growth is faster so spoilage will occur sooner. Another meat defect associated with post-mortem changes in muscle carbohydrates is known as pale, soft, exudative (PSE) condition. This occurs mainly in pigs and has no microbiological implications but does give rise to lower processing yields, increased cooking losses and reduced juiciness. The PSE condition results when normal non-exercised muscle is stimulated just before slaughter leading to a rapid post-mortem fall in pH while the muscle is still relatively warm. This denatures sarcoplasmic proteins, moisture is expelled from the tissues which assume a pale colour due to the open muscle texture and the oxidation of myoglobin to metmyoglobin.

CONTAMINATION, PRESERVATION AND SPOILAGE OF MEATS AND MEAT PRODUCTS

Contamination

The healthy inner flesh of meats has been reported to contain few or no micro-organisms, although they have been found in lymph nodes, bone marrow and even flesh. The important contamination, however, comes from external sources during bleeding, handling and processing. During bleeding, skinning and cutting, the main sources of micro-organisms are the exterior of the animal (hide, hoofs and hair) and the intestinal tract. Recently approved 'humane' methods of slaughter, mechanical, chemical or electrical, have little effect on contamination, but each method is followed by sticking and bleeding, which can introduce contamination. As with the older methods of use of a knife on hogs and poultry, any contaminating bacteria on the knife soon will be found in meat in various parts of the carcass, carried there by blood and lymph. The exterior of the animal harbours large numbers and many kinds of micro-organisms from soil, water, feed and manure, as well as its natural surface flora, and the intestinal contents contain the intestinal organisms (Table 14.2). Knives, cloths, air and hands and clothing of the workers can serve as intermediate sources of contaminants. During the handling of the meat thereafter, contamination can come from carts, boxes or other containers, from other contaminated meat, from air and from personnel. Especially undesirable is the addition of psychrophilic bacteria from any source, e.g. from other meats that have been in chilling storage. Special equipment such as grinders, sausage stuffers and casings, and ingredients in special products, e.g. fillers and spices, may add undesirable organisms in appreciable numbers, and sawdust on floors of processing rooms may contaminate meat with mould spores. Growth of micro-organisms on surfaces contacting the meats and on the meats themselves increase their numbers. According to European workers, numbers of micro-organisms contaminating meats may be reduced by treatment of the surface with hot water.

Table 14.2. Average numbers of micro-organisms contaminating beef in packing-plant slaughter room.

Sample	Bacteria	Yeasts	Molds
Beef, dressed, on floor	6400–830000/cm^2		
Soil from animals (dry)	110000000/g	50000/g	120000/g
Animal feces (fresh)	90000000/g	200000/g	60000/g
Rumen content	2000000,000/g	180000/g	1600/g
Room air	140/cm^2 of plate		2/cm^2
Water, washing beef	20–10000/ml		
Water, washing floor	1000–16000/ml		

Because of the varied sources, the kinds of micro-organisms likely to contaminate meats are many. Moulds of many genera may reach the surfaces of meats and grow there. Especially important are species of the genera *Cladosporium*, *Sporotrichum*, *Oöspora* (*Geotrichum*), *Thamnidium*, *Mucor*, *Penicillium*, *Alternaria* and *Monilia*. Yeasts, mostly asporogenous ones, often are present. Bacteria of many genera are found, among which some of the more important are *Pseudomonas*, *Achromobacter*, *Micrococcus*, *Streptococcus*, *Sarcina*, *Leuconostoc*, *Lactobacillus*, *Proteus*, *Flavobacterium*, *Bacillus*, *Clostridium*, *Escherichia*, *Salmonella*, and *Streptomyces*. Many of these bacteria can grow at chilling temperatures. There also is the possibility of the contamination of meat and meat products with human pathogens, especially those of the intestinal type.

In the retail market and in the home additional contamination usually takes place. In the market knives, saws, cleavers, slicers, grinders, chopping blocks, scales, sawdust, and containers, as well as the market operators, may be sources of organisms. In the home the refrigerator containers used previously to store meats can serve as sources of spoilage organisms.

FOOD BIOTECHNOLOGY OF MEAT AND FISH PRODUCTS

Meat tenderness appears as top rated issue to be solved concerning meat sensorial quality, which requires enhanced knowledge and further work to understand the processes involved. It is recognised that these processes are mainly enzymatic in nature and involve proteolytic systems. Two concepts based on enzymatic biochemistry during meat ageing are currently in explaining meat tenderness. Firstly, some researchers postulate that meat tenderisation is affected mainly or solely by µ-calpain, a proteinase responsible for myofibrilar protein degradation. Secondly, others have suggested a hypothesis that tenderisation is a multienzymatic process corresponding to apoptosis, where µ-calpain is an important enzyme. This process is common to all cells when damaged, being associated with muscle cells after animal bleeding. The µ-calpain activity explains many, but not all pathways of tenderisation and apoptosis appears be an attractive answer to explain the obscure processes within meat ageing.

Tenderness is one of or the most discussed features in meat. It is a real challenge for the scientific community and for the meat industry to achieve products with standardised and guaranteed tenderness, since these characteristics are exactly what consumers want in a meat product. The United States meat industry has identified solving the problem of inconsistent meat tenderness as a top priority. This requires a detailed understanding of the processes that affect meat tenderness and, perhaps more importantly, the utilisation of such information by the meat industry. Since the beginning of the 1990s, the North American meat industry has accelerated the adoption of new technologies to meet consumer expectations. Nowadays, we can find a product labelled as guaranteed tender such as marinades and case-ready products, and consumers are willing and able to pay more for these kinds of products.

It is well recognised that the biochemical post-mortem processes are key-steps for meat tenderisation. Actually, the tenderisation process is unanimously known as an enzymatic process of proteolytic systems, where two strong current ways of thinking appear to offer the most probable explanations. Some think that calpains are the only proteases responsible for meat tenderisation, while others propose a multienzymatic process implicating calpains and other enzymes which function is less clear (proteasomes, caspases). This chapter will discuss the fundamental biochemical aspects of meat tenderisation, as well as present both of the above-mentioned theories of meat tenderisation.

TENDERISATION PROCESS

The three factors that determine meat tenderness are background toughness, the toughening phase and the tenderisation phase. While the toughening and tenderisation phases take place during the post-mortem storage period, background toughness exists at the time of slaughter and does not change during the storage period.

The background toughness of meat is defined as the resistance to shearing of the unshortened muscle, and variation in the background toughness is due to the connective tissue component of muscle. In particular, the organisation of the perimysium appears to affect the background toughness, since a general correlation between the perimysium and the tenderness of muscles has been found for both chicken and beef. The toughening phase is caused by sarcomere shortening during *rigor mortis* development. It was shown that there is a strong negative relationship between sarcomere length and meat toughness, where shorter sarcomeres (less than 2 µm) result in tougher meat.

While the toughening phase is similar in all carcasses under similar processing conditions, the tenderisation phase is highly variable. There is a large variation in both rate and extent of post-mortem tenderisation of meat, and these result in the inconsistency of meat tenderness which are found at the consumer level. The tenderisation process begins just after slaughter. A measure of tenderness is the subjective consumer appreciation of meat's texture. On the other hand, an objective way to measure tenderness is the force required to shear a standardised piece of meat, with low shear values being desirable. During tenderisation, proteolysis affects all muscle proteins, including connective tissue, and it is now well established that post-mortem proteolysis of myofibrillar and myofibrillar-associated proteins is responsible for this process. Some non-enzymatic aspects also influence the meat tenderisation process, such as temperature, pH and Ca^{2+} concentration. Usually, meat ageing is done at low temperatures and some studies showed that a freeze-thaw-refrigeration cycle can prevent a high rate of sarcomere shortening leading to a more tender meat. Low temperature also decreases the glycolytic process by lowering enzymatic activity reducing the pH fall. A low pH can cause inhibition of proteolytic enzymes activity, denaturation of myofibrilar proteins and excessive shortening and consequently lead to toughness and loss of water-holding capacity.

The sarcoplasmic calcium ion concentration increases ultimately to 0.2 mM during tenderisation due to the loss of the ability of sarcoplasmic reticulum and mitochondria to accumulate calcium ions. This concentration is about 2000 times that in resting skeletal muscle. While calcium is necessary for muscle contraction, it is an activator of proteolytic enzymes, and many authors have shown that calcium injection as carbonate enhances the tenderisation process.

ENZYMATIC TENDERISATION

Breakage of myofibrilar and cytoskeletal proteins result from enzymatic proteolytic system activation. This degradation includes troponin-I, troponin-T, desmin, vinculin, meta-vinculin, dystrophin, nebulin and titin. Three major cytoskeletal structures are degraded when meat is tenderised: Z to Z-line attachments by intermediate filaments, Z- and M-line attachments to the sarcolemma by costameric proteins and the elastic filament protein titin.

Three proteolytic systems present in muscle have been investigated for their possible role in post-mortem proteolysis and tenderisation: the calpain system, the lysosomal cathepsins and the multicatalytic proteinase complex (MCP). The cathepsins were the first enzymatic system considered in the studies focusing on the mechanisms of meat tenderisation. Later, calpains received much more attention than cathepsins mainly because of their ability to alter the Z-line density, a modification often observed post-mortem, even if this change is not correlated with tenderness. More recently, many sets of the evidence showed the potential role of the 20S proteasome in the tenderisation process. Experiments based on different approaches reported results clearly showing the contribution of 20S proteasome onto the tenderisation of stored meat.

It has been reported that proteolytic systems must fulfill requirements to be considered involved in postmortem proteolysis in meat. First, the proteases must have access to the substrates, and secondly, they must be able to reproduce the proteolysis pattern observed after postmortem storage of meat. Incubation of myofibrillar proteins with cathepsins results in different degradation patterns than those that occur during postmortem storage of muscle, and it is doubtful that cathepsins are released from the lysosomes in post-mortem muscle. It has also been postulated that a significant role for MCP can be excluded, since myofibrils are very poor substrates for this protease system. Moreover, the degradation pattern of myofibrillar proteins by MCP does not mimic the degradation pattern observed in postmortem muscle. This leaves the calpain system or potentially another, not yet investigated, proteolytic system responsible for postmortem proteolysis of key myofibrillar proteins and the resultant meat tenderisation.

Calpains are calcium-activated proteases consisting of at least three proteases, μ-calpain, *m*-calpain and skeletal muscle-specific calpain or calpain 3, and an inhibitor of μ- and *m*-calpain, calpastatin. Calpains' importance during tenderisation is verified in many studies such as production of the same proteolytic pattern observed in post-mortem muscle when calpains are incubated with myofibrils. It was also found that injection of calcium (calpain activator), in muscles accelerates postmortem proteolysis and tenderisation, whereas infusion or injection of muscles with calpain inhibitors inhibits postmortem proteolysis and tenderisation. Moreover, the greatly reduced rate and extent of post-mortem proteolysis and tenderisation in callipyge lamb can be attributed to elevated levels of calpastatin in these animals, and therefore calpain ihibition. Overexpression of calpastatin in transgenic mice also resulted in a large reduction in post-mortem proteolysis of muscle proteins. According to Koohmaraie and Geesink, among calpains, μ-calpain appears as the only enzyme that fills all the requirements to work effectively on post-mortem proteolysis. These conclusions come from a study showing that post-mortem proteolysis was largely inhibited in μ-calpain knockout mice.

APOPTOSIS THEORY

Other researchers agree that the process of meat tenderisation results from the synergistic action of several endogenous enzymatic systems, even if the major peptidases of concern are not identified yet. They propose that the meat ageing process can be explained by programmed cell death or apoptosis.

Apoptosis is a physiological mechanism occurring in living organisms that eliminates excessive, damaged or potentially dangerous cells from an organism without damaging surrounding cells. This process is necessary for both the normal development of a multicellular organism during embryogenesis and the maintenance of tissue homeostasis in adults. Cell death occurs in an ordered manner, mediated by a particular group of cysteine peptidases called caspases. Among this group of enzymes, there are the caspases involved in apoptosis initiation (caspases 8–10), characterised by large prodomains often containing essential areas for their interactions with other proteins, regulating the cell death beginning. Moreover, there are the effectors caspases that disrupt the cell when activated (caspases 3, 6 and 7).

In meat animals, whatever the animal species and whatever the technology of stunning used, the last phase of slaughter is bleeding. Consequently, all cells and tissues will irreversibly be deprived of nutrients and oxygen. Under these very harmful environmental conditions, muscle cells will have no alternative but initiate apoptosis. Under similar conditions, such cell behaviour is currently observed in living organisms.

There are some features that are common in both apoptosis and meat tenderisation. Maybe, it is not only coincidence, but meat tenderisation is strongly correlated with apoptotic processes, such as inversion of membrane polarity. As mentioned above, injection of calcium in meat accelerates the process of tenderisation. The action of calcium is generally attributed to an activation of calpains, the calciumdependent peptidases. Because of the numerous roles of this cation in cell signalling pathways, other potential functions of calcium have received only little attention from meat scientists. However, if it is considered that, after slaughter, cells have no other alternative but engage towards suicide or apoptosis, we have to reconsider some of these functions. Calcium is indeed a crucial effector for triggering and controlling apoptosis. In post-mortem muscle, calcium concentration increases gradually in the cytoplasm during *rigor mortis* onset while the sarcoplasmic reticulum is emptied of its contents. It is known that this cation is a central element of the apoptotic process, inducing swelling and extensive alteration of mitochondria and the release of cytochrome *c* together with other proapoptotic proteins. This process ends by the activation of caspase 9, which in turn will activate the effectors caspases.

Moreover, acidification of muscle decreases protein charges and increases their hydrophobicity, thereby reducing water retention. This is confirmed by the very high correlation observed between the increase in extracellular space and muscle pH. The only point which remained unexplained was the early increase in extracellular space starting immediately after slaughter, while pH was still very close to neutrality. Events associated with cellular death provide an explanation since a cell entering in apoptosis is dissociated from others and 'shrinks'. The consequence will be a reduction in intracellular space and a parallel increase in extracellular space.

During stress, cells prepare their defense as quickly as possible. Among available means, the most described is the synthesis of various protective proteins known as heat shock proteins (HSPs). They have an anti-apoptotic activity, slowing down the cellular death process. When animals are stressed before or during slaughter, their meat does not follow a satisfactory ageing process that can be result not only from decreased pH, but also from apoptosis inhibition. If meat tenderisation is reconsidered through introduction of programmed cell death, the first active peptidases after animal bleeding would be undoubtedly caspases.

These peptidases are in a better position than others to alter cellular structures since this is their primary function *in vivo*. It is worthy to note that the implication of caspases can explain the often reported assertion that the first hours following slaughter are essential for the satisfactory meat ageing process.

Thus tenderness in meat is a well-known issue to be solved by the scientific community and industry. Nowadays, meat tenderisation process is primarily explained by two different theories. One explains this process as a result of μ-calpain action throughout the muscle cell. The other characterises tenderisation as a more complex procedure, dependent on many enzymatic processes, among them μ-calpain activity, called apoptosis. Apoptosis is known as programmed cell death that occurs in all kinds of tissues, including muscles. There are many more studies showing the importance of μ-calpain than apoptotic activities during meat ageing. However, there are still unclear gaps of understanding in the tenderisation process. Many recognised apoptotic steps coincide with tenderisation pathways, and maybe, with further studies, can explain these obscure gaps on meat ageing.

Biotechnology of Fish

INTRODUCTION

A fish is any aquatic vertebrate animal that is typically ectothermic (or cold-blooded), covered with scales, and equipped with two sets of paired fins and several unpaired fins. Fish are abundant in the sea and in freshwater, with species being known from mountain streams (e.g. char and gudgeon) as well as in the deepest depths of the ocean (e.g. gulpers and anglerfish).

Fish are of tremendous importance as food for people around the world, either collected from the wild or farmed in much the same way as cattle or chickens. Fish are also exploited for recreation, through angling and fish-keeping, and are commonly exhibited in public aquaria.

Here we are mainly concerned with what most people think of as fish principally the free swimming teleosts and elasmobranchs. The same term can also encompass all seafoods including crustaceans with a chitinous exoskeleton such as lobsters, crabs and shrimp, and molluscs such as mussels, cockles, clams and oysters. Microbiologically these have many common features with free swimming fish.

Historically the extreme perishability of fish has restricted its consumption in a reasonably fresh state to the immediate vicinity of where the catch was landed. This has detracted only slightly from it playing a significant role in human nutrition as, throughout the world, traditional curing techniques based on combinations of salting, drying and smoking were developed which allowed more widespread fish consumption. Poor keeping quality is a special feature of fish which sets it apart even from meat and milk. The biochemical and microbiological reasons for this are discussed later in this chapter.

STRUCTURE AND COMPOSITION

Although broadly similar in composition and structure to meat, fish has a number of distinctive features. Unlike meat, there are no visually obvious deposits of fat. Although the lipid content of fish can be up to 25 per cent, it is largely interspersed between the muscle fibres.

A further feature which contributes to the good eating quality of fish is the very low content of connective tissue, approximately 3 per cent of total weight compared with around 15 per cent in meat. This, and the lower proportion of body mass contributed by the skeleton, reflect the greater buoyancy in water compared with that in air.

202

Muscle structure also differs. Inland animals it is composed of very long fibres while in fish they form relatively short segments known as myotomes separated by sheets of connective tissue known as myocommata. This gives fish flesh its characteristically flaky texture. Fish flesh generally contains about 15–20 per cent protein and less than 1 per cent carbohydrate. In non-fatty fish such as the teleosts cod, haddock and whiting, fat levels are only about 0.5 per cent, while in fatty fish such as mackerel and herring, levels can vary between 3 and 25 per cent depending on factors such as the season and maturity.

CONTAMINATION, PRESERVATION AND SPOILAGE OF FISH AND OTHER SEAFOODS

Seafoods discussed in this section include fresh, frozen, dried, pickled and salted fish, as well as various shellfish. Freshwater fish also are considered.

Contamination

The flora of living fish depends upon the microbial content of the waters in which they live. The slime that covers the outer surface of fish has been found to contain bacteria of the genera *Pseudomonas*, *Achromobacter*, *Micrococcus*, *Flavobacterium*, *Corynebacterium*, *Sarcina*, *Serratia*, *Vibrio* and *Bacillus*. The bacteria on fish from northern waters are mostly psychrophiles, whereas fish from tropical waters carry more mesophiles. Freshwater fish carry freshwater bacteria, which include members of most genera found in salt water plus species of *Aeromonas*, *Lactobacillus*, *Brevibacterium*, *Alcaligenes* and *Streptococcus*. In the intestines of fish from both sources are found bacteria of the genera *Achromobacter*, *Pseudomonas*, *Flavobacterium*, *Vibrio*, *Bacillus*, *Clostridium* and *Escherichia*. Boats, boxes, bins, fish houses, and fishermen soon become heavily contaminated with these bacteria and transfer them to the fish during cleaning. The numbers of bacteria in slime and on the skin of newly caught ocean fish may be as low as 100 and as high as several million per square centimetre, and the intestinal fluid may contain from one thousand to 100 million per millilitre. Gill tissue may harbour a thousand to a million per gram. Washing reduces the surface count. Unopened fish or fish 'in the round', have been reported to keep better for a while than opened fish because contamination of the body cavity is avoided. Bacteria are supposed to spread through the fish flesh mostly via the gills.

Oysters and other shellfish that pass large amounts of water through their bodies pick up soil and water micro-organisms in this way, including pathogens if they are present. *Achromobacter* and *Flavobacterium* predominate. Shrimps, crabs, lobsters, and similar seafood have a bacteria-laden slime on their surfaces that probably resembles that of fish. Species of *Achromobacter*, *Bacillus*, *Micrococcus*, *Pseudomonas*, *Flavobacterium*, *Alcaligenes* and *Proteus* have been found on shrimp.

Since the surface flora of fish and other sea animals seems to consist chiefly of water bacteria, the icing of these foods would not be expected to add much in the way of contamination. Ices containing antiseptic or germicidal chemicals have been used, chlortetracycline may now be used as a dip or in ice for fish in many parts of the world and for scallops and unpeeled shrimp in the United States.

GENETICALLY ENGINEERED FISH

Biotechnology, the use of biological systems or living organisms in production process has a wide range of useful applications in fisheries and aquaculture. The potential of biotechnology to contribute to increasing agricultural, food and feed production, improving human and animal health, abating pollution and protecting the environment, has been acknowledged. Biotechnology makes it possible to achieve increased growth rate in farmed species, boost the nutritional value of aqua feeds, improve fish health, help restore and protect environment, extend the range of aquatic species and improve the management and conservation of wild stocks. Some biotechnologies are simple with a long history of application: e.g. fertilisation of ponds to increase feed availability. Others are more advanced and take advantage of increasing knowledge of molecular biology and genetics, e.g. genetic engineering and DNA disease diagnosis. The field of genetic biotechnology similarly ranges from simple techniques such as hybridisation, to more complex processes such as the transfer of specific genes between species to create Genetically Modified Organisms (GMOs).

FISH FEED BIOTECHNOLOGY

Biotechnology is helping to answer some of the technical and environmental concerns of fish farming. Presently, the most common protein source for many fish diets is fish meal. Fish meal a by-product of fish processing is used because of its high quality and high protein content. However, it has some disadvantages. One disadvantage for fish producers is that it is expensive. So any cheaper alternative protein source would be welcomed. Another concern regarding fish meal is the stability of supply. Fishmeal comes from by-products of wild fish, but world fish stocks are declining. The use of fish meal in aquaculture causes environmental problems. It contains levels of phosphorous far above the requirement for optimal growth in fish. The excess phosphorus goes into the water, causing problems such as eutrophication or excess algal growth.

As a result of these concerns with fish meal, researchers are using biotechnology to produce alternative plant-based protein source. Plant protein has the potential to address the problem of phosphorus pollution. For Prairie crops to be used as the main protein source for fish, they must be processed into a concentrate. Biotechnology is often used in this processing, plant protein also requires processing because plants contain what are called anti-nutritional compounds that serve as a defense mechanism. These compounds must be destroyed during processing or they could harm the fish or interfere with the fish's ability to utilise the feed. Researchers are also trying to deal with these anti-nutritional factors by producing feed enzymes to counter them. Phytase is one example. This enzyme can help fish make the best use of the phosphorus available in a plant protein based feed.

Bio-remediation: Farmed aquatic animals are much more sensitive to their immediate environment than land animals. The water, in which they depend for oxygen and a range of other important chemicals also takes up their waste products and may carry pollution from the nearby environment. The process of disease in aquaculture species is thus much more strongly connected to environmental factors than would be the case say, with cattle. A further biotechnology field that has developed in aquaculture, because of the nature of this relationship, is that of bio-remediation. This refers to the use of friendly bacteria or 'probiotics' to treat water or feeds and by natural processes, discourages the development of 'unfriendly' bacteria that potentially would cause disease.

TRANSGENIC FISH

Researchers are seeking to improve the genetic traits of the fish used in aquaculture by using different transgenic techniques. Researchers are trying to develop fish which are larger and grow faster, more efficient in converting their feed into muscle, resistant to disease, tolerant of low oxygen levels in the water and tolerant to freezing temperatures.

The exploitation of Tilapia in small scale aquaculture in developing countries is constrained by the performance characteristics of fish currently in use. There have been significant advances in the genetic improvement of Tilapia used in aquaculture in recent years. For example through the use of selective breeding and monosex techniques. However, limited growth rate and excessive reproduction resulting in fish that are small and variable in size, still pose a considerable constraint to the exploitation of Tilapia in developing countries. Transgenic techniques offer the means of producing immediate large quantum changes in performance, for example in growth rate, that far exceed those attainable with other approaches.

GENETICALLY ENGINEERED FISH AND SEAFOOD: ENVIRONMENTAL CONCERNS

Farmers and scientists have a history of modifying animals to maximise desirable traits. Genetic modification is one of the current approaches for modifying animals to increase their beneficial traits. In the broadest sense, genetic modification refers to changes in an organism's genetic makeup not occurring in nature, including the production of conventional hybrids. With the advent of modern biotechnology (e.g. genetic engineering or bioengineering), it is now possible to take the gene (or genes) for a specific trait either from an organism of the same species or from an entirely different one and transfer it to create an organism having a unique genetic code. This technique can add both speed and efficiency to the development of new foods and products. Genetically engineered plant varieties, such as herbicide-resistant corn and soyabeans, have already been widely adopted by US farmers, and some advocate using similar techniques to produce genetically engineered fish or seafood for the aquaculture industry.

A number of environmental concerns have been raised related to the development of genetically modified (GM) fish, including the potential for detrimental competition with wild fish, and possible interbreeding with wild fish so as to allow the modified genetic material to escape into the wild fish population. Sterilisation and bioconfinement have been proposed as means of isolating GM fish to minimise the potential for harming wild fish populations. In the process of congressional oversight of executive agency regulatory action, concerns have been raised about the adequacy of the US Food and Drug Administration's review of applications for approval of GM animals, with respect to the potential for environmental harm.

Genetic engineering and fish: Genetically engineered (also called transgenic) fish are those that carry and transmit one or more copies of a recombinant DNA sequence (i.e. a DNA sequence produced in a laboratory using *in vitro* techniques). Because genetic engineering is defined by the technology that is used to create and transfer the DNA sequence, and not the source species of the donor DNA, even fish that are engineered with DNA derived entirely from fish species are considered to be genetically engineered. Currently, no genetically engineered fish has been approved for food production in the United States. To date only one company, AquaBounty, has publicly announced that it has requested FDA approval to market a genetically engineered food animal, a growthenhanced Atlantic salmon that is capable of growing 4 to 6 times faster (but not larger) than standard salmon grown under the same conditions.

Science-based concerns associated with genetically engineered fish: The greatest science-based concerns associated with genetically engineered fish are those related to their inadvertent release or escape. Concerns range from interbreeding with native fish populations to ecosystem effects resulting from heightened competition for food and prey species. There is, in principle, no difference between the types of concerns associated with the escape of genetically engineered fish and those related to the escape of fish that differ from native populations in some other way, such as captively bred populations. Ecological risk assessment requires an evaluation of the fitness of the genetically engineered fish relative to non–genetically engineered fish in the receiving population in order to determine the probability that the transgene will spread into the native population. Ecological impacts are the result of the characteristics of the organism, regardless of whether the organism acquired those characteristics through natural selection, artificial selection, or genetic engineering. The presence of genetically engineered fish does not a *priori* have a negative effect on native populations. If genetically engineered fish are ill-suited to an environment or are physically unable to survive outside of containment, they may pose little risk to the native ecosystems. Regulators apply a scientifically derived, risk-based framework to assess the ecological risks involved with each transgene, species, and receiving ecosystem combination on a case-by-case basis. Risks will be quite specific to the gene, species, and site in question, and simple generalisations concerning the risks (and benefits) of genetically engineered fish are not scientifically meaningful.

Can containment be used to isolate genetically engineered fish: Commercialisation of genetically engineered fish will likely depend on the development of effective containment strategies. If genetically engineered fish are adequately contained, they pose little risk to native populations. The NRC recommended the simultaneous use of multiple containment strategies for genetically engineered fish (National Research Council). Physical containment is an obvious first line of defense to prevent the escape of genetically engineered fish. Examples of such measures may include building facilities on land or in locations removed from native populations, or ensuring that water chemistry (temperature, pH, salinity, and concentrations of certain chemicals) is lethal to one or more life stages of the genetically engineered fish, such as treating effluent water to prevent the release of viable gametes or fry. Biological containment or bioconfinement approaches such as sterilisation are also being developed. To ensuring food safety, the FDA also evaluates environmental risks posed by genetically engineered animals as directed by the National Environmental Policy Act (NEPA). Under NEPA, Federal agencies are obligated to cooperate with other involved federal agencies. In the case of the AquaBounty genetically engineered salmon, this cooperation includes involvement of the US Fish and Wildlife Service and the National Marine Fisheries Service in the development of a scientifically based environmental risk assessment.

Will consumers accept genetically engineered fish: Ultimately, it is the marketplace, and not science, that decides the fate of new technologies and acceptability of certain risks. Food retailers and even farmers may be unwilling to stock genetically engineered fish and risk having their market become the target of an organised antibiotech campaign. Such a scenario occurred in Europe, where activist campaigns targeted retailers stocking labelled genetically engineered food products, and attempts to differentiate brands resulted in the removal of these products from supermarket shelves altogether. Despite strong public support for medical applications of genetic engineering, there is less public support for agricultural biotechnology. However, market response and consumer behaviour may differ markedly between affluent Western countries and developing countries. Even if genetically engineered fish are approved by the FDA in the United States, it will likely be activist, food retailer, and consumer response in the marketplace that will ultimately decide whether genetically engineered food fish will sink or swim.

ENVIRONMENTAL CONCERNS AND CONTROL OPTIONS

In addition to its responsibility for assuring food safety, FDA is charged with assessing the potential environmental impacts of newly engineered plants and animals. To fully assess these potential impacts, FDA consults with the Fish and Wildlife Service and the National Marine Fisheries Service (NOAA Fisheries). Despite this consultation, critics question whether FDA has the mandate and sufficient expertise to identify and protect against all potential ecological damage that might result from the widespread use of transgenic fish. The possible impacts from the escape of GM organisms from aquaculture facilities are of great concern to some scientists and environmental groups. A National Research Council report stated that transgenic fish pose the 'greatest science-based concerns associated with animal biotechnology, in large part due to the uncertainty inherent in identifying environmental problems early on and the difficulty of remediation once a problem has been identified.'

Interbreeding with Wild Fish

Critics and scientists argue that GM fish could breed with wild populations of the same species and potentially spread undesirable genes. One study postulated a 'Trojan gene hypothesis' after observing that GM Japanese medaka, a fish commonly used as an experimental model, were able to out-compete nonaltered fish for mates in a laboratory environment. However, the resulting offspring of this mating between GM fish and wild fish were less fit, lacking certain physical or behavioural attributes that resulted in the eventual demise of the modified population. The ecological risks of stocking GM shellfish in the wild have not yet been thoroughly examined, since confining and isolating these organisms ᵢ more difficult than confinement of many fish species, due to their methods of reproduction and dispersa .

Competition with Wild Fish

Even if fast-growing GM fish do not spread their genes among their wild counterparts, critics fear ᴜM fish could disrupt the ecology of streams by competing with native fish for scarce resources. Escapeᴅ transgenic fish could harm wild fish through increased competition or predation. In addition, some argue that transgenic fish, especially if modified to improve their ability to withstand wider ranges of salinity or temperature, could be difficult or impossible to eradicate, similar to an invasive species. The consequences of such competition would depend on many factors, including the health of the wild population, the number and specific genetic strain of the escaped fish, and local environmental conditions. Critics maintain that an indication of the magnitude of this potential problem may be noted where non-GM Atlantic salmon from nearshore net pens in the northwest United States and British Columbia have escaped and entered streams, in some cases outnumbering their wild Pacific salmon counterparts.

However, it is not known whether GM fish could survive in the wild in sufficient numbers to inflict permanent population damage. One study indicated that, when food supplies were low, GM fish might have the ability to harm a wild population, although the researchers have caution that laboratory experiments may not reflect what would happen in the wild. Biotechnology proponents argue that GM fish, if they escape, would be less likely to survive in the wild, especially when they are reared in protected artificial habitats and have not learned to avoid predators.

Potential Control Options

A number of potential safeguards to address these environmental concerns exist and could be required.

Sterilisation

FDA could require that only sterile GM fish be approved for culture. Fertilised fish eggs that are subjected to a heat or pressure shoc`, retain an extra set of chromosomes. The resulting triploid fish do not develop normal sexual characteristics and, in general, the degree of sterility in triploid females is greater than males. Thus, all-female lines of triploid fish are considered to be one of the best current methods to insure nonbreeding populations of GM fish. Nonetheless, there are batch-to-batch variations, and it is uncertain whether this method could be effective for all species, it has not been successful for shrimp, for example. Also, critics question whether escaped triploid fish, which in some species have sufficient sex hormone levels to enable normal courtship behaviour, could mate with wild individuals, lowering reproductive success of the wild population. Other sterilisation methods are currently under study, and it is likely that research in this area will increase options. Critics of GM fish counter that the risks to native fish populations, however small, may outweigh the potential benefits of this technology, especially where native fish populations are already threatened or endangered.

POSSIBLE BENEFITS AND DISADVANTAGES OF GENETICALLY ENGINEERED FISH AND SEAFOOD

Potential Benefits

Biotechnology proponents maintain that genetic modification techniques have many advantages over traditional breeding methods, including faster and more specific selection of beneficial traits. Because scientists are able to directly select traits they wish to create or amplify, the desired change can be achieved in very few generations, making it faster and lower in cost than traditional methods, which may require many generations of selective breeding. Genetic modification techniques allow scientists to precisely select traits for improvement, enabling them to create an organism that is not just larger and faster-growing, but potentially improved, for example, by increasing nutritional content. Proponents claim that faster-growing fish could make fish farming more productive, increasing yields while reducing the amount of feed needed, which in turn could reduce waste. With intense exploitation of wild fish stocks, GM fish and seafood could be important means to meet increasing human nutrition needs and address food security concerns.

Shellfish and finfish, genetically modified to improve disease resistance, could reduce the use of antibiotics. Increased cold resistance in fish could lead to the ability to grow seafood in previously inhospitable environments, allowing aquaculture to expand into previously unsuitable areas. Research efforts are also under way to improve human health by genetically modifying fish to produce human drugs like a blood clotting factor and to create shellfish that will not provoke allergic reactions. Biotechnology proponents claim that these advantages could translate into a number of potential benefits, such as reduced costs to producers, lower prices for consumers for edible fish and pharmaceuticals, and environmental benefits, such as reduced water pollution from wastes. Food scientists and the aquaculture industry may support the introduction of genetic engineering, provided that issues of product safety, environmental concerns, ethics, and information are satisfactorily addressed.

Disadvantages

On the other hand, while the majority of consumers in the United States appear to have generally accepted GM food and feed crops, it is uncertain whether consumers will be as accepting of GM fish. Although such fish may taste the same and are expected, like their traditionally bred counterparts, to be less expensive than wild-caught fish, ethical concerns over the appropriate use of animals, in addition to environmental concerns, may affect public acceptance of GM fish as food.

In addition, the commercial fishing industry says that it has successfully educated the public to discriminate among fish from different sources, such wild and farmed salmon. It is possible that a publicised escape of GM fish could lead to reduced public acceptance of their wild product. Environmental and consumer groups are asking that genetically engineered products be specially labelled. However, industry groups are concerned that such labelling might lead consumers to believe that their products are unsafe for consumption.

FUTURE APPLICATIONS OF BIOTECHNOLOGY

In the future we can expect the application of biotechnology in aquaculture to increase as new methods to improve fish performance are developed and perfected. In the area of chromosome manipulation gynogenesis will become a practical first step in the generation of monosex female stocks when combined with endocrine masculinisation. Androgenesis, although technically more difficult, may see eventual use as a practical means of producing supermales in male heterogametic species thus facilitating the production of all male stocks. Current research on meiotic and mitotic gynogenesis may possibly lead to the feasibility of producing clones of identical high performance fish. Triploidy induction may be used increasing as a means of sterilizing genetically altered fish. Tetraploid fish will be further developed as a means of producing diploid gametes and thus allotriploids.

In the field of controlled sex differentiation non-aromatizable androgens will be used as masculinizing agents and non-steroidal aromatase inhibitors may also be utilised Y-specific-DNA probes will be developed and applied in additional fish species and many species where one sex is more desirable for culture or more valuable in the marketplace will be grown in monosex culture.

Acceptable means of accelerating growth will be applied which utilise somatotropin, placental lactogen, related peptides or their analogues applied by injection or implantation in slow release formulations or devices or via the diet. Biotechnology will have a major impact on fish nutrition. Practical diets will be developed where suitably modified plant protein sources totally replace fish meal. Astaxanthin from the yeast *Phaffia rhodozyma* will provide a natural source of pigment for salmonid diets. Dietary optimisation of omega-3 polyunsaturated fatty acid levels may be used to produce designer seafood for health conscious consumers.

The techniques of molecular biology will be utilised in many different ways to produce high performance fish. In the field of fish health further diagnostic tests based on the polymerase chain reaction, dot blot and *in situ* hybridisation, etc. will facilitate disease and parasite detection. Increasing knowledge of the relationship between stress and the immune response and the utilisation of glucans to stimulate non specific defense mechanisms will all contribute to minimising the effects of stress on performance. The utilisation of molecular biology to develop new or more effective vaccines will further contribute to the elimination of disease outbreaks in aquacultured fish.

In the area of genetics the introduction of marker assisted selection into broodstock development should assist in the genetic improvement of aquacultured stocks. The engineering of transgenic fish in which specific characteristics are enhanced or suppressed may provide a quantum leap in the development of high performance fish. While still in this infancy this technology may provide means of improving growth and food conversion efficiency, accelerating smoltification, modifying nutritional requirements, increasing disease resistance, inhibiting or redirecting sexual development and facilitating adaptation to new environments. Early results show considerable promise for acceleration of growth and smoltification and potentially improvement of food conversion efficiency in transgenic salmonids, however, selection over several generations will be required to generate true breeding transgenic stocks.

Chapter 16

Impact of Biotechnology on Poultry Nutrition

INTRODUCTION

Poultry is defined as economically important birds used for food. This includes chickens, turkeys, quail, ducks, geese, and guineas. Other birds, such as pheasants, partridges and peafowl, can be classified as poultry. The following is poultry-related vocabulary.

Pullet: A young female chicken.

Hen: A sexually mature female chicken, usually more than 10 months old, that has started to lay eggs.

Chick: A baby chicken of either gender.

Broiler or fryer: A young male or female chicken, tender-meated with flexible breastbone cartilage, marketed at 6 to 8 weeks of age.

Roaster: A young male or female chicken, tender-meated with breastbone cartilage somewhat less flexible than a broiler or fryer, usually marketed at 7 to 10 weeks of age.

Capon: A surgically unsexed male chicken, usually under 8 months old, used for specialty markets.

Cockerel: A male chicken less than one year old.

Rooster: A sexually mature male chicken.

Poult: A male or female baby turkey.

POULTRY FARM

Poultry farming is the raising of domesticated birds such as chickens, ducks, turkeys and geese for the purpose of farming meat or eggs for food. Poultry are farmed in great numbers. Chickens raised for eggs are usually called layers while chickens raised for meat are often called broilers.

The increase of productivity in the poultry industry has been accompanied by various impacts, including emergence of a large variety of pathogens and bacterial resistance. These impacts are in part due to the indiscriminate use of chemotherapeutic agents as a result of management practices in rearing cycles. This chapter summarises the use of probiotics for prevention of bacterial diseases in poultry, as well as demonstrating the potential role of probiotics in the growth performance and immune response of poultry, safety and wholesomeness of dressed poultry meat evidencing consumers protection.

The poultry industry has become an important economic activity in many countries. In large-scale rearing facilities, where poultry are exposed to stressful conditions, problems related to diseases and deterioration of environmental conditions often occur and result in serious economic losses. Prevention and control of diseases have led during recent decades to a substantial increase in the use of veterinary medicines. However, the utility of antimicrobial agents as a preventive measure has been questioned, given extensive documentation of the evolution of antimicrobial resistance among pathogenic bacteria. So, the possibility of antibiotics ceasing to be used as growth stimulants for poultry and the concern about the side-effects of their use as therapeutic agents has produced a climate in which both consumer and manufacturer are looking for alternatives. Probiotics are being considered to fill this gap and already some farmers are using them in preference to antibiotics.

SLAUGHTERING AND PROCESSING OF MEAT

Broiler chickens (the type raised for meat) generally take up to seven weeks to reach market weight. Once they've reached the proper size and weight, workers trained in humane care arrive to catch each chicken at the farm, by hand. During this process, chickens are transferred into holding cages or modular bins, specifically designed for transport to the processing plant, aimed to ensure that birds don't hurt themselves or other birds, and that air is able to circulate.

Steps involved to slaughter the chickens. Generally 10 steps are invloved and these are discussed below:

Step 1: Arrival at the Processing Plant

Just as careful attention is paid to the welfare of chickens while being raised on the farm, the same is true for their short trip to the processing plant.

Step 2: Stunning chickens

Once birds arrive at the processing plant, workers trained in humane handling carefully suspend them by their feet on a moving line. In a matter of seconds, the chickens become calm due to "rub bars," which provide a comforting sensation on the chicken's chest. This, combined with low lighting, is used to keep birds calm.

In modern poultry processing plants, every attempt is made so that chickens are processed quickly and painlessly. First, they are rendered unconscious and unaware of pain, prior to slaughter.

There is one primary method of stunning broilers prior to slaughter in the US and that is 'electrical stunning.' It is the predominant method of rendering birds unconscious. There are a limited number of facilities in the US that utilise controlled atmosphere stunning (CAS) systems for broilers. These systems utilise carbon dioxide to render birds insensible. Another CAS system utilises a reduction of atmospheric pressure to stun birds.

When operating properly, both systems are equally humane as both require monitoring, proper adjustment and management to ensure they are meeting humane care standards.

Step 3: Slaughter

Technology makes slaughter extremely quick to minimise discomfort. While making a single cut to the throat of an unconscious bird is largely effective, should the blade miss for any reason, trained workers stand by to quickly euthanise remaining birds. Proper maintenance of equipment and this back-up 'human' system is key to a fast and humane slaughter process.

Step 4: Evisceration

After slaughter, birds enter a process where their feathers are removed. This is necessary in order to prepare the bird for processing. This begins by putting the chicken through a bath of hot water, which is designed to help loosen feathers. Feather removal is performed by a machine called a 'picker,' which includes hundreds of little rubber 'fingers' that rotate around to remove the feathers. After feathers are removed, the birds are sent to an 'eviscerating' line which removes internal organs and feet, also known as 'paws.' Every single part of the bird is used —for example, chicken feet are considered a delicacy in Asian countries, and feathers are rendered and used as protein in some animal feed.

Step 5: Cleaning and Chilling

After the organs are removed, the carcasses are then cleaned before being inspected. As an added measure to further reduce bacteria, water and an organic rinse may be applied to each bird. Any substance used for this purpose is closely regulated by both the USDA and Food and Drug Administration (FDA) and has been approved for use in food production.

Research has confirmed that the use of these rinses do not pose human health concerns; rather their use does improve the wholesomeness of finished products. Before this process, which includes chilling the birds to a lower temperature to keep fresh and clean, company quality assurance and food safety personnel inspect them once again for quality, food safety and wholesomeness. They follow strict regulatory and company standards for each bird entering the chilling process.

Step 6: Inspection by the U.S. Department of Agriculture (USDA) USDA chicken inspection

During the evisceration process, each bird is inspected by both a member of the processing plant and a USDA inspector. USDA inspectors visually evaluate every inch of each chicken to look for diseases, fecal matter or bruising.

Any birds flagged with issues are removed from the line, condemned, and the issue addressed. It's important to remember that chickens today are the healthiest they've ever been – condemned parts are only a fraction of one percent of total production.

Step 7: Additional Testing

After chickens are chilled, microbiological tests to further ensure food safety are conducted on equipment and products at chicken plants by companies and by the USDA. This includes tests for micro-organisms such as *Salmonella*.

Due to the effectiveness of these processes, the industry experiences a very small percentage of positive Salmonella results, when compared to overall production. According to the most recent USDA data, large establishments (which reflects most of US chicken production) had 2.7 percent positive tests for *Salmonella* on whole birds. This is well below the 7.5 per cent standard set forth by the USDA.

As a reminder, all chicken is safe to eat when it is properly handled and cooked to an internal temperature of 165°F. While the industry goes to great lengths to control micro-organisms responsible for foodborne illness before chicken products leave the plant, it's equally important for consumers to adhere to these very simple cooking instructions to ensure overall food safety.

Step 8: Second Processing

After properly tested and chilled, the carcass is typically cut and deboned to accommodate a variety of different products. Depending on the processing plant, these products may include the fresh or frozen chicken sold in stores, chicken used in restaurants or exported. This includes convenience products sold in 'tray-packs' most commonly seen at your local grocery stores, such as drumsticks, thighs, leg quarters, wings, breasts and more. All-in-all, before reaching consumers, each piece of chicken is inspected for quality, wholesomeness and food safety with more than 300 safety checks throughout the entire process.

Step 9: Packaging Packaged Chicken

Once chicken is cut up into parts, it is packed in trays and wrapped. The wrapped product is then inspected again to ensure that it meets or exceeds both consumer and customer expectations.

Wrapped product is placed into baskets and sent through a 'blast tunnel' to receive a chill. This is done so that the product can have an extended shelf life by keeping it fresh longer. Though the product is significantly cooled during this process, it does not freeze.

After the product is properly chilled, it is weighed and price and safe handling instructions are affixed to the package. Labels on chicken packages must be approved by USDA prior to application on a product. The product then passes through a metal detector for one final check to ensure that there is nothing present in the package that doesn't belong there.

Finally, the product is packaged into boxes where a label is placed on the exterior of the box. This label displays the date packaged, USDA seal of approval and the establishment number of the plant, so that the product can be traced to the establishment where it was produced.

Step 10: Shipping

Finally, the chicken is on its way to your local market. Prior to loading the finished product on trucks, trailers are inspected to ensure they are functioning correctly, and are properly cooled and cleaned.

Once a shipment load is completed, the trailer is sealed with a tamper-evident seal. The seal is not broken until the product arrives at the customer, in order to ensure product safety and wholesomeness.

Retail products are usually delivered to a retailer's warehouse the day after leaving the production plant. Most often, chicken products get placed into company grocery stores the same day of delivery.

IMPACT OF BIOTECHNOLOGY ON POULTRY NUTRITION

The impact of biotechnology in poultry nutrition is of significant importance. Biotechnology plays a vital role in the poultry feed industry. Nutritionists are continually putting their efforts into producing better and more economical feed. Good feed alone will not serve the purpose but its better utilisation is also essential. Dietary changes as well as lack of a healthy diet can influence the balance of the microflora in the gut thus predisposing to digestion upsets. A well-balanced ration sufficient in energy and nutrients is also of great importance in maintaining a healthy gut. A great deal of attention has recently been received from nutritionists and veterinary experts for proper utilisation of nutrients and the use of probiotics for growth promotion of poultry.

In broiler nutrition, probiotic species belonging to *Lactobacillus, Streptococcus, Bacillus, Bifidobacterium, Enterococcus, Aspergillus, Candida,* and *Saccharomyces* have a beneficial effect on broiler performance, modulation of intestinal microflora and pathogen inhibition, intestinal histological changes, immunomodulation, certain haemato-biochemical parameters, improving sensory characteristics of dressed broiler meat and promoting microbiological meat quality of broilers.

Probiotics are 'live micro-organisms which when administered in adequate amounts confer a health benefit on the host'. More precisely, probiotics are live micro-organisms of nonpathogenic and nontoxic in nature, which when administered through the digestive route, are favourable to the host's health.

It is believed by most investigators that there is an unsteady balance of beneficial and non-beneficial bacteria in the tract of normal, healthy, non-stressed poultry. When a balance exists, the bird performs to its maximum efficiency, but if stress is imposed, the beneficial flora, especially *lactobacilli*, have a tendency to decrease in numbers and an overgrowth of the non-beneficial ones seems to occur. This occurrence may predispose frank disease, i.e. diarrhea, or be subclinical and reduce production parameters of growth, feed efficiency, etc. The protective flora which establishes itself in the gut is very stable, but it can be influenced by some dietary and environmental factors. The three most important are excessive hygiene, antibiotic therapy and stress. In the wild, the chicken would receive a complete gut flora from its mother's faeces and would consequently be protected against infection (Fig. 16.1). However, commercially reared chickens are hatched in incubators which are clean and do not usually contain organisms commonly found in the chicken gut. There is an effect of shell microbiological contamination which may influence gut microflora characteristics. Moreover, also HCl gastric secretion, which starts at 18 days of incubation, has a deep impact on microflora selection. Therefore, an immediate use of probiotics supplementation at birth is more important and useful in avian species than in other animals. The chicken is an extreme example of a young animal which is deprived of contact with its mother or other adults and which is, therefore, likely to benefit from supplementation with microbial preparations designed to restore the protective gut microflora.

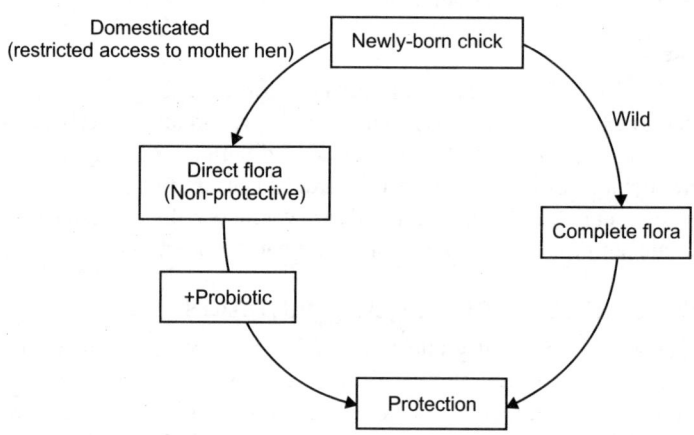

Fig. 16.1: Schematic representation of the concept of probiotics.

The species currently being used in probiotic preparations are varied and many. These are mostly *Lactobacillus bulgaricus*, *Lactobacillus acidophilus*, *Lactobacillus casei*, *Lactobacillus helveticus*, *Lactobacillus lactis*, *Lactobacillus salivarius*, *Lactobacillus plantarum*, *Streptococcus thermophilus*, *Enterococcus faecium*, *Enterococcus faecalis*, *Bifidobacterium* spp. and *Escherichia coli*. With two exceptions, these are all intestinal strains. The two exceptions, *Lactobacillus bulgaricus* and *Streptococcus thermophilus*, are yoghurt starter organisms. Some other probiotics are microscopic fungi such as strains of yeasts belonging to *Saccharomyces cerevisiae* species.

MECHANISMS OF ACTION

Enhancement of colonisation resistance and/or direct inhibitory effects against pathogens are important factors where probiotics have reduced the incidence and duration of diseases. Probiotic strains have been shown to inhibit pathogenic bacteria both *in vitro* and *in vivo* through several different mechanisms. The mode of action of probiotics in poultry includes: (i) maintaining normal intestinal microflora by competitive exclusion and antagonism, (ii) altering metabolism by increasing digestive enzyme activity and decreasing bacterial enzyme activity and ammonia production, (iii) improving feed intake and digestion and (iv) stimulating the immune system.

Probiotic and competitive exclusion approaches have been used as one method to control endemic and zoonotic agents in poultry. In traditional terms, competitive exclusion in poultry has implied the use of naturally occurring intestinal micro-organisms in chicks and poults that were ready to be placed in brooder house. Nurmi and Rantala first applied the concept when they attempted to control a severe outbreak of *S. infantis* in Finnish broiler flocks. In their studies, it was determined that very low challenge doses of *Salmonella* (1 to 10 cells into the crop) were sufficient to initiate *salmonellosis* in chickens. Additionally, they determined that it was during the 1st week post-hatch that the chick was most susceptible to *Salmonella* infections. Use of a *Lactobacillus* strain did not produce protection, and this forced them to evaluate an unmanipulated population of intestinal bacteria from adult chickens that were resistant to *S. infantis*. On oral administration of this undefined mixed culture, adult-type resistance to *Salmonella* was achieved. This procedure later became known as the Nurmi or competitive exclusion concept. The competitive exclusion approach of inoculating day-old chicks with an adult microflora successfully demonstrates the impact of the intestinal microbiota on intestinal function and disease resistance. Although competitive exclusion fits the definition of probiotics, the competitive exclusion approach instantaneously provides the chick with an adult intestinal microbiota instead of adding one or a few bacterial species to an established microbial population. Inoculating day-old chicks with competitive exclusion cultures or more classical probiotics serves as a nice model for determining the modes of action and efficacy of these micro-organisms. Because of the susceptibility of day-old chicks to infection, this practice is also of commercial importance. By using this model, a number of probiotics have been shown to reduce colonisation and shedding of *Salmonella* and *Campylobacter*. Competitive exclusion is a very effective measure to protect newly hatched chicks, turkey poults, quails and pheasants and possibly other game birds, too, against *Salmonella* and other enteropathogens.

Upon consumption, probiotics deliver many lactic acid bacteria into the gastrointestinal tract. These micro-organisms have been reputed to modify the intestinal milieu and to deliver enzymes and other beneficial substances into the intestines. Supplementation of *L. acidophilus* or a mixture of *Lactobacillus* cultures to chickens significantly increased ($P<0.05$) the levels of amylase after 40 days of feeding. This result is similar to the finding of Collington and others, who reported that inclusion of a probiotic (a mixture of multiple strains of *Lactobacillus* spp. and *Streptococcus faecium*) resulted in significantly higher carbohydrase enzyme activities in the small intestine of piglets. The *lactobacilli* colonising the intestine may secrete the enzyme, thus increasing the intestinal amylase activity. It is well established that probiotics alter gastrointestinal pH and flora to favour an increased activity of intestinal enzymes and digestibility of nutrients.

The effect of *Aspergillus oryzae* on macronutrients metabolism in laying hens was observed, of which findings might be of practical relevance. They postulated that active amylolytic and proteolytic enzymes residing in *Aspergillus oryzae* may influence the digested nutrients. Similarly, it was reported that an increase in the digestibility of dry matter was closely related to the enzymes released by yeast. In addition, probiotics may contribute to the improvement of health status of birds by reducing ammonia production in the intestines. Probiotic is a generic term, and products can contain yeast cells, bacterial cultures, or both that stimulate micro-organisms capable of modifying the gastrointestinal environment to favour health status and improve feed efficiency. Mechanisms by which probiotics improve feed conversion efficiency include alteration in intestinal flora, enhancement of growth of nonpathogenic facultative anaerobic and gram positive bacteria forming lactic acid and hydrogen peroxide, suppression of growth of intestinal pathogens, and enhancement of digestion and utilisation of nutrients. Therefore, the major outcomes from using probiotics include improvement in growth, reduction in mortality, and improvement in feed conversion efficiency. These results are consistent with previous experiment of Tortuero and Fernandez, who observed improved feed conversion efficiency with the supplementation of probiotic to the diet.

The manipulation of gut microbiota via the administration of probiotics influences the development of the immune response. The exact mechanisms that mediate the immunomodulatory activities of probiotics are not clear. However, it has been shown that probiotics stimulate different subsets of immune system cells to produce cytokines, which in turn play a role in the induction and regulation of the immune response. Stimulation of human peripheral blood mononuclear cells with *Lactobacillus rhamnosus* strain GG in vitro resulted in the production of interleukin 4 (IL-4), IL-6, IL-10, tumor necrosis factor alpha, and gamma interferon. Other studies have provided confirmatory evidence that Th2 cytokines, such as IL-4 and IL-10, are induced by lactobacilli. The outcome of the production of Th2 cytokines is the development of B cells and the immunoglobulin isotype switching required for the production of antibodies. The production of the mucosal IgA response is dependent on other cytokines, such as transforming growth factor β. Importantly, various species and strains of lactobacilli are able to induce the production of transforming growth factor β, albeit to various degrees. Probiotics, especially *lactobacilli*, could modulate the systemic antibody response to antigens in chickens.

CRITERIA FOR SELECTION OF PROBIOTICS IN THE POULTRY INDUSTRY

The perceived desirable traits for selection of functional probiotics are many. The probiotic bacteria must fulfill the following conditions: it must be a normal inhabitant of the gut, and it must be able to adhere to the intestinal epithelium to overcome potential hurdles, such as the low pH of the stomach, the presence of bile acids in the intestines, and the competition against other micro-organisms in the gastro-intestinal tract. The tentative ways for selection of probiotics as biocontrol agents in the poultry industry are illustrated in Fig. 16.2. Many in vitro assays have been developed for the pre-selection of probiotic strains. The competitiveness of the most promising strains selected by *in vitro* assays was evaluated in vivo for monitoring of their persistence in chickens. In addition, potential probiotics must exert its beneficial effects (e.g. enhanced nutrition and increased immune response) in the host. Finally, the probiotic must be viable under normal storage conditions and technologically suitable for industrial processes (e.g. lyophilised).

EVALUATING PROBIOTIC EFFECTS ON GROWTH PERFORMANCE

Studies on the beneficial impact on poultry performance have indicated that probiotic supplementation can have positive effects. It is clearly evident from the result of Kabir and other that the live weight gains were significantly ($P<0.01$) higher in experimental birds as compared to control ones at all levels during the period of 2nd, 4th, 5th and 6th weeks of age, both in vaccinated and nonvaccinated birds. This result is in agreement with many investigators who demonstrated increased live weight gain in probiotic fed birds. On the other hand, Lan and others found higher ($P<0.01$) weight gains in broilers subjected to two probiotic species. Huang and others demonstrated that inactivated probiotics, disrupted by a high-pressure homogeniser, have positive effects on the production performance of broiler chickens

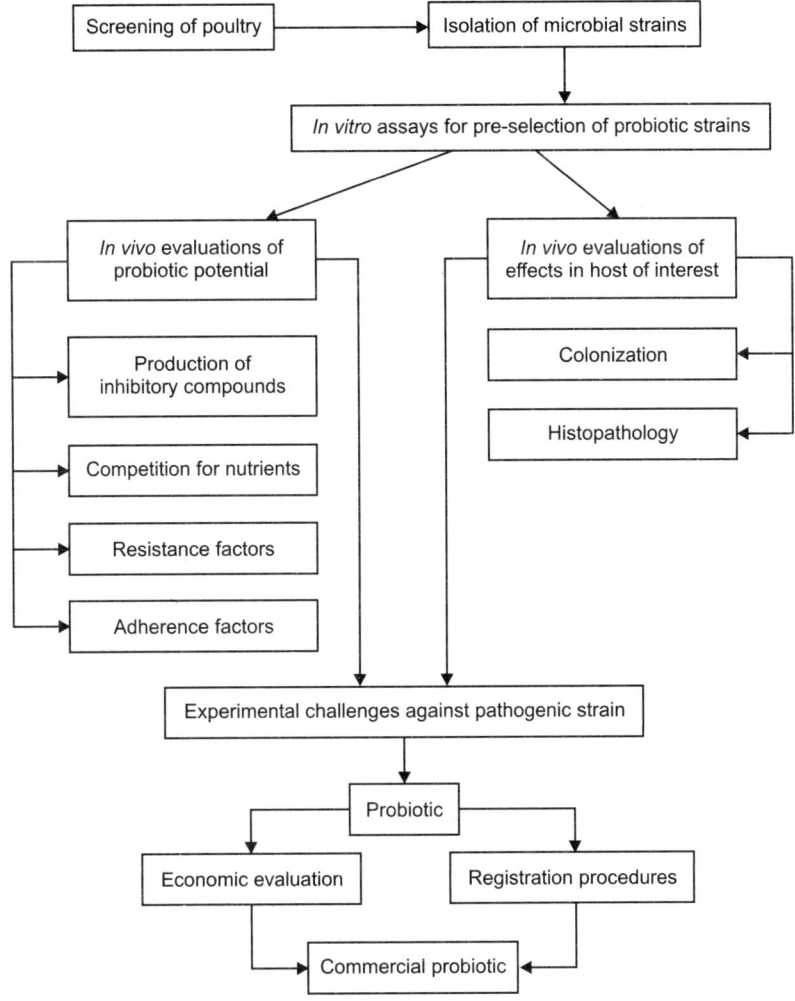

Fig. 16.2: Diagram for selection of probiotics in the poultry industry.

when used at certain concentrations. Kabir and others reported the occurrence of a significantly (P<0.01) higher carcass yield in broiler chicks fed with the probiotics on the 2nd, 4th and 6th week of age both in vaccinated and nonvaccinated birds. Although Mahajan and others recorded in their study that mean values of giblets, hot dress weight, cold dress weight and dressing percentage were significantly (P<0.05) higher for probiotic (Lacto-Sacc) fed broilers. On the other hand, Mutus and others investigated the effects of a dietary supplemental probiotic on morphometric parameters and yield stress of the tibia and they found that tibiotarsi weight, length, and weight/length index, robusticity index, diaphysis diameter, modulus of elasticity, yield stress parameters, and percentage Ca content were not affected by the dietary supplementation of probiotic, whereas thickness of the medial and lateral wall of the tibia, tibiotarsal index, percentage ash, and P content were significantly improved by the probiotic.

EVALUATING PROBIOTIC EFFECTS ON THE INTESTINAL MICROBIOTA AND INTESTINAL MORPHOLOGY

Kabir and others attempted to evaluate the effect of probiotics with regard to clearing bacterial infections and regulating intestinal flora by determining the total viable count (TVC) and total *lactobacillus* count (TLC) of the crop and cecum samples of probiotics and conventional fed groups at the 2nd, 4th and 6th week of age. Their result revealed competitive antagonism. The result of their study also evidenced that probiotic organisms inhibited some nonbeneficial pathogens by occupying intestinal wall space. They also demonstrated that broilers fed with probiotics had a tendency to display pronounced intestinal histological changes such as active impetus in cell mitosis and increased nuclear size of cells, than the controls.

Evaluating Probiotic Effects on Immune Response

Kabir and others evaluated the dynamics of probiotics on immune response of broilers and they reported significantly higher antibody production (P<0.01) in experimental birds as compared to control ones. They also demonstrated that the differences in the weight of spleen and bursa of probiotics and conventional fed broilers could be attributed to different level of antibody production in response to SRBC. They also suggested a positive impact of the probiotic in stimulating some of the early immune responses against *E. acervulina*, as characterised by early IFN-γ and IL-2 secretions, resulting in improved local immune defenses against coccidiosis. Brisbin and others investigated spatial and temporal expression of immune system genes in chicken cecal tonsil and spleen mononuclear cells in response to structural constituents of L. acidophilus and they found that cecal tonsil cells responded more rapidly than spleen cells to the bacterial stimuli, with the most potent stimulus for cecal tonsil cells being DNA and for splenocytes being the bacterial cell wall components.

Evaluating Probiotic Effects on Meat Quality

Kabir and others evaluated the effects of probiotics on the sensory characteristics and microbiological quality of dressed broiler meat and reported that supplementation of probiotics in broiler ration improved the meat quality both at prefreezing and postfreezing storage. Mahajan and others stated that the scores for the sensory attributes of the meat balls appearance, texture, juiciness and overall acceptability were significantly (p60.001) higher and those for flavour were lower in the probiotic (Lacto-Sacc) fed group. Simultaneously, Mahajan and others reported that meat from probiotic (Lacto-Sacc) fed birds showed lower total viable count as compared to the meat obtained from control birds. On the other hand, Loddi and others reported that neither probiotic nor antibiotic affected sensory characteristics (intensity of aroma, strange aroma, flavour, strange flavour, tenderness, juiciness, acceptability, characteristic colour and overall aspects) of breast and leg meats. On the other hand, Zhang and others conducted an experiment with 240, day-old, male broilers to investigate the effects of *Saccharomyces cerevisiae* (SC) cell components on the meat quality and they reported that meat tenderness could be improved by the whole yeast (WY) or *Saccharomyces cerevisiae* extract (YE).

Thus, the concept of probiotics constitutes an important aspect of applied biotechnological research and therefore as opposed to antibiotics and chemotherapeutic agents can be employed for growth promotion in poultry. In past years, men considered all bacteria as harmful, forgetting about the use of the organisms in food preparation and preservation, thus making probiotic concept somewhat difficult to accept. Scientists now are triggering effort to establish the delicate symbiotic relationship of poultry with their bacteria, especially in the digestive tract, where they are very important to the well being of man and poultry. Since probiotics do not result in the development and spread of microbial resistance, they offer immense potential to become an alternative to antibiotics.

SECTION VI

Genetically Modified Foods

SECTION IV

Cereals and Vegetables Biotechnology

Chapter 17

Fruits and Vegetables Biotechnology

INTRODUCTION

Fruits and vegetables contain important vitamins, minerals and fibres and are an important part of health diet, and as are all of great significance in agriculture. However, new varieties with traits achieved through biotechnology processes have played a great role today. Biotechnology of fruit and vegetable production are an aid to conventional breeding and its ability to transfer genes between different organisms. The ability to monitor the absence of such genes in plants is good for plant breeders. In order for the plants to transfer specific genes to one another, it can be added to one's crops by providing a majority of time and effort. Metabolic engineering can allow the genes transfer to fruits and vegetables by adding components to the nutritional values.

The integration of emerging biotechnologies with conventional breeding will greatly facilitate the modification of quality or 'value-added' attributes of fruits and vegetables, (i.e. appearance, organoleptic, nutrition, physiological benefits and safety. Beginning in 1994, the first wave of products from bio-technological applications to vegetables were introduced in pilot test markets. Vine-ripe tomatoes with extended shelf life, processing tomatoes with superior quality and deep red colour, squash with novel virus resistance, and potatoes genetically modified to produce an insect-killing protein are examples of the traits introduced into commercial vegetable varieties with the tools of biotechnology.

Biotechnology, in the simplest and broadest sense, is the utilisation of living organisms or their components to provide useful products or processes. This definition encompasses essentially all of agriculture, since agriculture is based on the production of plants and animals to provide food, fiber and other products for human use. It also includes familiar uses of micro-organisms, such as the yeast used in brewing or baking or the symbiosis of legume plants with nitrogen-fixing bacteria (such as commercial *Rhizobium* inoculants for peas). Thus, humans have been employing biotechnology for at least 10000 years since the origin of agriculture. Recently, however, the term 'biotechnology' has become synonymous with 'genetic engineering.' The cloning and movement of genes among life forms as diverse as bacteria, tomatoes or sheep is now possible. In this context, biotechnology refers to a wide range of enabling technologies that allow the alteration of heritable traits outside of the living organism and the subsequent reintroduction of the new trait into an organism for specific purposes. These new varieties are called transgenic because they require the recombination of genetic materials from different organisms to cause the desired change.

Some of these enabling techniques are applied without resulting in a transgenic plant. These advanced technologies are sophisticated enhancements of methods that have long been common practice. For example, potatoes are propagated primarily by planting the buds, or 'eyes,' present on the tubers. This vegetative reproduction ensures that the new plants are identical to their parents. As one tuber can give only a few new plants, one for each 'eye,' several rounds of seed increase are needed to produce commercial quantities of potato 'seed.' Unfortunately, this also allows the transmission of disease-causing bacteria or viruses that may have infected the parent plant.

Micropropagation, the mass production of identical plants from tiny buds of the parent plant, is a biotechnique that can eliminate these pathogens from the progeny plants while retaining the advantages of vegetative reproduction. Similarly, the individual cells of a plant can be separated, multiplied, and regenerated into whole plants through a process known as tissue culture. In this way, thousands of copies of a single plant can be made and certified pest-free. These techniques are valuable for propagating plants and are an essential component of more advanced methods of genetic modification.

While these tissue culture techniques are valuable, they do not, in principle, modify the characteristics that are encoded in the plant's genes.

For many crops, breeders have relied heavily on the introduction of genes from related wild, but interfertile, species to provide new characteristics. In tomatoes and potatoes, for example, resistance to diseases such as *Fusarium* and *Verticillium* and pests such as Root Knot Nematode have been introduced into commercial varieties from related wild relatives discovered in South America. Virtually all of our domesticated crops, including vegetables, are dependent upon genes derived from a wide array of germplasm, or genetically distinct variants, of each species.

DIRECT AND INDIRECT BENEFITS OF BIOTECHNOLOGY

The long-term beneficial impact of biotechnology on vegetable production will be realised both directly and indirectly. Many of the products will benefit producers and processors by improving the economic efficiency of production. California vegetable production faces many challenges in maintaining productivity while protecting the environment. Biotechnology will contribute to environmental quality protection by reducing the frequency of agricultural pesticide applications and by allowing more environmentally compatible materials and alternative methods to be employed. For example, plant-based pest resistance in both transgenic and conventionally selected varieties will reduce dependence on broad spectrum pesticides. To enjoy continuing benefits from this approach, producers will need to carefully manage their use of such varieties to prevent or delay the development of resistance in the target organisms. Alternative strategies for reducing herbicide use include introducing colour or leaf morphology modifications into seedlings to enhance the use of precision, robotic cultivation.

Biotechnology is also being applied to develop micro-organisms for biological control of pests. For example, an insect-attacking virus (baculovirus) has been modified using rDNA techniques to produce a protein toxin from a gene originally obtained from scorpions. Field trials in California and other states on several vegetable crops have been proposed. These types of products are much slower to be commercialised as they provoke far more public concern than plant-based pest resistance. On the other hand, DNA-based tools allow the detailed analysis of microbial interactions with plants, with other microbes and with the environment in ways never before possible. Two major categories of direct consumer benefit from biotechnology are sensory and nutritional quality and food safety.

CONSUMER BENEFITS

Biotechnology has been applied to improving the sensory properties and shelf life of vegetables. Some of the innovations, particularly in texture or flavour enhancement, have not been highly publicised as originating from plants modified by methods included in a broad definition of biotechnology. These include carrot, potato, celery, pepper, and melon varieties improved through applications of tissue culture. Public awareness of developments in agricultural biotechnology is primarily associated with genetic engineering, particularly of tomato.

In 1994, the first genetically engineered food product reached consumer markets. Calgene's Flavr Savr® tomato received widespread publicity during development and commercial introduction. The targeted benefit, shared by several companies developing new tomato varieties, was to deliver to the market tomatoes with improved flavour and reduced rates of softening or decay.

Although production issues have slowed large scale commercialisation, these enhanced sensory quality tomatoes were generally well received by consumers in test markets and limited retail distribution in United States.

Future products entering the commercial pipeline seek to specifically modify the accumulation and stability of sensory traits by increasing the sweetness and by maintaining the acid balance during maturation and ripening of tomatoes. Both plant and non-plant genes have been introduced into test tomato varieties to increase sucrose accumulation, increase the conversion of sucrose to fructose in the fruit, and sustain organic acids during ripening of tomatoes.

Food safety: Biotechnology is playing a key role in the development of rapid and sensitive diagnostic tools for food borne pathogens, microbial toxins, and other contaminants. Recent outbreaks of *Salmonella* and *Escherichia coli* (*E. coli*) 0157:H7 in fresh vegetables and processed fruit have heightened awareness of the need for proper sanitation and handling. Biotechnology is contributing new detection methods and information about the sources and persistence of foodborne pathogens in production, processing and distribution to the consumer.

Regulation of Biotechnology Products

The Federal agencies responsible for food, human, and environmental safety apply the same basic requirements to the products of biotechnology as to conventional vegetable production and consumption. A coordinated system of review and regulation has evolved and continues to be refined among the United States Department of Agriculture (USDA), Food and Drug Administration (FDA), and the Environmental Protection Agency (EPA).

Various resources are available which detail the review process, permitting, and product registration requirements for safety testing and approvals from research through commercialisation. In general for products related to vegetable production or consumption, the EPA is the lead agency if the product is a microbe or plant with genetically-engineered pesticidal traits.

Current regulations, nonetheless, require special conditions or considerations unique to biotechnology products. A non-transgenic product with the identical product trait will likely have reduced regulatory requirements for approval.

Labelling of Biotech Products

Essentially, if a foodstuff is produced using genetic engineering, this must be indicated on its label. Actual labelling practice, however, is far more complicated and must be planned and regulated with issues such as feasibility, legal responsibilities, coherence and standardisation in mind.

STRATEGIES FOR COMMERCIALISATION

The steps between concept, discovery, reduction to practice and commerciali-sation for biotechnology-based products are very complex. Several branch points in the path are reached at each key phase that affect both technical and business strategic decisions. Access to a broad array of enabling licenses and technologies must be obtained to have 'freedom to operate' in a commercial application. Very broad patents cover the primary tools of biotechnology and are the ground floor access points. A novel application may result in a new patent, but sale of the resultant product requires negotiation of license rights for those technologies utilised in the construction and introduction of any new gene to a plant or microbial product.

BIOTECHNOLOGICAL APPROACHES TO IMPROVE NUTRITIONAL QUALITY AND SHELF LIFE OF FRUITS AND VEGETABLES

The importance of fruits and vegetables (F&V) in the diet of mankind cannot be over emphasised. Many reviews have reported the wide range of determinants of desirable quality attributes in fruits and vegetables such as nutritional value, flavour, colour, texture, processing qualities and shelf-life. The understanding of the fundamental processes that influence fruit set, maturation, and ripening are required to manipulate fruits and vegetable yield and quality. Biotechnology has played a significant role in this respect. Typically, biotechnology technique such as genetic modification is used in F&V to enable plants tolerate the biotic and abiotic stresses, and plant resistances to problematic pests and disease, which may provide higher nutritional contents, and extend the shelf-life of the produce.

The broader definition of biotechnology refers to commercial techniques that use living organisms to make or modify a product, including techniques for improving the characteristics traits of plants and animals, and development of micro-organisms that act on the environment. Additionally, another view of traditional biotechnology covers well established and widely used technologies in brewing, food fermentation, conventional animal vaccine production, and many others based on the commercial use of living organisms.

In recent times, advanced biotechnology techniques involve the use of induced mutations, marker-assisted selection, homologous recombination, genomics, and genetic modifications. The major ones are the tissue or cell culture, cell fusion, embryo transfer, recombinant DNA, and age-old fermentation technique. In the mist of recent global challenges such as increasing population, increasing demand for food, climate change, and water scarcity, plant biotechnology has become a necessity tool for growth and yield performance to meet the food needs of today. The production of quality fruits and vegetables with improved shelf-life is no exception.

Need for Biotechnology in Fruits and Vegetable Production

A number of challenges have called for the application of biotechnology in the production of fruits and vegetables. These are population increase, water shortages, climate change, high perishability or postharvest decays, and short shelf-life associated with fruits and vegetables. Fruits and vegetables by their intrinsic properties require more water and in the face of water scarcity throughout the world, biotechnology will be required to develop fruits and vegetables that can withstand water stress and still be able to produce good crop of high quality and yield. For instance, during the last century, world population rose from 1.6 to 6 billion creating huge challenges for agriculture. However, new technologies increased crop yields drastically so the predicted catastrophic starvation and resulting conflicts did not occur. There are still serious challenges to be faced. Freshwater, vital for agricultural productivity, is becoming scarce and climate change could increase temperature, drought, and uncertainty. New crop varieties need to be developed quickly to meet these challenges and biotechnology will be needed to enhance existing technologies to achieve this. So far, biotechnology has been successfully used to develop insect and herbicide resistance in a limited number of crops such as corn. In the future, the actual metabolism of crop plants will be altered to produce new varieties or species that are tolerant against environmental stresses. In addition, the nutritional value of crops such as rice will be enhanced. Crops will also be used to harvest the sun for biofuels to replace fossil fuels and reduce the emission of CO_2. Some of these new crop plants are already in field trials and will be available to farmers in the near future.

Postharvest decay of fruits and vegetables are a major challenge throughout the world. The degree of postharvest loss through decay is well documented. In the industrialised countries, it is estimated that about 20–25% of the harvested fruits and vegetables are decayed by pathogens during postharvest handling. The situation is far more exasperating in the developing countries, where postharvest decays are often times over 35% , due to inadequate storage, processing and transportation facilities. The use of synthetic fungicides such as benomyl and iprodione to control postharvest diseases of fruits and vegetables is well known in scientific literature. The health and environmental concerns associated with the continuous use of synthetic fungicides have alarmed legal enforcers and consumers to demand greener technology and quality products from the food industry as well as the scientific community. In the past 40 years, microbial antagonists like yeasts, fungi, and bacteria have been used with limited successes to reduce postharvest decays in fruits and vegetables. For instance, fungal diseases like grey mold, powdery mildew, and downy mildew in grapes do notable only cause losses in yield but also reduce wine quality. However, the advances in biotechnology can be employed to develop fruits and vegetables with improved quality and shelf-life. The ability to maintain the quality of stored F&V during postharvest storage is highly related to the physiological, biochemical, and molecular traits of the plant from which they derive. These traits are genetically determined and can be manipulated using genetic breeding and/or biotechnology. Published research results have revealed potential genes, which when manipulated can be used to improve postharvest qualities of crop plants. The application of this biotechnological knowledge should not only lead to major improvements in postharvest storage of fresh fruits and vegetables but as well improved human food supply.

Dynamics of Ripening and Perishability in Fruits and Vegetables

Fruit ripening and softening are major attributes that contribute to perishability in both climacteric and non-climacteric fruits. Fruits and vegetables such as tomato, banana, mango, avocado etc. take about a few days after which it is considered inedible due to over-ripening. The spoilage includes excessive softening and changes in taste, aroma and skin colour. This unavoidable process brings significant losses to both farmers and consumers alike. Even though ripening in F&V can be delayed through several external procedures, the physiological and biochemical changes associated with ripening is an irreversible process and once started cannot be stopped. Ethylene has been identified as the major hormone that initiates and controls ripening in fleshy fruits and vegetables. Influencing ethylene biosynthesis during ripening in fleshy commodities has been the foremost attempt for combating postharvest deterioration.

Biotechnological Approaches applied to Fruits and Vegetables

The transfer of genetic material from one organism into the DNA of another called transgenic application has been widely used in fruits and vegetables. Tolerant plants to biotic and abiotic stress, higher nutritional contents and extended shelf-life are some of the advantages of transgenic plants. In addition, once a useful transformant is obtained, vegetative propagation, which is the normal method of multiplying in several fruit plants, provides unlimited production of the desired transgenic lines. Recently, reports indicate that recombinant DNA technology has been used by scientists to delay ripening in fruits and vegetables in order for farmers to have the flexibility in marketing their produce and ensure consumers good quality produce from their farms. Transgenic grapes were developed for modified auxin production, fungal and virus resistance as well as fruit quality and colour modifications. Costantini and others transformed grape cultivar Thompson Seedless with an ovule-specific auxin-synthesising (DefH9-iaaM) gene and observed that average number of inflorescence per shoot in transgenic grape lines was doubled as compared to control.

Binnie and McManus identified three ACO genes from apple and showed that all three genes express differently. MdACO1 is restricted to fruit tissues, with optimal expression during fruit ripening, MdACO2 expression occurs more predominantly in younger fruit tissue, with some expression in young leaf tissue, while MdACO3 is expressed predominantly in young and mature leaf tissue. Exposure of these fruit to different concentrations of exogenous ethylene showed that various ripening parameters like pulp softening, biosynthesis of volatile aroma compounds, and starch degradation, had different ethylene sensitivities. Their results suggested that the conversion of starch to sugars (an early ripening event) showed a low dependency on ethylene, but a high sensitivity to low concentrations of ethylene. On the other hand, late ripening events such as pulp softening and ester volatile production showed a high dependency on ethylene but were less sensitive to low ethylene concentrations. Nora and others constructed a gene having an antisense of apple 1-aminocyclopropane-1-carboxylic acid (ACC) oxidase (pAP4) and transformed melon leaves. The pAP4 gene detected in transformed leaves and fruits showed a low ethylene production.

Effect of Biotechnological Approaches on Nutritional Quality of Fruits and Vegetables

Many reviews have reported the wide range of determinants of desirable quality attributes in fresh fruits and vegetables such as nutritional value, flavour, colour, texture, processing qualities and shelf-life. Studies found that tomato plants transformed with yeast SAMDC gene under the control of E8 promoter showed improvement in tomato lycopene content, better fruit juice quality, and vine life. Fruit colouration and softening were essentially unaffected, and all the seedlings from first generation seed displayed a normal triple response to ethylene. Over-expression of Nr (wild-type) gene, in tomato using constitutive 35S promoter produced plants that were less sensitive to ethylene. As ethylene receptors belong to a multi-gene family, antisense reduction in expression of individual receptors did not show a major effect on ethylene sensitivity possibly due to redundancy except in case of LeETR4. Antisense plants developed using LeETR4 under the control of CaMV35S promoter exhibited a constitutive ethylene response and were severely affected.

In apples, Dandekar and others reported differential regulation of ethylene with respect to fruit quality components. A direct correlation between ethylene and aroma production during apple ripening has been reported. In a related study with two cultivars of apple, Zhu and others characterised the expression patterns of AAT and ACS gene family members in order to examine the relationship with volatile ester production during on-tree and post harvest ripening. They found that differential expression of AAT genes contributed to phenotypic variation of volatile ester biosynthesis in the apple cultivars.

Nishiyama and others found that there was expressed suppression of the ACO gene of transgenic melon fruit when they examined the cell wall polysaccharide depolymerisation and the expression of the wall metabolism-related genes. There was also a complete inhibition of softening in the transgenic melon fruits but were restored by exogenous ethylene treatment. Post harvest application of 1-MCP after the onset of ripening completely suppressed subsequent softening, suggesting that melon fruit softening is ethylene-dependent.

The transgenic melon fruits were 60% smaller in size and recorded increased sucrose and acidity invertase levels, with degraded chloroplast as a result of decreased photosynthetic rate than the control.

In another study involving avaocado fruits, Tateishi and others found that three cloned members of β-galactosidases (PaGAL2, PaGAL3 and PaGAL4) played a significant role in the cell wall metabolism during fruits growth and ripening. The study of expression pattern of the isozymes by the same authors during avocado ripening found that the accumulation pattern of the gene transcripts and the response to ethylene gave a correlation between AV-GAL1 transcript and isozyme AV-GAL III. This could be the reason why post harvest biotechnology of avocado has been strongly limited in spite of the fact that it provided early clues to the ripening mechanism in fleshy fruit.

Costantini and others transformed grape cultivar Thompson Seedless with an ovule-specific auxin-synthesising (DefH9-iaaM) gene and observed that average number of inflorescence per shoot in transgenic grape lines was doubled as compared to the control. In their studies, they reported that auxin enhanced fecundity in grapes, thus resulting in increased yield with lower production costs. Similarly, Symons and others have shown that brassinosteroids (steroidal hormones) might be implicated in ripening of non-climacteric fruits. The study showed that grape ripening was significantly promoted by exogenous application of brassinosteroids (BRs) and ripening could be delayed by brassinozole, an inhibitor of BR biosynthesis.

Recent advances in recombinant DNA technology and genetic engineering have opened up the possibility to manipulate ripening in fast perishable fruits like banana. Towards this, many genes involved in ripening have been cloned and characterised. Ripening in banana is characterised by a biphasic ethylene production with a sharp early peak followed by a post climacteric small peak. During banana fruit ripening ethylene production triggers a developmental cascade that is accompanied by a huge conversion of starch to sugars, an associated burst of respiratory activity and an increase in protein synthesis. Other changes include fruit softening. Banana fruit softening is attributed to activities of various cell wall hydrolases. Lohani and others reported participation of various cell wall hydrolases in banana softening during ripening.

Effect of Biotechnological approaches on the Shelf Life of Fruits and Vegetables

The shelf-life of transgenic tomato fruits was reported to last for at least 60 days at room temperature without significant change in hardiness and colour. After 15–20 days of treatment of the transgenic fruits with ethylene, most of the tomatoes reached the ripe stage. Antisense transgenic lines of tomato have also been raised with anti-ACO gene to alter ethylene biosynthesis. RNAi technique has also been used to produce tomato fruit with delayed ripening using ACO gene. According to Xiong and others transgenic tomato fruits had a prolonged shelf-life of at least 120 days. In another study with apples, Wang and others showed that null mutation in MdACS3 gene leads to longer shelf-life.

Challenges Associated with Commercialisation of Biotech Fruits and Vegetables

Many research findings revealed that even though biotechnological approaches is seen by the scientific community as a panacea to solve recent increased demands for fruits and vegetables, the technology is more of a scientific jargon than a commercially viable entity.

This is because:

- Dilemma and uncertainties remain up to today regarding the consumption of biotechnological fruits and vegetables. The impasse has created challenges with consumption of genetically modified fruits and vegetables in many countries and some continents mainly due to the complexities surrounding its use.

- Although the first biotech crop to be commercialised was a genetically modified tomato for processing as a consumer tomato paste, there have been comparatively few introductions of biotech fruits and vegetables since then. Reported cases with potential benefits for farmers in developing countries include virus resistant papaya in China, now commercially grown, and, more recently, the high profile case of *Bt* eggplant, or brinjal, in India. Because of the susceptibility of brinjal to the fruit and shoot borer insect, multiple insecticide applications are required to prevent uneconomic losses of yield in this crop.

Approvals of suitably developed and stewarded high-value vegetables and fruits could carry significant benefits for small farmers, because of the relatively high prices of these crops on the market.

The primary conclusion was that the traits did not reach the market not because of poor performance or lack of grower interest but because of regulatory approval uncertainty and prohibitively high and uneconomic development and regulatory costs-a de facto barrier for technology deployment for small holder farmers, even for high-value crops.

A review by Smith has argued that opinion and debate on acceptance of transgenic agricultural biotechnology remains polarised both 'for and against' and is often not aligned with rigorous review and balanced, empirically grounded assessment of socio-economic and community benefits, human safety, environmental considerations such as non-target safety, gene flow, biodiversity and associated risks. At a time when biotech crops have been grown extensively in the Americas and Asia for over 13 years, the precautionary principle prevails in many countries even for the traits embodied in these crops.

In the European Union (EU) for example, genetically modified fruits and vegetables are not allowed on the market and none of the GM plants currently authorised in the EU are intended for direct consumption. For instance, in the case of GM tomatoes, they are lurking in grocery stores in the USA and never received authorisation in the EU. The situation is the same for biotech bananas, apples, wine grapes, and papaya production. Recent reports in the EU member states indicate that whiles countries like Finland, Germany, and Greece have strongly opposed commercialisation of GM crops including fruits and vegetables, Spain and UK do not fundamentally oppose cultivating GM crops but have used the precautionary principle. So the question remains that 'is biotechnology in fruit and vegetable plant production a commercial activity or simply a research jargon?' A pragmatic approach proposed by Godfray and others move the debate forward saying 'genetic modification is a potentially valuable technology whose advantages and disadvantages need to be considered rigorously on an evidential, inclusive, case-by-case basis'. Genetic modification should neither be privileged nor automatically dismissed. In addition to governments' policy on regulation, key factors influencing future availability of biotech fruits and vegetables in developing countries are stewardship capability, and liability of technology providers. Excellence through stewardship reveals that stewardship biotechnology of fruits and vegetables includes not just management of biosafety and compliance with regulatory authorities' requirements but also product quality and integrity along the whole product life cycle right from early research ideas to the withdrawal of crop varieties. The need for steward- ship is fully founded. With the expected rise in numbers of commercialised events around the world, including potentially *Bt* rice in China, concern is growing about the potential for low-level presence of numerous events in trade channels and the food chain in countries without regulatory authorisations. International harmonised solutions need to be found. Otherwise the private sector will remain very cautious about supporting technology releases to the public sector to assist smallholder farmers in developing countries, especially for food crops that could cross national borders or enter international trade channels. Additionally, little impact has been realised to date with fruits and vegetables because of development timescales for molecular breeding and development and regulatory costs and political considerations facing biotech crops in many countries.

New Paradigm Shifts for Commercialisation of Biotech Fruits and Vegetables

Biotechnological approaches offer much potential to increase the development and introduction of improved varieties and as an enabler for greater genetic diversity, but the full benefits are yet to be established. Constraints to the development and adoption of technology-based solutions to reduce yield gaps need to be overcome.

The new paradigm shift proposed includes the: (i) integration of broader thrust that galvanises public and private investment for the development and provision of technology with the creation of seed systems and markets supported by agricultural extension and other services for farmers, (ii) a commitment to increase and sustain funding of agricultural R&D, (iii) the need to break barriers at the policy and operational level to enable formation of public–private partnerships for transformational change in research, product development, and the delivery of seed-embodied technology to farmers, (iv) the integration of low risk chemical fungicides, natural antimicrobial substances, and physical means such as hot water treatment, irradiation with ultraviolet light, microwave, and infrared treatment in the postharvest biocontrol process, (v) the enhancement in the expression of crucial recombinant DNA genes and/or combining genes from different agents in the mass production, formulation and storage, or in response to exposure and contact with parent plant tissue after application, (vi) the use of genetically modified organisms as biocontrol agents to enhance the postharvest quality and shelf-life of fruits and vegetables and (vii) the research towards discovering new DNA genes instead of the ones currently used in practices in that only a small portion of the earth micro flora has been identified and characterised.

FUTURE PROSPECTS

Thus, it is evident that developed biotechnological approaches have the potential to enhance the yield, quality, and shelf-life of fruits and vegetables to meet the demands of the 21st century. However, the developed biotech approaches for fruits and vegetables were more of academic jargon than a commercial reality. To make sure that the current debates and complexities surrounding the registration and the commercialisation of genetically modified fruits and vegetables are adequately addressed, various stake-holders in the industry (policy makers, private sectors, agriculturalists, biotechnologists, scientists, extension agents, farmers, and the general public must be engaged in policy formulations, seed embodiments, and products development. The full benefit of the knowledge can be reaped if there are total commitment by all stakeholders regarding increased and sustained funding, increase agricultural R&D, and less cost and time for registration and commercialisation of new traits.

Thus, the future strategic application of biotechnology is expected to follow a trend of increasing technological sophistication. During the pre-commercial phase, extensive research is required to develop the basic technologies for gene identification, manipulation of desired traits, understanding of plant developmental regulation of genes and the stable reintroduction into competitive varieties.

The progression of products will extend applications from single-gene modifications to multiple-gene introductions and developmentally or environmentally regulated gene expression. The level of participation of growers in the economic consequences of marketing of biotechnology products will be specific to the application.

As in the mainstream of the produce industry, strategic alliances between agricultural biotechnology companies, large grower/shipperssprocessors, and seed companies is a rapidly developing trend. Vertical integration to control all components of the production and marketing system is viewed as the best approach to ensure quality, safety, proprietary protection, and profitability for both conventional and biotechnologyderived products.

Chapter 18

Biotechnology of Mushrooms

INTRODUCTION

There are hundreds of identified species of fungi which, since time immemorial, have made a significant global contribution to human food and medicine. Some estimate that the total number of useful fungi – defined as having edible and medicinal value – are over 2300 species. Although this contribution has historically been made through the collection of wild edible fungi, there is a growing interest in cultivation to supplement, or replace, wild harvest.

This is a result of the increased recognition of the nutritional value of many species, coupled with the realisation of the income generating potential of fungi through trade. In addition, where knowledge about wild fungi is not passed on within families or throughout communities, people have become more reluctant to wild harvest and prefer to cultivate mushrooms instead.

Cultivated mushrooms have now become popular all over the world. There are over 200 genera of macrofungi which contain species of use to people. Twelve species are commonly grown for food and/ or medicinal purposes, across tropical and temperate zones, including the Common mushroom (*Agaricus*), Shiitake (*Lentinus*), Oyster (*Pleurotus*), Straw (*Volvariella*), Lion's Head or Pom Pom (*Hericium*), Ear (*Auricularis*), Ganoderma *(Reishi)*, Maitake (*Grifola frondosa*), Winter (*Flammulina*), White jelly (*Tremella*), Nameko (*Pholiota*), and Shaggy Mane mushrooms (*Coprinus*). Commercial markets are dominated by *Agaricus bisporus*, *Lentinula edodes* and *Pleurotus* spp, which represent three quarters of mushrooms cultivated globally.

IDENTIFICATION OF MUSHROOMS

Identifying mushrooms requires a basic understanding of their macroscopic structure. Most are Basidiomycetes and gilled. Their spores, called basidiospores, are produced on the gills and fall in a fine rain of powder from under the caps as a result. At the microscopic level, the basidiospores are shot off basidia and then fall between the gills in the dead air space. As a result, for most mushrooms, if the cap is cut off and placed gill-side-down overnight, a powdery impression reflecting the shape of the gills (or pores, or spines, etc.) is formed (when the fruit body is sporulating). The colour of the powdery print, called a spore print, is used to help classify mushrooms and can help to identify them. Spore print colours include white (most common), brown, black, purple-brown, pink, yellow, and creamy, but almost never blue, green, or red.

While modern identification of mushrooms is quickly becoming molecular, the standard methods for identification are still used by most and have developed into a fine art harking back to medieval times and the Victorian era, combined with microscopic examination. The presence of juices upon breaking, bruising reactions, odours, tastes, shades of colour, habitat, habit, and season are all considered by both amateur and professional mycologists. Tasting and smelling mushrooms carries its own hazards because of poisons and allergens. Chemical tests are also used for some genera.

In general, identification to genus can often be accomplished in the field using a local mushroom guide. Identification to species, however, requires more effort, one must remember that a mushroom develops from a button stage into a mature structure, and only the latter can provide certain characteristics needed for the identification of the species. However, over-mature specimens lose features and cease producing spores. Many novices have mistaken humid water marks on paper for white spore prints, or discoloured paper from oozing liquids on lamella edges for coloured spored prints.

Microscopic Features of Mushrooms

A hymenium is a layer of microscopic spore-bearing cells that covers the surface of gills. In the nongilled mushrooms, the hymenium lines the inner surfaces of the tubes of boletes and polypores, or covers the teeth of spine fungi and the branches of corals. In the Ascomycota, spores develop within microscopic elongated, sac-like cells called asci, which typically contain eight spores in each ascus. The Discomycetes, which contain the cup, sponge, brain, and some club-like fungi, develop an exposed layer of asci, as on the inner surfaces of cup fungi or within the pits of morels. The Pyrenomycetes, tiny dark-coloured fungi that live on a wide range of substrates including soil, dung, leaf litter, and decaying wood, as well as other fungi, produce minute, flask-shaped structures called perithecia, within which the asci develop.

In the Basidiomycetes, usually four spores develop on the tips of thin projections called sterigmata, which extend from club-shaped cells called a basidia. The fertile portion of the Gasteromycetes, called a gleba, may become powdery as in the puffballs or slimy as in the stinkhorns. Interspersed among the asci are threadlike sterile cells called paraphyses. Similar structures called cystidia often occur within the hymenium of the Basidiomycota. Many types of cystidia exist, and assessing their presence, shape, and size is often used to verify the identification of a mushroom.

The most important microscopic feature for identification of mushrooms is the spores. Their colour, shape, size, attachment, ornamentation, and reaction to chemical tests often can be the crux of an identification. A spore often has a protrusion at one end, called an apiculus, which is the point of attachment to the basidium, termed the apical germ pore, from which the hypha emerges when the spore germinates.

CONTRIBUTION TO LIVELIHOODS

Mushroom cultivation can help reduce vulnerability to poverty and strengthens livelihoods through the generation of a fast yielding and nutritious source of food and a reliable source of income. Since it does not require access to land, mushroom cultivation is a viable and attractive activity for both rural farmers and peri-urban dwellers. Small-scale growing does not include any significant capital investment: mushroom substrate can be prepared from any clean agricultural waste material, and mushrooms can be produced in temporary clean shelters. They can be cultivated on a part-time basis, and require little maintenance.

Indirectly, mushroom cultivation also provides opportunities for improving the sustainability of small farming systems through the recycling of organic matter, which can be used as a growing substrate, and then returned to the land as fertiliser. Through the provision of income and improved nutrition, successful cultivation and trade in mushrooms can strengthen livelihood assets, which can not only reduce vulnerability to shocks, but enhance an individual's and a community's capacity to act upon other economic opportunities.

NUTRITIONAL VALUE OF MUSHROOMS

Mushroom cultivation can directly improve livelihoods through economic, nutritional and medicinal contributions. However, it is essential to note that some mushrooms are poisonous and may even be lethal, thus the need for extra caution in identifying those species that can be consumed as food.

Mushrooms both add flavour to bland staple foods and are a valuable food in their own right: they are often considered to provide a fair substitute for meat, with at least a comparable nutritional value to many vegetables. The consumption of mushrooms can make a valuable addition to the often unbalanced diets of people in developing countries. Fresh mushrooms have a high water content, around 90 per cent, so drying them is an effective way to both prolong their shelf-life and preserve their flavour and nutrients.

Mushrooms are a good source of vitamin B, C and D, including niacin, riboflavin, thiamine, and folate, and various minerals including potassium, phosphorus, calcium, magnesium, iron and copper. They provide carbohydrates, but are low in fat and fibre, and contain no starch. Furthermore, edible mushrooms are an excellent source of high quality protein (reportedly between 19 and 35 per cent), and white button mushrooms contain more protein than kidney beans. In addition to all the essential amino acids, some mushrooms have medicinal benefits of certain polysaccharides, which are known to boost the immune system.

Medicinal Value of Mushrooms

Recently, there has been a spectacular growth in, and commercial activity associated with, dietary supplements, functional foods and other products that are 'more than just food'. Medicinal fungi have routinely been used in traditional Chinese medicine. Today, an estimated 6 per cent of edible mushrooms are known to have medicinal properties and can be found in health tonics, tinctures, teas, soups and herbal formulas. *Lentinula edodes* (shiitake) and *Volvariella volvacea* (Chinese or straw mushroom) are edible fungi with medicinal properties widely diffused and cultivated.

The medicinal properties of mushrooms depend on several bioactive compounds and their bioactivity depends on how mushrooms are prepared and eaten. Shiitake are said to have antitumour and antiviral properties and remove serum cholesterol from the blood stream. Other species, such as *Pleurotus* (oyster), *Auricularia* (mu-er), *Flammulina* (enokitake), *Termella* (yin-er) and *Grifola* (maitake), all have varying degrees of immune system boosting, lipidlowering, anti-tumour, microbial and viral properties, blood pressure regulating, and other therapeutic effects. Mushrooms represent a vast source of yet undiscovered potent pharmaceutical products and their biochemistry would merit further investigation.

Mushroom cultivation activities can play an important role in supporting the local economy by contributing to subsistence food security, nutrition, and medicine, generating additional employment and income through local, regional and national trade, and offering opportunities for processing enterprises (such as pickling and drying).

Income from mushrooms can supplement cash flow, providing either:

- A safety net during critical times, preventing people falling into greater poverty.
- A gap-filling activity which can help spread income and generally make poverty more bearable through improved nutrition and higher income.
- A stepping stone activity to help make people less poor, or even permanently lift them out of poverty.

Livelihood Opportunities

Trade in cultivated mushrooms can provide a readily available and important source of cash income - for men and women and the old, infirm and disabled alike. The role played by women in rural mushroom production can be very significant. Certain parts of the mushroom cultivation process, such as filling substrates in containers and harvesting, are ideally suited for women's participation. Several programmes have enhanced women's empowerement through mushroom production by giving them the opportunity to gain farming skills, financial independence and self-respect.

ESSENTIALS OF MUSHROOM CULTIVATION

Fungi come in many shapes, sizes and colours. Macrofungi is a general category used for species that have a visible structure that produces spores, which are generically referred to as fruiting bodies. Unlike the leaves of green plants, which contain chlorophyll to absorb light energy for photosynthesis (the process by which plants convert carbon dioxide and water into organic chemicals), mushrooms rely on other plant material (the substrate) for their food.

Life Cycle of a Mushroom

The key life cycle stages for fungi are as follows.

Vegetative growth of the mycelium in the substrate

As spores, released from the gills, germinate and develop they form hyphae, which are the main mode of vegetative growth in fungi. Collectively, these are referred to as mycelium, and these feed, grow and ultimately produce mushrooms (in most species). Mycelium appears as microscopic threads similar in appearance to the mould that sometimes grows on bread.

Reproductive growth when the fruit bodies are formed

The appearance of fruiting bodies or mushroom varies according to the species, but all have a vertical stalk (stipe) and a head (pileus or cap).

Production of spores by the mushroom fruit bodies

The underside of the cap has gills or pores from which mushroom spores are produced. The mushroom produces several million spores in its life, and this life cycle is repeated each time the spores germinate to form the mycelium.

Growing systems: Cultivated mushrooms are edible fungi that grow on decaying organic matter. Mushrooms obtain their nutrients in three basic ways:

1. Saprobic, growing on dead organic matter. Saprobic edible fungi can be wild harvested, but are most widely valued as a source of food and medicine in their cultivated forms. They need a constant supply of suitable organic matter to sustain production and, in the wild, this can be a limiting factor in production.

2. Symbiotic, growing in association with other organisms. The majority of wild edible fungi species (e.g. chanterelles - *Cantharellus* and *Amanita* species) are symbiotic and commonly form mycorrhizas with trees, where the fungus helps the tree gather water from a wider catchment and delivers nutrients from the soil that the tree cannot access and the tree provides the fungus with essential carbohydrates.

3. Pathogenic or parasitic, plant pathogenic fungi cause diseases of plants and a small number of these microfungi are eaten in the form of infected host material. Essentially, mushroom species can be cultivated in two ways:

Composted substrates: wheat and rice straw, corn cobs, hay, water hyacinth, composted manure, and various other agricultural by-products including coffee husks and banana leaves.

Woody substrates: logs or sawdust.

Generally, each mushroom species prefers a particular growing medium, although some species can grow on a wide range of materials. This booklet focuses on cultivating saprobic species. Some mushrooms - matsutakes and chanterelles - can also be cultivated by inoculation of tree roots with species that form mycorrhizae that then infect the roots, as with truffles, however this is not covered by this booklet.

Key steps in mushroom production

The basic concept in cultivation is to start with some mushroom spores, which grow into mycelium and expand into a mass sufficient in volume and stored up energy to support the final phase of the mushroom reproductive cycle, which is the formation of fruiting bodies or mushrooms.

The key generic steps in mushroom production – a cycle that takes between one to three months from start to finish depending on species are:

1. Identifying and cleaning a dedicated room or building in which temperature, moisture and sanitary conditions can be controlled to grow mushrooms in.

2. Choosing a growing medium and storing the raw ingredients in a clean place under cover and protected from rain.

3. Pasteurising or sterilising the medium and bags in which, or tables on which, mushrooms will be grown (to exclude other fungi that would compete for the same space - once the selected fungi has colonised the substrate it can fight off the competition).

4. Seeding the beds with spawn (spores from mature mushrooms grown on sterile media).

5. Maintaining optimal temperature, moisture, hygiene and other conditions for mycelium growth and fruiting, which is the most challenging step, adding water to the substrate to raise the moisture content since it helps ensure efficient sterilisation.

6. Harvesting and eating, or processing, packaging and selling the mushrooms.

7. Cleaning the facility and beginning again.

Spawn and inoculation

Mushroom spawn is purchased from specialist mushroom spawn producers, and there are several types or strains of spawn for each type of mushroom. It is not generally advisable for mushroom growers to make their own spawn because of the care needed to maintain the quality of spawn in the production process. Spawn is produced by inoculating a pasteurised medium, usually grain, with the sterile culture (grown from spores) of a particular mushroom species.

The cheapest cultivation system using composted substrate is one where mushrooms are grown in plastic bags (which can be sterilised and reused with new substrate) containing substrate or compost, in a simple building to provide controlled growing conditions. Bottles can also be used, and in other indoor low cost systems wooden trays of different sizes can be arranged in stacks to provide a useful cultivating space. Spawn is added to the sterilised/pasteurised substrate under hygienic conditions, in an enclosed space, and mixed thoroughly to ensure that the mushroom mycelium grows evenly throughout the substrate. Farmers with limited resources can overcome the need to purchase spawn each time a new crop is put down by removing a portion of the substrate colonised by the mushroom spawn from the new crop and using it for spawning the following crop. However, care must be taken to remove only healthy, uninfected substrate colonised fully by the mushroom spawn.

Maintaining suitable growing conditions

The inoculated substrate is put into bags, trays, etc. and transferred to an enclosed and darkened room or building to incubate for a period of up to 12 weeks, depending on the variety of mushroom. If space is limited, plastic bags can be suspended in darkened rooms. Humidity levels are important for the mycelium to colonise over the next two weeks, so water needs to be available, and the temperature controlled accordingly to the variety of mushroom. The crop should be protected from sunlight and strong winds at all times, which can cause the mushrooms to dry out. Humidity can be maintained in the growing room by hanging wet rags at several points around the walls, or watering the floor. Temperature can be regulated by a fire, (electric if available) and cooling could be assisted by using a table fan blowing over a container of water, and air circulating between the sacks should help assist with temperature regulation. It is essential to maintain hygienic conditions over the general cropping area, in order to protect the crop from contamination.

Harvesting cultivated mushrooms

The transition from fully-grown mycelium to produce mushroom fruiting bodies normally requires a change in the environmental conditions, such as temperature decrease and ventilation and humidity increase. Mushrooms fruit in breaks or flushes, and the type and size of mushrooms harvested depend on the type of mushrooms grown and market demand. Mushrooms should be harvested according to market demand, for example, there may be a price premium for small button mushrooms. Generally mushrooms are harvested by hand using sterilised knives to cut the ones that are ready. Pickers should be trained to recognise the appropriate stage for harvesting and be consistent in when the mushrooms are cropped. Handling such a perishable crop should be kept to a minimum to reduce the risk of damage.

Marketing mushrooms

Harvested mushrooms need to be carefully handled and should be kept in a container that allows for air circulation, such as a basket, and care needs to be taken to prevent bruising. The baskets containing mushrooms should be covered to keep flies out and protected from sunlight, high temperatures and draughts. High quality mushrooms that are healthy and clean fetch the best market price. Harvested mushrooms should be taken to market without delay in order to maintain their freshness and quality, or stored in a refrigerated environment or processed. Getting fresh specimens to market is considerably difficult, both for wild fungi and cultivated mushrooms. The physical appearance of fruiting bodies is obviously important and customer preferences must be observed. Some species discolour if the gills or cap are damaged and they must be handled with care. Depending on the soil where the fungi grow, some preliminary cleaning of gills and gaps may be needed to remove particles. Picking fruiting bodies at the correct stage of development is important. As they mature, some species become woody and much less desirable, while others rot away.

Pest and disease management

The basic principle in protecting the mushroom crop from pests and diseases is prevention, largely achieved through good hygiene. As mushrooms are grown mostly in an enclosed environment, the risk of pests and diseases spreading rapidly within the crop is high, so it is important to monitor the crop on a daily basis for incidence of pests and diseases, to prevent losing at least some of the crop. It is also important to sterilise the growing room and the preparation areas on a regular basis. If and when pests or diseases are detected, control measures should be applied immediately. This may involve removing infected mushrooms by carefully picking them off without spreading the disease, then applying a pesticide. The type of pesticide required should be carefully chosen from a list of registered chemicals and used strictly in accordance with the directions given on the label.

Scale of production: Growing systems should be selected that are best suited to local conditions and based on the assets available. Many species of mushrooms can be successfully cultivated on a small-scale, by farmers and other growers who have limited access to resources and vulnerable to risk. It is quite possible for growers to gradually shift from a low-cost system to a higher cost production process, with greater output, when they have gained sufficient skills and income. Large-scale commercial methods of mushroom cultivation require significant financial investment to purchase steam sterilisers, and technical equipment for sterilisation such as auto claves, and often have laboratory facilities to produce spores.

Species selection: Although there has been a great amount of research into mushrooms and their cultivation in temperate climates, there has unfortunately been comparatively little on varieties suitable for tropical climates. Many commercial mushrooms only fruit at around 20°C and are therefore not suitable for tropical regions. Suitable tropical strains are harder to obtain, but some commercial strains can be ordered which fruit at higher temperatures and local laboratories which manufacture spore will be best placed to advise on appropriate varieties and in providing advice on best planting practices.

Key species and their cultivation methods

Detailed in the following pages are a few species of commonly cultivated edible mushrooms that are of global relevance.

Agaricus bisporus: The white button mushroom is the most cultivated mushroom in the world, of particular importance in temperate regions. It is grown in composted substrate and is commonly cultivated in higher technology systems, requiring a low temperature of between 14 to 18°C to provide optimal fruiting conditions for the mushroom and for best results in cultivation.

Pleurotus ostreatus: Oyster mushrooms are a good choice for inexperienced cultivators because they are easier to grow than many other species. In addition, they can become an integral part of a sustainable agriculture system utilising organic waste, can be grown on a small-scale with a moderate initial investment, and convert high amounts of substrate to fruiting bodies thereby increasing potential profitability. Oyster mushrooms were first cultivated on tree logs, and are now commonly grown on sawdust, wheat or rice straw and a variety of high cellulose waste materials, which has shortened the fruiting period to about two months. Cultivation merely involves placing the sterilised and inoculated substrate in plastic bags, and keeping them in the cool and dark. Once the mycelium has grown throughout the substrate, openings are cut through the bag to allow fruiting bodies to develop. Nevertheless, they have some drawbacks. These mushrooms have a soft and fragile structure, the shortest shelf-life of any cultivated mushroom, often displaying bacterial or fungal contamination within a day or two of arriving at the market place. Some people are allergic to the spores, which are produced in profusion when the fruiting bodies start to emerge from growing bags, requiring at minimum a face mask to work in production areas (and aircleaning equipment or respirators in more high technological systems).

Lentinus edodes: Shiitake mushrooms are well suited as a low-input alternative enterprise because they can also be grown on a small-scale with a moderate initial investment. Shiitake are grown outside on logs, or inside and outside on compressed sawdust or in bottles or bags. A cultivation system using compressed sawdust and bags allows for a much faster fruiting cycle and a high level of return, but requires more skilful management than log production. The smaller the diameter of substrate logs, the quicker fruiting bodies appear, although production lasts for a shorter time, and the denser the wood, the longer the production will last. In the same way as substrate, logs are inoculated with spawn from a suitable and locally sourced strain and, as the spawn develops and the mycelium grow throughout the log, it must be kept shaded, moist, and out of the wind. When the mycelium has fully occupied the logs and the temperature and humidity are right for fruiting, the mycelium will initiate tiny 'pinheads' on the surface of the log, which will grow into mushrooms within a few days.

Volvariella volvacea: Paddy straw mushroom cultivation is often integrated with rice production across much of Southeast Asia, including Viet Nam. The mushrooms also grow on substrates in addition to paddy straw, including rice straw, cotton waste, dried banana leaves and oil palm bunch waste, but yields are lower than with paddy straw, where cultivation methods are similar to that of common or oyster mushrooms. Throughout many rural areas, including Indonesia and Malaysia, mushroom growers just leave thoroughly moistened paddy straw under trees and wait for mushrooms to appear.

OPPORTUNITIES AND CHALLENGES

Opportunities

Mushrooms can play an important role contributing to the livelihoods of rural and peri-urban dwellers, through food security and income generation. Mushrooms can make a valuable dietary addition through protein and various micronutrients and, coupled with their medicinal properties, mushroom cultivation can represent a valuable small-scale enterprise option.

Mushrooms can be successfully grown without access to land, and can provide a regular income throughout the year. Growing mushrooms also helps avoid some of the challenges facing collectors of wild fungi, including species identification, obtaining access and permits for collecting, and practicing sustainable harvest. Cultivation is also independent of weather, and can recycle agricultural by-products as composted substrate which, in turn, can be used as organic mulch in growing other horticultural crops, including vegetables.

Mushroom cultivation is highly combinable with a variety of other traditional agricultural and domestic activities, and can make a particularly important contribution to the livelihoods of the disabled, of women and the landless poor who, with appropriate training and access to inputs, can increase their independence and self-esteem through income generation. However, any interventions to promote livelihood activities should be carefully planned, and it is important at the outset to agree with potential mushroom growers: cultivation objectives and the skills, assets and resources available, as well as to identify what market opportunities exist, should they wish to trade their harvested crop. Successful mushroom cultivation for trade requires a good level of individual or collective organisation, and although mushroom cultivation can be a viable small-scale business, any investment in a growing scheme can be risky. Cooperatives and community groups can collaborate in set-up and production costs, harvesting and marketing. Working in joint ventures or partnerships with regional agroindustries, universities or wholesalers can help reduce vulnerability and risk for small-scale producers, and provide access to training and other forms of support.

Challenges in Cultivation Arger Scale Mushroom

Establishing larger scale mushroom cultivation systems can be more labour and management intensive. All production systems, to some extent, are vulnerable to sporadic yields, invasions of 'weed' fungi, insect pests, and unreliable market prices for traded goods. Moving from cultivating mushrooms for subsistence use to commercial production and marketing can be quite challenging to local growers. One of the most important aspects of growing mushrooms for commercial purposes is the ability to maintain a continuous supply for chosen market outlets, and if the mushroom enterprise is one of many livelihood activities, producers need to become multi-skilled to manage several enterprises successfully. The initial challenges which mushroom growers have to face include determining the most suitable mushroom to grow and identifying a spawn supplier, organising available resources to develop a growing system, and assessing requirements for supplying different marketing outlets. In spite of these, starting with home production is an advisable approach.

Some mushrooms have been given bad press because of poisonings, which fortunately are generally rare and have been associated with events, including: young children collecting indiscriminately and eating raw mushrooms, immigrants arriving in a new country and incorrectly identifying a local species that turns out to be poisonous, food shortages and economic hardship forcing people to hunt for food, and different physiological responses to an 'edible' fungus. Other health risks can include allergies to different mushroom spores.

Mushrooms have not often been actively promoted in the past by agricultural ministries of developing countries. Various reasons have been cited for this neglect, including: a lack of technical capacity in production techniques with poorly equipped government supported advisory services resulting in interested farmers having to seek technology on their own, comparatively few studies on tropical mushrooms, and a lack of technical skills to produce spawn with suitable strains often hard to find. The market can present an additional constraint in some regions as the prices of mushrooms are out of the range of most local consumers and unable to compete with other protein sources like beef, beans or eggs for a place in the average family diet. In conclusion, many of the challenges which face mushroom cultivation activities are not uncommon to other challenges still faced by small-scale rural producers. As a livelihood diversification option, mushroom cultivation has enormous potential to improve food security and income generation, which in turn can help boost rural and peri-urban economic growth.

Edible Mushrooms

Mushrooms are used extensively in cooking, in many cuisines (notably Chinese, Korean, European, and Japanese). Though neither meat nor vegetable, mushrooms are known as the 'meat' of the vegetable world. Most mushrooms sold in supermarkets have been commercially grown on mushroom farms. The most popular of these, Agaricus bisporus, is considered safe for most people to eat because it is grown in controlled, sterilised environments. Several varieties of *A. bisporus* are grown commercially, including whites, crimini, and portobello. Other cultivated species available at many grocers include Hericium erinaceus, shiitake, maitake (hen-of-the-woods), Pleurotus, and enoki. In recent years, increasing affluence in developing countries has led to a considerable growth in interest in mushroom cultivation, which is now seen as a potentially important economic activity for small farmers.

Separating edible from poisonous species requires meticulous attention to detail, there is no single trait by which all toxic mushrooms can be identified, nor one by which all edible mushrooms can be identified. People who collect mushrooms for consumption are known as mycophagists, and the act of collecting them for such is known as mushroom hunting, or simply 'mushrooming'. Even edible mushrooms may produce allergic reactions in susceptible individuals, from a mild asthmatic response to severe anaphylactic shock. Even the cultivated A. bisporus contains small amounts of hydrazines, the most abundant of which is agaritine (a mycotoxin and carcinogen). However, the hydrazines are destroyed by moderate heat when cooking.

A number of species of mushrooms are poisonous, although some resemble certain edible species, consuming them could be fatal. Eating mushrooms gathered in the wild is risky and should only be undertaken by individuals knowledgeable in mushroom identification. Common best practice is for wild mushroom pickers to focus on collecting a small number of visually distinctive, edible mushroom species that cannot be easily confused with poisonous varieties.

Toxic Mushrooms

Many mushroom species produce secondary metabolites that can be toxic, mind-altering, antibiotic, antiviral, or bioluminescent. Although there are only a small number of deadly species, several others can cause particularly severe and unpleasant symptoms. Toxicity likely plays a role in protecting the function of the basidiocarp: the mycelium has expended considerable energy and protoplasmic material to develop a structure to efficiently distribute its spores. One defense against consumption and premature destruction is the evolution of chemicals that render the mushroom inedible, either causing the consumer to vomit the meal, or to learn to avoid consumption altogether. In addition, due to the propensity of mushrooms to absorb heavy metals, including those that are radioactive.

Medicinal Properties of Mushrooms

Some mushrooms are used or studied as possible treatments for diseases, particularly their extracts, including polysaccharides, glycoproteins and proteoglycans. In some countries, extracts of polysaccharide-K, schizophyllan, polysaccharide peptide, or lentinan are government-registered adjuvant cancer therapies, even though clinical evidence of efficacy in humans has not been confirmed. Historically in traditional Chinese medicine, mushrooms are believed to have medicinal value, although there is no evidence for such uses.

Other Uses of Mushrooms

Mushrooms can be used for dyeing wool and other natural fibres. The chromophores of mushroom dyes are organic compounds and produce strong and vivid colours, and all colours of the spectrum can be achieved with mushroom dyes. Before the invention of synthetic dyes, mushrooms were the source of many textile dyes.

Some fungi, types of polypores loosely called mushrooms, have been used as fire starters (known as tinder fungi). Mushrooms and other fungi play a role in the development of new biological remediation techniques (e.g. using mycorrhizae to spur plant growth) and filtration technologies (e.g. using fungi to lower bacterial levels in contaminated water).

Chapter 19

Genetically Modified Fruits

INTRODUCTION

Fruits are major ingredients of human diet and provide several nutritional ingredients including carbohydrates, vitamins and functional food ingredients such as soluble and insoluble fibers, polyphenols and carotenoids. Biochemical changes during fruit ripening make the fruit edible by making them soft, changing the texture through the breakdown of cell wall, converting acids or stored starch into sugars and causing the biosynthesis of pigments and flavour components. Fruits are processed into several products to preserve these qualities. Because of various health benefits associated with the consumption of fruits and various products derived from fruits, these are at the centre stage of human dietary choices in recent days.

The selection of trees that produce fruits with ideal edible quality has been a common process throughout human history. Fruits are developmental manifestations of the seed-bearing structures in plants, the ovary. After fertilisation, the hormonal changes induced in the ovary result in the development of the characteristic fruit that may vary in ontogeny, form, structure and quality. Pome fruits such as apple and pear are developed from the development of the thalamus in the flower. Drupe fruits, such as cherry, peach, plum, apricot and so on, are developed from the ovary wall (mesocarp) enclosing a single seed.

Some of the important fruits are discussed below:

APPLES

Apple fruits turn brown quickly after being sliced or bitten. Although this is a natural phenomenon, browning has been considered as an undesirable trait that often discourages consumption and causes unnecessary waste. Apple browning is caused by a polymer compound of pigment that primarily consists of quinones. Quinones are produced from phenols, which are common in apple fruit cells, through the action of a class of enzymes called polyphenol oxidases (PPOs). In intact fruit cells, phenols and PPOs are separated in different compartments. When cells are damaged by slicing or biting, phenols and PPOs come into contact and react to produce quinones, which will eventually become the pigmented polymer compound (together with amino acids and proteins). Therefore, PPOs are the key enzymes responsible for apple browning. To stop or minimise fruit browning, reducing the activities of PPOs has

241

long been considered as an effective approach. The 'non-browning' apple is genetically engineered to keep from going brown after being cut. When apple flesh is cut and exposed to oxygen, it begins to brown but the GM apple or 'Arctic Apple' will not brown for 15 to 18 days.

'Foreign genes': There are no 'foreign genes' in Arctic Apples. All that is different about them is that certain apple genes are 'turned off.' Most genes in cells of an apple or any other organism are turned off most of the time. The genes to grow roots or leaves are turned off in apple fruits just as our genes to develop the features of an eye are turned off in all the other parts of our body. The scientists that developed the Arctic Apple simply used a natural mechanism to turn off the genes that make the enzymes that turn apples brown when they are cut.

Organic growers: For thousand of years, people have known that if you find a fruit that is good, you need to propagate it from cuttings or buds, not seeds. These are grafted onto rootstocks that have been selected for pest resistance and in most cases for dwarfing so the trees don't get too tall to safely pick. Actually, to get a good crop, growers need to make sure that the bees that visit each tree have typically gotten pollen from some other variety of apples, even crab apples, because otherwise fruit set is less likely.

There is a small chance that a bee will carry pollen from a GMO apple to a few flowers in an organic block. A few of the seeds in those organic apples might be pollinated with the GMO pollen. Only a small part of the seeds (the embryo) in those apples will have that extremely minor change. Apple seeds shouldn't be eaten. They contain compounds that can generate cyanide a rather potent poison. Consumers routinely discard them.

Bees and Pollen: Apples are pollinated by bees. The bees will stay very close to their hive when there is enough food (like when an orchard is in bloom) and so the risk of contamination from bees is small.

Apple seeds: Apple seeds do not breed true (they may retain some traits of their parents but the resulting trees are not an exact copy of their parent) so grafting is used to propagate apple trees rather than seeds. However if apple seeds pollinated with GM pollen germinate they will result in GM apple trees (trees expressing the GM trait or carrying the new gene sequence). There are many ways that GM apple seeds can spread in our environment such as by humans discarding apple cores, cores in compost piles, seeds scattered by animals, and deliberate plantings.

GM pollen and fate of apples: If an apple tree is pollinated with GM pollen, the genes would be present in the resulting apple seeds, not the apple flesh. If pollen from GM apple trees moves into a non-GE apple orchard, some seeds in apples from the non-GM trees that were pollinated could carry the new gene sequence and could express the new GM trait.

Brand protection: Technology opponents have long known that they have the most leverage if they can threaten an entity with a valuable consumer brand with potential, controversial attention. The apple industry is concerned that even though the Arctic apples would be clearly labelled as such, activists could smear the entire apple industry 'brand' and make consumers afraid that they might be eating something the don't want.

GM apple is not necessary: The GM technology is unnecessary as there are already techniques that slow browning in apples in our kitchens we use lemon juice, and the food service industry uses ascorbic acid (vitamin C). There are also varieties of naturally slow-browning apples, such as Ambrosia.

CITRUS

Citrus is one of the most important commercial and nutritional fruit crops in the world, hence it needs to be improved to cater to the diverse needs of consumers and crop breeders. Genetic manipulation through conventional techniques in this genus is invariably a difficult task for plant breeders as it poses various biological limitations comprising long juvenile period, high heterozygosity, sexual incompatibility, nucellar polyembryony and large plant size that greatly hinder cultivar improvement. Hence, several attempts were made to improve Citrus sps. by using various *in vitro* techniques. Citrus sps. are widely known for their recalcitrance to transformation and subsequent rooting, but constant research has led to the establishment of improved protocols to ensure the production of uniformly transformed plants, albeit with relatively low efficiency, depending upon the genotype.

Genetic modification through *Agrobacterium*-mediated transformation has emerged as an important tool for introducing agronomically important genes into Citrus sps. Somatic hybridisation has been applied to overcome self and cross-incompatibility barriers and generated inter-specific and inter-generic hybrids. Encouraging results have been achieved through transgenics for resistance against viruses and bacteria, thereby augmenting the yield and quality of the fruit. Now, when major transformation and regeneration protocols have sufficiently been standardised for important cultivars, ongoing citrus research focuses mainly on incorporating such genes in citrus genotypes that can combat different biotic and abiotic stresses.

Citrus are subject to many biotic and abiotic stresses and markets are continuously requiring fruits of higher quality. These pose important problems in most citrus growing areas that can only be solved with the establishment of citrus improvement programs to recover new and healthy genotypes to be used as rootstocks and varieties. Success of traditional citrus breeding strategies is limited by the peculiar genetic and reproductive characteristics of citrus. Biotechnology offers different approaches that can overcome many of these limitations. Some of the technologies used in Spain are:

1. Shoot tip grafting *in vitro*: It is used worldwide to recover plants free of all known citrus pathogens, and already hundreds of millions of trees originated in this technique have been planted in the field.

2. Embryo rescue: It is used in large programs to recover triploid seedless hybrid varieties from aborted seeds from $2n \times 4n$ and $2n \times 2n$ crosses.

3. Protoplast fusion: It allows the production of allotetraploid hybrids among sexually compatible and incompatible parents to be used as rootstocks or as parentals for interploid breeding. Fusions between diploid and haploid cell lines are also used to produce triploid hybrids.

4. Genetic transformation. Efficient protocols for transformation of several genotypes have been developed and introduction in citrus of genes of potential agronomical interest has been accomplished.

CHERRY

While worldwide production of genetically modified crops has been increasing dramatically since 1996, the use of transformation technologies for Prunus crops has been rather modest when compared to the major commercialised genetically engineered crops. To date, only transgenic plum 'HoneySweet' with PPV (plum pox virus) resistance has received USDA non-regulated status. The lack of efficient biotechnology platforms is considered to be the 'bottleneck' preventing the improvement of Prunus species through genetic transformation. To overcome this limitation for cherry species, previous efforts have focused on optimising shoot micropropagation, plant regeneration, gene delivery, and stable transformation. To date, plant regeneration has been reported for 19 sweet cherry (*Prunus avium L.*) cultivars, 12 sour cherry (*Prunus cerasus L.*) cultivars, and 24 genotypes for other cherry species such as cherry hybrids, black cherries (*Prunus serotina Ehrh.*), and wild cherries (*Prunus avium L.*). Stable transgenic plants have been reported for sour cherry 'Montmorency', cherry rootstocks 'Gisela 6' and

'Colt', and *Prunus subhirtella* Miq. 'Rosa', but only reporter genes were used in these successful transformations. Most recently, an interfering RNAs (RNAi) vector targeted to prunus necrotic ringspot virus (PNRSV) has been transformed into the hybrid cherry rootstocks. This RNAi strategy, utilising transformed rootstocks to achieve virus-resistance in the scion, minimises concerns about 'transgene-flow' and 'foreign protein production' in commercial scion cultivars, and it is therefore a potential approach for engineering virus resistance for fruit-bearing cherry genotypes without direct transformation of the fruiting genotype itself. As more genomic resources for Prunus species become available, more genes of interest will be identified and isolated. In the future, success in cherry genetic engineering will still depend on availability of desirable target genes and reliable transformation systems with efficient plant regeneration, efficient gene delivery, and effective selection without using antibiotics or herbicides.

GUAVA

Fruits and vegetables have specific healthy attributes. Guava is the most nutritious of all edible fruits. High pectin contents make guava suitable for jelly making. Guava has antidiarrheal, antibacterial, antiamoebic and antispasmodic activities. High concentrations of several vitamins, dietary fiber, carotenoids, lectins, saponins, tannins, phenols, triterpenes, and flavonoids altogether make guava therapeutically an important fruit. Ascorbic acid in guava fruit far exceeds that in citrus. Consumption of guava fruit is reported to lower blood cholesterol, triglycerides, hypertension and some cardiac problems. Due to high quercetin contents, guava leaves are used to develop a phytomedicine for diarrhea. Since guava plant is susceptible to frost, genetic improvement to enhance cold hardiness of guava germplasm is necessary for adapting and cultivating it in temperate climates. Efficient and reliable biotechnology using *in vitro* plant regeneration amenable to genetic transformation will ensure cold hardiness enhancement in guava. Development of biotechnology protocols for morphogenesis and somatic embryogenesis along with production of edible quality guava fruits and their nutritional analysis has been achieved using various tissue explants from mature guava trees growing at Fort Valley State University. For guava micropropagation, researchers have developed and germinated 'synseeds' from somatic embryos. Successful preliminary work on genetic transformation has paved the way for incorporating cold hardy genes into the guava genome. Guava (*Psidium guajava* L.), also called 'apple of tropics,' is immensely nutraceutical and horticulturally important. Being a tropical plant, it cannot stand temperatures below 25° F and needs frost protection to grow in temperate regions. To adapt in cold climate, cold hardy guava cultivars are needed. Conventional ways are uneconomic in time and efforts. Still, transgenic plants developed using biotechnological approaches of tissue culture and rDNA technology, appear to have great potential. Thus, protocols for *in vitro* propagation of guava were developed via organogenesis and somatic embryogenesis using nodal explants from mature trees and young zygotic embryos, respectively.

Nodal explants induced multiple shoots when cultured on MS medium fortified with KIN, BAP and Ad.S. Adding a $(NO_3)_2$ to medium was useful to prevent *in vitro* shoot tip browning of adventitious shoots. Rocker liquid culture greatly increased growth of multiple shoots compared to the agar-based medium. It appears to be a good tool for woody plant tissue culture. Induction of somatic embryos in guava was also achieved on MS medium supplemented with IAA auxin. About 80% to 90% somatic embryos germinated normally. To achieve Agro-bacterium-mediated gene transfer in guava, on-going co-cultivation of organogenic tissues of guava is to optimise protocols for freeze tolerance gene (CBF1, CBF2, CBF3) transfer. Plasmid vectors containing selectable markers (nptII gene for antibiotic selection and GUS reporter gene as scorable gene mediated selection), with CaMV 35S promoter gene has been introduced into guava tissues and the resultant plants showed antibiotic resistance.

APRICOTS

Apricot (*Prunus armeniaca L.*) is a species originated in China and Central Asia, from Tien Shan to Kashmir. Finally, different studies are being performed to the identification of genomic regions involved in fruit quality traits by joining the phenotypic data with the molecular characterisation of different apricot progenies. The related QTL identification linked to quality traits is the first step in order to develop specific molecular markers which will offer the opportunity of optimise apricot breeding programs for fruit quality by introducing an early marker assisted selection (MAS) strategy.

Molecular breeding strategies and genome sequencing: Characterisation of genetic variability in apricot species has been traditionally based on morphological traits. However, such traits are not always available for analysis, and do not provide sufficient information to trace the expansion of apricot from the centers of origin, to expose the major genetic events in the evolution of cultivars or to identify and characterise genotypes during the breeding process. Molecular marker technology offers several advantages over the sole use of morphological markers, it have proved to be a powerful tool for solving the above problems, and have become an essential tool for the molecular studies.

Isoenzymes were the first molecular markers to be utilised in apricot. Nevertheless, the utilisation of these markers is limited due to the small number of *loci* that can be analysed with staining methods and the low variation observed. In this sense, the availability of DNA markers provides a new opportunity to evaluate plant diversity. Restriction fragment length polymorphism (RFLP) markers are based on the differential hybridisation of DNA fragments from restriction-enzyme digestion. RFLP markers are codominant and can detect a virtually unlimited number of markers. These markers have been used in the molecular characterisation of apricot genotypes and also in the development of the first genetic linkage map in apricot.

More recently, the utilisation of PCR-based markers has increased the opportunities for molecular characterisation and mapping of populations. Random amplified polymorphic DNA (RAPD) markers are based on the PCR amplification of random locations in the genome. RAPDs are characterised by using arbitrary primers. A single oligonucleotide is utilised for the amplification of genomic DNA. RAPD techniques have been successfully used in apricot for identifying cultivars and also for mapping population. In contrast to isoenzymes and RFLPs, RAPDs are dominant markers. This feature, as well as their variable degree of repeatability and problems in transferring across populations, limits their utilisation to map construction. These difficulties can be overcome by converting RAPDs to sequence-characterised amplified regions or SCARs.

On the other hand, amplified restriction fragment length polymorphism (AFLP) technology is a powerful DNA fingerprinting technology based on the selective amplification of a subset of genomic restriction fragments using PCR developed later than RAPD. These markers have been mainly used in apricot for molecular characterisation of varieties and genetic linkage mapping. AFLP has a number of advantages over the RAPD: more *loci* analysed and better reproducibility of banding although presents the inconvenient of the difficulty to the use in routine.

Presently, simple sequence repeat (SSR or microsatellite) markers, also based on the PCR technique but with specific primers, are the best suited markers for the assessment of genetic variability within crop species because of their high polymorphism, abundance, and codominant inheritance. SSRs are extremely abundant and dispersed relatively evenly throughout the genome. DNA flanking SSRs is often well conserved in related species, which allows the cross-species amplification with the same primer pairs in related species. This situation gave the first application of SSR in apricot using primers developed in other *Prunus* species such as peach. More recently, primer sets using apricot sequence information have been developed. These markers have been mainly used in apricot for molecular characterisation of apricot varieties, identification of accidental pollination in apricot progenies and genetic mapping.

Genetic linkage analysis: Genetic linkage analysis was initially performed in apricot using the combination of different molecular makers including RFLPs, RAPDs, AFLPs and SSRs. The use of RFLPs provides a virtually unlimited source of high quality markers, located all over the genome. In this sense, the most recent utilisation of PCR-based markers has increased the opportunities for mapping and tagging a wide range of traits. New genetic linkage maps in apricot has been developed using a reduced type of markers. AFLPs have been other wide used markers allowing the detection of a higher level of polymorphism in apricot than RFLPs or RAPDs. Finally, SSR markers are currently becoming the markers of choice for genetic mapping in apricot.

The genotype-dependency of apricot transformation could be overcome by transforming meristematic cells and therefore avoiding the bottleneck of most transformation procedures. Such a procedure has been developed for apricot allowing the transformation of four different cultivars. Although only marker genes have been introduced in apricot commercial cultivars, transgenic apricots plants expressing the *Plum pox virus* (PPV) coat protein gene were successfully developed from seed derived tissue. In this case, the coat protein gene of PPV was used to introduce coat protein mediated resistance. In plum (*Prunus domestica* L.), PPV resistance has been successfully achieved by post trascripcional gene silencing.

Future work perspective: A key point for apricot breeding is to maintain production and consumers confidence by assuring acceptable production and quality levels through the introduction of new apricot cultivars on the market. Until now, apricot cultivars have been mainly generated through controlled crosses and open pollination. Additional advantages encouraging the utilisation of the new biotechnologies to apricot breeding include high levels of synteny between genomes and a well established international network of cooperation among researchers. In this sense, future works regarding marker assisted selection (MAS) of apricot breeding must include the comparative mapping of different progenies. Genomic methodologies including expressed sequences tags (ESTs) cloned gene analogs (CGAs) and single point mutations (single nucleotide polymorphisms, SNPs) may make it possible to discover genes of interest in quality selection in apricot. More recent efforts are being oriented to the elaboration of physical maps, the development of quick gene sequencing and cloning tools, and the complete sequencing of the peach genome to develop efficient molecular markers applicable to assistant selection in peach breeding programs. Finally, the increasing availability of biotechnological techniques such as genetic transformation further complements *in vitro* culture opportunities. In this sense, at this time several apricot genotypes genetically modified are being assayed although to date there is no commercial varieties.

PAPAYA

Papaya is cultivated primarily in tropical and sub-tropical regions. It is an important part of the diet in many developing countries. It is the source of the enzyme papain that is used in the pharmaceutical and food industries. For several decades the global production of papaya has been threatened by ringspot disease caused by the papaya ringspot virus (PRSV).

PRSV was identified for the first time in 1945 in the American state of Hawaii. Since then there have been outbreaks of ringspot disease in other areas of the world where papaya is cultivated. Hawaii is responsible for just 0.1% of global papaya production. Researchers from Cornell University (USA) and the University of Hawaii developed a genetically altered variety of papaya that is resistant to PRSV. A remarkably efficient development and approval procedure allowed the GMO papayas to be introduced as soon as in 1998. In less than four years' time papaya production was back to pre-PRSV levels. Hawaiian papaya cultivation had been saved. In the meantime, Hawaii has started exporting its biotech papayas to Canada and Japan. Due to the specificity of the virus and the virus resistance mechanism Hawaiian GMO papayas are resistant to the Hawaiian PRSV in particular. However, the developments on Hawaii have inspired and stimulated other papaya cultivating countries to use the same methods to develop virus resistant papayas for their local markets. Resistant papaya varieties have now been developed in Brazil, Taiwan, Jamaica, Indonesia, Malaysia, Thailand, Venezuela, Australia and the Philippines.

Role of biotechnology: The development of the virus resistant GMO papayas shows that in certain cases using biotechnology can improve plants more efficiently than traditional plant breeding. By using traditional crossing techniques breeders are limited to the genetic information that is available within a certain plant species. In the case of PRSV, the genetic information required for resistance is not present in papaya (*Caricapapaya*). This resistance is known to be present in the *Vasconcellea*papaya. The *Vasconcellea* papayabelongs to the same family as the *Carica papaya*, but a different genus. Attempts have been made since 1958 to transfer the PRSV resistance in *Vasconcellea*to *Carica* using intergenetic crossing (between genera). Such hybridisations are exceedingly inefficient and not only do they require the development of special procedure and laboratory techniques, they also rely on a great deal of luck. Over the course of more than 50 years only a small number of infertile offspring have been generated. Filipino and Australian researchers have recently reported the first fertile, classically bred *Carica papaya*plant that is resistant to PRSV. Despite the very significant contribution that classical breeding has made, and will continue to make in the future, the classically bred plant was far too late to save the damaged Hawaiian papaya plantations.

Following evaluation of all the food safety and environmental factors, the import of the Rainbow papayawas approved in Japan on 1 December 2011.

Thus, the genetically modified crops are often associated with multinationals and large scale cultivation. The story of the GMO papaya demonstrates that this does not necessarily always have to be the case. The Hawaiian GMO papaya was developed by the public sector and the intellectual property rights were transferred to the local papaya industry. The GMO papaya has been cultivated since 1998 to deal with the papaya ringspot virus, a pathogen that does not just cause serious losses in yield but can also make commercial papaya cultivation impossible. Thanks to the use of GMO papayas cultivation has been able to continue on Hawaii. This success story inspired many papaya producing countries to look for a similar solution to deal with local papaya viruses. Until now virus resistant GMO papayas are only being cultivated in Hawaii and China but more countries are ready to make use of biotechpapayasin the fight against the papaya ringspot virus.

EGG PLANT

Eggplant is a member of the nightshade family as are tomatoes, peppers, potatoes and sweet potatoes. As such, eggplant is technically a fruit. As with tomatoes and other of nightshade veggies there are a number of varieties of eggplant. The large dark purple pear shaped eggplant is the most common. The smaller version of the larger purple skinned eggplant is often called Italian or baby eggplant. These have a somewhat more intense flavour and the flesh is much more tender. The straight thin eggplants known as Japanese or Asian eggplant have thin delicate skins like Italian eggplant but the flesh is sweeter. The colour ranges from dark purple to a striped purple as well as a light amethyst.

There is a wide genetic diversity in the cultivated as well as the wild species of eggplant. Cultivated varieties of eggplant are susceptible to a wide array of pests and pathogens as well as to various abiotic stress conditions. In contrast, the majority of wild species are resistant to nearly all known pests and pathogens of eggplant and thereby are a source of desirable traits for crop improvement. Transgenic eggplants for insect resistance, for the production of parthenocarpic fruits and abiotic stress tolerance have been accomplished. However, transgenics of eggplant are yet to be developed for improvement of other agronomic traits, including disease and pest resistance, and quality and shelf life of fruits.

FSB resistant eggplant: FSB resistant (FSBR) eggplant is an insect resistant eggplant developed with the help of biotechnology. Also called *Bt* eggplant or *Bt* brinjal, it produces a natural protein that makes it resistant to FSB. Once the FSB caterpillar feed on plant leaves, shoots and fruits, they stop eating and eventually die. The *Bt* protein in the biotech eggplant only affects FSB and does not affect humans, farm animals and other non-target organisms.

Biotechnology has opened up new vistas for crop improvement. Bio-technological tools like *in vitro* propagation, genetic engineering and molecular biology has helped overcoming constraints of conventional breeding, and identification and introduction of useful genes that confer resistance to pests and diseases, and tolerance to abiotic stresses in eggplant. Although some developments in the field of biotechnological applications has taken place, the full potential is yet to be exploited for improvement of eggplant. The understanding of specific metabolic pathways directly or indirectly involved in plant morphogenesis has helped in understanding and improving the regeneration potential of eggplant genotypes.

Plant transformation procedures have been well established and are being utilised for producing eggplant transgenics. However, genetic engineering has not yet been utilised to its full potential for eggplant improvement as the existing studies have been confined mainly to the development of insect resistant transgenics and transgenics for parthenocarpic fruit development as well as tolerance to abiotic stresses. Thus, such studies need to be further exploited so as to address the development of transgenic plants tolerant to the wide array of environmental stresses, and fungal and bacterial pathogens. The prospects of eggplant improvement appear brighter with the advent of biotechnology tools. The areas that need to be strengthened in eggplant research are genetic engineering and molecular biology.

Chapter 20

Genetically Modified Vegetables

INTRODUCTION

Vegetables are very important components of diet. Vegetables are highly perishable, and transportation, storage, and distribution require low-temperature conditions. Vegetables are processed into juice, sauce, and canned under aseptic conditions. This enables long-term storage of processed products. Tomato is the major produce that is processed into juice and sauce. Preservation of nutritional components is compromised during processing. Stability of juice is influenced by the particle size distribution.

Vegetables and fruits have many similarities with respect to their composition, harvesting, storage properties, and processing. In the true botanical sense, many vegetables are considered as fruits. Thus, tomatoes, cucumbers, eggplant, peppers, and so on could be considered as fruits, since they develop from ovaries or flower parts and are functionally designed to help the development and maturation of seeds. However, the important distinction between fruits and vegetables is based on their use. In general, most vegetables are immature or partially mature and are consumed with the main course of a meal, whereas fruits are generally eaten alone or as a dessert.

Some of the important vegetables are discussed below:

TOMATOES

Tomato is a major vegetable crop that has achieved tremendous popularity over the last century and it is grown in almost every country of the world. Tomatoes breeding programs can highly benefit of biotechnological tools, such as gene transfer technology, which allows the introduction of foreign genes into a germplasm, without modifying the genetic background of elite varieties. However, a breeding program associated to biotechnological tools depends upon the development of an efficient *in vitro* plant regeneration system. *In vitro* plant regeneration of tomatoes using protocols for adventitious shoot regeneration from cotyledon segments has been reported. The system is based on three culture steps: a bud induction phase, culturing the explants in medium supplemented with cytokinin; an elongation phase, transferring the shoot buds to medium with a lower concentration of cytokinin; and a rooting phase, using a culture medium supplemented with auxin. A genetically modified tomato, or transgenic tomato, is a tomato that has had its genes modified, using genetic engineering. The first commercially available genetically modified food was a tomato engineered to have a longer shelf life (the Flavr Savr). Currently there are no genetically modified tomatoes available commercially, but scientists are developing tomatoes with new traits like increased resistance to pests or environmental stresses. Other projects aim to enrich tomatoes with substances that may offer health benefits or be more nutritious. As well as aiming to produce novel crops, scientists produce genetically modified tomatoes to understand the function of genes naturally present in tomatoes.

Genetically Engineered Traits

At present, tomatoes are the only food that has been marketed with GE delayed-ripening traits. Delaying the ripening process in tomatoes is of interest to producers because it allows more time for shipment of tomatoes from the farmer's fields to the grocer's shelf, and increases the shelf life of the tomatoes for consumers. Although ripening makes tomatoes edible and flavourful, it also begins the gradual decline towards softening and rot, causing losses for producers and consumers. Tomatoes that are genetically engineered to have delayed ripening can be left longer on the plant to mature, will have longer shelf-life in shipping, and may last longer for consumers.

Some tomatoes have been genetically engineered to alter one particular aspect of tomato ripening: softening. The process of fruit softening is caused in part by the breakdown of pectins—compounds which give support to the walls of tomato cells. Tomatoes have been engineered to have reduced levels of a pectin breakdown enzyme called polygalacturonase. This not only increases shelf life, it makes the tomato products thicker (higher pectin to water ratio), which is of interest to tomato processors. This is the technique used in the well-known 'FlavrSavr' tomato. Tomato plants naturally produce the compound ethylene to trigger the ripening of tomatoes on the plant. Several genetic engineering strategies involve the reduction or prevention of ethylene production to slow or stop the ripening process.

Methods of Transformation

There are various methods of transformation available which are followed by different laboratories. To create the transgenic tomato, a gene from *E. coli* (a bacterium which occurs naturally in the mammalian gut) called kan(r) and the FLAVR SAVR gene (from a tomato) were inserted into a plasmid (a circular ring of DNA) and plasmids like these were inserted into a group of tomato cells in a growth medium containing an antibiotic (Engel 77). The Kan(r) gene, when established in the cell, produced a substance called APH (3') II that gave the cell resistance to the antibiotic.

The antibiotic killed cells that did not receive the plasmid. The purpose of the bacterial gene was, therefore, to identify the cells that were genetically transformed. The FLAVR SAVR gene coded for a strand of RNA that was the reverse of a strand of RNA that naturally occurs in the plant. The original RNA strand in the plant is responsible for the production of the enzyme polygalacturonase. Polygalacturonase breaks down pectin in the cell walls of the tomato during the ripening process and causes the entire tomato to become soft. The complementary strand of RNA from the FLAVR SAVR gene binds to the polygalacturonase RNA and the two strands 'cancel each other out,' preventing the production of polygalacturonase and the softening of the tomato. Tomato transformation and regeneration were analysed and optimised by Carolina Cortina and coworkers. They infected Cotyledon explants from *Lycopersicon esculentum cv.* UC82B with *Agrobacterium tumefaciens* strain LBA4404 harbouring the neomycin phosphotransferase (*NPTII*) reporter gene. They found that on increasing concentration of thiamine(vitamin) from 0.1 mg l^{-1} in standard medium to 0.4 mg l^{-1} decreased the chlorophyll lost that accompanied the expansion of necrotic areas in cotyledon explants. They observed Optimal shoot regeneration rate with a balanced concentration of 0.5 mg l^{-1} auxin indolelacetic acid (IAA) and 0.5 mg l^{-1} cytokinin zeatin riboside. Finally, when the phenolic acetosyringone was present in the co-culture medium at 200 µM, they confirmed transgenic lines reached 50% of antibiotic resistant shoots. The efficiency of transformation reached 12.5% with this protocol.

Chyi Y.S. and coworkers investigated the genetic behaviour of DNA sequences in the backcross progeny of 10 transformed *Lycopersicon esculentum* x *L. pennellii hybrids*. They used Isozyme and restriction fragment length polymorphism (RFLP) markers to test linkage relationships of the insertion in each backcross family. The TDNA inserts in 9 of the 10 transformants were mapped in relation to one or more of these markers, and each mapped to a different chromosomal location. Because only one insertion did not show linkage with the markers employed, it must be located somewhere other than the genomic regions covered by the markers assayed. They concluded that *Agrobacterium*-mediated insertion in the *Lycopersicon* genome appears to be random at the chromosomal level.

Backcross progeny of two nopaline negative transformants showed incomplete correspondence between the T-DNA genotype and the kanamycin resistance phenotype. Two kanamycin resistant progeny plants of one of these two transformants possessed altered T-DNA restriction patterns, indicated genetic instability of the T-DNA in this transformant.

McCormick S. and others modified leaf disc transformation system in tomato. They used leaf explants and hypocotyls sections to regenerate transformed plants. They found evidences for both single and multicopy insertions of the T-DNA, and have demonstrated inheritance of the T-DNA insert in the expected Mendelian ratios. A reduced efficiency of transformation was observed with binary T-DNA vectors as compared to co-integrate T-DNA vectors. Table 20.1 shows comparison of ranges of nutrients between transgenic and normal tomatoes.

Table 20.1: Comparison of ranges of nutrients between transgenic and normal tomatoes (per 100 g fruit).

Nutrient	*Normal range*	*Transgenics*	*Controls*
Protein	0.85 g	0.75–1.14 g	0.53–1.05 g
Vitamin A	192–1667 IU	330–1660 IU	420–2200 IU
Thiamin	16–80 µg	38–72 µg	39–64 µg
Riboflavin	20–78 µg	24–36 µg	24–36 µg
Vitamin B6	50–150 µg	86–150 µg	10–140 µg
Vitamin C	8.4–59 mg	15.3–29.2 mg	12.3–29.2 mg
Niacin	0.3–0.85 mg	0.43–0.70 mg	0.43–0.76 mg
Calcium	4.0–21 mg	9–13 mg	10–12 mg
Phosphorus	7.7–53 mg	25–37 mg	29–38mg
Sodium	1.2–32.7 mg	2–5 mg	2–3 mg

Note: For Table 20.1, the 'normal range' represents values that the researchers looked up in standard references. The 'controls' column represents actual amounts of nutrients found in nontransgenic (traditional) varieties grown by the researchers alongside the transgenic varieties.

Various benefits of transformed plants

The increased consumption of fruits and vegetables is associated with reduced cardiovascular disease. D. Rein and coworkers studied the health effects of wildtype tomato (wtTom) and flavonoid-enriched tomato (flTom). Human C-reactive protein transgenic (CRPtg) mice express markers of cardiovascular risk. They analysed markers of general health (bodyweight, food intake, and plasma alanine aminotransferase activities) and of cardiovascular risk (plasma CRP, fibrinogen, E-selectin, and cholesterol levels). CRPtg mice were fed a diet containing 4 g/kg wtTom, flTom peel, vehicle, or 1 g/kg fenofibrate for 7 weeks which reduced cardiovascular risk. A.L.E.Lopez and coworker showed that transgenic tomato expressing interleukin-12 has a therapeutic effect on progressive pulmonary tuberculosis. They observed that transgenic tomato L-12 administration resulted in a reduction of bacterial loads and tissue damage compared with wild-type tomato (non-TT). In the late infection, a longterm treatment with TT–IL-12 was essential. They successfully demonstrated that TT–IL-12 increases resistance to infection and reduce lung tissue damage during early and late drug-sensitive and drug-resistant mycobacterial infection.

Konijeti R and coworkers found dietary lycopene combined with other constituents from whole tomatoes have greater chemopreventive effects against prostate cancer as compared to pure lycopene provided in a beadlet formulation in mice. They fed mice with lycopene in form of tomato paste and lycopene beadlets. The incidence of prostate cancer was significantly decreased in the lycopene beadlets LB group relative to the control group up to 95%.

Smith and coworkers investigated the expression of coat protein gene of Tomato leaf curl virus (TLCV) into an expression vector and mobilised to *Agrobacterium tumefaciens* through triparental mating. Cotyledon leaf explants of Pusa Ruby tomato were transformed by co-cultivation with Agrobacterium containing TLCV–CP constructs. Kanamycin-resistant transformants were regenerated and established in glasshouse. They observed that in TI generation transformed plants showed disease tolerance when compared to non transformed ones.

Thus, the tomato has developed into an excellent model system for the plant molecular biologist. Plants can be easily regenerated from single cells or protoplasts. Selection of single cells have yielded lines genetically resistant to pathogens, herbicides, and other abiotic environs. Tomato protoplasts can be fused with protoplasts of other tomato lines, or species, as well as with protoplasts of other plant genera.

Tomato DNA has been isolated, cloned, and engineered. Successful transformation of tomato has been effected, using a variety of methods to introduce foreign DNA. The critical road block to improvement of tomato varieties through genetic engineering is the discovery and manipulation of useful single genes that can be isolated by molecular geneticists.

SOYABEANS

Soyabean is the oil crop of greatest economic relevance in the world. Its beans contain proportionally more essential amino acids than meat, thus making it one of the most important food crops today. Processed soyabeans are important ingredients in many food products.

To help meet the challenges of increased soyabean demand, biotechnology tools are being used to develop soyabeans with improved nutritional value and greater resistance to disease, herbicides, and drought. Producers are increasingly turning to biotech soyabeans because of the cost and time savings and reasonable yield enhancement these soyabeans offer. Future traits offer the promise of further crop protection benefits, higher yield, and grain value enhancement through oil and protein modification. Despite all the opportunities, biotech soyabeans face numerous challenges. Because of the cost of technology and regulatory clearance, it is challenging for developers to capture an acceptable return on biotechnology investments. In order for the full benefits of biotechnology to be realised by the world's farmers and consumers, global acceptance of biotech crops and grain is critical.

Biotechnology can be defined broadly as a set of tools that allows scientists to genetically characterise or improve living organisms. Several emerging technologies, led by transformation and molecular characterisation, are already being used extensively for the purpose of plant improvement. Other emerging sciences, including genomics and proteomics, also are starting to impact plant improvement. Looking forward, biotechnology promises to deliver products with improved nutritional value and yield enhancement through greater resistance to disease, herbicides, and drought. Enhanced productivity is increasingly important, because per-capita soyabean consumption is growing. Soyabean consumption and production are increasing. Soyabean consumption is driven primarily by meat consumption in human diets, as soyabeans are used primarily for animal feed.

Impact of Technology

The soyabean is particularly difficult to transform. Despite the wide-ranging impact and notoriety of biotechnology solutions, there are additional biotechnology tools that are quietly, yet effectively, improving the productivity of soyabeans. One of these is the use of molecular markers, which help scientists track and select key genes during breeding. The story of how Pioneer is tackling a serious soyabean pest is an example of the role this biotechnology tool can play in agriculture.

The soyabean cyst nematode (SCN) is the most destructive soyabean pest in the United States, causing yield losses estimated at more than $1 billion a year. Nematodes attach to roots, causing significant damage, plant stunting and yellowing, and yield loss. While attached, female nematodes are fertilised by male nematodes and produce a large number of eggs. At season's end, the female dies and the eggs remain in her body, which forms a protective shell or cyst. Nematode cysts may remain in infested fields for more than a decade.

Gene mapping has been used to identify the location of SCN resistance genes on specific chromosomes. Another emerging tool of biotechnology is genomics, which refers to the study of the function and structure of genes. The soyabean genome, as with genomes of other species, holds a vast resource of blueprints that determine what this great plant can provide. Genomics is helping researchers understand soyabean DNA structure and function to change traits that affect pest resistance, yield, and grain composition.

Herbicide-tolerant soyabean

Herbicide tolerant soyabean varieties contain a gene that provides resistance to one of two broad spectrum herbicides. This modified soyabean provides better weed control and reduces crop injury. It also improves farm efficiency by optimising yield, using arable land more efficiently, saving time for the farmer, and increasing the flexibility of crop rotation. It also encourages the adoption of no-till farming-an important part of soil conservation practice.

Insect resistant soyabean

This biotech soyabean exhibits resistance to lepidopteran pests through the production of Cry1Ac protein. Insect resistant soyabean was developed to reduce or replace high insecticide applications and at the same time maintain soyabean yield potential.

Oleic acid soyabean

This modified soyabean contains high levels of oleic acid, a monounsaturated fat. According to health nutritionists, monounsaturated fats are considered 'good' fats compared with saturated fats found in beef, pork, cheese, and other dairy products. Oil processed from these varieties is similar to that of peanut and olive oils. Conventional soyabeans have an oleic acid content of 24%. These new varieties have an oleic acid content that exceeds 80%.

Challenges

Capturing value for quality-enhancing traits is one of the biggest challenges facing biotech soyabeans. Put simply, to capture value, value must be created. To ensure adoption, this value must be sufficient to reward adequately all participants in the value chain (developer, grain producer, grain handler, and end user). Biotech solutions will likely be used to expand the use of soyabean protein and oil in the food market. This might include improvement of soyabean isolate flavor and functionality and improved oil/ protein health characteristics for consumers. Biotechnology also will be used to create traits that lead to soyabeans with increased industrial and energy uses (e.g. biodiesel, lubricants).

However, capturing value also hinges on acceptance of the technology used to create the value. To help ensure acceptance, soyabeans derived through biotechnology must be developed while following stringent safety and regulatory guidelines established by industry and government agencies. These guidelines must be followed from product concept to postmarket.

Today, every product is subject to extensive regulatory scrutiny. Approval costs for a product can reach into the millions of dollars. The value of new soyabean traits, however beneficial they may be for producers and consumers, could be derailed by increased regulatory cost. The biotech industry will likely face more and tighter regulations with increased scrutiny of field research and more data requirements. There will be increased pressure on technology globally. If biotech soyabeans are going to provide the hoped-for benefits, the industry and others will need to work to help consumers and regulators understand the impact of additional regulatory proposals and the safety procedures that are already in place.

Soyabeans Pros and Cons

There are some potential disadvantages to growing GMO soyabeans as well.

Pros of genetically modified soyabeans

Higher yields are a real possibility. Genetic modifications may make it possible for farmers to grow multiple crops of soyabeans on the same land during the year. This would help to make a farm more profitable every year and enhance the food supply that is available to everyone in a region, country, or even the world. More farmers could plant soyabeans. Although soyabeans are a rather hearty crop, there are places in the world where they just won't grow. Through the use of genetic modifications, the plants could be designed to respond to more difficult environments and this would help farmers anywhere in the world to use their land more efficiently. They are easy to care for on a daily basis. GMO soyabeans cannot only be engineered to be pest resistant, but also weed resistant. This allows farmers to spend more of their time working on other crops or livestock instead of focusing all of their attention and energy on their soyabeans.

Cons of genetically modified soyabeans

They may increase the rate of food allergies. More people than ever before are allergic to soyabeans. It's become so prevalent that it has become a listed food allergy on grocery products. Because soya protein is extensively used in many food products due to its affordability, the development of a food allergy makes it difficult for people to find foods to eat at times. The crops are hard on the soil. Many farmers have found that they can grow GMO soyabeans on their land for a limited amount of time before they'll have to do a crop rotation. Unlike other crops, it may take more than one year for land to recover from the growing of genetically modified soyabeans. Sometimes the land isn't usable ever again.

There is no real consumer benefit. Although more soyabeans can potentially be grown, the costs of growing GMO soyabeans is the same. This means that consumers are going to pay the same amount for their food products, whether modifications were made to it or not. Genetically modified soyabeans may be able to change the world's food supply, but is the cost of that alteration a price that is too high to pay?

CORN

The large productivity gains in corn production made during the last several decades have come primarily from advanced plant breeding techniques and improved corn management of the crop. GM corn products have increased corn production over a wide range of growing conditions, helping promote yield stability and reduce production risks. Corn products with biotechnology traits have the latest technology integrated with seed genetics to help protect yield potential.

Biotechnology benefits for corn production: There are many options to consider for corn product selection including products which are conventional, herbicide-tolerant, or insect-protected and herbicide-tolerant. Each farmer must determine what product offers the best opportunity to help maximise profitability on his or her farm. Field research trials from a variety of sources show that corn products with genetically modified (GM) traits have a number of benefits over conventional corn to help manage risk under variable yield conditions and protect yield potential.

- GM corn provided more yield than conventional corn.
- GM corn responded to higher plant densities more than conventional corn.

Weed management benefits:
- Reduced plant stress due to weed infestations to protect yield potential and plant health.
- Removes hosts for insects, diseases, and nematodes.
- Facilitates the use of reduced-tillage for soil and water conservation.

Insect management benefits:
- Damage from the multi-pest complex (corn earworm, European corn borer, western bean cutworm, fall armyworm) causes stress and injury to plant tissue, reducing yield potential. Insect damage can allow fungi to infect, proliferate, and produce mycotoxins which have the potential to cause health problems in animals and humans.
- GM corn rootworm protection has been shown to have agronomic benefits in addition to insect management. Improved root growth and activity can allow corn products to export more cytokinins from the roots and utilise nitrogen more effectively after flowering to promote higher kernel weight and yield potential.
- Higher corn plant densities are important for maximising grain yield potential.

GM products with insect and herbicide tolerance protect corn yield potential and provide other benefits researchers found that GM crops are credited with decreasing pesticide and fuel use, facilitating conservation tillage practices that reduce soil erosion, improving carbon retention, and lowering greenhouse gas emissions.

To sum up farmers planting GM corn products with herbicide resistance and multiple mode of action insect protection traits can realise higher yield potential by using intensive corn management practices to:

- Reduce plant stress from corn borers, ear feeding insects, stalk boring insects, and rootworm root damage.
- Plant corn-intensive crop rotations.

- Maintain higher plant densities to help maximise corn production.
- Harvest better quality grain by preventing insect damage that can lead to stalk and ear rot diseases. Mycotoxins produced by these diseases have the potential to cause health problems in animals and humans.
- Reap the economic benefits of higher yield potential in feedstuffs for cattle.

Pros and Cons of Genetically Modified Corn

Pros of GM corn

GM corn are genetically engineered to last longer and be in a better condition than regular crops, this can bring down the price of corn. GM corn are genetically engineered to resist harsh conditions like weed killers (Roundup) and weather, allowing more of the crops to survive. Since corn are used in so many products as well as feeding livestock (are huge building blocks of the food industry), the extra GM corn and soyabeans produced can be used to help the impoverished.

Cons of GM corn

Farmers are free to use chemically-rich weed killers to kill all other plants except the GM corn leaving these chemicals on the surface of the crops. GM corn mate just like all other corn and soybeans, with all other corn and soybeans. This causes problems by widely spreading the GM corn and soybean DNA before we know its effects on the environment and human health.

CARROT

All varieties of carrots contain valuable amounts of antioxidant nutrients. Included here are traditional antioxidants like vitamin C, as well as phytonutrient antioxidants like beta-carotene. Different varieties of carrots contain differing amounts of these antioxidant phytonutrients. Red and purple carrots, for example, are best known for the rich anthocyanin content. Oranges are particularly outstanding in terms of beta-carotene, which accounts for 65% of their total carotenoid content. In yellow carrots, 50% of the total carotenoids come from lutein. Carrot DNA apparently contains a gene that causes carrots to express orange pigmentation. The orange colour in carrots is caused by beta-carotene, a precursor to vitamin A. Morris, together with Kendal Hirschi and other Texan colleagues, has found a way to double the calcium content of carrots through genetic modification, making them a rich source of the element that is so vital for bones.

Morris and others loaded their super-carrots with a protein called sCAX1, which pumps calcium into the plant's cells. The protein originally hailed from the plant-of-choice for geneticists, *Arabidopsis thaliana*, where it exists in a larger version. Morris's team lopped off a small piece from its tip that stops the protein from funnelling in more calcium once a certain amount has been reached.

In this shortened form, sCAX1 is relentless in its import of calcium and the researchers have found that it can greatly increase the calcium content of several vegetables including tomatoes, potatoes and carrots. These super-charged vegetables could help to reduce the risk of osteoporosis, one of the world's leading nutritional disorders, where a lack of calcium leads to brittle bones.

Benefits and risks: Morris's tinkered carrots are not only more nutritious but also more long-lasting and more productive. Again, calcium is the key. Farmers have known this for some time. Many growers soak apples in calcium solutions to keep them firm during shipping and fresh on the shelf. Potato crops are also sprayed with extra calcium, which helps them to tolerate hot conditions and ward off infections. There's good reason to believe that the modified carrots would also enjoy similar benefits.

Biotechnological Applications in Carrot Improvement

Trait mapping

The genetics of two multigenic traits of carrot, carotenoids and restoration of CMS were evaluated in the 1960s–1980s. Several individual genes were and the heritabilities of several carrot traits have been measured. Multigene trait mapping only began with the development of molecular markers and linkage maps. To date, seven monogenic traits have been mapped for carrot: yel, cola, *Rs*, *Mj*-1, *Y*, *Y2* and *P1*. Work is underway to localise genes for flavour compounds and pigments, and to localise genes to specific chromosomes using fluorescent *in situ* methods. Quantitative trait loci (QTL) have been mapped for carrot total carotenoids and five component carotenoids; phytoene, α-carotene, Beta-carotene, zeta-carotene, and lycopene and the majority of the structural genes of the carotenoid pathway is now placed into this map. Marker-assisted selection has been reported for *Rs* gene.

Gene mapping

Bradeen and Simon studied 103 F2 individuals of the cross B9304 X YC7262, which segregated for core colour. Using bulked segregant analysis combined with F2 mapping, they identified six AFLP markers linked to and flanking the Y2 locus. Markers were located between 3.8 and 15.8 cM from the gene. Using the same F2 mapping population, Vivek and Simon subsequently identified a single AFLP marker located 2.2 cM from the Y2 locus, assigning the locus to one end of linkage group B. Anthocyanin accumulation in the carrot phloem is conditioned by the *P1* locus, with purple (*P1*) dominant to nonpurple (*p1*). Simon studied the inheritance of *P1* and *Y2* in *F2* and BC populations originating from Eastern carrot germplasm and concluded that the two loci are unlinked. Consistent with this, Vivek and Simon mapped *P1* to linkage group A, independent of *Y2*. *P1* is flanked by AFLP markers mapping 1.7 and 8.1 cM away from the gene.

The rs allele has recently been found to be a naturally occurring knockout mutant of a carrot invertase isozyme which produces no functional enzyme. Vivek and Simon mapped *rs* to one end of linkage group C, 8.1 cM away from an AFLP marker. Mapping results are consistent with inheritance data indicating that *rs* is genetically unlinked to *Y2* and *P1*.

Quantitative trait loci (QTL) detected

In addition to single genes conditioning important traits, several QTL have been identified in carrot through segregation analysis. To date, QTL conditioning synthesis in carrot roots of provitamin A α-and β-carotenes, the carotene lycopene, and precursors in the carotene pathway have been mapped. Among orange carrots, heritability of 0.40 and approximately 20 major QTL have been reported to control carotenoid content.

Most modern carrot breeding effort has exclusively involved intercrosses among orange carrots and the numerous QTL involved in that colour class. A major exception to this generalisation has been the use of white wild carrot as a source of CMS. As yellow, red, and even white cultivated carrots become more popular, the major genes and eventually QTL conditioning these colours will be better described.

Smith and others reported that two major interacting loci, *Y* and *Y2* on linkage groups 2 and 5, respectively, control much variation for carotenoid accumulation in carrot roots. These two QTLs are associated with carotenoid biosynthetic genes zeaxanthin epoxidase and carotene hydroxylase and carotenoid dioxygenase gene family members as positional candidate genes. Dominant Y allele inhibits carotenoid accumulation. When Y is homozygous recessive, carotenoids that accumulate are either only xanthophylls in Y2 plants, or both carotenes and xanthophylls, in y2y2 plants. These two genes played a major role in carrot domestication and account for the significant role that modern carrot plays in vitamin A nutrition.

Future directions: Genes conditioning root pigmentation and sugar and terpenoid content are candidates for gene mapping in the near future. MAS may be more widely integrated into carrot breeding programs as converted markers linked to quality traits are generated. Genomics resources such as expressed sequence tag libraries, microarrays, *in situ* hybridisation methodologies, and transposon-tagging systems are likely to be developed which would further accelerate the generation and identification of useful variation for carrot quality improvement.

POTATOES

Potato has always been a close companion to biotechnology. Potato, being vegetatively propagated crop, is highly amenable to asexual clonal propagation techniques *in vitro* and consequently genetic engineering. Besides, potato has the distinct advantage of possessing a commercially viable carbon sink in the form of tuber. Therefore, it has also been looked upon as a potential bioreactor for the production of novel compounds of therapeutic and industrial values.

Potato is the third most important global food crop and the most widely grown noncereal crop. Transformation is highly effective for adding single genes to existing elite potato clones with no, or minimal, disturbances to their genetic background and represents the only effective way to produce isogenic lines of specific genotypes/cultivars. This is virtually impossible via traditional breeding as, due to the high heterozygosity in the tetraploid potato genome, the genetic integrity of potato clones is lost upon sexual reproduction as a result of allele segregation. These genetic attributes have also provided challenges for the development of genetic maps and applications of molecular markers and genomics in potato breeding. Various molecular approaches used to characterise loci, (candidate) genes and alleles in potato, and associating phenotype with genotype.

New molecular biology and plant cell culture tools have enabled scientists to understand better how potato plants reproduce, grow and yield their tubers, how they interact with pests and diseases, and how they cope with environmental stresses. Those advances have unlocked new opportunities for the potato industry by boosting potato yields, improving the tuber's nutritional value, and opening the way to a variety of non-food uses of potato starch, such as the production of plastic polymers.

Producing High-quality Propagation Material

Unlike other major field crops, potatoes are vegetatively reproduced as clones, ensuring stable, 'true-to-type' propagation. However, tubers taken from diseased plants also transmit the disease to their progenies. To avoid that, potato tuber 'seed' needs to be produced under strict disease control conditions, which adds to the cost of propagation material and therefore limits its availability to farmers in developing countries.

Micropropagation or propagation *in vitro* offers a low-cost solution to the problem of pathogens in seed potato. Plantlets can be multiplied an unlimited number of times, by cutting them into single-node pieces and cultivating the cuttings. The plantlets can either be induced to produce small tubers directly within containers or transplanted to the field, where they grow and yield low-cost, disease free tuber 'seed'. This technique is very popular and routinely used commercially in a number of developing and transition countries. For example, in Viet Nam micropropagation directly managed by farmers contributed to the doubling of potato yields in a few years.

Protecting and exploring potato diversity

The potato has the richest genetic diversity of any cultivated plant. Potato genetic resources in the South American Andes include wild relatives, native cultivated species, local farmer-developed varieties, and hybrids of cultivated and wild plants. They contain a wealth of valuable traits, such as resistance to pests and diseases, nutrition value, taste and adaptation to extreme climatic conditions. Continuous efforts are being made to collect, characterise and conserve them in gene banks, and some of their traits have been transferred to commercial potato lines through cross-breeding.

To protect collections of potato varieties and wild and cultivated relatives from possible diseases and pest outbreaks, scientists use a variation of micropropagation techniques to maintain potato samples *in vitro*, under sterile conditions. Accessions are intensively studied using molecular markers, the identifiable DNA sequences found at specific chromosomal locations on the genome and transmitted by the standard laws of inheritance.

Obtaining improved varieties

Potato genetics and inheritance are complex, and developing improved varieties through conventional cross breeding is difficult and time consuming. Molecular-marker based screening and other molecular techniques are now widely used to enhance and expand the traditional approaches to potato in food production. Molecular markers for characteristics of interest help identify desired traits and simplify the selection of improved varieties. Such techniques are currently applied in a number of developing and transition countries.

Through the Potato Genome Sequencing Consortium, significant progress is being made in mapping the complete DNA sequence of the potato genome, which will enhance our knowledge of the plant's genes and proteins, and of their functional traits. Technical advances in the fields of structural and functional potato genomics - and the ability to integrate genes of interest into the potato genome - have expanded the possibility of genetic transformation of the potato using recombinant DNA technologies. Transgenic varieties with resistance to Colorado Potato Beetle and viral diseases were released for commercial production in the early 1990s in Canada and the USA, and more commercial releases can be expected in the future.

Transgenic potato varieties offer the possibility of increasing potato productivity and production, as well as creating new opportunities for non-food industrial use. However, all biosafety and food safety aspects must be carefully assessed and addressed before their release.

Organic Potatoes

Organic potatoes can be grown on a large-scale without commercial pesticides and standard fertilisers. However, production costs for organic potatoes are higher and their yields are lower than for conventionally produced potatoes. Whether prices for organic potatoes can be high enough to offset these costs remains a question. To demonstrate the production and profitability of commercial organic potatoes, the researchers used two 25- to 30-acre potato fields in Coloma, Wisconsin. The researchers wanted not only to demonstrate commercial organic potato production, but also to determine crop rotation systems best suited to potato production.

In Wisconsin, farmers grow potatoes on light, sandy soils under irrigation and rely heavily on purchased chemical pesticides and fertilisers. Since sandy soils are typically more permeable than heavier soils, many people worry about these chemicals passing through the soil and contaminating the groundwater.

The research team substituted cultural and biological inputs for synthetic chemical inputs. By not putting the chemicals on the land, they hoped to eliminate the negative effects on the environment. It was found that organic potato production removes the environmental concerns associated with pesticides and nutrients.

Superior, Red Norland, and Norkotah Russet varieties were grown in 10 acre blocks according to organic labeling requirements. Researchers closely monitored all crop inputs and operations.

The team used mechanical and cultural practices to control weeds, insects, and potato diseases. Early planting and harvest eliminated or reduced such problems as potato leaf hopper infestation, green peach and potato aphid infestation, and early and late blight.

Also, they planted the potatoes in fields that were planted to alfalfa or corn the year before to reduce nitrogen purchases (alfalfa) and potato diseases (alfalfa and corn) that carry over from year to year.

Chapter 21

Genetically Modified Foods

INTRODUCTION

Genetically modified organisms (GMO) can be defined as organisms (i.e. plants, animals or micro-organisms) in which the genetic material (DNA) has been altered in a way that does not occur naturally by mating and/or natural recombination. The technology is often called 'biotechnology' or 'genetic engineering'. It allows selected individual genes to be transferred from one organism into another, also between nonrelated species. Foods produced from or using GM organisms are often referred to as GM foods. GM foods are developed – and marketed – because there is some perceived advantage either to the producer or consumer of these foods. The genetic modifications are meant to translate into a product with a lower price, greater benefit (in terms of durability or nutritional value) or both. Initially, GM seed developers wanted their products to be accepted by producers and have concentrated on innovations that bring direct benefit to farmers (and the food industry generally).

GM foods assessed differently from conventional foods: Generally consumers consider that conventional foods (that have an established record of safe consumption over the history) are safe. It is understandable that novel varieties of organisms for food use, whether they are developed using the traditional breeding methods or newer technologies such as gene technology, may encounter resistance from consumers. After all, some or all of the usual characteristics may be altered, either in a positive or a negative way. National food authorities may be called upon to examine the safety of conventional foods obtained from novel varieties of organisms. In contrast, most national authorities consider that specific assessments are necessary for GM foods. Specific systems have been set up for the rigorous evaluation of GM organisms and GM foods relative to both human health and the environment. Similar evaluations are generally not performed for conventional foods. Hence, there currently exists a significant difference in the evaluation process prior to marketing for these two groups of food.

Genetically-modified foods have the potential to solve many of the world's hunger and malnutrition problems, and to help protect and preserve the environment by increasing yield and reducing reliance upon chemical pesticides and herbicides. Yet there are many challenges ahead for governments, especially in the areas of safety testing, regulation, international policy and food labelling. Many people feel that genetic engineering is the inevitable wave of the future and that we cannot afford to ignore a technology that has such enormous potential benefits. However, we must proceed with caution to avoid causing unintended harm to human health and the environment as a result of our enthusiasm for this powerful technology. The term GM foods or GMOs (genetically-modified organisms) is most commonly used to refer to crop plants created for human or animal consumption using the latest molecular biology techniques. These plants have been modified in the laboratory to enhance desired traits such as increased resistance to herbicides or improved nutritional content.

The enhancement of desired traits has traditionally been undertaken through breeding, but conventional plant breeding methods can be very time consuming and are often not very accurate. Genetic engineering, on the other hand, can create plants with the exact desired trait very rapidly and with great accuracy. For example, plant geneticists can isolate a gene responsible for drought tolerance and insert that gene into a different plant. The new genetically-modified plant will gain drought tolerance as well. Not only can genes be transferred from one plant to another, but genes from non-plant organisms also can be used. The best known example of this is the use of *Bt* genes in corn and other crops. *Bt* or *Bacillus thuringiensis*, is a naturally occurring bacterium that produces crystal proteins that are lethal to insect larvae. *Bt* crystal protein genes have been transferred into corn, enabling the corn to produce its own pesticides against insects such as the European corn borer.

ISSUES RELATED TO POTENTIAL HEALTH EFFECTS OF FOODS DERIVED FROM GENETICALLY MODIFIED PLANTS

The basic tenet of the biotechnology industry engaged in the production of genetically modified (GM) crop plants and foods is that no 'credible' evidence exists that GM crops damage the environment or that GM foods harm human/animal health. Accordingly, they are as safe as their 'substantially equivalent conventional counterparts' and need no safety testing. The general acceptance of such a view could, of course, save a great deal of money for the biotechnology industry that otherwise would have to be spent on very expensive environmental and health risk assessments of their GM products.

However, practically all recent reviews that have critically assessed the results of GM crop/food safety research data published in peer-reviewed science journals have come to the conclusion that, at best, their safety has not yet been adequately established, or at worst, that the results of risk assessment studies, particularly (but not exclusively) those carried out independently of the biotechnology industry, have raised important safety concerns which have not been properly settled. Thus, one review concluded that the most pertinent questions on environmental safety of GM crops have not yet been asked. A more recent update came up with a long list of important questions that regulatory authorities should ask before any GM crops are released into the environment. Unfortunately, few of these questions have been addressed in the biotechnology companies' submissions to the regulatory authorities.

The situation is not much better with the results of studies in which the potential health effects of GM foods have been investigated. Thus, an early review found only eight peer-reviewed papers published on the potential health aspects of GM food. Pryme and Lembcke reported a rather curious aspect of the results of health risk assessment studies using laboratory animals. It appeared that most independently funded research scientists who performed animal testing of GM crops reported some potential health problems, while the results of the studies sponsored by the industry indicated none.

Further reviews confirmed the scarcity of GM risk assessment research, particularly research carried out independently of the biotechnology industry. Thus, there were just over a dozen academic research papers on the health aspects of GM crops published by 2003 and this number had increased to approximately 20 by 2005.

A report by the Canadian Royal Society stated that without in-depth biological testing of GM crops, 'substantial equivalence' is a fatally flawed concept and regulation based on it exposes Canadians to potential health risks of toxic and allergic reactions. Neither did the British Medical Association accept that all GM crops/foods are safe, and therefore no testing is needed. In their report (The Medical Research Council 2000, recently updated) it was stated that 'any conclusion upon the safety of introducing GM material into the UK is premature as there is insufficient evidence to inform the decision making process at present'. It is, therefore, not surprising that the majority of British consumers think that GM foods are unsafe. As there is no demand for them most supermarkets in the UK have phased them out. Most consumers in Europe demand, as a minimum, the labelling and rigorous, transparent and independent safety testing of all GM foods.

Most GM crops are grown in America, the bulk in the USA. It is therefore regrettable that effectively there is no regulation in the USA that would guarantee their safety. The food regulatory agency in the USA, the Food and Drug Administration (FDA), almost totally relies on voluntary notification by the biotechnology companies that they carried out their own safety assessment of the GM crops they want to release commercially and found them to be safe. The FDA has no laboratory of its own and never, in fact, underwrites the safety of GM crops/foods. It only accepts the assurances of the biotechnology companies that their product is safe. This, in most instances, relies on a safety assessment that is based on the poorly defined and not legally binding concept of substantial equivalence. However, similarity in composition is no guarantee that GM food is as safe as conventional food. Thus, the content of proteins, lipids and carbohydrate components of a BSE cow (a cow suffering from a condition known as bovine spongiform encephalopathy) will be similar to that of a healthy cow but, obviously, these two cows cannot be regarded as substantially equivalent for consumer health. True, compositional analysis is an obligatory starting point in risk assessment but it cannot be its endpoint. Whether GM food is toxic or allergenic cannot be decided on the basis of chemical analyses but only by biological testing with animals.

Furthermore, the biotechnology companies try to claim as much 'confidential business information' concerning their risk assessments as possible, and therefore most of the time these are unavailable in full for public or independent scrutiny or even for some national regulatory bodies.

ALIMENTARY TRACT AS THE FIRST TARGET OF GM FOOD RISK ASSESSMENT

To show by chemical methods the presence of new toxins/allergens in GM food products is, at best, difficult. In contrast, the presence of even minute amounts of unexpected but harmful potent bioagents in GM foods could be more easily established from their possibly disproportionally large effect on health. Thus, exposure of individuals to biologically active transgenic proteins can have major effects on their gastrointestinal tract. As most proteins are immunogenic their consumption may trigger immune/allergic effects both in the mucosal immune system of the gut and the body. It is also likely that, in addition to the effects on the gastrointestinal tract, the size, structure, and function of other internal organs will be affected, particularly in young and rapidly growing humans or animals. According to some recent unconfirmed reports, the dietary exposure to GM foods may also have harmful effects on reproduction. In addition, the risks will also have to be investigated as to whether measurable amounts of the transgenic DNA constructs in GM crops/foods survive in a functionally active state/size in the gastrointestinal tract of the human/animal ingesting them, and whether they can incorporate into the genome of the cells of their gut and body organs and what will be the consequences, if any, for the individual. The GM risk assessment protocol presented in the following chapter outlines a gradual, step-by-step course of investigation by reliable and up-to-date methodology that addresses all these possible effects. These steps must be regarded as a minimum before any foods/feeds based on GM crops are allowed into the human/animal food chain.

SUGGESTED PROTOCOL FOR GM CROP/FOOD HEALTH RISK ASSESSMENT

Before any new GM crop could be made potentially safe transgenes must be identified and selected in preliminary model studies. The main criterion of the selection should be that the selected transgene and its protein product must have no toxic effects on humans or animals when given orally. However, the process of selection must be taken a step further by verifying that the selected transgene does function in the GM plant as intended.

The transgene product must therefore be isolated from the GM plant and show unequivocally that its chemical and biological properties are the same as those of the gene product expressed in the original source from which the transgene was taken. It is absolutely essential that all safety studies be carried out on this isolated transgene product and not on *E. coli* recombinant surrogates.

In the GM safety studies performed by the biotechnology industry great emphasis is laid on the assertion that, according to their *in vitro* tests, all transgene products rapidly break down in simulated intestinal proteolytic digestion tests. Obviously, should a transgenic protein quickly break down to amino acids and small peptides in the alimentary tract its toxic effects or allergenicity could not be more than minimal and thus the safety of the GM crop should apparently be assured. However, in contrast to the protocols used in the biotechnology industry's safety assessment, true proteolytic digestibility must be established in the gut *in vivo* and not in a test tube *in vitro*. Clearly, one of the most important differences between the digestion of a protein in the alimentary canal and in a test tube using only pancreatic proteases is that *in vivo*, the binding of the transgene product to the intestinal wall and/or to the food matrix reduces the availability of the transgene protein (particularly in the case of the widely used transgenic lectins, such as the various *Bacillus thuringiensis*, Bt-toxins) to the action of the proteases. Thus, an *in vitro* assay may give a false assurance of safety. In addition, as the structure, conformation and stability of a transgenic protein expressed in and isolated from *E. coli* is very different from that expressed in GM plants, no scientifically valid conclusions may be drawn from the results of experiments in which the assessment of the digestibility of a plant transgenic protein is attempted with an *E. coli* recombinant. Plants and eukaryotic bacteria are aeons apart on an evolutionary scale and therefore no bacterial recombinants should be used in tests aimed at establishing the true properties of transgenic proteins expressed in GM plants even though they are coded for by the same DNA.

Chemical Composition

One of the first steps in any proper risk assessment protocol should be the characterisation of the GM plant using well-authenticated and up-to-date methods of chemical analysis to estimate the contents of its major and minor components and to compare their amounts to those of the corresponding parent line. Although the results of such analysis and comparison can also be used to establish whether the GM and non-GM plants are 'substantially equivalent', first and foremost, this is an obligatory step that will allow us to carry out further biological risk assessment tests.

However, for such a comparison to be scientifically valid large numbers of the GM and the isogenic lines grown side-by-side and harvested at the same time are needed to be tested for the measurement of their major and minor constituents in parallel by classical and new analytical methods (proteomics, finger-printing, DNA/metabolic profiling, microarray analysis of all novel RNA species, full molecular biological examination with particular attention to the possibility of secondary DNA insertions into the plant genome, obligatory metabolomic NMR analysis of the transformed plant, stability of expression of foreign DNA, including the gene construct, promoter, antibiotic resistance marker gene, etc.).

Nutritional/Toxicological Testing with Animals

As outlined, GM crops/foods will need to be examined in obligatory short- and long-term nutritional/ toxicological tests with laboratory animals under controlled conditions. The intention is to find out whether there are any toxic effects in the animals fed on diets containing GM foods that would make the progression to human clinical trials unsafe. The animal tests are therefore designed to establish the effects of the GM crop/food on growth, metabolism, organ development, immune and endocrine functions, with particular emphasis on how diets based on GM food will affect the structure, function and bacterial flora of the animal gut. As the normality of these functions determines the development of young animals into healthy adults, the absence of significant differences between the health statuses of animals fed on GM and non-GM diets may possibly indicate that the GM crop is not unsafe, at least in animal nutrition.

Diet

It is of paramount importance that the conditions of nutritional testing are rigorously standardised. Thus, all diets must be *iso*-proteinic and *iso*-energetic (i.e. contain the same amounts of protein and energy) and are fully supplemented with vitamins and essential minerals. The composition of the control diet containing the parent line should be as close to the GM diet as possible. Diet formulation is therefore – particularly when there are significant compositional differences between the GM and its corresponding non-GM parent-line crops – not an easy task and supplementation with pure ingredients may be necessary to make good the compositional differences. In a second control diet, the parent line should be supplemented with the gene product isolated from the GM crop whose concentration should be the same as in the GM crop. All crops/foods should be fed both raw and after heat-treatment.

Experimental Protocol

Groups of young rapidly growing animals (5–6 in each group) closely matched in weight (less than ± 2% w/w), housed separately, should be strictly pair-fed these diets in short- and long-term experiments. Both males and females should be tested. The progress of the animals should be closely monitored, urine and faecal samples collected throughout the experiment and the nutritional performance of the animals and the nutritional value of the diets assessed by Net Protein Utilisation (NPU), and with measurements of nitrogen and dry weight balances and feed utilisation ratios. The animals should be weighed daily and any possible abnormalities observed. Blood samples should be taken before, during and at the end of the feeding experiments for immune studies (immune responsiveness assays, Elispot, etc.), hormone assays (insulin, CCK, etc.) and determination of blood constituents. At the end of the experiments the animals should be killed, dissected, and their guts rinsed and the contents saved for further studies (enzyme contents, GM products, DNA, etc.), gut sections taken for histology, the wet and dry weights (after freeze-drying of the tissues) of organs recorded, and the organs subjected to compositional analyses. All these data could be used to comprehensively characterise the health and metabolic status of the animals and the behaviour of the GM-fed animals could be directly compared with that of the controls. The results could then be evaluated by appropriate methods of statistics.

If any of the effects of the diet containing the GM crop on the rats is significantly different from that of the non-GM parental line control diet, the inclusion of the GM crop in food is unsafe and therefore not recommended. If the effects of feeding rats with the parent line control diet are significantly changed when this is spiked with the isolated transgene product, the *transgene is unsafe*. Most importantly, if the effects of the diets containing the GM plant and the parent line control spiked with the gene product differ, the harm is likely to be due to the use of the particular construct vector or caused by an unintended and unforeseen effect of the *transgene insertion or position* in the plant genome. Accordingly, this method of gene transfer and the resulting GM crop is unacceptable. Thus, further research is needed to find other, more precise and safer methods of genetic modification.

DIFFERENCES IN NUTRITIONAL PERFORMANCE USEFUL FOR DIAGNOSIS OF HARM

Organ weight changes are useful indicators of metabolic events after feeding laboratory animals with diets containing GM foodstuffs, particularly if followed up by histological examinations as part of the safety assessment of GM crops. Assessment of potential deviations in the normal development of key organs is of great diagnostic value, as shown in one of our GM-potato rat feeding studies. Sections of the various compartments of the gut taken for histology indicated a strong trophic effect of the GM potatoes on the rats' small intestine and, to a lesser extent, on their stomach. This hyperplastic gut growth was of particular significance because the jejunum was not enlarged when the parent line diet was supplemented with the gene product, GNA (*Galanthus nivalis* lectin), confirming previous observations which showed that the gene product had negligible growth factor effect on the jejunum, even when included in the diet at a several hundredfold concentration in comparison with that expressed in the GM potato lines. This was, in fact, one of the main reasons for selecting the gene of the natural insecticidal GNA for the genetic transformation of potatoes to make them pest-resistant but nutritionally safe. As similar hypertrophic and other similar changes in gut ultrastructure in the ileum of mice fed GM potatoes expressing *Bacillus thuringiensis* var. *kurstaki* Cry 1 toxin gene or the toxin itself were shown in a different study, GM potatoes of different origins may have common trophic effects on the gut. Changes in the ultrastructure of other organs, such as the liver, pancreas, etc., on feeding with GM crop containing diets, as shown by the work of the Malatesta group, may also be taken as a first indication of possible harmful effects that should make follow-up studies mandatory.

Changes in blood cells and blood protein levels in GM-fed animals may also suggest serious health problems, including disturbances in erithropoiesis, blood protein synthesis and the immune system. Thus, measurement of immune responsiveness could be a useful follow-up study when blood cell counts show significant differences in lymphocyte numbers that may point to one of the potentially serious hazards of the ingestion of GM foodstuffs. This is a particularly useful method because it is in general clinical use and could therefore be easily carried out with humans. Although no hormone assays were performed on rats fed GM or non-GM diets in our GM potato study, the consistently strong pancreatic growth stimulated by GM potato diets in the feeding studies suggests that this possibly was the result of the release of CCK (cholecystokinin) or some other humoral growth factor from the duodenum by an unknown growth/proliferative signal only found in the GM potatoes. Again, GNA (*Galanthus nivalis* lectin) could not be responsible for this because it does not stimulate the enlargement of the pancreas when fed to rats in its original source.

The measurement of circulating insulin levels after ingestion of GM diets would also be a good indicator for possible disturbances in the general metabolic state of the animals, particularly as insulin assays can be easily done on humans. Changes in blood basophile counts may also suggest possible problems of allergenicity that need to be followed up by more dynamic studies. Although the recommended decision-tree approach is a useful start to look at the allergenic potential of the GM crop, the criteria used in this, such as the lack of structural similarities of the GM protein to known allergens, the lack of glycosylation, small molecular size, or the *in vitro* digestibility of the GM protein, etc., are not sufficiently decisive to exclude the possibility that the GM protein is an allergen. The development of delayed hyper-sensitivity reaction found recently in GM peas expressing the kidney bean ~-amylase inhibitor gene has demonstrated that proteins that are not 21 known to be allergens in the original plant source can develop allergenic reactivity when their genes are transferred to other plant species by genetic engineering, even in the case of closely related species. Finding immune-reactive antibodies to GM proteins in blood circulation, particularly of IgE-type, in humans or animals should, of course, be strong evidence for the occurrence of immune/allergenic reactions. Although there is at present no satisfactory animal model for allergenicity testing of GM proteins, immunisation studies in brown Norway rats (*Rattus norvegicus*) show some promise.

PROBLEMS AND PERSPECTIVES

Compositional studies and animal tests are but the first steps in GM risk assessment. Next, long-term, preferably lifetime-long metabolic, immune and reproduction studies with both male and female laboratory and other animal species should also be conducted under controlled conditions. However, setting up proper protocols for these is a task that has not been accomplished yet. If none of the short- or long-term risk assessment tests on animals show harm, only then could the safety of the GM food be further tested in double-blind placebo-controlled clinical studies with human volunteers. However, it should be pointed out that most clinical studies rely on volunteers in a reasonably good state of health even though any possibly harmful effects of GM foods are expected to be more serious with the old, young and the diseased. Thus, even the results of human clinical investigations may not be representative for the whole population, particularly when it is considered that, according to some estimates, up to 40% of the population may suffer from some sort of disease of the gastrointestinal tract. It also has to be taken into consideration that because it is an irreversible technology once a GM crop is generally grown on the land and foods based on these are released into the human food chain and included in animal rations, its removal or recall will become nearly impossible.

Effects of Transgenic Plant DNA

In addition to the changes in protein/metabolite profiles and the possible formation of new toxins and allergens in the plant resulting from the unanticipated effects of transgene insertion and the destabilisation of the recipient genome and the interference with the expression of the plant's own genes, the effects of transgenic plant DNA should also be considered. Thus, it is essential in any risk assessment protocol to determine in humans/animals ingesting GM foods whether appreciable amounts of the DNA vector construct used for developing the GM plant survive in the gut in functional form, whether they are taken up and integrated into the genome of the individual, and what, if any, effects the foreign transgenic DNA will have on them.

GM DNA SAFETY STUDIES IN THE GASTROINTESTINAL TRACT

The first task is to trace the GM DNA used for the development of the GM crop, such as the Bt toxin-expressing maize lines, through the intestinal tract, measure the proportion of the construct DNA surviving in functional form, establish by appropriate methods whether it is absorbed by the gut epithelial cells or by gut bacteria and integrated into the genome of these cells and whether they will express the transgene. Next, it has to be shown whether the GM DNA is absorbed into the systemic circulation and taken up by cells of body organs. In addition, it has to be investigated whether the GM DNA can pass into the placenta in pregnant females, foetus and brain, and, if so, what the biological consequences are. In these investigations, special emphasis should be laid on whether parts of the DNA constructs, particularly the promoter, such as the cauliflower mosaic virus 35s (CaMV 35s) are taken up by the gut and have biological effects. Obviously, as discussed in previous sections, it is of particular relevance whether the Bt toxin expressed in the GM plant has any harmful effect on the gut, body organs and the immune system. When an antibiotic resistance gene is used in the DNA construct as a selection marker gene, one of the most important questions that the risk assessment protocol will have to answer is whether this antibiotic resistance gene can transform gut bacteria *in vivo*. This has become highly pertinent since it was shown that functional DNA constructs used in the development of GM soybean survived in sufficient quantities in human volunteers and were found to be taken up by the bacteria in the gut.

GM SAFETY STUDIES

In the absence of safety studies, the lack of evidence that GM food is unsafe cannot be interpreted as proof that it is safe, particularly as all well-designed GM safety studies published to date and carried out independently of the biotechnology industry have demonstrated potentially worrisome biological effects of GM food as referred to in this paper and recently documented by Smith. Unfortunately, the regulators have largely ignored these.

In the light of these problems one can ask whether the future of the present generation of GM crops/ foods rests on solid scientific foundations. If not, as it appears, the question is whether it is needed at all, particularly as according to the FAO apparently there is sufficient food for feeding the world population, providing that it is evenly and properly distributed. It is possible that GM foods may be needed in future but should such a need arise we ought to first find more reliable and safer genetic transformation techniques for the development of GM crops. However, even then, their safety must be rigorously tested with biological methods, as without proper, transparent, inclusive, and independent testing the sceptical public is unlikely to be convinced of their safety and accept any present-day or future GM foods.

Effectiveness of GM Crops

According to the FDA and the United States Department of Agriculture (USDA), there are over 40 plant varieties that have completed all of the Federal requirements for commercialisation. Some examples of these plants include tomatoes and cantaloupes that have modified ripening characteristics, soyabeans and sugarbeets that are resistant to herbicides, and corn and cotton plants with increased resistance to insect pests. While there are very, very few genetically-modified whole fruits and vegetables available on produce stands, highly processed foods, such as vegetable oils or breakfast cereals, most likely contain some tiny percentage of genetically-modified ingredients because the raw ingredients have been pooled into one processing stream from many different sources. Also, the ubiquity of soyabean derivatives as food additives in the modern American diet virtually ensures that all US consumers have been exposed to GM food products.

CRITICISMS AGAINST GM FOODS

Environmental activists, religious organisations, public interest groups, professional associations and other scientists and government officials have all raised concerns about GM foods, and criticised agribusiness for pursuing profit without concern for potential hazards, and the government for failing to exercise adequate regulatory oversight. It seems that everyone has a strong opinion about GM foods. Most concerns about GM foods fall into three categories: environmental hazards, human health risks, and economic concerns.

Environmental Hazards

- *Unintended harm to other organisms:* Recently a research study was published Transgenic pollen harms monarch larvae (Nature, Vol 399, No 6733, p 214, May 1999) showing that pollen from *Bt* corn caused high mortality rates in monarch butterfly caterpillars. Monarch caterpillars consume milkweed plants, not corn, but the fear is that if pollen from *Bt* corn is blown by the wind onto milkweed plants in neighbouring fields, the caterpillars could eat the pollen and perish. The results seemed to support this viewpoint. Unfortunately, *Bt* toxins kill many species of insect larvae indiscriminately, it is not possible to design a *Bt* toxin that would only kill cropdamaging pests and remain harmless to all other insects.
- *Reduced effectiveness of pesticides:* Just as some populations of mosquitoes developed resistance to the now-banned pesticide DDT, many people are concerned that insects will become resistant to *Bt* or other crops that have been genetically modified to produce their own pesticides.
- *Gene transfer to non-target species:* Another concern is that crop plants engineered for herbicide tolerance and weeds will cross-breed, resulting in the transfer of the herbicide resistance genes from the crops into the weeds. These 'superweeds' would then be herbicide tolerant as well. Other introduced genes may cross over into nonmodified crops planted next to GM crops.

There are several possible solutions to the three problems mentioned above. Genes are exchanged between plants via pollen. Two ways to ensure that non-target species will not receive introduced genes from GM plants are to create GM plants that are male sterile (do not produce pollen) or to modify the GM plant so that the pollen does not contain the introduced gene. Cross-pollination would not occur, and if harmless insects such as monarch caterpillars were to eat pollen from GM plants, the caterpillars would survive.

Another possible solution is to create buffer zones around fields of GM crops. For example, non-GM corn would be planted to surround a field of *Bt* GM corn, and the non-GM corn would not be harvested. Beneficial or harmless insects would have a refuge in the non-GM corn, and insect pests could be allowed to destroy the non-GM corn and would not develop resistance to *Bt* pesticides. Gene transfer to weeds and other crops would not occur because the wind-blown pollen would not travel beyond the buffer zone. Estimates of the necessary width of buffer zones range from 6 meters to 30 meters or more. This planting method may not be feasible if too much acreage is required for the buffer zones.

Human Health Risks

- *Allergenicity:* Many children in the US and Europe have developed life-threatening allergies to peanuts and other foods. There is a possibility that introducing a gene into a plant may create a new allergen or cause an allergic reaction in susceptible individuals.
- *Unknown effects on human health:* There is a growing concern that introducing foreign genes into food plants may have an unexpected and negative impact on human health.

On the whole, with the exception of possible allergenicity, scientists believe that GM foods do not present a risk to human health.

Economic Concerns

Bringing a GM food to market is a lengthy and costly process, and of course agri-biotech companies wish to ensure a profitable return on their investment. Many new plant genetic engineering technologies and GM plants have been patented, and patent infringement is a big concern of agribusiness. Yet consumer advocates are worried that patenting these new plant varieties will raise the price of seeds so high that small farmers and third world countries will not be able to afford seeds for GM crops, thus widening the gap between the wealthy and the poor.

Patent enforcement may also be difficult, as the contention of the farmers that they involuntarily grew Monsanto-engineered strains when their crops were cross-pollinated shows. One way to combat possible patent infringement is to introduce a 'suicide gene' into GM plants. These plants would be viable for only one growing season and would produce sterile seeds that do not germinate. Farmers would need to buy a fresh supply of seeds each year. However, this would be financially disastrous for farmers in third world countries who cannot afford to buy seed each year and traditionally set aside a portion of their harvest to plant in the next growing season. In an open letter to the public, Monsanto has pledged to abandon all research using this suicide gene technology.

REGULATION OF GM FOODS AND ROLE OF GOVERNMENT

Governments around the world are hard at work to establish a regulatory process to monitor the effects of and approve new varieties of GM plants. Yet depending on the political, social and economic climate within a region or country, different governments are responding in different ways.

In Japan, the Ministry of Health and Welfare has announced that health testing of GM foods will be mandatory as of April 2001. Currently, testing of GM foods is voluntary. Japanese supermarkets are offering both GM foods and unmodified foods, and customers are beginning to show a strong preference for unmodified fruits and vegetables.

India's government has not yet announced a policy on GM foods because no GM crops are grown in India and no products are commercially available in supermarkets yet. India is, however, very supportive of transgenic plant research. It is highly likely that India will decide that the benefits of GM foods outweigh the risks because Indian agriculture will need to adopt drastic new measures to counteract the country's endemic poverty and feed its exploding population.

Some states in Brazil have banned GM crops entirely, and the Brazilian Institute for the Defense of Consumers, in collaboration with Greenpeace, has filed suit to prevent the importation of GM crops. Brazilian farmers, however, have resorted to smuggling GM soyabean seeds into the country because they fear economic harm if they are unable to compete in the global marketplace with other grain-exporting countries.

In Europe, anti-GM food protestors have been especially active. In the last few years Europe has experienced two major foods scares: bovine spongiform encephalopathy (mad cow disease) in Great Britain and dioxin-tainted foods originating from Belgium. These food scares have undermined consumer confidence about the European food supply, and citizens are disinclined to trust government information about GM foods. In response to the public outcry, Europe now requires mandatory food labelling of GM foods in stores, and the European Commission (EC) has established a 1% threshold for contamination of unmodified foods with GM food products.

In the United States, the regulatory process is confused because there are three different government agencies that have jurisdiction over GM foods. To put it very simply, the EPA evaluates GM plants for environmental safety, the USDA evaluates whether the plant is safe to grow, and the FDA evaluates whether the plant is safe to eat. The EPA is responsible for regulating substances such as pesticides or toxins that may cause harm to the environment. GM crops such as *Bt* pesticide-laced corn or herbicide-tolerant crops but not foods modified for their nutritional value fall under the purview of the EPA. The USDA is responsible for GM crops that do not fall under the umbrella of the EPA such as drought-tolerant or disease-tolerant crops, crops grown for animal feeds, or whole fruits, vegetables and grains for human consumption. The FDA historically has been concerned with pharmaceuticals, cosmetics and food products and additives, not whole foods. Under current guidelines, a genetically-modified ear of corn sold at a produce stand is not regulated by the FDA because it is a whole food, but a box of cornflakes is regulated because it is a food product. The FDA's stance is that GM foods are substantially equivalent to unmodified, 'natural' foods, and therefore not subject to FDA regulation.

The EPA conducts risk assessment studies on pesticides that could potentially cause harm to human health and the environment, and establishes tolerance and residue levels for pesticides. There are strict limits on the amount of pesticides that may be applied to crops during growth and production, as well as the amount that remains in the food after processing. Growers using pesticides must have a license for each pesticide and must follow the directions on the label to accord with the EPA's safety standards. Government inspectors may periodically visit farms and conduct investigations to ensure compliance. Violation of government regulations may result in steep fines, loss of license and even jail sentences.

The USDA has many internal divisions that share responsibility for assessing GM foods. Among these divisions are the Animal Health and Plant Inspection Service (APHIS) which conducts field tests and issues permits to grow GM crops, the Agricultural Research Service which performs in-house GM food research, and the Cooperative State Research, Education and Extension Service which oversees the USDA risk assessment program. The USDA is concerned with potential hazards of the plant itself. The USDA has the power to impose quarantines on problem regions to prevent movement of suspected plants, restrict import or export of suspected plants, and can even destroy plants cultivated in violation of USDA regulations. Many GM plants do not require USDA permits from APHIS.

A GM plant does not require a permit if it meets these 6 criteria: (i) the plant is not a noxious weed, (ii) the genetic material introduced into the GM plant is stably integrated into the plants own genome, (iii) the function of the introduced gene is known and does not cause plant disease, (iv) the GM plant is not toxic to non-target organisms, (v) the introduced gene will not cause the creation of new plant viruses and (vi) the GM plant cannot contain genetic material from animal or human pathogens

LABELLING OF GM FOODS

Labelling of GM foods and food products is also a contentious issue. On the whole, agribusiness industries believe that labelling should be voluntary and influenced by the demands of the free market. If consumers show preference for labelled foods over nonlabelled foods, then industry will have the incentive to regulate itself or risk alienating the customer. Consumer interest groups, on the other hand, are demanding mandatory labelling. People have the right to know what they are eating, argue the interest groups, and historically industry has proven itself to be unreliable at self-compliance with existing safety regulations. The FDA's current position on food labelling is governed by the Food, Drug and Cosmetic Act which is only concerned with food additives, not whole foods or food products that are considered Generally recognised as safe (GRAS).

There are many questions that must be answered if labelling of GM foods becomes mandatory. First, are consumers willing to absorb the cost of such an initiative? If the food production industry is required to label GM foods, factories will need to construct two separate processing streams and monitor the production lines accordingly. Farmers must be able to keep GM crops and non-GM crops from mixing during planting, harvesting and shipping. It is almost assured that industry will pass along these additional costs to consumers in the form of higher prices. Secondly, what are the acceptable limits of GM contamination in non-GM products? The EC has determined that 1% is an acceptable limit of cross-contamination, yet many consumer interest groups argue that only 0% is acceptable. Some companies such as Gerber baby foods and Frito-Lay have pledged to avoid use of GM foods in any of their products. But who is going to monitor these companies for compliance and what is the penalty if they fail? Once again, the FDA does not have the resources to carry out testing to ensure compliance.

What is the level of detectability of GM food cross-contamination? Scientists agree that current technology is unable to detect minute quantities of contamination, so ensuring 0% contamination using existing methodologies is not guaranteed. Yet researchers disagree on what level of contamination really is detectable, especially in highly processed food products such as vegetable oils or breakfast cereals where the vegetables used to make these products have been pooled from many different sources. A 1% threshold may already be below current levels of detectability.

Finally, who is to be responsible for educating the public about GM food labels and how costly will that education be? Food labels must be designed to clearly convey accurate information about the product in simple language that everyone can understand. This may be the greatest challenge faced by a new food labelling policy: how to educate and inform the public without damaging the public trust and causing alarm or fear of GM food products.

In January 2000, an international trade agreement for labelling GM foods was established. More than 130 countries, including the US, the worlds largest producer of GM foods, signed the agreement. The policy states that exporters must be required to label all GM foods and that importing countries have the right to judge for themselves the potential risks and reject GM foods, if they so choose. This new agreement may spur the US government to resolve the domestic food labelling dilemma more rapidly.

Labelling of GMO Products: Freedom of Choice for Consumers

Exactly what must be labelled and how it is to be labelled, and why - is explained in the following.

Labelling guide

A basic principle applies to most food products: if genetically modified plants or micro-organisms have been used in production, this must be clearly indicated.

Labelling: Yes

However, under certain conditions, numerous products are exempt from labelling obligations. These exemptions primarily concern additives and processing aids, but also apply to meat, milk and eggs.

Labelling: No

The status of flavours, additives and enzymes in regard to labelling obligations is complex. The use of genetic engineering is common, but there is no general labelling practice.

Labelling: Flavours, additives and enzymes

Labelling is also required for foods which are offered by restaurants, canteens and takeaways although there are exceptions.

Organic products without genetic engineering

By law, the use of genetical engineering is prohibited for products defined as 'organic'. Nevertheless, these products are permitted in certain cases to contain slight traces of genetically modified organisms.

Labelling: Organic products without genetic engineering

GMO labelling in the European Union: basic principles

All food, and any ingredients, directly produced from a GMO must be labelled, even if this GMO is undetectable in the final product.

Labelling requirements: For neutral information only not for warning

Labelling empowers the buyer: In order to choose between products with or without genetically modified organisms, consumers need transparent, controllable and straightforward labelling regulations. However, the extent and breadth of these regulations are decided politically.

MOST COMMON GENETICALLY MODIFIED FOODS

Genetically modified material sounds a little bit like science fiction territory, but in reality, much of what we eat on a daily basis is a genetically modified organism (GMO). Whether or not these modified foods are actually healthy is still up for debate—and many times, you don't even know that you are buying something genetically modified.

It is not required to label GMOs in the US and Canada, but there are substantial restrictions, and even outright bans, on GMOs in many other countries.

However, by 2018, Whole Foods Market will start labelling GMOs in the U.S. This grocery chains' locations in Britain already provide GMO labelled products, as required by the European Union. According to the EU, GMO refers to plants and animals 'in which the genetic material has been altered in a way that does not occur naturally by mating and/or natural recombination.'

Some of the most common genetically modified foods are briefly discussed below:

Corn: Almost 85 per cent of corn grown in the US is genetically modified. Even Whole Foods's brand of corn flakes was found to contain genetically modified corn. Many producers modify corn and soya so that they are resistant to the herbicide glyphosate, which is used to kill weeds.

Soya: Soya is the most heavily genetically modified food in US. The largest US producer of hybrid seeds for agriculture, Pioneer Hi-Bred International, created a genetically engineered soyabean, which was approved in 2010. It is modified to have a high level of oleic acid, which is naturally found in olive oil. Oleic acid is a monounsaturated omega-9 fatty acid that may lower LDL cholesterol (traditionally thought of as 'bad' cholesterol) when used to replace other fats.

Alfalfa: Cultivation of genetically engineered alfalfa was approved in 2011, and consists of a gene that makes it resistant to the herbicide Roundup, allowing farmers to spray the chemical without damaging the alfalfa.

Canola: Canola is genetically engineered form was approved in 1996, and as of 2006, around 90 percent of US canola crops are genetically modified.

Sugar Beets: A very controversial vegetable, sugar beets were approved in 2005, banned in 2010, then officially deregulated in 2012. Genetically modified sugar beets make up half of the US sugar production, and 95 per cent of the country's sugar beet market.

Milk: To increase the quantity of milk produced, cows are often given rBGH (recombinant bovine growth hormone), which is also banned in the European Union, as well as in Japan, Canada, New Zealand and Australia.

Zucchini: Genetically modified zucchini contains a toxic protein that helps make it more resistant to insects. This introduced insecticide, has recently been found in human blood, including that of pregnant women and fetuses. This indicates that some of the insecticide is making its way into our bodies rather than being broken down and excreted.

Yellow squash: Yellow squash has also been modified with the toxic proteins to make it insect resistant. This plant is very similar to zucchini, and both have also been modified to resist viruses.

Papaya: Genetically modified papaya trees have been grown in Hawaii since 1999. These Papayas are sold in the United States and Canada for human consumption. These papayas have been modified to be naturally resistant to Papaya Ringspot virus, and also to delay the maturity of the fruit. Delaying maturity gives suppliers more time to ship the fruit to supermarkets.

PROS AND CONS OF GENETICALLY MODIFIED FOODS

Many people today take for granted exactly where the foods they eat come from. In fact, genetically modified foods have become a commonplace thing in America, even though few people understand just what 'genetically modified' means. While there are some benefits that genetically modified foods may offer, there are also some risks and negative effects that these foods can cause as well.

When the term 'genetically modified' is used to describe a food, it means that the genetic makeup of one of the ingredients in that food has been altered. This is achieved by a very special set of technologies that combine the genes from different organisms, with the resulting organism being called a genetically modified food. In most cases, the specific genes that are combined have been hand-picked for the specific traits that they have. Those traits could include everything from the resistance to insects to specific nutritional value. These genetically modified foods can be in anything from corn to canola oil, which are quite common ingredients in many foods found on the market today such as snacks, cereals and sodas.

Pros Of Genetically Modified Foods

There are several benefits that have been linked to genetically modified foods, including:

- *Resistance to disease:* Genes can be modified to make crops more resilient when it comes to disease, especially those spread through insects. This can lead to higher crop yields, which many experts argue can help to feed people in developing countries.
- *Cost:* Because foods can be more resistant to disease, it reduces the cost necessary for pesticides and herbicides. And although genetically modified seeds are a more costly investment initially, this reduction in cost along with fewer lost crops leads to more profits. In many cases, that lower cost is passed onto the consumer through lower food prices.
- *Quality:* Some genetically modified foods, particularly fruits and veggies, have a longer shelf life than natural products.
- *Taste:* Some people claim that genetically modified foods have a better taste. In some cases, the genes can be altered in order to improve taste, although this is still one factor that varies from person to person.
- *Nutritional content:* Foods are often genetically modified in order to increase their nutritional content. This is especially helpful for certain populations where a specific nutrient is lacking in the local diet.

Cons of Genetically Modified Foods

Although there are some benefits to genetically modified foods, there are some risks that have been associated with these foods.

Some of these risks include:

- *Allergens and toxins:* Some genetically modified foods may contain higher levels of allergens and toxins, which can have negative effects on the personal health of those who eat them. This may be especially dangerous for people with serious food allergies.
- *Antibiotic resistance:* Because genetically modified foods are often developed to fight off certain pesticides and herbicides, there may be an increased risk that people who eat those foods may be more resistant to antibiotics.
- *New diseases:* Viruses and bacteria are used in the process of modifying foods, which means that there is a possibility that they could cause the development of a new disease.
- *Nutritional content:* Not all genetically modified foods are changed to increase their nutritional content. Instead, these foods may actually lose nutritional content in the process of altering their genetic makeup.
- *Loss of biodiversity:* Genetically modified foods could potentially cause damage to other organisms in the ecosystems where they are grown. If these organisms are killed off, it leads to a loss of biodiversity in the environment while also putting other organisms at risk by creating an unstable ecosystem.

Public Debate on GMO

The release of GMO into the environment and the marketing of GM foods have resulted in a public debate in many parts of the world. Even though the issues under debate are usually very similar (costs and benefits, safety issues), the outcome of the debate differs from country to country. On issues such as labelling and traceability of GM foods as a way to address consumer preferences, there is no worldwide consensus to date. Despite the lack of consensus on these topics, the Codex Alimentarius Commission has made significant progress and developed Codex texts relevant to labelling of foods derived from modern biotechnology in 2011 to ensure consistency on any approach on labelling implemented by Codex members with already adopted Codex provisions.

ADVANTAGES OF GM FOODS

The world population has topped 7.3 billion people and is predicted to double in the next 50 years. Ensuring an adequate food supply for this booming population is going to be a major challenge in the years to come. GM foods promise to meet this need in a number of ways:

- *Pest resistance:* Crop losses from insect pests can be staggering, resulting in devastating financial loss for farmers and starvation in developing countries. Farmers typically use many tons of chemical pesticides annually. Consumers do not wish to eat food that has been treated with pesticides because of potential health hazards, and run-off of agricultural wastes from excessive use of pesticides and fertilisers can poison the water supply and cause harm to the environment. Growing GM foods such as *Bt* corn can help eliminate the application of chemical pesticides and reduce the cost of bringing a crop to market.
- *Herbicide tolerance:* For some crops, it is not cost-effective to remove weeds by physical means such as tilling, so farmers will often spray large quantities of different herbicides (weed-killer) to destroy weeds, a time-consuming and expensive process, that requires care so that the herbicide doesn't harm the crop plant or the environment. Crop plants genetically-engineered to be resistant to one very powerful herbicide could help prevent environmental damage by reducing the amount

of herbicides needed. For example, Monsanto has created a strain of soyabeans genetically modified to be not affected by their herbicide product Roundup ready soyabeans. A farmer grows these soyabeans which then only require one application of weed-killer instead of multiple applications, reducing production cost and limiting the dangers of agricultural waste run-off.

- *Disease resistance:* There are many viruses, fungi and bacteria that cause plant diseases. Plant biologists are working to create plants with genetically-engineered resistance to these diseases.

- *Cold tolerance:* Unexpected frost can destroy sensitive seedlings. An antifreeze gene from cold water fish has been introduced into plants such as tobacco and potato. With this antifreeze gene, these plants are able to tolerate cold temperatures that normally would kill unmodified seedlings.

- *Drought tolerance/salinity tolerance:* As the world population grows and more land is utilised for housing instead of food production, farmers will need to grow crops in locations previously unsuited for plant cultivation. Creating plants that can withstand long periods of drought or high salt content in soil and groundwater will help people to grow crops in formerly inhospitable places.

- *Nutrition:* Malnutrition is common in third world countries where impoverished peoples rely on a single crop such as rice for the main staple of their diet. However, rice does not contain adequate amounts of all necessary nutrients to prevent malnutrition. If rice could be genetically engineered to contain additional vitamins and minerals, nutrient deficiencies could be alleviated. For example, blindness due to vitamin A deficiency is a common problem in third world countries. Researchers at the Swiss Federal Institute of Technology Institute for Plant Sciences have created a strain of 'golden' rice containing an unusually high content of beta-carotene (vitamin A).

- *Pharmaceuticals:* Medicines and vaccines often are costly to produce and sometimes require special storage conditions not readily available in third world countries. Researchers have developed edible vaccines in tomatoes and potatoes. These vaccines are much easier to ship, store and administer than traditional injectable vaccines.

- *Phytoremediation:* Not all GM plants are grown as crops. Soil and groundwater pollution continues to be a problem in all parts of the world. Plants such as poplar trees have been genetically engineered to clean up heavy metal pollution from contaminated soil.

SECTION VII

Production, Purification and Application of Enzymes in Food Industry

Chapter 22

Microbial Enzymes, Production, Purification and Isolation

INTRODUCTION

Enzymes occur in every living cell, hence in all microorganisms. Each single strain of organism produces a large number of enzymes, hydrolysing, oxidising or reducing, and metabolic in nature. But the absolute and relative amounts of the various individual enzymes produced vary markedly between species and even between strains of the same species. Hence, it is customary to select strains for the commercial production of specific enzymes which have the capacity for producing highest amounts of the particular enzymes desired. Commercial enzymes are produced from strains of molds, bacteria, and yeasts.

Microbial enzymes: Production of a new microbial enzyme starts with screening of micro-organisms for desirable activity using appropriate selection procedures. The harsh environment to which several enzymes are subjected during process applications has given impetus to screening of extremophiles for enzymes having desirable features of activity and stability. The level of enzyme activity produced by an organism from a natural environment is often low and needs to be elevated for industrial production. Increase in enzyme levels is often achieved by mutation of the organism. An alternative strategy that has gained favour is production of the enzyme in a recombinant organism of choice whose growth conditions are well optimised and whose GRAS status is established. Random or site-directed mutagenesis with the purpose of engineering the activity and stability properties of an enzyme prior to its production is becoming a common practice. The micro-organisms used for enzyme production are grown in fermenters using an optimised growth medium. Both solid state- and submerged fermentation are applied commercially, however the latter is preferred in many countries because of a better handle on aseptic conditions and process control. The enzymes produced by the micro-organism may be intracellular or secreted into the extracellular medium.

Fungal enzymes: For fungal enzymes, modifications of Dr. Takamine's original mold bran process have usually been employed. In this process, the mold is cultivated on the surface of a solid substrate. Takamine used wheat bran and this has come to be recognised as the most satisfactory basic substrate although other fibrous materials can be employed. Other ingredients may be added, such as nutrient salts, acid or buffer to regulate the pH, soyabean meal or beet cosettes to stimulate enzyme production. In one modification of the bran process, the bran is steamed for sterilisation, cooled, inoculated with the mold spores, and spread out on trays. Incubation takes place in chambers where the temperature and humidity are controlled within limits by circulated air. It may be stated that instead of trays for incubation, Takamine, as well as other producers, at one time used slowly rotating drums. Generally tray incubation gives more rapid growth and enzyme production.

Bacterial enzymes: Bacterial enzymes have been and are also produced by the bran process. However, until recently the process originally invented by Boidin and Effront was most extensively employed. In this process, the bacteria are cultivated in special culture vessels as a pellicle on the surface of thin layers of liquid medium, the composition of which is adjusted for maximum production of the desired enzyme. Different strains of *Bacillus subtilis* and different media are employed, depending on whether bacterial amylase or protease is desired.

The submerged method was originally developed and first extensively employed for production of penicillin and other antibiotics. In the laboratory, submerged cultures are grown in shake flasks or in aerated tubes or flasks. Commercially, deep tanks are employed which have provision for introduction of sterile air and for vigorous agitation. The amount of air, degree of dispersion of air, and amount of agitation are dependent variables. For effective results the air must be dispersed in very fine bubbles throughout the mass of culture liquid. Fine aeration through porous substances may be used to produce high dispersion. Most manufacturers, however, depend upon efficient agitators to break up the air into the requisite small bubbles.

Either surface or submerged culture methods currently may be employed for most microbial enzymes production. Usually different cultures must be used for maximum enzyme yields by the two methods, although there are exceptions to this rule. There are advantages and disadvantages to each method, some of which are shown in Table 22.1. Which method is used for a particular commercial enzyme will be dictated by plant equipment, convenience, relative yields, and application.

Table 22.1: Comparison of surface and submerged processes.

Surface	*Submerged*
Requires much space for trays	Uses compact closed fermentors
Requires much hand labour	Requires minimum of labour
Uses low pressure air blower	Requires high pressure air
Little power requirement	Needs considerable power for air compressors and agitators
Minimum control necessary	Requires careful control
Little contamination problem	Contamination frequently a serious problem
Recovery involves extraction with aqueous solution, filtration or centrifugation and perhaps evaporation and/or precipitation	Recovery involves filtration or centrifugation, and perhaps evaporation and/or preceipitation

Recovery of the enzyme generally depends upon precipitation from an aqueous solution, although some enzymes may be marketed as stabilised solutions. In the bran process, the enzyme is extracted from the *koji* (the name given to the mass of material permeated with the mold mycelium) into an aqueous solution by percolation. In the liquid processes, the microbial cells are filtered from the beer. The enzyme may be precipitated by addition of solvents, such as acetone or aliphatic alcohols, to the aqueous enzyme solution, either directly or after concentration by vacuum evaporation at low temperature. The precipitated enzyme may be filtered and dried at low temperature, for example in a vacuum shelf dryer. The dry enzyme powders may be sold as undiluted concentrates on a potency basis or, for most applications, may be diluted to an established standard potency with an acceptable diluent. Some common diluents are salt, sugar, starch, and wheat flour. Most commercial enzymes are quite stable in the dry form, but some require the presence of stabilisers and activators for maximum stability and efficiency in use.

In theory, the fermentative production of microbial enzymes is a simple matter, requiring an appropriate organism grown on a medium of optimum composition under optimum conditions. The stocks in trade of microbial enzyme manufacturers are thus the selected cultures, the composition of media, and the cultural conditions, all of which are usually held confidential.

In practice, enzyme manufacturers suffer the same difficulties in fermentation, frequently in even greater degree, as antibiotics producers. Total loss of fermentation batches may result from contamination, culture variation, failure of cultural control, and other like causes. Furthermore, knowledge and careful application of the best methods for recovery, stabilisation, and storage of such delicate biological entities as the labile enzymes presents a constant challenge.

Isolation and purification: Isolation and purification, i.e. downstream processing of enzyme from the raw material constitutes the subsequent key stage in the production process. The desired level of purification depends on the ultimate application of the enzyme product. The industrial bulk enzymes are relatively crude formulations while speciality enzymes undergo a thorough purification to yield a homogeneous product. A traditional downstream processing scheme involves stages of clarification for separation of the enzyme from the solids comprising the raw material, concentration to reduce the process volumes, and purification to separate it from other soluble contaminants. In case of the intracellular enzymes, disruption of cells or tissue for release of the product is among the primary separation steps. There is a choice of different separation techniques for each stage. Chromatography is the major technique for high-resolution purification of enzymes. Some separation techniques allow integration of the downstream processing stages required for purification thus reducing the number of steps and hence the production costs. The enzyme is finally formulated as a liquid or solid product. In either case, stabilising additives are added for rendering long shelf life to the product. Some enzymes are immobilised to solid supports or enzyme crystals are cross-linked to render them insoluble and stable for repeated or long term use in a process application. Large scale production of enzymes has to comply with the standards set by International Organisation of Standardisation for ensuring quality and production efficiency, and also environmental management control, whenever applicable.

The production of enzymes is central to the modern biotechnology industry. The traditional industrial enzymes continue to have expanding markets, and the recognition of potential to use biocatalysis in various industrial sectors for new applications generates demand for enzymes with novel activities and/ or improved stability. Man has utilised enzymes throughout the ages either in the form of vegetables rich in enzymes, or as micro-organisms used for a variety of purposes, for instance in brewing, baking, and in cheese production. However, it was only in the 19th century that the various biological conversions were ascribed to the action of enzymes. This was initiated by the isolation of an enzyme complex from malt by Payen and Persoz in 1833, termed 'diastase' that converts gelatinised starch into sugars, primarily maltose. The history of modern enzyme production really began in 1874 when the Danish chemist Christian Hansen first produced rennet by extracting it from dried calves' stomachs with saline solution. Apparently, this was the first enzyme preparation of relatively high purity used for industrial purposes. During the early part of the last century, in the Far East, an age-old tradition involving the use of mould fungi called *koji* in the production of certain foodstuffs and flavour additives based on soya protein and fermented beverages, formed the basis on which the Japanese scientist Takamine developed a fermentation process for the industrial production of fungal amylase. The process included the culture of *Aspergillus oryzae* on moist rice or wheat bran, and the product was called 'Takadiastase' which is still used as a digestive aid. The industrial or bulk enzymes include proteases, amylases, lipases, etc. which are required in large volumes, but have an inherently low unit value so that they demand significantly lower manufacturing costs. On the other end of the scale is the therapeutics sector with products such as urokinase, which are produced in lower volumes and at inherently greater manufacturing cost. In between these two lie the diagnostic enzymes.

The technology for producing and using commercially important enzyme products combines the disciplines of microbiology, genetics, biochemistry and engineering, which have developed and matured through time both singly and in an interactive manner. Demands for new enzymes arise from the development of new processes or from the unsatisfactory performance of known enzymes in established processes. The revolution in gene technology over the last two decades has had a big impact on enzyme industry. Genetic engineering techniques have enabled enzyme manufacturers to produce sufficient quantities of almost any enzyme no matter what the source, while protein engineering allows the properties of the enzymes to be adjusted prior to production. This chapter provides an overview of enzyme production processes starting from raw material to the finished product, and gives an insight of the various alternative technologies available for different stages of production.

ENZYME SOURCE

The primary consideration in the production of any enzyme relates to the choice of source. In most cases, the desired activity can be obtained from several sources. Traditionally, however, the choice of source has been more restricted for some enzymes. For example, the enzyme rennet was until recently obtained from the stomach of suckling calves, the corresponding microbial enzyme led to an off-flavour in the cheese produced. Today, recombinant DNA technology is used to produce the calf enzyme in micro-organisms. Micro-organisms represent an attractive source of enzymes as they can be cultured in large quantities in a relatively short period by established methods of fermentation. However, the level of production of a particular enzyme varies in different micro-organisms, and moreover the enzymes often differ in composition and properties. One usually finds that the closely related organisms have enzymes with nearly similar properties, while unrelated organisms have enzyme systems that differ widely. The most critical feature of the organisms for producing industrially significant enzymes is their GRAS (generally regarded as safe) status, which implies that they must be non-toxic, non-pathogenic and generally should not produce antibiotics.

The GRAS listed micro-organisms include fewer than 50 bacteria and fungi. Examples are the bacteria including *Bacillus subtilis*, *B. licheniformis*, and various other *bacilli*, *lactobacilli*, *Streptomyces* species, the yeast *Saccharomyces cerevisiae*, and the filamentous fungi belonging to the genera *Aspergillus*, *Mucor*, *Rhizopus*, etc. In case of *Bacillus*, mutants are selected that can no longer form spores.

Since *Aspergillus* cultures are frequently inoculated with conidia, enzyme production using these fungi relies on good spore formation. Most of the bulk enzymes (hydrolases) are secreted by the micro-organisms directly into the culture medium, while some enzymes, e.g. penicillin acylase and glucose isomerase are intracellular. For some applications, it may not be necessary to isolate the enzymes but the microbial cells themselves are used as enzyme source.

The organism is preferred which gives high yields of enzyme in shortest possible fermentation time. The production strains used in industry are normally modified by genetic manipulation to have high levels of production. A common trend in the industry today is that the gene coding for the enzyme with desired characteristics is transferred into one of the selected microbial production strains which have all the required features of safety and high expression levels and for which the growth medium has been optimised, hence avoiding the need for optimisation of individual enzyme producing strains.

A further short cut in the search for the right enzyme has been made in eliminating the step of screening, isolation and cultivation of micro-organisms which may either be present in low number or produce low levels of the activity. Instead, DNA is directly isolated from an environmental sample and the possession of the desired activity is located using an appropriate gene probe. The gene is cloned and expressed in the desired production organism. Despite the advantages of micro-organisms as enzyme source, some enzymes are still economically produced from plant and animal sources. This is possible because of sufficiently high amounts of these enzymes in such sources and also as a means to convert inexpensive, renewable material like agricultural and slaughter waste into value added products.

Examples of the enzymes isolated from plants include several proteases such as papain, ficin and bromelain, and peroxidase. As mentioned above, rennet has been among the most industrially significant enzymes obtained from animal tissue. The other enzymes obtained from animal sources, e.g. proteases like trypsin, chymotrypsin and urokinase, lactate dehydrogenase, lysozyme, etc., have diverse applications in industry, analysis, and therapy. In recent years, protein production in transgenic animals and –plants has attracted attention. Focus on transgenic animals (e.g. sheep, cattle) has been for the production of therapeutic proteins. The expression of the foreign gene is targeted to the mammary gland so that the protein is secreted directly into the milk.

Although both pharmaceutical and industrial proteins have been expressed in transgenic plants, they are suggested to be ideal bioreactors for production of the latter category of proteins. Production of bulk enzymes like α-amylase, xylanase, phytase, etc. combines the advantages of low production costs of plant biomass with the minimal purification requirements for such products.

PURIFICATION OF ENZYMES

Downstream Processing

Downstream processing is a very important step in biotechnology because costs for collection, concentration and purification of the final product are substantial. High product concentrations in the supernatant or inside the cells and efficient purification are therefore important aspects in the overall economy of enzyme manufacture. The degree of purity of commercial enzymes ranges from raw enzymes to highly purified forms and depends on the application. Raw materials for the isolation of enzymes are animal organs, plant material and micro-organisms.

Often enzymes may be purified several hundred-fold but the yield of the enzyme may be very poor, frequently below 10% of the activity of the original material. In contrast, industrial enzymes will be purified as little as possible, only other enzymes and material likely to interfere with the process which the enzyme is to catalyse, will be removed. Unnecessary purification will be avoided as each additional stage is costly in terms of equipment, manpower and loss of enzyme activity. As a result, some commercial enzyme preparations consist essentially of concentrated fermentation broth, plus additives to stabilise the enzymes activity.

However, the content of the required enzyme should be as high as possible (e.g. 10% w/w of the protein) in order to ease the downstream processing task. This may be achieved by developing the fermentation conditions or, often more dramatically, by genetic engineering. It may well be economically viable, e.g. to spend some time cloning extra copies of the required gene together with a powerful promoter back into the producing organism in order to get over-producers. Downstream processing involves isolation and purification steps and ends up in the formulation of the enzyme preparation.

Enzymes are universally present in living organisms, each cell synthesises a large number of different enzymes to maintain its metabolic reactions. The choice of procedures for enzyme purification depends on their location. On the one hand, isolation of intracellular enzymes often involves the separation of complex biological mixtures. On the other hand, extracellular enzymes are generally released into the medium with only a few other components

Enzymes are very complex proteins and their high degree of specificity as catalysts is manifest only in their native state. The native conformation is attained under specific conditions of pH, temperature and ionic strength. Hence, only mild and specific methods can be used for enzyme isolation. Figure 22.1 shows the sequence of steps involved in the recovery of enzymes.

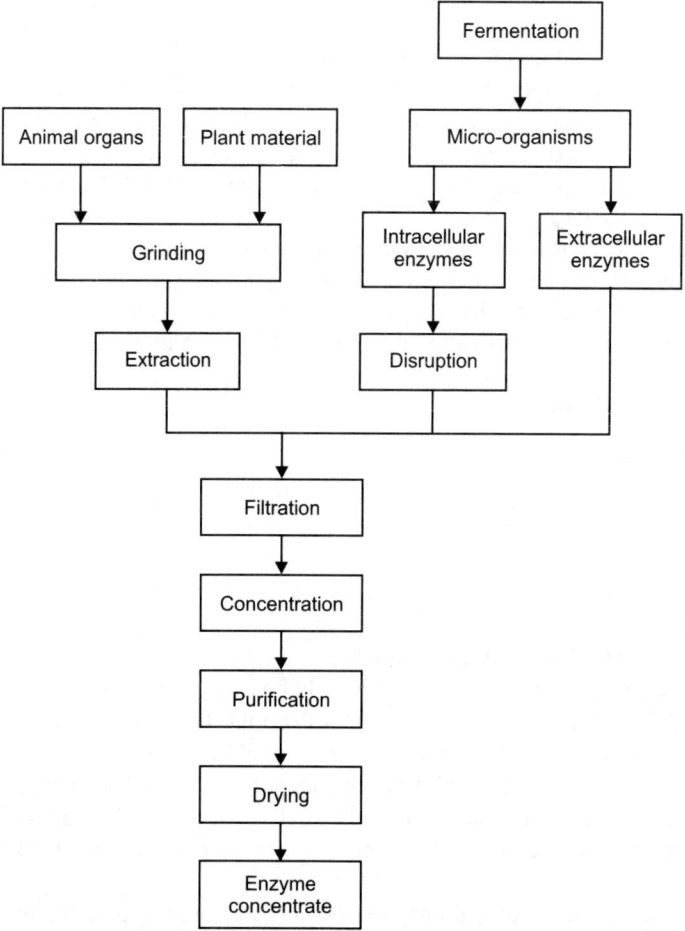

Fig. 22.1: Flowchart of the downstream processing of enzymes.

Preparation of Biological Starting Materials

Animal organs: Animal organs must be transported and stored at low temperature to retain enzymatic activity. Frozen organs can be minced with machines generally used in the meat industry and the enzymes can be extracted with a buffer solution. Besides mechanical grinding, enzymatic digestion can also be employed.

Plant material: Plant material can be ground with various crushers or grinders and the desired enzymes can be extracted with buffer solutions. The cells can also be disrupted by previous treatment with lytic enzymes.

Micro-organisms: Enzymes might accumulate inside the cells or be released into the medium. Most enzymes used commercially are extracellular enzymes and the first step in their isolation is the separation of the cells from the solution. For intracellular enzymes, which are being isolated today in increasing amounts, the first step involves grinding to rupture the cells. A number of methods for the disruption of cells are known, corresponding to the different types of cells and the problems involved in isolating intracellular enzymes. However, only a few of these methods are used on an industrial scale.

Cell disruption by mechanical methods

High-pressure homogenisation is the most common method of cell disruption. The cell suspension is pressed through a valve and hits an impact ring. The cells are ruptured by shearing forces and simultaneous decompression. The wet grinding of cells in a high-speed bead mill is another effective method of cell disruption. Glass balls with a diameter of 0.2–1 mm are used to break the cells. The efficiency of this method depends on the geometry of the stirrer system.

Cell disruption by non-mechanical methods

Cells may frequently be disrupted by chemical, thermal, or enzymatic lysis. The drying of micro-organisms and the preparation of acetone powders are standard procedures in which the structure of the cell wall is altered to permit subsequent extraction of the cell contents.

Separation of Solid Matter

After cell disruption, the next step is separation of extracellular or intracellular enzymes from cells or cellular fragments, respectively. This operation is rather difficult because of the small size of bacterial cells and the slight difference between the density of the cells and that of the fermentation medium. Continuous filtration is used in industry. Large cells, e.g. yeast cells, can be removed by decantation. Today, efficient centrifuges have been developed to separate cells and cellular fragments in a continuous process. Residual plant and organ matter can be separated with simpler centrifuges or filters.

Filtration

Pressure filters: A filter press (plate filter, chamber filter) is used to filtrate small volumes or to remove precipitates formed during purification. The capacity to retain solid matter is limited and the method is rather work-intensive. However, these filters are highly suitable for the fine filtration of enzyme solutions.

Vacuum filters: Vacuum filtration is generally the method of choice because biological materials are easily compressible. A rotary vacuum filter is used in the continuous filtration of large volumes. The suspension is usually mixed with a filter aid (e.g. kieselguhr) before being applied to the filter.

Cross-flow filtration: In recent years, a new method of filtration, cross-flow filtration, has been devised. In conventional methods, the suspension flows perpendicular to the filtering material. In cross-flow filtration, the input stream flows parallel to the filter area, thus preventing the accumulation of filter cake and an increased resistance to filtration. Cross-flow filtration can be conveniently used in recombinant DNA techniques to separate organisms in a closed system.

Centrifugation

The sedimentation rate of a bacterial cell with a diameter of 0.5 mm is less than 1 mm/h. An economical separation can be achieved only by sedimentation in a centrifugal field. The range of applications of centrifuges depends on the particle size and the content of the solids. Decanters (scroll-type centrifuges) work with low centrifugal forces and are used in the separation of large cells or protein precipitates.

Solid matter is discharged continuously by a screw conveyer moving at a differential rotational speed. Tubular bowl centrifuges are built for very high centrifugal forces and can be used to sediment very small particles. However, these centrifuges cannot be operated in a continuous process. Moreover, solid matter must be removed by hand after the centrifuge has come to a stop. A further disadvantage is the appearance of aerosols.

Separators (disk stack centrifuges) can be used in the continuous removal of solid matter from suspensions. Solids are discharged by a hydraulically operated discharge port (intermittent discharge) or by an arrangement of nozzles (continuous discharge). Bacteria and cellular fragments can be separated by a combination of high centrifugal forces. Disk stack centrifuges that can be sterilised with steam are used for recombinant DNA techniques in a closed system.

Extraction

An elegant method used to isolate intracellular enzymes is liquid-liquid extraction in an aqueous two-phase system. This method is based on the incomplete mixing of different polymers, e.g. dextran and poly(ethylene glycol), or a polymer and a salt in an aqueous solution. The first extraction step separates cellular fragments. Subsequent purification can be accomplished by extraction or, if high purity is required, by other methods. The extractability can be improved by using affinity ligands or modified chromatography gels, e.g. phenyl-sepharose.

Flocculation and flotation

The flocculation of bacterial cells to form larger particles can be achieved by the addition of mineral colloids, salts, or organic polymers. The neutralisation of charges on the cell surface and the formation of bridges between individual cells lead to agglomeration. Agglomerates can then be removed by filtration or centrifugation.

Concentration

The enzyme concentration in starting material is often very low. The volume of material to be processed is generally very large and substantial amounts of waste material must be removed. Thus, if economic purification is to be achieved, the volume of starting material must be decreased by concentration. Only mild concentration procedures which do not inactivate enzymes can be employed. These procedures include thermal methods, precipitation and to an increasing extent, membrane filtration.

Thermal methods

Only brief heat treatment can be used for concentration because enzymes are thermolabile. Evaporators with rotating components which achieve a thin liquid film (thin-layer evaporator, centrifugal thin-layer evaporator) or circulation evaporators (long-tube evaporator) can be employed.

Precipitation

Enzymes are very complex protein molecules possessing both ionisable and hydrophobic groups which interact with the solvent. Indeed, proteins can be made to agglomerate and, finally, precipitate by changing their environment. Precipitation is actually a simple procedure for concentrating enzymes.

Precipitation with salts: High salt concentrations act on the water molecules surrounding the protein and change the electrostatic forces responsible for solubility. Ammonium sulphate is commonly used for precipitation, hence, it is an effective agent for concentrating enzymes. Enzymes can also be fractionated, to a limited extent, by using different concentrations of ammonium sulphate. Sodium sulphate is another precipitating agent used.

Precipitation with organic solvents: Organic solvents influence the solubility of enzymes by reducing the dielectric constant of the medium. The solvation effect of water molecules surrounding the enzyme is changed, the interaction of protein molecules is increased and therefore, agglomeration and precipitation occur. Commonly used solvents are ethanol and acetone.

Precipitation with polymers: The polymers generally used are polyethyl-enimines and polyethylene glycols of different molecular masses. The mechanism of this precipitation is similar to that of organic solvents and results from a change in the solvation effect of the water molecules surrounding the enzyme. Most enzymes precipitate at polymer concentrations ranging from 15 to 20%.

Precipitation at the isoelectric point: Proteins are ampholytes and carry both acidic and basic groups. The solubility of proteins is markedly influenced by pH and is minimal at the isoelectric point at which the net charge is zero. Because most proteins have isoelectric points in the acidic range, this process is also called acid precipitation.

Ultrafiltration

A semipermeable membrane permits the separation of solvent molecules from larger enzyme molecules, because only the smaller molecules can penetrate the membrane when the osmotic pressure is exceeded. This is the principle of all membrane separation processes including ultrafiltration. In reverse osmosis, used to separate materials with low molecular mass, solubility and diffusion phenomena influence the process, whereas ultrafiltration and cross-flow filtration are based solely on the sieve effect. In processing enzymes, cross-flow filtration is used to harvest cells, whereas ultrafiltration is employed for concentrating and desalting.

Desalting: The desalting of enzyme solutions can be carried out conveniently by diafiltration. The small salt molecules are driven through a membrane with the water molecules. The permeate is continuously replaced by fresh water.

Purification

For many industrial applications, partially purified enzyme preparations will suffice, however, enzymes for analytical purposes and for medical use must be highly purified. Special procedures employed for enzyme purification are crystallisation, electrophoresis and chromatography. However crystallisation and electrophoresis are not relevant for large scale purifications. Chromatography, in contrast, is of fundamental importance to enzyme purification. Molecules are separated according to their physical properties (size, shape, charge, hydrophobic interactions), chemical properties (covalent binding), or biological properties (biospecific affinity). In gel chromatography (also called gel filtration), hydrophilic, cross-linked gels with pores of finite size are used in columns to separate biomolecules. In gel filtration, molecules are separated according to size and shape. Molecules larger than the largest pores in the gel beads, i.e. above the exclusion limit, cannot enter the gel and are eluted first. Smaller molecules, which enter the gel beads to varying extent depending on their size and shape, are retarded in their passage through the column and eluted in order of decreasing molecular mass. Gel filtration is used commercially for both separation and desalting of enzyme solutions.

Ion-exchange chromatography is a separation technique based on the charge of protein molecules. Enzyme molecules possess positive and negative charges. The net charge is influenced by pH and this property is used to separate proteins by chromatography on anion exchangers (positively charged) or cation exchangers (negatively charged). The ability to process large volumes and the elution of dilute sample components in concentrated form make ion exchange very useful.

For hydrophobic chromatography, media derived from the reaction of CNBr-activated Sepharose with aminoalkanes of varying chain length are suitable. This method is based on the interaction of hydrophobic areas of protein molecules with hydrophobic groups on the matrix. Adsorption occurs at high salt concentrations and fractionation of bound substances is achieved by eluting with a negative salt gradient. This method is ideally suited for further purification of enzymes after concentration by precipitation with such salts as ammonium sulphate. In affinity chromatography the enzyme to be purified is specifically and reversibly adsorbed on an effector attached to an insoluble support matrix. Suitable effectors are substrate analogues, enzyme inhibitors, dyes, metal chelates, or antibodies. The insoluble matrix is contained in a column. The biospecific effector, e.g. an enzyme inhibitor, is attached to the matrix. A mixture of different enzymes is applied to the column.

The immobilised effector specifically binds the complementary enzyme. Unbound substances are washed out and the enzyme of interest is recovered by changing the experimental conditions, for example by altering pH or ionic strength. Immunoaffinity chromatography occupies an unique place in purification technology. In this procedure, monoclonal antibodies are used as effectors. Hence, the isolation of a specific substance from a complex biological mixture in one step is possible. In this procedure, enzymes can be purified by immobilising antibodies specific for the desired enzyme. In this way, enzymes that usually do not bind to an antibody can be purified by immunoaffinity chromatography. Covalent chromatography differs from other types of chromatography by forming a covalent bond in between the required protein and the stationary phases.

FORMULATION OF THE FINAL ENZYME PRODUCT

Once the enzyme has been purified to the desired extent and concentrated, the manufacturer's main objective is to retain the activity. Enzymes for industrial use are sold on the basis of overall activity. Often a freshly supplied enzyme sample will have a higher activity than that stated by the manufacturer. This is done to ensure that the enzyme preparation has the guaranteed storage life. The manufacturer will usually recommend storage conditions and quote the expected rate of loss of activity under those conditions. It is of primary importance to the enzyme producer and customer that the enzymes retain their activity during storage and use.

Some enzymes retain their activity under operational conditions for weeks or even months. However, most enzymes do not. Most industrial enzyme preparations contain a relatively little amount of active enzyme, the rest being due to inactive protein, stabilisers, preservatives, salts and the diluent which allows standardisation between production batches of different specific activities.

The key to maintaining enzyme activity is maintenance of conformation, so preventing unfolding aggregation of the enzyme molecules and changes in the covalent structure. Three approaches are possible:

1. Use of additives.

2. The controlled use of covalent modification.

3. Enzyme immobilisation.

In general, proteins are stabilised by increasing their concentration and the ionic strength of their environment. Neutral salts compete with proteins for water and bind to charged groups or dipoles. This may result in the interactions between an enzyme's hydrophobic areas being strengthened causing the enzyme molecules to compress and making them more resistant to thermal unfolding reactions. Not all salts are equally effective in stabilising hydrophobic interactions, some are much more effective at their destabilisation by binding to them and disrupting the localised structure of water (the chaotropic effect). From this it can be seen why ammonium sulphate and potassium hydrogen phosphate are powerful enzyme stabilisers whereas sodium thiosulphate and calcium chloride destabilise enzymes. Many enzymes are specifically stabilised by low concentrations of cations which may or may not form part of the active site, for example Ca^{2+} stabilises α-amylases and Co^{2+} stabilises glucose isomerases. At high concentrations (e.g. 20% NaCl) salt discourages microbial growth due to its osmotic effect. In addition, ions can offer some protection against oxidation to groups such as thiols by salting-out the dissolved oxygen from solution.

Low molecular weight polyols (e.g. glycerol, sorbitol and mannitol) are also useful for stabilising enzymes, by repressing microbial growth due to the reduction in the water activity and by the formation of protective shells which prevent unfolding processes. Glycerol may be used to protect enzymes against denaturation due to ice-crystal formation at sub-zero temperatures. Some hydrophilic polymers (e.g. polyvinyl alcohol, polyvinylpyrrolidone and hydroxypropylcelluloses) stabilise enzymes by a process of compartmentalisation, whereby the enzyme-enzyme and the enzyme-water interactions are somewhat replaced by less potentially denaturing enzyme-polymer interactions. They may also act by stabilising the hydrophobic effect within the enzymes.

Many specific chemical modifications of amino acid side chains are possible which may (or, more commonly, may not) result in stabilisation. An useful example of this is the derivatisation of lysine side chains in proteases with *N*-carboxyamino acid anhydrides.

These derivatissation result in polyaminoacylated enzymes with various degrees of substitution and length of amide-linked side chains. This derivatisation is sufficient to disguise the proteinaceous nature of the protease and prevent autolysis. Enzymes are much more stable in the dry state than in solution. Solid enzyme preparations sometimes consist of freeze-dried protein. More usually they are bulked out with inert materials such as starch, lactose, carboxymethylcellulose and other poly-electrolytes which protect the enzyme during a cheaper spray-drying stage. Other materials which are added to enzymes before sale may consist of substrates, thiols to create a reducing environment, antibiotics, benzoic acid esters as preservatives for liquid enzyme preparations, inhibitors of contaminating enzyme activities and chelating agents. Additives of these types must, of course, be compatible with the final use of the enzymes product. In order to ensure safe handling, stability, suitable mixing, functionality, etc. in the various applications, most enzyme preparations are formulated in a variety of liquid and granular forms. Some enzyme preparations are immobilised. Often the precise details of the methods used to stabilise enzyme preparations are kept secret or revealed to customers as a confidential information only.

Nature of Enzyme Products

The nature of enzyme products is influenced by: (i) the enzyme itself and its properties as active compound, (ii) the enzyme source, the fermentation media and conditions and the purification steps (resulting in the enzyme concentrate), (iii) the additives used to formulate the final enzyme preparation. Thus, one has to differentiate between enzyme, enzyme concentrate and enzyme preparation.

Enzyme Concentrate

Impurities in the enzyme isolate/concentrate result from the fermentation broth and subsequent purification and consist of proteins, peptides and amino acids, carbohydrates, minerals and other minor components. The relative amounts of these components vary considerably within and between categories of enzyme concentrates. Enzyme content and purity is similar for technical, food and feed enzymes. Enzymes used in personal care products, for therapeutic and analytical or diagnostic application may generally be of higher purity and concentration. Ash constituents comprise small amounts of minerals and diluents are additives for granulation, liquid formulation, stabilisation, preservation, etc.

Enzymes used for food and feed have to comply with purity specifications comprising limits for heavy metals and contaminating micro-organisms and absence of mycotoxins, antibiotics and the production strain. The type and range of impurities, which are present in the enzyme concentrate, is highly varying and depend on the production strain, the media used, the fermentation conditions and the subsequent purification steps. Some of the impurities may also fulfil a technical function during enzyme application. This is particularly true for side activities.

An enzyme concentrate will typically contain other enzymes usually referred to as side activities (not to be confused with side activities of a particular enzyme protein). The type and range of these side activities are largely depending on the enzyme manufacturing conditions and the production strain. These side activities may be more or less important in any given application. An example of this is the use of amylases in baking. Fungal alpha-amylase is added to dough as part of a flour improver. The objective is to hydrolyse starch to provide more fermentable sugars for the yeast. Different enzymes, however, result in different effects on the rheology of the dough and the final product quality. The reason for this is that the enzyme preparation contains a side activity of endo-beta-xylanase which cleaves the hemicellulose in the flour. This enzyme is now manufactured solely for this application. Side activities can partly be inactivated during the enzyme manufacturing process. Nevertheless, it is quite normal to find a variety of different activities in an industrial enzyme preparation. If a preparation is marketed as a protease, e.g. glycosidases and lipases might be present as well.

Factors Affecting the Purity of Enzyme Concentrates

The purity of the enzyme concentrate is largely influenced by: (i) the fermentation and purification processing applied, (ii) the production organism used and (iii) the media used for fermentation.

Production organisms

The particular production strain is obviously affecting the nature as well as the range of by-products present in the enzyme concentrate. As micro-organisms are known to produce toxins, the production organisms may themselves be sources of hazardous materials and have therefore been a chief focus of attention by the regulatory authorities. Some of these toxins are quite well known. However, there is always the possibility of introducing new toxins. Therefore, production strains which are investigated not to produce toxins and which do have a long history of safe use are preferred.

Media for enzyme production

Media used have a bearing on the cost of the enzyme and media components often find their way into commercial enzyme preparations. Details of components used in industrial scale fermentation broths for enzyme production are not readily obtained. Not surprisingly, as manufacturers do not wish to reveal information that may be of technical or commercial value to their competitors. Also some components of media may be changed from batch to batch as availability and cost of, e.g. carbohydrate feedstock change. Such changes reveal themselves in often quite profound differences in appearance from batch to batch of a single enzyme from a single producer. The effects of changing feedstock must be considered in relation to downstream processing. If such variability is likely to reduce the efficiency of the standard methodology significantly, it might be economical to use a more expensive defined medium of easily reproducible composition.

Clearly defined media are usually out of question for large scale use on cost grounds but may be perfectly acceptable when enzymes are to be produced for high value uses, such as analysis or medical therapy where very pure preparations are essential. Less-defined complex media are composed of ingredients selected on the basis of cost and availability as well as composition. Waste materials and by-products from the food and agricultural industries are often major ingredients. Thus, molasses, corn steep liquor, distillers solubles and wheat bran are important components of fermentation media providing carbohydrate, minerals, nitrogen and some vitamins. Extra carbohydrate is usually supplied as starch, sometimes refined but often simply as ground cereal grains. Soyabean meal and ammonium salts are frequently used sources of additional nitrogen. Most of these materials will vary in quality and composition from batch to batch causing changes in enzyme productivity.

Additives for Stabilising the Activity of Enzyme

Enzyme preparations contain a varying amount of additives for the purpose of stabilising the enzyme activity, preservation, granulation, coating or as (de)colouring aids. The choice of formulation ingredients (additives) is not specific for a given application category, but certain substances used for, e.g. technical enzymes may not comply with food, feed and cosmetics regulations, specifications, etc. Some applications may require special substances and technologies, e.g. in the case of immobilised enzymes or enteric coating of digestive aid preparations.

ENZYME RECOVERY

In enzyme production there is a very unfavourable ratio between input of raw material and output of product. This requires the installation of concentration procedures. For economic reasons of enzyme application a concentration up to 10-fold is usually satisfactory for industrial enzyme preparations. For example, enzyme products employed in detergents contain about 5–10 per cent protease while amylase preparations for use in flour treatment contain only about 0.1 per cent pure α-amylase. However, in applications where high purity enzymes are required, e.g. in enzymie analysis, 1000-fold purification is quite common. In some applications, such as baking and dextrose manufacture, the presence of contaminating enzymes must be very low or rigidly controlled. Moreover, the raw enzyme solutions obtained from microbial cultures contain—independent of their source—different types of by-products.

Separation of all these substances may be necessary because of the possibility of undesired effects. Considering enzyme stability there is another reason for treatment of crude enzyme preparations. Since the trend in enzyme applications is toward use of liquid preparations, stabilisation is an important procedure. Figure 22.2 is a presentation of some treatments which are used in the preparation of enzymes on a commercial scale. Techniques for the large-scale isolation and (partial) purification of enzymes from microbial sources make use mainly of traditional procedures. Most of the equipment can be found in food-processing plants. Large-scale equipment specific for enzyme isolation is not marketed. Nearly all process operations are carried out at low temperatures (preferably 0–10°C), with the exception of drying. Separation processes are usually conducted in batches rather than continuously.

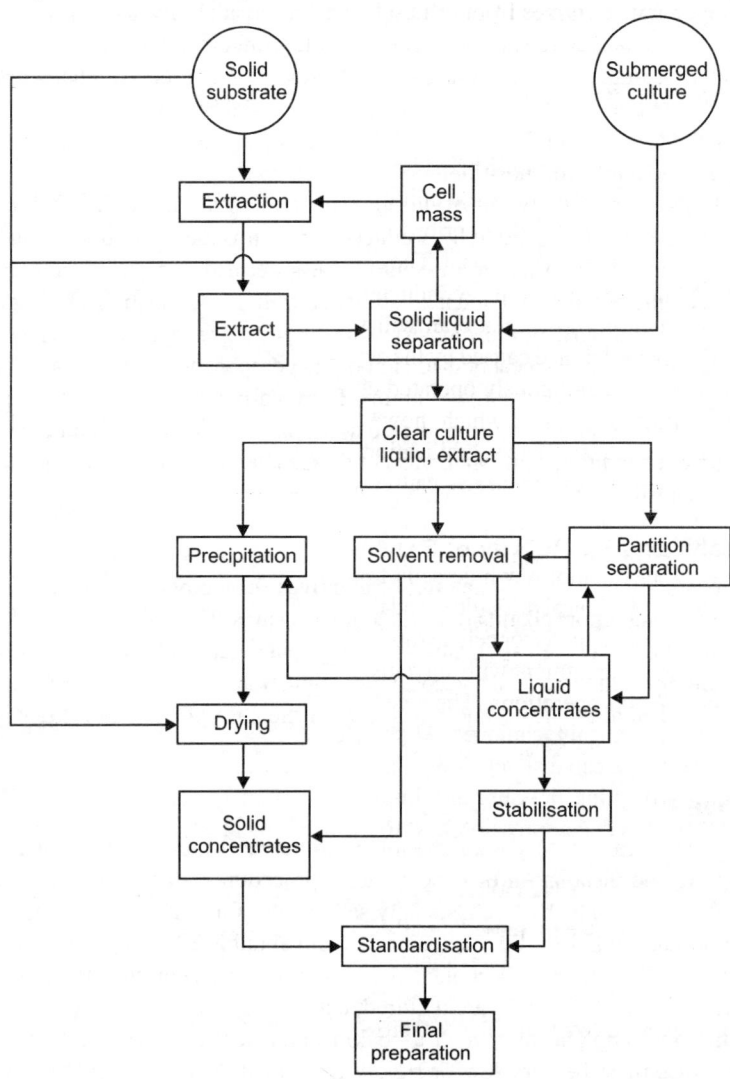

Fig. 22.2: Concentration and purification of enzymes.

However, the scale-up of batch operations inherently causes extended processing times which for many enzymes result in increased losses of activity due to denaturation of the enzyme protein. For this reason the application of continuous operations seems to be useful, but the necessity for highly reliable machines and ingenious process control delays introduction of continuous methods. In addition the value of continuous processing is lost when a single process step is conducted batchwise, perhaps during precipitation.

Extraction Methods

The first step in the isolation of enzymes is their extraction. Techniques that fall into this group are employed either to separate enzymes from solid substrate culture or to release enzymes from the interior of microbial cells.

Extraction of solid substrate cultures

Enzymes produced by solid substrate cultivation used to be of the extracellular type. It is therefore easily conceived that extraction of mould brans is rather a washing out process. Countercurrent techniques of percolation are the most frequently used unit operation.

In many cases the mould bran is dried prior to extraction. This is convenient when the utilisation of the particular enzyme preparation is seasonal. The cultures can be produced in relatively small equipment all the year round, while the extraction is conducted in times of enzyme demand. On the other hand, it is easily seen that extraction from dried bran will yield solutions with higher enzyme concentrations. And last, drying avoids interference caused by the activity of living cells of fresh cultures. This argument, however may not apply in continuously operated culture plants.

In all cases the extractant is water which, however, may contain acids (inorganic or organic), salts, buffer, or other substances to facilitate solubilisation of the enzyme or to improve its stability in solution, or to exclude or minimise undesired effects caused by contaminating by-products or micro-organisms.

Extraction of cells

The decision on whether to employ whole cells for a biochemical process or to use isolated enzymes depends on many factors. Technical difficulties and the related cost of large-scale isolation play an important role. There are a number of methods for cell disruption, as reviewed by Hughes. Chemical and biochemical methods, such as autolysis, treatment with solvents, detergents, or lytic enzymes, have the disadvantage of being in principle batch operations. Their conduct is difficult to standardise and optimise. More recommendable are mechanical techniques. At present, the APV-Manton-Gaulin homogeniser seems to be the most versatile type for cell disintegration. In this machine the cell suspension passes a homogenising valve at the selected operating pressure and impinges on an impact ring. The strong shearing forces combined with the sudden decompression lead to a disruption of the cell wall.

FUTURE OF INDUSTRIAL ENZYMES

Industrial uses of enzymes have increased greatly during the past few years. Prospects are excellent for continuing increased usage of presently available enzymes in present applications, and in new uses, and of new enzymes for many purposes.

Enzymes have several distinct advantages for use in industrial processes:

1. They are of natural origin and are nontoxic.
2. They are easily inactivated when reaction has gone as far as desired.

3. They have great specificity of action, hence can bring about reactions not otherwise easily carried out.

4. They work best under mild conditions of moderate temperature and near neutral pH, thus not requiring drastic conditions of high temperature, high pressure, high acidity, and the like, which necessitate special expensive equipment.

5. They act rapidly at relatively low concentrations, and the rate of reaction can be readily controlled by adjusting temperature, pH, and amount of enzyme employed.

Because of these inherent advantages, many industries are keenly interested in adapting enzymatic methods to the requirements of their processes. Examples of some applications under intensive investigation include unhairing of hides for leather, protection of foods and other materials against oxidation, resolution of racemic mixtures of amino acids, and restoration of flavor to dehydrated or canned foods. Another recent application of enzymes has been in clinical test reagents. Additional developments in this field can be expected.

Clinical application of enzymes has been developing also. Proteolytic enzymes are used for debridement of wounds, and promising clinical results have been reported by injection of certain enzymes such as streptokinase, crystalline trypsin, and chymotrypsin. Since many physical ailments result from derangement of metabolic enzyme systems, increased therapeutic use of enzymes, presently unpredictable, may be expected. For clinical and therapeutic uses, highly purified and perhaps crystalline enzymes will be necessary. Availability of high purity enzymes on an industrial scale is just beginning, and rapid advances in this field may be expected.

Currently much enzyme research is underway by various industries including enzyme manufacturers. Such research is devoted to finding new and improved methods for using enzymes, to improving yields of industrial microbial enzymes, and to finding new enzymes for industrial purposes. Continually increasing usage of old and new enzymes will result from such research.

Chapter 23

Applications of Enzymes in Food Industry

INTRODUCTION

Enzymes are natures sustainable catalysts. They are biocompatible, biodegradable and are derived from renewable resources. Enzymes constitute a large biological globular protein molecule responsible for thousands of metabolic processes that sustain life and function as catalysts to facilitate specific chemical reactions within the cell. These reactions are essential for the life of the organism. The living cell is the site of tremendous biochemical activity called metabolism. This is the process of chemical and physical changes which go on continually in the living organism, enzyme facilitate life processes in essentially all life-forms from viruses to man. Enzymes have been naturally tailored to perform under different physiological conditions. Build-up of new tissues, replacement of old tissues, conversion of food into energy, disposal of toxic materials, reproduction almost all the activities that can be characterised as 'life.' Enzymes act as life catalysts, substances that accelerate the rate of a chemical reaction. By reducing the activation energy (E_a) necessary to initiate the reaction, thus dramatically increasing the rate of reaction. Enzymes do not initiate reactions that would not naturally occur but they accelerate any reaction that is already underway. Enzymes enable the reaction to take place more rapidly at a safer, relatively low temperature that is consistent with living systems. During an enzyme-mediated reaction, the substrate physically attaches to the enzyme at its active site, allowing the substrates to be converted to new product molecules. Most enzyme reaction rates are millions of times faster than those of comparable uncatalysed reactions. Enzymes are neither consumed by the reactions they catalyse, nor do they alter the equilibrium of these reactions.

Enzymes are proteins and are natures own biocatalyst and their function is determined by their complex structure. With the exception of a small group of catalytic RNA molecules, all enzymes are proteins which are made up of amino acids linked together by peptide bonds.

SOURCES OF ENZYMES

Enzymes occur in all living organisms and catalyse biochemical reactions necessary to support life. A wide array of enzymes are extracted from plant sources, they have many advantages including cost of production and stability of products. An ample range of sources are used for commercial enzyme production from a broad spectrum of plant species. Non-microbial sources provide a larger proportion of these, at the present time. Microbes are preferred to plants and animals as sources of enzymes because:

1. They are generally cheaper to produce.
2. Their enzyme contents are more predictable and controllable.
3. Regular supply due to absence of seasonal fluctuations and rapid growth of micro-organisms on inexpensive media.
4. Plant and animal tissues contain more potentially harmful materials than microbes, including phenolic compounds (from plants), endogenous enzyme inhibitors and proteases.
5. Microbial enzymes are also more stable than their corresponding plant and animal enzymes and their production is more convenient and safer.

About fifty years ago, enzymes were being extracted strictly from animals like pig and cow from their pancreases. Animal enzymes were multifold, they were not very stable at the low pH environment so that the enzyme product was destroyed before doing the job. To overcome this problem plant enzymes were discovered, most important one is extraction of peroxidase from horseradish roots occurs on a relatively large scale because of the commercial uses of the enzyme. Peroxidase can also be extracted from soyabean, it is also having the common features with horseradish peroxidase. Some plants like *Cruciferous* vegetables, including *broccoli*, cabbage, kale and collard and turnip greens and papaya are rich in catalase. Wheat sprouts contain high levels of catalase and vegetarian sources of catalase include apricots, avocados, carrots. Catalase is also present in some microbes and bacteria, *Aspergillus niger* culture also produces catalase enzyme.

MECHANISM OF ENZYMES

Enzymes are macro-molecules that help to accelerate (catalyse) chemical reactions in biological systems. Some biological reactions in the absence of enzymes may be as much as a million times slower. Any chemical reaction converts one or more molecules, called the substrate, into different molecules, called the product. Most of the reactions in biochemical processes require chemical events that are unfavourable or unlikely in the cellular environment, such as the transient formation of unstable charged intermediates or the collision of two or more molecules in the precise orientation required for reaction. In some of the reactions like, digestion of food, send nerve signals, or contract a muscle simply do not occur at a useful rate without catalysis.

Enzyme overcomes these problems by providing a specific environment within which a given reaction can occur more rapidly. Enzymes are usually proteins – each has a very specific shape or conformation. Within this large molecule is a region called an active site, which has properties allowing it to bind tightly to the substrate molecules.

FUNCTIONS OF ENZYMES

Enzymes allow many chemical reactions to occur within the homeostasis constraints of a living system. Enzymes function as organic catalysts. A catalyst is a chemical involved in, but not changed by, a chemical reaction. Many enzymes function by lowering the activation energy of reactions. By bringing the reactants closer together, chemical bonds may be weakened and reactions will proceed faster than without the catalyst. The use of enzymes can lower the activation energy of a reaction. Enzymes can act rapidly, as in the case of carbonic anhydrase (enzymes typically end in the -ase suffix), which causes the chemicals to react 10^7 times faster than without the enzyme present. Carbonic anhydrase speeds up the transfer of carbon dioxide from cells to the blood. There are over 2000 known enzymes, each of which is involved with one specific chemical reaction. Enzymes are substrate specific. The enzyme peptidase (which breaks peptide bonds in proteins) will not work on starch (which is broken down by human-produced amylase in the mouth).

Enzymes are proteins. The functioning of the enzyme is determined by the shape of the protein. The arrangement of molecules on the enzyme produces an area known as the active site within which the specific substrates will 'fit'. It recognises, confines and orients the substrate in a particular direction.

IMMOBILISED

Immobilised enzymes are widely used for variety of applications. Based on the type of application, the method of immobilisation and support material can be selected. The immobilised enzymes can be separated from the reaction mixture and reused and also immobilised in order to prevent the enzyme from being exposed to harsh conditions, high temperature, surfactants and oxidising agents, etc. The immobilised enzymes are also widely used in food industry, pharmaceutical industry, bioremediation, detergent industry, textile industry, etc. Enzyme immobilisation improves the operational stability and is also due to the increased enzyme loading which causes the controlled diffusion. Several hundreds of enzymes are immobilised and used for various large scale industries. Immobilisation technique reduces the effluent treatment costs.

The enzymes have various other limitations such as low stability, highly sensitive to the process conditions and these problems can be overcome by the immobilisation techniques. The repeated assay can be done with the immobilised enzyme which reduces the cost of assay and the reuse of enzyme process is also very simple and it can be attained through ultrafiltration technique. Irreversible enzyme immobilisation includes covalent binding and entrapment. Reversible enzyme immobilisation includes adsorption, ionic binding, affinity binding and metal binding.

Application of the Immobilised Enzymes

Food industry

In food industry, the purified enzymes are used but during the purification the enzymes will denature. Hence the immobilisation technique makes the enzymes stable. The immobilised enzymes are used for the production of syrups. Immobilised beta-galactosidase used for lactose hydrolysis in whey for the production of bakers yeast. The enzyme is linked to porous silica matrix through covalent linkage. This method is not preferably used due to its cost and the other technique developed by Valio in 1980, the enzyme galactosidase was linked to resin (food grade) through cross-linking. This method was used for the various purposes such as confectionaries and icecreams.

Advantages and Disadvantages of Enzyme Immobilisation

Advantages of enzyme immobilisation

1. Multiple or repetitive use of a single batch of enzymes.
2. Immobilised enzymes are usually more stable.
3. Ability to stop the reaction rapidly by removing the enzyme from the reaction solution.
4. Product is not contaminated with the enzyme.
5. Easy separation of the enzyme from the product.

6. Allows development of a multienzyme reaction system.

7. Reduces effluent disposal problems.

Disadvantages of enzyme immobilisation

1. It gives rise to an additional bearing on cost.

2. It invariably affects the stability and activity of enzymes.

3. The technique may not prove to be of any advantage when one of the substrate is found to be insoluble.

4. Certain immobilisation protocols offer serious problems with respect to the diffusion of the substrate to have an access to the enzyme.

AMYLASES

Amylases are starch degrading enzymes. They are widely distributed in microbial, plant and animal kingdoms. They degrade starch and related polymers to yield products characteristic of individual amylolytic enzymes. Initially the term amylase was used originally to designate enzymes capable of hydrolysing α-1,4-glucosidic bonds of amylose, amylopectin, glycogen and their degradation products. They act by hydrolysing bonds between adjacent glucose units, yielding products characteristic of the particular enzyme involved.

In recent years a number of new enzymes associated with degradation of starch and related polysaccharides structures have been detected and studied. The enzymes of actual or potential commercial importance of microbial origin that split α-1,4 and α-1,6 bonds in these structures, may be divided in the following six classes.

1. Enzymes that hydrolyse α-1,4 bonds and by-pass α-1,6 linkages, e.g. α-amylase (endoacting amylases).

2. Enzymes that hydrolyse α-1,4 and cannot by-pass α-1,6 linkages, e.g. β-amylase (exoacting amylases producing maltose as a major end product).

3. Enzymes that hydrolyse α-1,4 and α-1,6 linkages, e.g. amyloglucosidase (glucoamylase) and exoacting amylase.

4. Enzymes that hydrolyse only α-1,6 linkages, e.g. pullulanase and other debranching enzymes.

5. Enzymes that hydrolyse preferentially α-1,4 linkages in short chain oligosaccharides produced by the action of other enzymes on amylose and amylopectin, e.g. α-glucosidases.

6. Enzymes that hydrolyse starch to a series of non-reducing cyclic D-glucosyl polymers called cyclodextrins or Sachardinger dextrins, e.g. *Bacillus macerans* amylase (cyclodextrin producing enzyme).

Starch

Before describing the action pattern and properties of amylolytic enzymes, it is essential to discuss the features of the natural substrate, starch. Starch is a major reserve carbohydrate of all higher plants. In some cases it accounts for as high as 70% of the undried plant material. It occurs in the form of water insoluble granules. The size and shape of the granules are often characteristic of the plant species from which they are extracted. When heated in water the hydrogen bonds holding the granules together begin to weaken and this permits them to swell and gelatinise. Ultimately they form paste or dispersion, depending on the concentration of polysaccharide.

Starches are produced commercially from the seeds of plants, such as corn, wheat, sorghum or rice, from the tubers and roots of plants such as cassava, potato, arrowroot and the pith of sago palm. The major commercial source of starch is corn from which it is extracted by a wet milling process. Starch is a heterogeneous polysaccharide composed of two high molecular weight entities called amylose and amylopectin. These two polymers have different structures and physical properties.

Starch may be separated into its two components by addition of a polar solvent, e.g. *n*-butanol, to a dispersion of starch. The insoluble amylose complex can then be separated from soluble amylopectin fraction. Amylose is composed of linear chains of α-1,4 linked D-glucose residues. Hence it is extensively degraded by α-amylase. Some amylose is not totally degraded to maltose by this enzyme. Amylose has a degree of polymerisation of several thousands of glucose units. Because of the molecular shape and structure of amylase, it is not stable in aqueous solution and retrogrades (precipitates spontaneously). This is because linear chains align themselves by hydrogen bonding and thus forms aggregates. This process is irreversible. Retrograded amylase will only dissolve in alkaline solution. Amylose has considerable viscosity in alkaline solutions due to its molecular shape. Amylase forms complex with iodine to form intense blue colour and this forms the basis of a method for quantitative determination of amylase.

Amylopectin may account for 75 to 85% of most starches. It has molecular weight in excess on 10^7–10^8 and has a branched structure composed of chains of about 20–25 α-1,4 linked D-glucose residues. Amylopectin which is branched by α-1,6 linkages may contain 4 to 5% α-1,6-D-glucosidic bonds. In aqueous solutions, amylopectins are relatively stable due to branched molecules and are not able to form compact aggregates. There is no apparent relationship between the limiting viscosity number and the degree of polymerisation. Due to the nature of branched structure, the iodine binding power is reduced. The branched components of starch is amylopectin which has different types of chains referred to as A, B and C chains. The hydrolysis of starch may be carried out using either acid or enzyme as catalyst. Enzyme hydrolysis has several advantages: it is more specific, therefore fewer by-products are formed and hence yields are higher. Conditions for enzyme hydrolysis are milder therefore refining stages to remove ash and colour is minimised. The enzymatic hydrolysis of starchy has been practiced on an industrial scale for many years and is gradually replacing the traditional acid hydrolysis process.

Applications of Amylases

Industrial production of dextrose powder and dextrose crystals from starch using α-amylase and glucoamylase began in 1959. Since then, amylases are being used for various purposes. Conversion of starch into sugar, syrups and dextrins forms the major part of the starch processing industry. The hydrolysates are used as carbon sources in fermentation as well as sources of sweetness in a range of manufactured food products and beverages. Hydrolysis of starch to products containing glucose, maltose, etc. is brought about by controlled degradation. Some of the applications of amylase are discussed below.

Liquefaction

Liquefaction is a process of dispersion of insoluble starch granules in aqueous solution followed by partial hydrolysis using thermostable amylases. In industrial processes, the starch suspension for liquefaction is generally in excess of 35% (w/v). Therefore the viscosity is extremely high following gelatinisation. Thermostable α-amylase is used as a thinning agent, which brings about reduction in viscosity and partial hydrolysis of starch. Retrogradation of starch is thus avoided during subsequent cooling. The traditional thinning agent used in starch technology was acid (hydrochloric or oxalic acids, pH 2 and 140–150°C for 5 min). The introduction of thermostable amylases has meant milder processing conditions. The formation of by-products is reduced and refining and recovery costs are lowed.

In the enzymatic process the hydrolytic action is terminated when the average degree of polymerisation is about 10–12. Two distinct types of thermostable α-amylases are commercially available and used extensively in starch processing technology. The amylase of *Bacillus amyloliquefaciens* was the first liquefying α-amylase used on a large scale. Later, a more heat stable enzyme from *Bacillus licheniformis* was introduced commercially. Liquefaction can be done by two methods:

Single stage enzyme liquefaction: In this process, starch slurry containing 30–40% dry solids is prepared in the feed tank. The pH is adjusted to about 6–6.5 with sodium hydroxide. Calcium salts may be added if the level of free calcium ions is below 50 ppm. The liquefying enzyme is then added. The slurry is then pumped continuously through a jet cooker where the temperature is raised to 105°C by direct injection of live steam. Tremendous shearing forces are exerted on the slurry as it is pumped through the jet cooker. So in addition to the viscosity reduction action of the enzyme, some mechanical thinning also occurs. The slurry is maintained at this high temperature in the pressurised holding cell for about 5 min, after which it is discharged via a spring loaded release valve into a reaction, where enzyme action is allowed to continue for about 2 hr at 95°C. After this treatment the liquefied starch will have dextrose equivalent (DE) of 10–20 depending on the amount of enzyme used. DE is defined as reducing sugars expressed as dextrose and calculated as a percentage of dry substance. This process is simple energy consumption is relatively low because the maximum operating temperature is only 105°C as compared to 140–150°C normally used.

Acid enzyme liquefaction: This is another process which takes advantage of the thermostability of *B. licheniformis* amylase. The enzyme is added after the starch has been cooked and cooled to 100–95°C. A starch slurry containing 30–40% dry solids is cooked at a high temperature for about 5 min. A jet cooker is used so that sufficient mechanical thinning, due to shearing takes place. The pH may be in the range 2–5, but if it is too low, by-product formation will be significant. If it is too high there will be no thinning effect from the acid and there will be an increased colour formation. After cooking, the slurry is flash cooled to about 100°C and the pH is set to 6 to 6.5 before the addition of enzyme. By this process the enzyme consumption is slightly reduced. The filtration properties are also improved because better fat/protein separation is achieved. There is an increase in steam consumption and hence fuel costs due to high temperature cooking. Liquefaction is the first and most important step in starch processing. The purpose is to provide a partially hydrolysed starch suspension of relatively low viscosity which is free from by products, stable to retro gradation and suitable for further processing, i.e. saccharification. If the liquefaction process does not go well, problems like poor filtration and turbidity of the processed solution occurs. The most important factor for ideal liquefaction of starch is that the starch slurry which contains suitable amount of α-amylase is treated at 105 to 107°C as quickly and uniformly as possible. Thermostable amylase are not sufficiently heat stable to be used during liquefaction process, but they can be used as saccharifying enzymes. The most widely used enzymes in this group are the maltogenic enzymes.

Manufacture of maltose

Maltose is a naturally occurring disaccharide. Its chemical structure has 4-O-α-D-glucopyranosyl-D-glucopyranose. It is the main component of maltosugar syrup. Maltose is widely used as sweetener and also as intravenous sugar supplement. It is used in food industries because of low tendency to be crystallised and is relatively non-hygroscopic. Corn, potato, sweet potato and cassava starches are used for maltose manufacture. The concentration of starch slurry is adjusted to be 10–20% for production of medical grade maltose and 20–40% for food grade. Thermostable α-amylase from *B. licheniformis* and *B. amyloliquefaciens* are used.

Manufacture of high fructose containing syrups

High fructose containing syrups (HFCS) (Fructose content = 42%) is prepared by enzymic isomerisation of glucose with glucose isomerase. The starch is first converted to glucose by enzymic liquefaction and saccharification.

Manufacture of oligosaccharides mixture

Oligosaccharides mixture (malto-oligomer mix) is obtained by digestion of corn starch with α-amylase, β-amylase and pullulanase. Malto-oligomer mix is a new commercial product. Its composition is usually as follows: glucose, 2.2%, maltose, 37.5%, maltotriose, 46.4% and maltotetraose and larger malto-oligosaccharides, 14%. Malto-oligomer mix powder obtained by spray drying is highly hygroscopic. Therefore it serves as a moisture regulator of the food with which it is mixed. Malto-oligomer mix tastes less sweet than sucrose. Its solution shows a lower viscosity than corn syrup because of its low content of glucose.

Malto-oligomer mix is mainly used as a substitute for sucrose and other saccharides. It is also used for preventing crystallisation of sucrose in foods and keeping a certain level of hardness of the texture during storage.

Manufacture of maltotetraose syrup

Maltotetraose syrup (G4 syrup) is produced by subjecting starch to the action of maltotetraose forming amylase (G4 amylase). The sweetness of the syrup is as low as 20% of sucrose. Therefore a partial replacement of sucrose with G4 syrup reduces the sweetness of foods without affecting their inherent taste and flavour. It has high moisture retention power which serves to prevent retrogradation of starch ingredient and retains suitable moisture in foods. G4 syrup shows less Millard reaction as it has less glucose and maltose content. It has higher viscosity than sucrose thus improving the food texture. G4 syrups depresses the freezing point of water more moderately than sucrose or high fructose syrup. Therefore, G4 syrup can be used to control the freezing points of frozen foods. G4 syrup imparts gloss and can be used in industry such as a paper sizer. Commercial thermostable a-amylase of *B. licheniformis* or *B. subtilis* is used to make G4 syrups.

Production of anomalously linked oligosaccharides mixture (Alo mixture)

'Alo mixture' is a mixture of isomaltose, panose, isomaltotriose and branched oligosaccharide composed of 4 and 5 glucose residues. The 'Alo mixture' has properties that are favourable to food industry. It is mildly sweet, has low viscosity, high moisture retaining capacity and low water activity convenient in controlling microbial contamination. For the manufacturing of 'Alo mixture', starch is dextrinised using thermostable bacterial α-amylase. The degree of hydrolysis (DE) of starch is kept between 6 to 10. The simultaneous reaction of sacharification and transglucosidation of dextrin is done by using soyabean β-amylase and *Aspergillus niger* transglucosidase. The reaction mixture is finally purified and concentrated to 25% moisture.

Manufacture of high molecular weight branched dextrins

Branched dextrins of high molecular weight are prepared by hydrolysis of corn starch with α-amylase. The extent of starch degradation to be carried out depends on the type of starch and the physical properties desired. The branched dextrins are obtained as powder after chromatography and spray drying. These are used as extender and a glozing agent for production of powdery foods and rice cakes, respectively.

PECTIC ENZYMES

Pectic enzyme, also known as pectinase, is a protein that is used to break down pectin, a jelly like glue that holds plant cells together. In wines pectin can cause troublesome 'pectin haze' that is not easily cleared without the use of pectic enzymes. While this enzyme does occur naturally in grapes as well as yeast there is not enough of it to overcome the amount of pectin present in the must. Other sources of pectic enzyme include plants, bacteria and fungus.

It turns out that fungus produces a special kind of pectic enzyme that is particularly adept at breaking down pectin even in the harsh environment created during fermentation. Most commercially sold pectic enzymes come from fungus. Pectic enzymes may be purchased in a liquid form or as a powder at any home brewing supply store. Pectins are polysaccharides ubiquitous in the plant kingdom and constitute the major component of plant cell walls. The pectinases are a group of related enzymes capable of degrading pectin. Therefore, this group of enzymes have been used for decades in the food and wine making industry for the processing of fruit juices.

The pectinases are synthesised by plants and micro-organisms, the latter being used for industrial production. Micro-organisms are used to produce many enzymes of industrial interest in processes relatively inexpensive and environmentally friendly. Moreover, enzymatic catalysis is preferred over other chemical methods since it is more specific, less aggressive and generates less toxicity. Advances in biotechnology, especially in the fields of molecular biology and microbial genetics, have led to major advances in enzyme technology and have allowed, in many cases, the development of new producing strains and microbial enzymes. The production of pectinases may also benefit from these technologies.

This section reviews the characteristics of pectic substances, the types and mode of action of enzymes which degrade them and the main applications of commercial preparations of microbial pectinases in the food and wine making industry, followed by a review of new micro-organisms and pectolytic enzymes, evaluating new approaches to their production, marketing and use.

Pectic substances are polysaccharides of high molecular weight, with a negative charge, appearing mostly in the middle lamella and the primary cell wall of higher plants, found in the form of calcium pectate and magnesium pectate. They are formed by a central chain containing a variable amount although in high proportion of galacturonic acid residues linked through α-(1–4) glycosidic bonds partially esterified with methyl groups.

The generic name of pectic substances is used for referring to four types of molecules: protopectin (pectic substance in intact tissue), pectinic acids (polygalacturonan containing >0–75% methylated galacturonate units), pectic acids (polygalacturonan that contains negligible amount of methoxyl groups) and pectins (pectinic acid with at least 75% methylated galacturonate units). Protopectines are insoluble in water, while the rest are wholly or partially soluble in water.

Pectic substances represent between 0.5–4% of fresh weight plant material. In addition to their role as cementing and lubricating agents in the cell walls of higher plants, they are responsible for the texture of fruits and vegetables during growth, maturation and their storage. Furthermore, pectic substances are involved in the interaction between plant hosts and their pathogens.

Pectins have numerous and important applications in the food and pharmaceutical industries. In the food sector, it is primarily used as a gelling agent, replacing sugars and/or fats in low-calorie food and as nutritional fibre. The pharmaceutical industry offers them as preparations to reduce cholesterol or to act as a lubricant in the intestines thus promoting normal peristaltic movement without causing irritation. In addition, these polysaccharides are used as drug delivery systems, which can also reduce the toxicity of these and make their activity longer lasting without altering their therapeutic effects.

Pectinases in the Food Industry

Extraction of fruit and vegetable juices

The main industrial application of pectinases is the extraction and clarification of fruit and vegetable juices. Most of the microbial pectinases produced by the industry are dedicated to this purpose. Pectins are responsible for the turbidity and consistency of the juice causing an increase in their viscosity which hinders its clarification, filtration and concentration. The degradation of pectic substances in mashed fruit purees is achieved through the addition of pectolytic enzymes resulting in an increase in juice yield and its clarification as well as a decrease in viscosity. Treatment with pectinases also provides filtering of the product.

The fruit and vegetable juice industry uses mainly acidic pectinases of fungal origin, principally from *Aspergillus* spp. Commercial preparations are mixtures of polygalacturonases, pectate lyases and pectin esterases. Pectate lyases can act on the esterified pectin while the polygalacturonases act on the desesterified pectin thus it might require previous action of the pectin esterases. Pectic enzymes treatments vary depending on the type of juice.

Clear juices (i.e. apple, pear, grape): In this type of product, pectolytic enzymes are added to increase the yield in juice during the pressing and for removing matter in suspension. In the case of apple juice, the most commonly used enzymes are those that can depolymerise the highly esterified apple pectin. Apple juice can be obtained through a two-step process consisting of a first treatment of the crushed apple mush with pectinases to obtain the premium juice followed by pomace liquefaction treatment made with a mixture of different pectinases and cellulases for the complete extraction of the juice.

After washing and crushing, the apples are pressed to obtain the juice. Pectic enzymes are used to facilitate the pressing and juice extraction and to assist in the separation of a flocculant precipitate by means of sedimentation, filtration or centrifugation. If a cloudy product is required, the juice is pasteurised after pressing to inactivate residual enzymes. Centrifugation removes the large-size remains leaving small particles in suspension.

However, if a clear juice is required, these suspended particles have to be withdrawn. In order to do this a treatment with mixtures of commercial enzymes is carried out containing pectinases, cellulases and hemicellulases. Finally, the fluid is centrifuged to obtain the clear juice. Although it has been noted that the proper clarification of apple juice requires mixtures of polygalacturonase and pectin methyl esterase, subsequent studies have shown that it can be clarified by only using a pure pectin lyase.

Cloudy juices (i.e. citrus, tomato, nectars): In the case of orange juice, where natural pectin esterases are present, pectin is only partially methylated. Polygalacturonases are the pectic enzymes which are most commonly used and of great interest for this type of fruit juice.

In the process of orange juice extraction, pectinases can be added at the end of the pulp wash extraction to reduce viscosity or, preferably, at the end of the first finisher. This leads to higher yield in juice, a better extraction of soluble solids and to a lower viscosity. The action of these enzymes just reduces the viscosity without attacking the insoluble pectin that maintains the stability of the cloud. Enzyme preparations should lack or have the least possible amount of pectin methyl esterases to avoid the clarification of the product. It has been suggested that the best enzyme might be pure pectin lyase.

Coffee, cocoa and tea fermentation

Traditionally, fermentation of coffee is made with pectolytic micro-organisms in order to remove the layer of mucilage from the coffee beans. With the same purpose, commercial enzyme preparations containing pectinase is sprayed onto the beans to ferment. A cheaper alternative is to use, with the same purpose, the filtrate of inoculated fermentations. The enzyme treatment significantly reduces the fermentation time. Cocoa fermentation is essential to develop the chocolate flavour. This fermentation is carried out by a succession of different micro-organisms, some of them pectolytic. Pectic enzymes allow the degradation of the cocoa pulp and are indispensable for the fermentation process and the good quality of fermented beans. Similarly, treatment of tea leaves with pectic enzymes of fungal origin (at a dose adjusted to avoid damaging the leaf), facilitates and accelerates the fermentation.

Pectinases in the Wine Industry

Wine is the result of the fermentation of grape juice. Pectinases are the most important enzymes used by the wine making industry although commercial preparations may contain other enzymatic activities such as hemicellulases, glucanases and glycosidases.

Pectic enzymes are synthesised naturally by the plant and are present in the grape. However, they have low activity during the wine producing process. Microbial pectolytic enzymes especially of fungal origin are resistant to the conditions of fermentation and can be used to facilitate processes, improve quality and diversify products. So far, commercial enzymes are produced all from fungi, mainly of the genus *Aspergillus*. Although all enzymes are produced by *Aspergillus*, pectinase preparations currently available for the wine market are very different. Both the type of activities as well as their concentration in the preparations depend on the strain of *Aspergillus* used, the fermentation conditions for production, the nature of the fermented substrate and the degree of partial purification.

Various research studies have shown that the addition of pectolytic enzymes leads to increased levels of methanol in wine due to the activity of pectin methyl esterase. Methanol is toxic and its maximum concentration in wine is regulated. Therefore, pectin methyl esterase activity should be at low concentrations in commercial mixtures. The functions of pectic enzymes in the wine making process are to support the extraction process, maximise juice yield, facilitate filtration and intensify the flavour and colour.

Extraction, clarification and filtration

The addition of pectinase to the must reduces its viscosity and causes the grouping of suspended particles in larger aggregates that can be removed by sedimentation. If the enzymes are added to the pulp before pressing, must yield increases, facilitating the pressing and enhancing the colour. Macerating enzyme preparations for this purpose contain pectinases as well as cellulases. A high level of polygalacturonase is very effective for clarification but may require the prior action of pectin lyase activity. For this reason the enzyme preparations with a high content of pectin lyase are desirable when very fast racking is required to prevent any problems related to must oxidation, development of endogenous microbiota and nutrient loss. In addition, wines made with pectic-enzyme-treated grape must significantly reduces filtration times.

Intensification and stabilisation of colour

In the case of musts obtained from red grapes, the degradation of cell walls in the skin of the grapes through pectolytic enzyme treatment results in an increased release of phenolic compounds responsible for colour. Early work related to the use of pectinases to enhance the colour of the wines were a bit confusing because Ough and others were able to intensify the colour of wine by the enzyme treatment, but Wightman and others found that some enzyme preparations containing pectic enzymes reduced red wine colour. Subsequent studies performed with two different enzyme preparations confirmed that both allow to produce wines with higher concentrations of anthocyanins and total phenols and have a higher colour intensity.

Boosting aroma and flavour

The aromatic profile of wines consists of two components: the varietal aromas characteristic of the variety of grape used and the aromas originated by the yeast during fermentation. In many cases the grape variety used completely determines the aroma of the wine, especially in young wines. The volatile compounds of grapes include monoterpenes, C13 nor-isoprenoides, benzene derivatives and aliphatic alcohols. The aromatic components of the grape may appear as free forms, which contribute directly to its scent or bound forms, of greater concentration, to sugars and nonvolatile. Nonodorous glycoside flavour precursors accumulate in the grape especially in the skin during the ripening process. The hydrolysis of these precursors via beta-glucosidases releases the olfactory active aglycones. Pectic enzymes help break down the cell walls of the grapes and thus to extract the aromatic precursors. The addition of pectic enzymes during the extraction or fermentation of the must results in an increase in aromatic precursors susceptible to being attacked by beta-glucosidases from the must, those produced by yeasts and bacteria during fermentation or those which are included in commercial enzyme preparations, thereby enhancing the aroma of wines.

Biotechnological Aspects for Production of Microbial Pectic Enzymes

As discussed until now, there are numerous applications of microbial pectic enzymes in food and wine-making industry. Not surprisingly, the sales volume of these enzymes represents 25% of the enzymes that are commercialised in the food and alcoholic beverages industry. These many applications require one or more types of pectinases that must act in very different condition according to the process in which they are involved. For example, while in the extraction and clarification of fruit juice enzymes can be used at temperatures between 45–95°C, the wine industry employs temperatures below 15–10°C. Although most food applications pectinases have to act in a medium acid, in others, such as oil extraction, they should perform in a medium alkaline.

So far, commercial pectic enzymes are prepared only from cultures of filamentous fungi, mainly of the genus *Aspergillus*. Commercial products contain mainly a mix of polygalacturonase, pectin lyase and pectin methyl esterase. The use of different strains of *Aspergillus* and modification of substrates and culture conditions can lead to mixtures enriched in one type of enzyme.

However, these commercial pectic preparations are not always optimal for each process and its use is not without side effects and controversy. Thus, while pectin methyl esterase activity present in the samples may be necessary for the action of polygalacturonase in the case of pectin with a high degree of esterification, its action may lead to an undesirable increase of methanol in the products. Moreover, although the pectic enzymes are the most abundant in commercial mixtures, they can also contain other undesirable activities, such as in the case of making wine, polyphenoloxidases or cinnamyl esterases.

At present especially in recent years, the accumulated knowledge of new microbial pectolytic enzymes as well as methodological and technological advances can address the production and use of these enzymes with a different approach. The diversity of applications and conditions in which these enzymes must work also demand a large number of different enzymes capable of acting in such conditions. Even more interesting would be having more robust, broad spectrum enzymes which allow for a more versatile use in different applications.

Considering the variety of enzymes, traditionally the majority of studies refer to the pectinases of *Erwinia* and *Bacillus* within the bacteria and various fungi especially *Aspergillus* although there has been a major advance in the description and characterisation of pectic enzymes produced by yeast in the last 15 years. In recent years there has been also a growing interest in studying pectic enzymes with very interesting properties from the point of view of their application. These include thermostable pectinases or pectinases with optimal activity at low temperatures.

The possibility of producing different types of pectolytic enzymes separately and for later preparation of their mixtures in the proper proportions would allow to provide more suitable commercial preparations for each application and lacking undesirable activities. In this sense, micro-organisms such as some strains of yeast *Saccharomyces cerevisiae* and *Kluyveromyces marxianus* which produce only one type of enzyme or that constitutively synthesise it are of great interest. Similarly, obtaining constitutive mutants from producing strains allow the optimisation of production and contribute to making cost-effective production processes. Furthermore, these strains can be used to produce pectinases that accumulate together with other metabolites in the culture broth, which contributes favourable to the overall economy of the process. Heterologous expression of enzymes in prokaryotic or eukaryotic systems is a technique of great interest for the production of a single type of enzyme. The many pectic enzymes, both from bacteria, yeast and fungi, which are rightly expressed in *Escherichia coli* and different species of yeast. Although *E. coli* is unable to carry out post-translational modifications of proteins, fungal pectolytic enzymes expressed in these bacteria are active. However, one of the most interesting strategies seems to be the expression of pectinase genes in yeast, particularly in *Pichia pastoris*, in which very high levels of constitutive expression has been achieved. In some cases changes in glycosylation patterns conducted by yeast did not affect the activity and characteristics of recombinant pectinases, while other changes do occur with respect to the characteristics of the native protein, which even lead to enzymes with interesting properties for certain applications.

Thus, the enzymes that degrade the pectic substances play an essential role in the food and wine making industries because they are used to degrade the pectins that interfere with the extraction and clarification of fruit juices and oils as well as being important in the fermentation of coffee, cocoa and tea. Also, in the wine industry they play an important role by contributing to the release of the molecules responsible for aroma and colour, two of the major components that characterise a wine.

LACTASE ENZYME

For those with lactose intolerance, we may be all too familiar with what we picked from our digestive enzymes, lactase. Lactase digestion helps you process the lactose in dairy, helping to prevent bloating, gas, constipation, diarrhea and cramping that happens to many who ingest milk products. Lactose intolerance is not an allergy, it is about a missing enzyme necessary for the proper digestion of dairy. For those who can digest dairy, lactase is produced by cells in the small intestine. For those who are intolerant, it is not—or it is not produced in sufficient quantity.

Lactase's primary function is to break down lactose, a type of sugar found in milk and other dairy products. Our body cannot naturally absorb lactose, so lactase enzymes help digest this milk sugar. This in done in order for our body to metabolise this form of sugar by breaking down the lactose into smaller, more digestible sugars called glucose and galactose. When our body is unable to produce enough lactase to function properly, a condition known as lactose intolerance develops. Lactase deficiency is the most common and well-known form of carbohydrate intolerance. Most mammals, including humans, have high intestinal lactase activity at birth (which makes sense when you consider all mammals nurse during their first few months). But, in many cases, this activity declines to low levels during childhood and drops even further (or completely disappears) in adulthood. The low lactase levels cause incomplete digestion of milk and other foods containing lactose. Undigested lactose in the bowel then is subject to fermentation which causes the bloating, gas, constipation, diarrhea and cramping associated with lactose intolerance. It is estimated that approximately 70% of the world's population is deficient in intestinal lactase with more than one-third of the US population presumed to be lactose intolerant and unable to digest dairy products, causing sometimes severe digestion problems. According to MedlinePlus, a service of the National Institutes of Health, symptoms of lactase deficiency begin 30 min to 2 hr after ingesting milk or a similar dairy product. Symptoms can include abdominal bloating, abdominal cramps, diarrhea, flatulence and diarrhea. Supplemental lactase enzymes have been found to decrease the symptoms of lactose intolerance associated with the consumption of dairy foods. One way of treating lactose intolerance is by adding lactase enzymes to regular milk, or take digestive enzymes that contain lactase in capsule or chewable tablet form.

INVERTASE

Invertase, also called beta-fructofuranosidase cleaving the terminal non-reducing beta-fructofuranoside residues, is a glycoprotein with an optimum pH 4.5 and stability at 50°C. Invertase is used for the inversion of sucrose in the preparation of invert sugar and high fructose syrup (HFS). It is one of the most widely used enzymes in food industry where fructose is preferred than sucrose especially in the preparation of jams and candies, because it is sweeter and does not crystallise easily. The enzymatic activity of Invertase has been characterised mainly in plants and micro-organisms. Among micro-organisms, *Saccharomyces cerevisae, Candida utilis, Aspergillus niger, Thermomyces lanuginosus* and *Penicillium chrisogenum* has been widely studied.

Invertase exhibits marked stability towards temperature, pH changes and denaturants. Temperature of the reaction mixture determines the rate of sucrose inversion by the active enzyme.

Sources of invertase: The official name for Invertase is beta-fructo-furanosidase (EC.3.2.1.26), which implies that the reaction catalysed by the enzyme, is the hydrolysis of the terminal non-reducing beta-fructofuranoside residues in beta-fructofuranosides. Invertaseis widely distributed among the biosphere. It is mainly characterised in plants and micro-organisms. *Saccharomyces cerevisiae* commonly called Baker's yeast is the chief strain used for the production of Invertase commercially. They are found in wild growing, on the skin of grapes and other fruits. Though plants like Japanese Pear fruit (*Pyrus pyrifolia*), Pea (*Pisum sativum*), Oat (*Avena sativa*) can also be used, but generally micro-organisms like *S. cerevisiae, Candida utilis, A. niger* are considered ideal for their study.

Isozymes of Invertase

Isozymes in baker's yeast

Invertase exists in more than one form in yeasts generally, either extracellular Invertase or intracellular Invertase. The external yeast Invertase is a glycoprotein containing about 50% carbohydrate, 5% mannose, 3% glucosamine, whereas internal Invertase contains no carbohydrate. The former one has a molecular weight of 135 KDa whereas the latter variety has a molecular weight of 270 KDa. It has been established that in depressed cells most of the Invertase is external whereas in fully repressed state all the Invertase is intracellular.

Both differ in amino acid sequences particularly the internal Invertase does not contain cysteine. Both the enzymes are inhibited by Iodine and reactivated by mercaptoethanol. Both require an acid with pK_a about 6.8 in its protonated form. Both are inhibited by cyanogen bromide in a biphasic reaction.

PROTEASE

Proteolytic enzyme, also called protease, proteinase, or peptidase, any of a group of enzymes that break the long chainlike molecules of proteins into shorter fragments (peptides) and eventually into their components, amino acids. Proteolytic enzymes are present in bacteria, archaea, certain types of algae, some viruses and plants, they are most abundant, however, in animals. There are different types of proteolytic enzymes, which are classified according to sites at which they catalyse the cleavage of proteins. The two major groups are the exopeptidases, which target the terminal ends of proteins and the endopeptidases, which target sites within proteins. Endopeptidases employ various catalytic mechanisms, within this group are the aspartic endopeptidases, cysteine endopeptidases, glutamic endopeptidases, metalloendopeptidases, serine endopeptidases and threonine endopeptidases. The term oligopeptidase is reserved for those enzymes that act specifically on peptides.

Among the best-known proteolytic enzymes are those that reside in the digestive tract. In the stomach, protein materials are attacked initially by a gastric endopeptidase known as pepsin. When the protein material is passed to the small intestine, proteins, which are only partially digested in the stomach, are further attacked by proteolytic enzymes secreted by the pancreas. These enzymes are liberated in the small intestine from inactive precursors produced by the acinar cells in the pancreas. The precursors are called trypsinogen, chymotrypsinogen, proelastase and procarboxypeptidase. Trypsinogen is transformed to an endopeptidase called trypsin by an enzyme (enterokinase) secreted from the walls of the small intestine. Trypsin then activates the precursors of chymotrypsin, elastase and carboxypeptidase. When the pancreatic enzymes become activated in the intestine, they convert proteins into free amino acids, which are easily absorbed by the cells of the intestinal wall. The pancreas also produces a protein that inhibits trypsin. It is thought that in this manner the pancreas protects itself from autodigestion.

Proteases occur in all organisms, from prokaryotes to eukaryotes to viruses. These enzymes are involved in a multitude of physiological reactions from simple digestion of food proteins to highly regulated cascades (e.g. the blood-clotting cascade, the complement system, apoptosis pathways and the invertebrate prophenoloxidase-activating cascade).

Proteases constitute one of the most important groups of industrial enzymes, accounting for about 60% of the total enzyme market. A protease is an enzyme that conducts proteolysis, that is, begins protein catabolism by hydrolysis of the peptide bonds that link amino acids together in the polypeptide chain forming the protein. For several physiological processes the action of the proteolytic enzyme is essential, e.g. in digestion of food proteins, protein turnover, cell division, blood clotting cascade, signal transduction, processing of polypeptide hormones, apoptosis and also in the life cycle of disease-causing organisms including the replication of retrovirus. With special reference to their key role in life-cycle of many hosts and pathogens they have great medical, pharmaceutical and academic importance. Protease is of commercial value and various industrial applications. They are widely used as detergent, in food, pharmaceutical and leather tanning industries. The vast variety of proteases, with their specificity of their action and application have attracted worldwide attention to exploit their physiological as well as biotechnological applications. It has been considered as eco-friendly because the appropriate producers of these enzymes for commercial exploitation are non-toxic and non- pathogenic that are designated a safe.

Classification of Proteases

The physiological function of proteases is essential for all living organism, from viruses to humans and the enzymes can be classified based on their origin: microbial (bacterial, fungal and viral), plant, animal and human enzymes can be disguished. On the basis of the site of action on protein substrates, proteases are broadly classified as endopeptidases or exopeptidases enzymes. Exopeptidases cleave the peptide bond proximal to the amino or carboxy termini of the substrate. Based on the site of action at the nitrogen (N) or carbon (C) terminus, they are classified as aminopeptidases and cabroxy-petidases. Endopeptidases cleave peptide bonds distant from the termini of the substrate. Based on the functional group present at the active site, endo-peptidases are further classified into four prominent groups, i.e. serine proteases, aspartic proteases, cysteine proteases and metallo-proteases.

Based on the pH optima, they are referred to as acidic, neutral, or alkaline proteases.

Protease from different sources

The proteases are physiologically necessary for living organisms, they are ubiquitous, being constituted in a wide diversity of sources such as plants, animals and micro-organism. The use of plants for the production of proteases is dependent on the availability of land for agriculture and certain climatic conditions.

Papain, bromelain, keratinases are some of the well-known proteases of plant origin. The most familiar proteases of animal origin are pancreatic trypsin, chymotrypsin, pepsin and rennins. These are prepared in pure form in bulk quantities. However, their production depends on the availability of livestock for slaughter, which in turn is governed by political and agricultural policies. The inability of the plant and animal proteases to meet current world demands has led to an increased interest in microbial proteases.

Micro-organisms represent an excellent source of enzymes owing to their broad biochemical diversity and their susceptibility to genetic manipulation. Proteases from microbial sources are preferred to the enzymes from plant and animal sources since they possess almost all the characteristics desired for their biotechnological applications.

Protease Functionalised Nanoparticles

In recent years, nanotechnology has showed a significant attraction to the preparation of immobilised enzymes. Extensive studies have been carried out on the use of micrometer-sized particles for the immobilisation of enzymes by using nanoporous, nanofibrous and nanoparticles. Overall, nanoparticles provide an ideal remedy to the usually contradictory issues encountered in the optimisation of immobilised enzymes, such as minimum diffusional limitation, maximum surface area per unit mass and high enzyme loading. Protein/peptide oral drug delivery is a complicated process mainly due to poor intrinsic protein permeability. This is due to large molecular weight, degradation by proteolytic enzymes in the stomach and in the small intestine and chemical instability. Hence protein functionalised nanoparticles which are also pH sensitive were made to overcome this problem.

The preparation of biocatalytic ultra micro and nanoparticles of low molecular weight chitosan can be done using nanoprecipitation method. This process involves a non solvent, chemical cross-linking with tripoly phosphate, stabilised by a anionic surfactant and the enzyme was loaded on to the nanoparticle by glutaraldehyde cross linking chemistry. TEM and SEM micrographs confirms the presence of enzyme functionalised nanoparticles. The nanoparticles released the enzyme at the physiological pH of 7.4. The molecular weight of chitosan, concentration of the polymer, chitosan-TPP ratio and the pH of the solution are important parameters for the size, surface charge and aggregation of the nanoparticles.

Compared with conventional solid-phase supports, nanoparticulate matrices have a higher catalyst loading capacity due to their very large surface areas. Important application of enzyme or protein functionalised nanoparticle is in the controlled release of protein and peptide drugs, because they show excellent muco-adhesiveness and permeation enhancing effect across the biological surfaces.

Applications of Protease in Food Industry

Proteases execute a large variety of functions, extending from the cellular level to the organ and organism level, to produce cascade systems such as haemostasis and inflammation, which are responsible for the complex processes involved in the normal physiology of the cell as well as in abnormal pathophysiological conditions. Their involvement in the life cycle of disease- causing organisms has led them to become a potential target for developing therapeutic agents against fatal diseases such as cancer and AIDS. Microbial proteases are increasingly used in treatment of various disorders namely cancer, inflammation, cardiovascular disorders, necrotic wounds, etc. Proteases are used an immune – stimulatory agents. Increased antibiotic concentration at a target site when protease was concomitantly used with an antibiotic. Proteases are used extensively in the pharmaceutical industry for preparation of medicines such as ointments for debridement of wounds. It is also used in denture cleaners and as contact-lens enzyme cleaners.

Proteases have a large variety of applications, mainly in the detergent and food industries. Proteases are envisaged to have extensive applications in leather treatment and in several bioremediation processes. Proteases that are used in the food industries are prepared in bulk quantities and used as crude preparations, whereas those that are used in medicine are produced in small amounts but require extensive purification before they can be used.

The use of proteases in the food industry dates back to antiquity. They have been routinely used for various purposes such as cheese making, baking, preparation of soya hydrolysates and meat tenderisation.

The major application of proteases in the dairy industry is in the manufacture of cheese. The milk-coagulating enzymes fall into three main categories: (i) animal rennets, (ii) microbial milk coagulants and (iii) genetically engineered chymosin. Both animal and microbial milkcoagulating proteases belong to a class of acid aspartate proteases and have molecular weights between 30000 to 40000. Rennet extracted from the fourth stomach of unweaned calves contains the highest ratio of chymosin to pepsin activity. A world shortage of calf rennet due to the increased demand for cheese production has intensified the search for alternative microbial milk coagulants.

In cheese making, the primary function of proteases is to hydrolyse the specific peptide bond to generate para-k-casein and macro-peptides. Chymosin is preferred due to its high specificity for casein, which is responsible for its excellent performance in cheese making.

Wheat flour is a major component of baking processes. It contains an insoluble protein called gluten, which determines the properties of the bakery doughs. Endo and exo-proteinases from *Aspergillus oryzae* have been used to modify wheat gluten by limited proteolysis. Enzymatic treatment of the dough facilitates its handling and machining and permits the production of a wider range of products. The addition of proteases reduces the mixing time and results in increased loaf volumes. Bacterial proteases are used to improve the extensibility and strength of the dough.

Protease Inhibitors

Soyabeans and other legumes

Legumes and especially soyabeans, are an abundant source of natural protease inhibitors. Some research suggests that protease inhibitors found in soyabeans may play a role in soya's anti-cancer properties and may help protect against certain cancers such as breast and colon. The particular protease inhibitor found in soya, called Bowman-Birk, appears to help prevent tumour formation, according to 'Pathology for the Health Professions.'

Potato tubers

Another natural source of protease inhibitors is the potato. Protease inhibitors from potatoes are shown to suppress tumour growth in clinical research settings. In addition, potato protease inhibitors suppress a hormone responsible for stimulating appetite. Recently researchers found that potato protease inhibitors may be useful at suppressing appetite. More research is needed to understand how well it works.

Green tea

Antioxidant compounds in green tea called catechins produce protease-inhibiting activity. The major catechin in green tea is epigallocatechin gallate (EGCG). The protease-inhibiting activity of green tea catechins is also reportedly involved in green tea's anti-cancer benefits. The natural protease inhibitors in green tea may suppress infection by other pathogens, such as the flu virus and the adenovirus. These pathogens commonly cause respiratory infection in children.

Health Benefits of Protease

Proteolytic enzymes are extremely important for the digestion of many foods. But their intestinal duties are not solely limited to digesting food. They also digest the cell walls of unwanted harmful organisms in the body and break down unwanted wastes such as toxins, cellular debris and undigested proteins. In this way, protease helps digest the small stuff, so that our immune system can work hard to avoid toxin overload. In this way digestion plays a huge role in overall health and enzymes are a big part of digestive health. With the distinct ability to breakdown peptide bonds and liberate amino acids, proteolytic enzymes are now being studied by modern science and medicine for their clinical and therapeutic use in the realms of general oncology and overall immune function.

The following list offers some health benefits of supplementing with protease, as well some of the exciting research being done on protease and its applications to human health and disease prevention/management.

Inflammatory bowel disease: A 2010 US study on Inflammatory Bowel Diseases found that the proteolytic enzyme bromelain could help reduce indications and problems in the colon.

Skin burns and stomach ulcers: A 2010 Brazilian study in 'Burns' journal found that intake of protease can help cellular repair in cases of skin burns and stomach ulcers in laboratory mice.

Sprains: One double-blind, placebo-controlled trial involving 30 individuals with chronic neck discomfort found that those patients using a protease enzyme experienced a moderate reduction in discomfort when compared to a placebo. Clinical trials show that protease enzymes can speed the healing time of sprains, bruises, fractures and tissue injuries.

Slow or stop irritation: Protease enzymes slow or stop irritation by neutralising the biochemicals associated with the response (i.e. bradykinins and eicosanoids). This may support cardiovascular and brain health.

Osteoarthritis discomfort: Although only preliminary and not to be interpreted as a new therapy, one study done on over 400 people found that protease enzymes could reduce osteoarthritis discomfort.

Sports-related injuries: Research suggests that protease enzyme combinations may aid in the recovery of sports injuries. A double-blind, placebo-controlled study of over 40 people with sports-related injuries found that taking protease enzymes could offer faster recovery, as well as reducing the time away from practicing sports again by up to 50%. Another research has found that taking protease enzymes could significantly increase recovery time after injury, particularly in cases involving mild fractures.

Beneficial gut bacteria: Research shows that protease activities give our cells critical amino acids and are essential for the growth of the good forms of gut flora that break compounds down.

Circulatory and lymph system: Protease enzymes help to cleanse organic debris out of our circulatory and lymph system.

Clots: Some forms of protease can boost the 'quality' of blood cells, allowing a better circulatory response and reducing the risk of clots.

Future Scope of Protease

Proteases are a unique class of enzymes, since they are of immense physiological as well as commercial importance. They possess both degradative and synthetic properties. Since proteases are physiologically necessary, they occur ubiquitously in animals, plants and microbes. However, microbes are a goldmine of proteases and represent the preferred source of enzymes in view of their rapid growth, limited space required for cultivation and ready accessibility to genetic manipulation. Microbial proteases have been extensively used in the food, dairy and detergent industries since ancient times. There is a renewed interest in proteases as targets for developing therapeutic agents against relentlessly spreading fatal diseases such as cancer, malaria and AIDS. The development of recombinant rennin and its commercialisation by Pfizer and Genencor is an excellent example of the successful application of modern biology to biotechnology. Analysis of sequences for acidic, alkaline and neutral proteases has provided new insights into the evolutionary relationships of proteases. Despite the systematic application of recombinant technology and protein engineering to alter the properties of proteases, it has not been possible to obtain microbial proteases that are ideal for their biotechnological applications. Industrial applications of proteases have posed several problems and challenges for their further improvements. The biodiversity represents an invaluable resource for biotechnological innovations and plays an important role in the search for improved strains of micro-organisms used in the industry. A recent trend has involved conducting industrial reactions with enzymes reaped from exotic micro-organisms that inhabit hot waters, freezing Arctic waters, saline waters, or extremely acidic or alkaline habitats.

The proteases isolated from extremophilic organisms are likely to mimic some of the unnatural properties of the enzymes that are desirable for their commercial applications. The existing knowledge about the structure-function relationship of proteases, coupled with gene-shuffling techniques, promises a fair chance of success, in the near future, in evolving proteases that were never made in nature and that would meet the requirements of the multitude of protease applications.

LIPASE

Lipase is a physiologically necessary enzyme. It occurs in many plants and animal organisms, as well as in micro-organisms. However, its richest source are bacteria, fungi and yeast. Lipolytic enzymes are widespread in the plant world, however, the knowledge and experience pertaining to plant-originating lipases is still limited in comparison to the information on lipases from mammals and micro-organisms and this is especially connected to the difficulties with their isolation.

Microbiological lipases, especially those originating from bacteria, are more stable than those from plants or animals. They possess unique qualities and are used more often for industrial purposes. The knowledge about lipolytic bacterial enzymes is developing with surprising rate. The classification of bacterial lipases has been based mainly on the sequence of amino acids and some basic biological properties. Bacterial lipases from the *Pseudomonas* kind constitute the main group from which two types are distinguished – one is connected with the auxiliary protein in order to show proper activity and the other one is not. Lipases (triacylglycerol acylhydrolases, EC 3.1.1.3) catalyse the hydrolysis and the synthesis of esters formed from glycerol and long-chain fatty acids. Lipases occur widely in nature, but only microbial lipases are commercially significant. The many applications of lipases include speciality organic synthesis, hydrolysis of fats and oils, modification of fats, flavour enhancement in food processing, resolution of racemic mixtures and chemical analyses. This chapter discusses the production, recovery and use of microbial lipases. Immobilised preparations of lipases are also discussed. In view of the increasing understanding of lipases and their many applications in high-value synthesis and as bulk enzymes, these enzymes are having an increasing impact on bioprocessing.

Isolation and Purification of Lipase

Lipases available on the market are usually isolated from higher eukaryotic organisms and micro-organisms which include bacteria, fungi, yeast and actinomycetes. Lipase-producing micro-organisms occur in various environments such as industrial waste-collection plants, plants producing vegetable oils, dairies, oil-polluted soils, decomposing food, composting plants, coal heaps or hot springs.

Microbiological lipases are obtained mainly by a submerged fermentation. Its production is influenced by: the concentration of coal and nitrogen source, environmental pH, temperature and concentration of dissolved oxygen. The majority of microbiological lipases are active in a basic habitat (pH 7.0–9.0). For this reason, the activity of these enzymes is so strongly dependent on pH changes in the reaction mixture which influence the catalytic potential of lipases. Heavy metal ions are also effective lipase antagonists and modulators. On the other hand, chelating agents, e.g. EDTA, inhibit only the activity of metal lipases. Because of the possibility of intensive production, good stability and numerous stereospecific properties, microbiological lipases still draw scientists' attention. Lipases have been isolated and studied from such micro-organisms as *Pseudomonas*, *Geotrichum*, *Candida rugosa*, *Aspergillus*, *Rhizophus*, *Mucor hirmalis f. hiemalis*, *Streptomyces rimosus*, *Penicillinum*. These lipases not only differ from each other molecularly, but they also possess different catalytic properties.

Lipase Immobilisation

Lipases are readily and often used in biotechnological processes. Their activity may be improved by immobilisation. Because of this, new studies are conducted which pertain to the activity and application of immobilised lipase enzymes.

There exist many methods of immobilising lipase, starting from adsorption or precipitation on hydrophobic materials, covalent bonding to functional groups, trapping in polymer gels, adsorption on macroporous, anionic ion-exchange resins, or microencapsulation in lipid membranes. Among the mentioned methods, covalent bonding of lipase is dominating. On non-organic substrates (aluminium oxide, silica and porous glass) Moreno and others covalently immobilised the Candida cylindracea lipase, previously activating the carriers with cyanuric acid. Derivative immobilising on silica and aluminium oxide were characterised by a higher activity and thermal resistance than pure enzymes or even those immobilised on porous glass.

Applications of Lipases

Because of their significance, for a long time lipases have been the subject of numerous studies which concentrate mainly on the structure of this enzyme, the mechanism of functioning, kinetics, sequencing and cloning genes, as well as on the general characteristics of their occurring. Lipases are commonly used in processing fats and oils, as additions to detergents and degreasing agents, in food processing, chemical and pharmaceutical synthesis, in paper production and the cosmetic industry. They may also be used in increasing the speed of fatty and polyurethane waste decomposition.

Major applications of lipases are summarised in Table 23.1.

Table 23.1: Industrial applications of microbial lipases.

Industry	*Action*	*Product or application*
Detergents	Hydrolysis of fats	Removal of oil stains from fabrics
Dairy foods	Hydrolysis of milk fat, cheese ripening, modification of butter fat	Development of flavouring agents in milk, cheese and butter
Bakery foods	Flavour improvement	Shelf-life prolongation
Beverages	Improved aroma	Beverages
Food dressings	Quality improvement	Mayonnaise, dressings and whippings
Health foods	Trans-esterification	Health foods
Meat and fish	Flavour development	Meat and fish products, fat removal
Fats and oils	Trans-esterification, hydrolysis	Cocoa butter, margarine, fatty acids, glycerol, mono- and diglycerides
Chemicals	Enantioselectivity, synthesis	Chiral building blocks, chemicals
Pharmaceuticals	Trans-esterification, hydrolysis	Specialty lipids, digestive aids
Cosmetics	Synthesis	Emulsifiers, moisturisers
Leather	Hydrolysis	Leather products
Paper	Hydrolysis	Paper with improved quality
Cleaning	Hydrolysis	Removal of fats

Lipases in Food Industry

Fats and oils are important constituents of foods. The nutritional and sensory value and the physical properties of a triglyceride are greatly influenced by factors such as the position of the fatty acid in the glycerol backbone, the chain length of the fatty acid and its degree of unsaturation. Lipases allow us to modify the properties of lipids by altering the location of fatty acid chains in the glyceride and replacing one or more of the fatty acids with new ones. This way, a relatively inexpensive and less desirable lipid can be modified to a higher value fat.

Cocoa butter, a high-value fat, contains palmitic and stearic acids and has a melting point of approximately 37°C. Melting of cocoa butter in the mouth produces a desirable cooling sensation in products such as chocolate. Lipase-based technology involving mixed hydrolysis and synthesis reactions is used commercially to upgrade some of the less desirable fats to cocoa butter substitutes. One version of this process uses immobilised *Rhizomucor miehei* lipase for the trans-esterification reaction that replaces the palmitic acid in palm oil with stearic acid. Similarly, Pabai and others described a lipase-catalysed interesterification of butter fat that resulted in a considerable decrease in the long-chain saturated fatty acids and a corresponding increase in C18:0 and C18:1 acids at position 2 of the selected triacylglycerol.

Micro-organisms Producing Lipases

Lipases are produced by many micro-organisms and higher eukaryotes. Most commercially useful lipases are of microbial origin. Some of the lipase-producing micro-organisms are listed in Table 23.2.

Table 23.2: Some lipase-producing micro-organisms.

Source	Genus	Species
Bacteria (Gram-positive)	Bacillus	B. megaterium
		B. cereus
		B. stearothermophilus
		B. subtilis
		Recombinant B. subtilis 168
		B. brevis
		B. thermocatenulatus
		Bacillus sp. IHI-91
		Bacillus strain WAI 28A5
		Bacillus sp.
		B. coagulans
		B. acidocaldarius
		Bacillus sp. RS-12
		B. thermoleovorans ID-1
		Bacillus sp. J 33
	Staphylococcus	S. canosus
		S. aureus
		S. hyicus
		S. epidermidis
		S. warneri
	Lactobacillus	Lactobacillus delbruckii sub sp. bulgaricus
		Lactobacillus sp.
	Streptococcus	Streptococcus lactis
	Micrococcus	Micrococcus freudenreichii
	Propionibacterium	Propionibacterium acne
		Pr. granulosum
	Burkholderia	Burkholderia sp.
		Bu. glumae
Bacteria (Gram-negative)	Pseudomonas	P. aeruginosa
		P. fragi
		P. mendocina
		P. putida 3SK
		P. glumae
		P. cepacia
		P. fluorescens
		P. aeruginosa KKA-5
		P. pseudoalcaligenes F-111
		Pseudomonas sp.
		P. fluorescens MF0

(Cont'd...)

Source	Genus	Species
		Pseudomonas sp. KWI56
	Chromobacterium	*Ch. viscosum*
	Acinetobacter	*Aci. pseudoalcaligenes*
		Aci. radioresistens
	Aeromonas	*Ae. hydrophila*
		Ae. sorbia LP004
Fungi	*Rhizopus*	*Rhizop. delemar*
		Rhizop. oryzae
		Rhizop. arrhizus
		Rhizop. nigricans
		Rhizop. nodosus
		Rhizop. microsporous
		Rhizop. chinensis
		Rhizop. japonicus
		Rhizop. niveus
		Aspergillus A. flavus
		A. niger
		A. japonicus
		A. awamori
		A. fumigatus
		A. oryzae
		A. carneus
		A. repens
		A. nidulans
	Penicillium	*Pe. cyclopium*
		Pe. citrinum
		Pe. roqueforti
		Pe. fumiculosum
		Penicillium sp.
		Pe. camambertii
		Pe. wortmanii
		Mucor Mucor miehei
		Mu. javanicus
		Mu. circinelloides
		Mu. hiemalis
		Mu. racemosus
	Ashbya	*Ashbya gossypii*
	Geotrichum	*G. candidum*
		Geotrichum sp.
	Beauveria	*Beauveria bassiana*

(Cont'd...)

Source	Genus	Species
	Humicola	H. lanuginosa
	Rhizomucor	R. miehei
	Fusarium	Fusarium oxysporum
		F. heterosporum
	Acremonium	Ac. strictum
	Alternaria	Alternaria brassicicola
	Eurotrium	Eu. herbanorium
	Ophiostoma	O. piliferum
Yeasts	Candida	C. rugosa
		C. tropicalis
		C. antarctica
		C. cylindracea
		C. parapsilosis
		C. deformans
		C. curvata
		C. valida
	Yarrowia	Y. lipolytica
	Rhodotorula	Rho. glutinis
		Rho. pilimornae
	Pichia	Pi. bispora
		Pi. maxicana
		Pi. sivicola
		Pi. xylosa
		Pi. burtonii
	Saccharomyces	Sa. lipolytica
		Sa. crataegenesis
	Torulospora	Torulospora globora
	Trichosporon	Trichosporon asteroides
Actinomycetes	Streptomyces	Streptomyces fradiae NCIB 8233
		Streptomyces sp. PCB27
		Streptomyces sp. CCM 33
		Str. coelicolour
		Str. cinnamomeus

Purification and Kinetic Characterisation Oflipases

Many lipases have been extensively purified and characterised in terms of their activity and stability profiles relative to pH, temperature and effects of metal ions and chelating agents. In many cases, lipases have been purified to homogeneity and crystallised. Purification methods used have generally depended on nonspecific techniques such as precipitation, hydrophobic interaction chromatography, gel filtration and ion exchange chromatography. Affinity chromatography has been used in some cases to reduce the number of individual purification steps needed.

OXIDOREDUCTASE

Oxidoreductases are a class of enzymes that catalyse oxidoreduction reactions. Oxidoreductases catalyse the transfer of electrons from one molecule (the oxidant) to another molecule (the reductant).

Polyphenol Oxidase

Polyphenol oxidase (PPO or monophenol mono-oxygenase or polyphenol oxidase I, chloroplastic) is a tetramer that contains four atoms of copper per molecule and binding sites for two aromatic compounds and oxygen. In fact, browning by PPO is not always an undesirable reaction, the familiar brown colour of tea (especially black tea) and cocoa is developed by PPO enzymatic browning during product processing. Grape reaction product (2-S glutathionyl caftaric acid) is an oxidation compound produced by action of PPO on caftaric acid and found in wine. This compound production is responsible for the lower level of browning in certain white wines. Arctic Apples are a suite of trademarked apples that contain a nonbrowning trait. Specifically, gene silencing is used to turn down the expression of polyphenol oxidase (PPO), thus preventing the fruit from browning. They are therefore genetically modified food. Polyphenol oxidase has been given two entries in the International Union of Biochemistry (IUB) classification, as EC 1.14.18.1, monophenol monoxygenase and EC 1.10.3.1, catechol oxidase. Common trivial names for monophenol monoxygenase are tyrosinase, phenolase, monophenol oxidase and cresolase. The enzyme complexes from higher plants and fungi are virtually non-specific and oxidise a wide variety of mono-phenolic and *o*-diphenolic compounds.

PEROXIDASE

Peroxidases are distributed ubiquitously in both the plant and animal kingdoms. Perhaps the one most widely known to enzymologists is horseradish peroxidase (EC 1.11.1.7), because of its wide use as an indicator enzyme in standard spectrophotometric and immunoassay techniques. All peroxidases act on hydrogen peroxide as an electron acceptor and oxidise a multitude of donor compounds, many times to coloured end products. Horseradish peroxidase has a heme prosthetic group and is a hemoprotein. NADH peroxidase (EC 1.11.1.1) is a flavoprotein. Glutathione peroxidase (EC 1.11.1.9) is a selenium-containing protein. Chloride peroxidase (EC 1.11.1.10) is probably a heme-thiolate protein.

LACTOPEROXIDASE

Lactoperoxidase is a peroxidase enzyme secreted from mammary, salivary and other mucosal glands that functions as a natural antibacterial agent. Lactoperoxidase is a member of the heme peroxidase family of enzymes. In humans, lactoperoxidase is encoded by the LPO gene. Lactoperoxidse catalyses the oxidation of a number of inorganic and organic substrates by hydrogen peroxide. These substrates include bromide and iodide and therefore lactoperoxidase can be categorised as a haloperoxidase. Another important substrate is thiocyanate. The oxidised products produced through the action of this enzyme have potent bactericidal activities. Lactoperoxidase together with its inorganic ion substrates, hydrogen peroxide and oxidised products is known as the lactoperoxidase system. The lactoperoxidase system plays an important role in the innate immune system by killing bacteria in milk and mucosal secretions hence augmentation of the lactoperoxidase system may have therapeutic applications.

Applications of Lactoperoxidase

Furthermore lactoperoxidase have found application in dental and wound treatment. Finally lactoperoxidase may find application as anti-tumor and anti-viral agents.

Dairy products

Lactoperoxidase is an effective antimicrobial agent and is used as an antibacterial agent in reducing bacterial microflora in milk and milk products. Activation of the lactoperoxidase system by addition of hydrogen peroxide and thiocyanate extends the shelf life of refrigerated raw milk. It is fairly heat resistant and is used as an indicator of over pasteurisation of milk.

CATALASE

Catalase is an enzyme that brings about (catalyses) the reaction by which hydrogen peroxide is decomposed to water and oxygen. Found extensively in organisms that live in the presence of oxygen, catalase prevents the accumulation of and protects cellular organelles and tissues from damage by peroxide, which is continuously produced by numerous metabolic reactions. In mammals, catalase is found predominantly in the liver. Catalase has various industrial applications. In the food industry, it is used in combination with other enzymes in the preservation of foodstuffs and in the manufacture of beverages and certain food items. Commercial catalases also are used to break down hydrogen peroxide in wastewater.

Applications of Catalase

Catalase is used in the food industry for removing hydrogen peroxide from milk prior to cheese production. Another use is in food wrappers where it prevents food from oxidising.

SULPHYDRYL OXIDASE

Sulphydryl oxidase catalyses the oxidation of thiols to their corresponding disulphides using molecular oxygen as the electron acceptor. The reaction requires that two moles of thiol are consumed for each mole of oxygen. A FAD-activated sulphydryl oxidase was isolated and identified from whey obtained from bovine skim milk. The enzyme also has been identified in the spores of the fungus *Myrothecium werrucaria*. More recently, the enzyme was isolated and characterised from *Aspergillus sojae* and *Aspergillus niger*.

Mammalian: Sulphydryl oxidase from bovine milk was isolated and characterised. Cream was separated from the skim milk by centrifugation. Chymosin was added to the skim milk to precipitate the casein and the precipitate was separated from the whey by centrifugation. The whey was saturated to 50% with ammonium sulphate and the precipitate collected by centrifugation. The precipitate was dissolved in phosphate buffer at pH 7.0 and dialysed against the same buffer. The dialysed enzyme was concentrated to about 3% protein and allowed to stand overnight at 4°C. The sedimented enzyme was centrifuged and the precipitate was dissolved in buffer, dialysed against deionised water and lyophilised. The sulphydryl oxidase was purified further by gel filtration and differential centrifugation. The enzyme was purified about 3200-fold with a 41% recovery. The specific activity was 104 μmol oxygen consumed/min/mg N. The reaction catalysed by bovine milk sulphydryl oxidase involves oxidation of sulphydryl groups to the disulphide in the presence of molecular oxygen. The sulphydryl groups in cysteine, glutathione, or even protein can be oxidised. Glutathione was chosen as the substrate because it is more stable than cysteine and reacts more quickly than protein-bound sulphydryl groups. This enzyme can be assayed by three techniques: oxygen consumption, disappearance of sulphydryl groups and hydrogen peroxide production. The oxygen consumption method usually is used because of its greater sensitivity and ease of use. The enzyme exhibited a sharp peak at 35°C and had only 10% activity at the relatively low temperature of 50°C. It had an optimal pH of 7.0 and showed very little activity at pH 4.5 and 8.5. The K_m for glutathione was 9.0×10^{-5} at 35°C and pH 7.0.

GLUCOSE OXIDASE

The glucose oxidase enzyme (GOx) (EC 1.1.3.4) is an oxido-reductase that catalyses the oxidation of glucose to hydrogen peroxide and D-glucono-δ-lactone. In cells, it aids in breaking the sugar down into its metabolites. Glucose oxidase is widely used for the determination of free glucose in body fluids (diagnostics), in vegetal raw material and in the food industry. It also has many applications in biotechnologies, typically enzyme assays for biochemistry including biosensors in nanotechnologies. It is often extracted from *Aspergillus niger*.

PYRANOSE OXIDASE

Pyranose oxidase (EC 1.1.3.10) is an enzyme that catalyses the chemical reaction. Pyranose oxidase free from glucosone utilising impurities was obtained from strains of *Coriolus versicolour, Lenzites betulinus* and *Polyporus obtusus*. The enzyme can be used to make pure glucosone for use in making high purity food sugars, for example, fructose, mannitol and sorbitol. When pyranose oxidase is prepared from *P. obtusus*, it is a tetramer with a molecular mass of about 200 kDa with a sub-unit mass of about 69 kDa.

XANTHINE OXIDASE

Xanthine oxidase (XO, sometimes 'XAO') is a form of xanthine oxidoreductase, a type of enzyme that generates reactive oxygen species. These enzymes catalyse the oxidation of hypoxanthine to xanthine and can further catalyse the oxidation of xanthine to uric acid. These enzymes play an important role in the catabolism of purines in some species, including humans.

LIPOXYGENASE

Lipoxygenases (EC 1.13.11) are a family of iron-containing enzymes that catalyse the dioxygenation of polyunsaturated fatty acids in lipids containing a *cis, cis*-1,4-pentadiene structure. Lipoxygenases are found in plants, animals and fungi. Products of lipoxygenases are involved in diverse cell functions. These enzymes are most common in plants where they may be involved in a number of diverse aspects of plant physiology including growth and development, pest resistance and senescence or responses to wounding. In mammals a number of lipoxygenases isozymes are involved in the metabolism of eicosanoids (such as prostaglandins, leukotrienes and nonclassic eicosanoids).

DEHYDROGENASE

A dehydrogenase is an enzyme belonging to the group of Oxidoreductases that oxidises a substrate by a reduction reaction that transfers one or more hydrides (H$^-$) to an electron acceptor, usually NAD$^+$/NADP$^+$ or a flavin coenzyme such as FAD or FMN. These oxidoreductases are categorised mostly as dehydrogenases, although 37 are categorised as reductases. This class of enzyme is very important to the metabolic functions of plants and animals but, in foods or food processing, only a few might be considered important. These enzymes are alcohol dehydrogenases, shikimate dehydrogenases, malate dehydrogenases, isocitrate dehydrogenases and lactate dehydrogenases.

Meat lactate dehydrogenase: Lactate dehydrogenase may be important in meat research. There is a relationship between the lactate dehydrogenase isoenzyme pattern and different types of muscle fibres. Differences exist in muscle lactate dehydrogenase activity and in isoenzymes among different species of animal.

SECTION VIII

Brewing of Yeast, Beer and Wine Industry

Chapter 24

Yeast as a Versatile Tool in Biotechnology

INTRODUCTION

Yeasts represent a very diverse group of micro-organisms, and even strains that are classified as the same species often show a high level of genetic divergence. Yeasts biodiversity is closely related to their applicability. Biotechnological importance of yeast is almost immeasurable. For centuries, people have exploited its enzymatic potential to produce fermented food as bread or alcoholic beverages. Admittedly, yeasts application was initially instinctual, but with science and technology development, these micro-organisms got the object of thorough scientific investigations. It must be recognised that yeast represents an excellent scientific model because of its eukaryotic origin and knowledge of genetics of yeast cells as well as metabolism examined in detail. In 1996, the genome of baker yeast *Saccharomyces cerevisiae* has been elucidated, what opened the opportunity for the global study of the expression and functioning of the eukaryotic genome.

YEASTS – COMMERCIAL APPLICATIONS

Yeasts have a wide range of applications mainly in food industry (wine making, brewing, distilled spirits production, and baking) and in biomass production [single-cell protein (SCP)]. More recently, yeast has also been used in the biofuel industry and for the production of heterologous compounds. Obviously, their main application arises from the metabolic capacity to carry out the transformation of sugars into ethyl alcohol and carbon dioxide under anaerobic conditions. Moreover, a large number of secondary flavour compounds are created what implies on organoleptic attributes of particular food products. However, it would be misguided to trivialise their metabolic capacities only to fermentative activity. The main factors influencing yeast metabolism are the oxygen availability and the type of carbon source. Many yeast strains can function under both anaerobic as well as aerobic conditions of environment, switching their metabolism types easily. Obviously, the courses of main metabolic pathways are conserved, but some regulative mechanisms attract the attention, denoting unusual metabolism flexibility. In food industry, *Saccharomyces cerevisiae* is the genus of yeast most frequently used, whereas Candida, Endomycopsis, and Kluyveromyces are crucial for SCP production. Yeast strains of industrial importance are carefully selected from the immense natural biodiversity and their properties improved according to process outcome. Both classical approaches as well as modern strategies of gene manipulations are applied to generate variants relevant to work under industrial specific conditions. Strains alterations concern not only the route of fermentations step and its direct yield but also facilitation of product recovery procedures and finally the best quality of particular end product. These nonpathogenic strains with both genotypic and phenotypic stability should have short-generation time and low nutritional requirements. They should perform the fermentation process quickly to minimise the contamination risk. Additionally, they should be tolerant of a wide range of physiological stresses, such as low pH, high ethanol concentration, and high osmotic stress.

The enzymatic power of yeast is central to beer manufacturing. Industrial species are carefully selected, nonsporulated polyploides that do not perform sexual reproduction process. Brewer's yeasts are divided into two separate categories: top-fermenting yeast (ale) and bottom-fermenting one (lager). Both yeast types have similar cell morphology, but they differ in some physiological and metabolic features, what closely corresponds to the process conditions and type of end product. Fermentation of ale yeast is carried out at room temperature and results in beers with a characteristic fruity aroma. In the case of lager yeast, the fermentation temperature is lower, and therefore, this step takes longer than fermentation with ale strains.

The share of aerobic respiration in yeast metabolism is higher in case of ale strains than in lager yeast. Both top-fermenting and bottom-fermenting yeast belong to the genus *Saccharomyces*. Ale-brewing yeasts are genetically more diverse and classified as *Saccharomyces cerevisiae*, whereas the taxonomy of the lager strains has undergone several changes. Initially, bottom-fermenting yeast was classified as *Saccharomyces carlsbergensis*, but due to the application of modern taxonomic approaches, the species *S. carlsbergensis* were included as a part of *Saccharomyces pastorianus* taxon. On the basis of genetic studies, *S. pastorianus* strains are now considered as allopolyploid interspecies hybrids of *S. cerevisiae* and *S. bayanus*. From technological point of view, beside the high fermentative activity of yeast cells and tolerance for several environmental stresses, the aptitude for asexual aggregation known as flocculation seemed to be especially important, because this ability causes the yeast to sediment to the bottom of the fermenter at the end of fermentation step, what simplifies downstream processing.

Various yeast species constitute the predominant microbial group of natural microbiota of fruits ecosystems, what is the reason of fast and spontaneous fermentation of juices or musts resulted in wine production. Yeast colonising various fruits belongs to the genera *Saccharomyces*, *Brettanomyces* (its sexual form (teleomorph) *Dekkera*), *Candida*, *Cryptococcus*, *Debaromyces*, *Hanseniaspora* (anamorph *Kloeckera*), *Hansenula*, *Kluyveromyces*, *Pichia*, *Rhodotorula*, *Torulaspora*, *Schizosaccharomyces*, and *Zygosaccharomyces*. It goes without saying that most fermented products are generated by a mixture of microbes. These microbial consortia perform various biological activities responsible for the nutritional, hygienic, and aromatic qualities of the product. Doubtlessly, yeast plays a principal role in wine making both in domestic environment as well as in commercial scale. Nonconventional yeast (non-*Saccharomyces* species) dominates during early and middle stages of fermentation process, whereas the latter phase of natural process is mediated by *Saccharomyces cerevisiae*. The raw material of particular importance in global wine making is grapes that are harvested at specific stages of ripeness depending on the style of wine to be made.

Once again yeast enzymatic power is crucial for the product chemical profile—except for ethanol biosynthesis, the creation of flavour compounds and fragrances from substrates abundant in fruits implies the organoleptic features of particular wine product. During the degradation of grape sugars, amino acids, fatty acids, terpenes, and thiols, some by-products like glycerol, carboxylic acids, aldehydes, higher alcohols, esters, and sulphides are formed, and their synthesis is largely dependent on the peculiarities of the strains used. In 1965, the first two commercially active dried wine yeasts called Montrachet and Pasteur Champagne were produced for a large Californian winery. Now-a-days, modern winegrowers routinely use selected yeast starters in practice. These micro-organisms dominate native yeast species and give desired direction of chemical transformations occurring in musts and allow to obtain product of predictable quality. Presently exploited commercially available starters have been created as a result of naturally occurring phenomenon called 'genome renewal' as well as planned processes of genetic improvement followed by a careful selection for their good fermentation performance. Genome renewal hypothesis in the standard version assumes that infrequent sexual cycles, characterised by a high degree of selfing, can help to purge deleterious alleles and fix beneficial alleles, thus helping to facilitate adaptation in yeast. This hypothesis had to be re-evaluated due to the fact that in the case of many environmental isolates, very high levels of genomic heterozygosity had been observed. Presently, the majority of commercial wine yeast comprises strains of *Saccharomyces cerevisiae*, including those described by enologists as *S. bayanus*, which has been re-identified in most cases as *S. cerevisiae*.

The growing demand for more diversified wines or for specific characteristics has led to the exploration of new species for wine making. This nonconventional yeast may contribute to the wine's organoleptic characteristic by producing a broad range of unique secondary metabolites and secreting particular enzymes or exhibiting others substantial features (release of mannoproteins, contributions to color stability, etc.). The wine industry currently proposes starters of a few nonconventional yeast (*Torulaspora delbrueckii, Metschnikowia pulcherrima, Pichia kluyveri, Lachancea thermotolerans*, etc.). Yeast variants used as starters must qualify for commercial application. The list of desired properties is very long and includes both fermentation characteristics as well as flavour characteristics (e.g. low sulphide/dimethyl sulphide/thiol formation, liberation of glycosylated flavour precursors, low higher alcohol production, etc.) metabolic properties with health implications (low formation of sulphite, biogenic amines, or urea) and technological properties (e.g. high sulphite tolerance, low foam formation, flocculation properties, etc.).

Another example of *Saccharomyces cerevisiae* strains of commercial interest encompasses distiller's yeasts, applied in industry where alcohol fermentation is followed by distillation. In this industry type, spontaneous fermentations are not practiced, and specific yeast is inoculated into fermenter. Starters used for distilled beverages (aquavit, gin, vodka, whisky, rum, tequila, cognac, brandy, and kirsch) commercial production should exhibit exquisitely intensive anaerobic metabolism, what should result not only in high ethanol productivity but also in the high level of cellular tolerance to this product. Additionally, thermostable variants tolerant to acids and increased osmotic pressure are in greater demand for distillers. Since many years, there has been a rising interest in new variants that are able to degrade starch. Many attempts have been made to produce ethanol from starch using recombinant haploid strains that express amylolytic enzymes, due to the simplicity of genetic manipulation of these strains. However, it is difficult to use laboratory haploid strains in practice because their fermentation characteristics are not as good as those of industrial polyploidy strains.

In addition to alcoholic beverages production, enzymatic power of yeasts is also essential for baking industry, where concentrated yeast biomass is used as a starter in dough fermentation in order to produce bread and other bakery products. Commercially available baker's yeast forms include fresh compressed biomass, dehydrated cells (dry yeast), and lyophilised cells (instant). Fermentation of dough substrates leads to ethanol production as well as number of volatile and nonvolatile compounds that have an important contribution to the flavour of bread. As a result of carbohydrates (maltose mainly), fermentation carbon dioxide is generated what increases the dough volume and is responsible for crumb texture. Baker's yeast is simply brewery yeast produced via submerged fermentation process carried out in the presence of oxygen. Aerobic conditions favour yeast cells production, which is not of interest to ethanol producers, but is important when large amount of cells mass must be produced.

The main ingredient of industrial production medium used in yeast production factories are beet or cane molasses, mainly because of the low cost of this waste products and high sucrose content. In most cases, the industrial production is a multistage process carried out under batch or fed-batch conditions with sequential stages differing in fermenter size, performed under controlled intense aeration. Aeration is generally considered as the most important single factor to increase yeast yield and numerous studies have been carried out to investigate the optimisation of particular technological solutions. It should be underlined, the particular uniqueness of yeast metabolism—baker's yeast must exhibit efficient respiratory metabolism during yeast manufacturing, which determines biomass yield, but at the same time, cells must possess strong fermentative potential in order to produce excellent bakery products. During the fermentation of dough yeast cells is exposed to numerous environmental stresses (baking associated stresses) such as freeze thaw, high sugar concentrations, air drying, and oxidative stresses. Nor should it be surprising that industrial starters should be characterised by appropriate stress tolerance. Yeasts certainly have evolved some mechanisms of adaptation, but if the level of stress will increase too much, their enzymatic potential will be restricted. It has been demonstrated that two molecules: trehalose and proline are extremely important for yeast stress tolerance, so the engineering of their metabolism is a promising approach to the development of stress-tolerant yeast strains relevant to industrial use.

Next long-standing industrial processes involving yeast are the production of single-cell protein (SCP)—alternative source of high nutritional value proteins used as a food or feed supplements. Idea of such protein concentrate production was born in response to growing human population in the world and worldwide protein deficiency. SCP manufacture includes simply the cell mass obtaining by way of the application of cheap, waste raw materials to cultivate various nonpathogenic micro-organisms (bacteria, fungi, algae) under conditions of submerged (rarely solid-state) fermentation. Besides its high protein content (about 60–82% of dry matter), SCP also contains fats, carbohydrates, nucleic acids, vitamins, and minerals. In practice, several technologies were evaluated, products commercialised and currently obtained SCP found the application primarily in animal feeding. However, baker's yeast mentioned above can be considered also as an example of particular SCP preparation. Many fungal species are used as protein-rich food. Most popular among them are the yeast species Candida, Hansenula, Pichia, Torulopsis, and *Saccharomyces*, but also, marine yeast is considered now as a valuable source of protein. Yeast-derived SCP has desired nutritional value and contains essential amino acids—above all sulphur containing amino acids. It is also significant that people accept yeast as a food source, because of its historical impact. The main disadvantage of this preparation limiting the utilisation as food is the nucleic acid content what is associated with additional purification step in downstream processing as well as with external mannoprotein layer of yeast cell wall what impedes the digestion.

Yeast can also be considered as an alternative source of lipids. Some species are capable of synthesis and accumulation of over 20% of biomass in form of neutral lipids and for that reason are called 'oleaginous.' Under optimal growth conditions and/or as a result of genetic improvement, the level of lipid accumulation can reach even 70%. Oleaginous yeast includes species of *Candida, Cryptococcus, Pichia (Hansenula), Lipomyces, Pseudozyma, Rhodosporidium, Rhodotorula, Trichosporon, Trigonopsis, Yarrowia,* and *Saccharomycopsis*. The ability of these yeast species to accumulate high quantities of lipids offers the commercial potential for production of single-cell oil (SCO). Microbial oils can serve both as alternative edible oils for food industry as well as substrates used in synthesis of the oleochemicals such as fuels, soaps, plastics, paints, detergents, textiles, rubber, surfactants, lubricants, additives for the food and cosmetic industry, and many other chemicals. Micro-organisms regarded as an alternative oils source cannot presently compete directly with plants, but their use has many advantages like shorter process cycle, independence of season and climate, facility for genetic improvement and the possibility to manufacture lipid of unique structure, and nonsynthesised by plants. It should be mentioned that the composition of yeast oils is similar to vegetable products, and predominant fraction of them is made of triacylglycerols (TGA) rich in polyunsaturated fatty acids. Lipids are stored intracellularly mainly as granular forms called 'lipid body,' and their content and profile of fatty acids differs between species. The main fatty acids formed by oleaginous yeast are the myristic (C14:0), palmitic (C16:0), stearic (C18:0), oleic (C18:1), palmitoleic (C16:1), and linoleic (C18:2) acids. Two different pathways are involved in lipid accumulation by oleaginous yeast: *de novo* lipid synthesis and ex novo lipid accumulation. Fundamental differences at the biochemical level exist between *de novo* lipid accumulation from hydrophilic substrates and ex novo lipid accumulation from hydrophobic substrates. In contrast, ex novo lipid production is a growth-associated process that occurs simultaneously with cell growth and is entirely independent of nitrogen exhaustion of the culture medium. The synthesis of ex novo lipids is the modification of fats and oils by oleaginous micro-organisms. The production of microbial oils became more important in the light of biodiesel production. The oleaginous yeast is considered as potential candidate for the production of '2nd generation' biodiesel deriving from lipid produced by oleaginous micro-organisms growing on wastes or agro-industrial residues like sewage sludge, hemicelluloses, hydrolysates, waste glycerol, cheese whey, etc. SCO may be produced via submerged fermentation performed under aerobic conditions through the mode of batch, fed batch, or continuous operation. To achieve both sufficient biomass formation and proper lipid accumulation level, yeast must be cultivated under carefully evaluated conditions. Many factors have been described as influencing lipid accumulation process and proper medium composition seemed to be one of the most important ones. After biomass had been formed, to promote lipid accumulation process, stress conditions must be induced. Lipid storage is triggered by a nutrient limitation combined with an excess of carbon. Mostly nitrogen limitation is used to trigger lipid accumulation, but also other nutrients as phosphorus and sulphur have been shown to induce lipid storage. In spite of the applicative potential, the commercialisation of microbial products is limited to oils containing polyunsaturated fatty acids like docosahexanoic acid (DHA), arachidonic acid (ARA), and eicosapentaenoic acid (EPA). Producing other microbial oils either for human consumption, industrial use, or for biofuel production is still cost inhibitory—currently, the production of microbial oils is more expensive than that of vegetable oils. To extend the industrialisation of yeast as alternative source of oils, all efforts should be focused on genetic improvement of cell factories combined with optimisation of nutrient supply and cultivation conditions and modifications of downstream technology.

YEASTS AS WHOLE-CELL BIOCATALYSTS

Different genera of yeasts are convenient biocatalysts applied in many fields of chemistry, especially for the synthesis of chiral building blocks and fine chemicals. They are interesting catalytic tools, not only for their varied enzymatic activities but also for their microbiological features such as simplicity of cultivation, low nutritional requirements, and adaptive capacities. These capacities result from their flexible metabolism, which responds to the environmental impacts, so the direction (also stereoselectivity) and the effectiveness of the biotransformations can be driven by the physical chemical parameters of the process. What is also important, they are susceptible to the engineering of the reaction media (e.g. water, organic) and to the biocatalysts form (e.g. permeabilisation, immobilisation). This significantly broadens the field of their application by overcoming the limitations such as low solubility of the bioconversion substrates. It can be said that yeasts are used for decades and are one of the first whole-cell biocatalysts applied in industrial processes.

Literature data proved that the core applications of yeasts are connected with their extraordinary reductive abilities. Since it was proven that whole-cell biocatalysis is (as enzymatic one) chemoselective, regioselective, and stereoselective and able to regenerate dehydrogenases cofactors under biocatalytic conditions, yeasts were extensively examined as reductive catalysts for chiral building blocks synthesis—especially chiral alcohols of defined absolute configuration. The activity of a number of yeasts genera has been tested toward structurally different ketones, and in the most cases desired, alcohols have been obtained as pure enantiomers. Alcohols drawn below represent both, simple, and more complicated—unusual structures obtained thanks to the whole-cell biocatalysis driven by yeasts.

Pichia methanolica is one of the most important biocatalysts employed in industry because of its reductive ability and low substrate specificity. Thus, Pichia methanolica mediated reduction of ethyl-5-oxo-hexanoate and 5-oxohexanenitrile led to S-alcohols: 1a and 1b obtained with conversion degree up to 90 and 97% of ee (enantiomeric excess). What is also important, bioreduction of compound 1a was performed successfully on the gram scale with the similar yield. Pichia methanolica is employed for the reduction purposes by Bristol-Mayers Squibb Company (USA). Alcohol 2 is important for pharmaceutical industry, this molecule is considered as supporting drug in the diabetes type 2 therapy by elevating the rate of metabolisms. Application of Candida sorbophila MY 1833 allowed receiving pure product on the large scale level with up to 60% of conversion degree and 98% of ee. Optically pure diols are also important, mainly as chiral building blocks for pharmaceuticals and fine chemicals synthesis. As an example—compound 3—2S, 5S–hexanediol can be received by diketone reduction performed with *Saccharomyces cerevisiae* on preparative scale, with the complete substrate reduction and with the de (distereomeric excess) of 96%.

Except to mentioned diol, such procedure can be applied also for 5-hydroxyhexane-2-on formation, which is a substrate for chiral tetrahydrofuranes synthesis—chemicals of significant meaning for obtaining of the biodegradable polymers, drugs, and perfumes. Among different chiral compounds with hydroxyl functionalities, there are some of special interest—they are building platforms for the synthesis of series compounds. Such importance can be attributed to the (S)-4-chloro-3-hydroxybutanoate ester (S)-CHBE. A number of fungal catalysts, e.g. *Geotrichum* sp., *Candida* sp., *Aureobasidium* sp. are active toward appropriate ketone, but the reaction proceeds with an average stereoselectivity. Among others, *Pichia stipites* CBS 6054 was found as one allowed obtaining chiral product with ee of 80%. The other products of yeast-mediated enantioselective bioreduction are chiral bicyclic alcohols, e.g. synthons of (S)-pramipexole (anti-Parkinson drug) and also its (R) isomer—considered as an anti-amyotrophic lateral sclerosis (ALS) agent. Chirality is introduced into these molecules by the prochiral bicyclic ketone reduction mediated by *Sachcaromyces cerevisiae*, and this is the crucial step of the sequences of the reaction leading to the final enantiomers of pramipexole. Discussed objective is also an example of the yeast biotransformation of the compounds with heteroatoms in their structures—here, sulphur atom is an element of the bicyclic.

Reductive activity of *Saccharomyces cerevisiae* is also applied for some nontypical reactions such as geraniol into citronellol hydrogenation achieved with resting yeast cells on preparative level. Process productivity reaches 2.38 g/L for the reaction carried out in the continuous-closed-gas-loop bioreactor.

Entirely a different activity of yeasts as biocatalysts found some practical application, lately. Successful experiments with *Saccharomyces cerevisiae* were performed according to the addition of diverse 3-substituted indol derivatives to nitroolefines. Chemical synthesis of such substituted indols requires some hazardous organic solvents, while biological synthesis significantly reduces this problem and offers easy operated and effective system (yield of the reaction range between 72 and 90%). Such indol motifs are crucial part of the biologically active compounds such as arbidol (influenza A and B virus treatment and prophylactic), golotimod (immunostimulating, antimicrobial, and antineoplastic agent), and panobinostat (acute myeloid leukemia treatment).

YEAST'S ENZYMES APPLICATIONS

Biocatalysis includes both biotransformations (e.g. the conversion of xenobiotics using whole cells or resting cell systems) and enzyme catalysis (e.g. the conversion of xenobiotics using cell-free extracts or purified enzymes). Although both whole cells and isolated enzymes can be used as biocatalysts, whole cells are very often preferable because they are more stable and cheap sources than purified enzymes, without the need for purification and coenzyme addition.

However, in the case of single-step biotransformation, isolated enzymes can be considered as a better choice and can be used as a free or immobilised biocatalyst either in aqueous or organic media. Yeasts, especially *Saccharomyces* species, are primarily known from whole cell reductive activity and are used in the food industry for the production of alcoholic beverages as well as for bread fermentation. However, yeasts are a source of enzymes such as: lipases, dehydrogenases, or invertase.

Yeast's Lipases

Lipases are widely distributed in nature and are produced by plants, animals, and micro-organisms. Microbial enzymes are more useful than the other ones because of the diversity of catalytic activities, simplicity of manipulations, and low cost production (extracellularly during rapid growth on inexpensive media). Additionally, microbial enzymes are free from problems associated with contamination with hormones, viruses, and can be used in food processing and pharmaceutics productions for vegetarian or kosher diets. Microbial lipases (EC 3.1.1.3) are suitable enzymes for organic synthesis because they are active toward broad range of nonphysiological substrates and are stable in biphasic systems or pure organic media. Lipases can be applied for either of lipid modifications and synthesis of special compounds: pharmaceuticals, polymers, biodiesels, and biosurfactants. Under physiological conditions, lipases catalyse hydrolysis of ester bond in triacylglycerol to glycerol and free fatty acids. Under nonaqueous conditions, they catalyse the reverse process—esterification. The term transesterification refers to exchange the group between an ester and acid, ester and alcohol, or at least between two esters.

Mentioned features make them significant biocatalyst for various applications. There are a certain number of yeast species able to produce lipases, most of them belong to Candida genus: *Candida utilis*, *C. rugosa* (*cylindracea*), *C. antarctica*, *C. viswanathii*, and additionally, *Yarrowia lipolytica*.

Hydrolysis

For pharmaceutical industry, lipases are used to resolve racemic mixtures of alcohols or carboxylic acids through asymmetric hydrolysis of acyl derivatives. Candida antarctica lipase, isoform B (CAL-B) is one of the most employed psychrophilic lipases for many different applications (kinetic resolutions, desymmetrisation, aminolysis, etc.). Commercial CAL-B is available either in free, lyophilised, and immobilised forms (onto Lewatit VP OC 1600 (poly(methyl methacrylate-co-divinylbenzene)—

Novozyme 435, Chirazyme L2-C2). Novozym 435 is a suitable catalyst for both small organic molecules and for polymerisation reactions. Also, the immobilisation of CAL-B onto different supports may result in different activity and enantioselectivity and may be a tool of control of selectivity of the hydrolysis. This feature was used for the resolution of racemic mixture of 2-O-butyryl-2-phenylacetic acid—precursor of both enantiomers of mandelic acid and for the enantioselective hydrolysis of 3-phenylglutaric dimethyl diester—precursor in the drug synthesis (e.g. HIV inhibitor). It is possible to change the enantioselectivity of biocatalyst just by simple replacement of one support material to another one.

Another possible way to change the enantioselectivity of hydrolysis is the addition of the organic co-solvent to reaction medium. In organic media, the conformation of enzyme appears to be more rigid which may influence the enantioselectivity of the reaction.

For the Candida rugosa, lipase-catalysed hydrolysis of various substituted phenoxypropionates, the addition of 30–70% dimethyl sulphoxide (DMSO) or sodium dodecyl sulfate (SDS) improved the enantioselectivity, (E = 4 to >100). For the same lipase (*C. rugose*), complete reversal of enantioselectivity of hydrolysis of 1,4-dihydropyridines was observed in different organic solvents saturated with water, this allowed to obtain both enantiomers of 1,4-dihydropiridine.

Candida antarctica produces two isoforms of lipases (A and B). However, more attention has been directed to the application of CAL-B, but in the last few years, also CAL-A has found remarkable applications. The most surprising biochemical property of CAL-A is its high termostability of over 90°C. It is quite strange that rather psychrophilic micro-organism produces thermostable enzyme. This feature allows using the CAL-A under unique reaction conditions and towards unusual substrates, especially sterically hindered tertiary alcohols and their derivatives or bulky cyclic compounds.

Bioconversion of tertiary alcohols can be useful for the removal of tert-butyl protecting group even under high temperature. Another interesting application of CAL-A is the regioselective hydrolysis of cyclic diacetates, useful building blocks for the synthesis of vitamin D3 derivatives, or hydrolysis of different sterically hindered carboxylic acids.

Esterification and transesterification

The most important application of lipases in organic synthesis is esterification important for the resolution of racemic mixtures of secondary alcohols and carboxylic acids. Chiral secondary alcohols serve as intermediates for pharmaceutical synthesis. Lipase-catalysed methods available for the preparation of enantiopure compounds are kinetic resolution (KR), dynamic kinetic resolution (DKR), and desymmetrisation. Enzymatic kinetic resolution is based on the difference between the reaction rates of the enantiomers of a racemate at the presence of chiral catalyst—enzyme. Dynamic kinetic resolution combines kinetic resolution with the in-situ racemisation of the unreacted enantiomer. Racemisation can be performed chemically or enzymatically. The kinetic resolution of secondary alcohols and esters is carried out in organic solvents with lipase catalysed acylation and alcoholysis. It leads to the formation of one enantiomer obtained as an alcohol and the other one as an ester. The maximum theoretical yield for each enantiomer is 50%. *C. antarctica* enzymes are often used for the resolution of secondary alcohols, mostly with bulky group but also for resolution of aliphatic compounds. As an example— CAL-B was applied for the resolution of racemic mixture of 2-pentanol in heptane—media (yield (49,6%) and ee >99%). S-(+)-2-pentanol is a key chiral intermediate for synthesis of anti-Alzheimer drugs, which inhibit β-amyloid peptide release and/or its synthesis. *C. rugosa* is the producer of lipase employed for the resolution of profens (2-aryl propionic acids) in enantioselective transesterification process. Profens are an important group of nonsteroidal anti-inflammatory drugs, and their biological activity depends on the optical purity of the compounds, mainly (S)-enantiomer. For instance, (S)-ibuprofen ((S)-2(4-isobutylphenyl) propionic acid) is 160 times more effective than (R)- isomer in the inhibition of prostaglandins synthesis. Optically, pure profens can be synthesised by asymmetric chemical synthesis, catalytic kinetic resolution, and chiral chromatography, but enzymatic enantioselective esterification seems to be the best method. Discussed reaction was carried out in saturated cyclohexane with 1-propanol or 2-propanol as acyl agents and completed with good conversion degree and excellent enantiomeric excess of (S)-ibuprofen.

Also, enantiomerically pure amines constitute a class of compounds with possible biological properties and industrial applications. Candida antarctica lipase B is one of the most effective catalyst in the preparation of enantiomerically pure nitrogenated compounds (e.g. amines, amides, amino acids, amino alcohols, etc.). This is achieved by enantioselective acetylation. For example, resolution of amino-alkylpyridines was most effective (conversion 50%, time 4h, ee of product, and substrate >99%) with the use CAL-B and ethyl acetate as an acyl donor in the tert-butyl methyl ether (TBME-medium).

CAL-A isoform of C. *antarctica* lipase is able to selectively acylate cyclic, sterically hindered structures via kinetic resolution alicyclic β-aminocarboxylic acids esters—building blocks for the synthesis of various pharmaceutical important heterocycles are synthesised this way. The best activity and enantioselectivity were observed in diethyl ether or diisopropyl ether with 2,2,2-trifluoroethyl hexanoate as an acyl donor. CAL-A is also active towards sterically hindered tertiary alcohols. This feature is quite unique among hydrolases. The first example of enantioselective kinetic resolution of racemic mixture of tertiary alcohol was acylation of 2-phenylbut-3-yn-2-ol. The reaction was quite enantioselective, but the yield was rather moderate (25%) because of the steric hindrance. Another interesting application of CAL-A is selective acylation of sterols, furyl substituted allyl alcohol, or cyanohydrins.

Yeast's Invertase

Invertase (β-fructofuranosidase- EC 3.2.1.26) catalyses hydrolysis of the glycoside bond from the terminal nonreducing beta-fructofuranoside side in disaccharide. It is also widely distributed in the environment, mainly in plants and micro-organisms. The most important application of invertase is production of invert syrup—equimolar mixture of fructose and glucose, released from sucrose, which is used in food and beverage industries. Monosaccharides mixture is sweeter than sucrose and hygroscopic, it mainly is used for production of soft-centered candies and fondants. Invertase is also applied for the manufacture of artificial honey, plasticising agents for cosmetics, pharmaceutical and paper industries, and enzyme electrodes for the detection of sucrose. Additionally, it can be applied for the synthesis of probiotic oligosaccharides like non-digestible oligosaccharides (NDO), e.g. lactosucrose. Commercially invertase is produced mainly by *Saccharomyces cerevisiae* (Baker's yeast) or *Saccharomyces* carlsbergensis. In yeast cells, invertase is produced either in intracellular or extracellular form.

Yeast's Oxidoreductases

Enantiometrically pure alcohols including α- and β-hydroxyesters are important and valuable intermediates in the synthesis of pharmaceuticals and other fine chemicals. Enantioselective ketone reductions are one of the most common methods applied for optically pure alcohols productions. Because reactions catalysed by dehydrogenases/reductases require cofactors (NADH or NADPH), the use of whole cells rather than isolated enzymes is preferred, to decrease the cost of enzyme purification and cofactor regeneration. However, isolated dehydrogenases employment decreased product purification problems.

Generally, α-ketoesters are reduced with lower enantioselectivities by whole yeast cells, so pair of purified reductases are selected to produce both enantiomers of (S)-ethyl-3-hydroxybutyrate (pharmaceutical building block) in optically pure forms on preparative scale. Reduction of β-ketoesters depends on both structure of substrate and specificity of the enzyme and usually yields in desired enantiomer of high optical purity.

Saccharomyces are also known as a producers of old yellow enzyme (OYE), the first discovered and characterised flavoprotein, which can be used for double bond reduction or for dismutation reactions toward cyclic substrates and also for the reduction of nitrate esters with the addition of coenzyme—NADPH. Other example of reductase is selective carbonyl reductase from Candida magnolia, active toward the structurally different ketones, reduced to the corresponding optically pure (R)-aryl and aliphatic alcohols. This enzyme also catalysis reduction of ketones with anti-Prelog enantioselectivity which is an unusual feature of bioreductions. Configuration of obtained aryl alcohols is mostly R—enantiomers but also strongly dependent on R group structure.

YEAST'S APPLICATIONS IN MOLECULAR BIOLOGY

Yeasts of the *Saccharomyces* genus, in particular *S. cerevisiae*, are one of the fundamental models for eukaryotic organisms, commonly used in genetic and molecular biology studies. *S. cerevisiae* is a unicellular organism that can be grown on defined media, which gives the complete control over its chemical and physical environment. Culturing yeast is simple, economical, and rapid and can be conducted under aerobic and anaerobic conditions. As a nonpathogenic and nontoxic organism, they are safe for laboratory work, without any special precautions. Big accessibility as well as easy culturing on both liquid and solid medium makes yeast cheap and handy organism with significant biotechnological capabilities.

Although yeast and humans have been evolving along separate paths for 1 billion years, still a substantial amount of yeast genes exhibit high homology to mammalian ones. Since the basic cellular mechanics of replication, recombination, cell division, and metabolism are generally conserved between yeast and larger eukaryotes, they constitute a good model for studying different processes such as aging, regulation of gene expression, signal transduction, cell cycle, metabolism, apoptosis, neuro-degenerative disorders, and many more. Furthermore, its protein expression systems have more in common with higher organisms than with prokaryotic ones, mainly due to the post-transcriptional and post-translational processing, which makes it a great candidate for acquiring a number of industrially or medically significant biomolecules, such as recombinant proteins for pharmaceutical purposes.

Life cycle of *Saccharomyces cerevisiae* strains include haploid and diploid phase, both of which typically grow asexually by budding. The cell cycle consists of four distinct phases (G1, S, G2, and M) and is regulated in a similar way to that of the cell cycle in larger eukaryotes. Haploid yeast cells can be either mating type a or a and under normal condition can mate together to generate a/a diploids. The diploid cells cannot mate but can reproduce asexually by budding like haploids. However, under specific circumstances, like unfavourable environment conditions (lack of nutrients), diploid cell can undergo meiosis to produce haploid spores. Subsequently, the newly produced haploid nuclei are packaged into four spores that contain modified cell walls, resulting in structures that are very resistant to environmental stress. Each single haploidic spore from tetrad arising after meiosis can be isolated and analysed by various micromanipulation methods. It provides a unique opportunity to study the coupling between genes among many others. Haploid states cell can be also used for recessive mutation studies, while diploid strains can be exploited for complementation tests.

S. cerevisiae have also been first eukaryotes whose genome has been fully sequenced and published in 1996. Its nucleus genome constitutes of 12068 kb organised into the haploid set of 16 chromosomes ranging in size from 200 to 1600 kb. The characteristic feature is that yeast genome is much more compact in comparison to other eukaryotic relatives, with genes representing 70% of total sequence. It possesses around 5885 potential protein-encoding genes, approximately 140 genes specifying ribosomal RNA, 40 genes for small nuclear RNA molecules, and 275 transfer RNA genes. Currently, an international, multidisciplinary team is involved in the production of 16 chromosomes of *S. cerevisiae* by synthetic biology tools, and the results are expected at the end of the year. Another highly unique and unusual, as for eukaryotes, feature of *S. cerevisiae* genome is the presence of DNA plasmids that enables a variety of genetic manipulations and are of great importance for modern molecular biology.

Techniques used for yeast transformation and specific selection have been well described. For this purpose, shuttle vectors are commonly used due to the fact that they can transform both yeast and bacteria, such as *Escherichia coli*. Various yeast strains carry different auxotrophic markers that can be generated by genetic engineering methods, for instance, by gene deletion in amino acid biosynthesis pathways. Scientists have developed a number of bifunctional vectors that are easy to isolate and can autonomously replicate in each yeast and bacteria cells.

Yeast's Plasmid Vectors

Saccharomyces cerevisiae is a very important micro-organism in modern biotechnology, not only for its contribution to brewing and bread-making industry, but also for showing great potential in the field of molecular biology and biomedicine due to its unique form of genetic material and protein expression systems. *S. cerevisiae* is one of very few eukaryotic organisms that contain circular DNA in the form of plasmids. Almost every strain of this yeast has the 2-μm plasmid, which can constitute the outstanding basis for cloning vectors, as it is 6 kb in size and is equipped with four following elements: origin of replication, genes REP1 and REP2 that code for proteins involved in replication process, FLP gene that is utilised by the plasmid to switch between isoforms, and gene D which role is not established yet. In order for 2-μm plasmid to work as a fully functional cloning vector, there has to be an incorporated element, called a selective marker, which allows for the transformed cells to be identified after cloning experiment.

Most of the bacterial vectors are provided with genes-encoding resistance to various kinds of antibiotics, such as ampicillin (ampR) or tetracycline (tetR). Therefore, upon culturing transformed cells in the medium with the addition of such antibiotic, the colonies that were unable to grow did not carry the resistance gene (hence, the uptake of the vector did not occur), thus, the only cells that survive are the transformed ones carrying the properly inserted vector.

Cloning techniques with yeast differ mostly in the strategic approach of the selective markers. In this case, usually a special kind of organism is required as the host, namely an auxotrophic mutant that is unable to obtain or synthesise a pivotal compound of one of its metabolic pathways. A good example is leu2- yeast that has an inactive form of LEU2 gene, hence, it cannot synthesise leucine and can only grow in a medium that is supplied with this amino acid. To properly use that organism as the host, a vector with LEU2 active gene has to be prepared. The cells are then transformed with the plasmid and cultured in minimal medium that lacks leucine. This way the only colonies that will grow will be the transformed cells.

There are few kinds of yeast cloning vectors, but all of them are so-called shuttle vectors, which means that they can replicate and be selected in both bacteria and yeast. Shuttle vectors were developed mostly because plasmid preparation from yeast only is highly ineffective, hence, the large-scale DNA propagation and convenient genetic manipulation are performed in bacterial organism, such as *E. coli*.

Yeast cloning vectors based on 2-μm plasmid are called yeast episomal plasmids (YEps). Depending on the kind of YEp, they can either contain most of the 2-μm plasmid or just the origin of replication, their backbone is usually constructed from *E. coli* vectors, such as pBR322 that contains genes encoding resistance for antibiotics, like ampicillin or tetracycline. As the name suggests, YEps can replicate independently, or they can be integrated into one of the yeast chromosomes. The most common reason for YEps integration is that they carry selective marker gene, which is very similar to the mutant chromosomal DNA of the host organism, for example, the already mentioned LEU2 gene or others, such as URA3, HIS3, TRP1, or LYS2 all involved in biosynthesis pathways of pyrimidine nucleotides, histidine, tryptophan, or lysine, respectively. YEps are considered as high copy number vectors, yielding up to 200 copies per cell with the transformation frequency between 10,000 and 100,000 transformed cells per μg. Unfortunately, the major drawback is that their recombinants are highly unstable, which makes it very difficult and time consuming to achieve conclusive and reliable results when working with YEps. Genetic structure of exemplary yeast vectors. AmpR and tetR are the antibiotic resinstance genes from *E. coli* pBR322 plasmid, whereas LEU2, URA3, TRP1 fragments represent yeast chromosomal DNA: (a) yeast episomal plasmid YEp13, (b) yeast integrative plasmid YIp5, (c) yeast replicative plasmid YRp7.

The other important type of yeast vectors are yeast integrative plasmids (YIps). They mainly consist of *E. coli* plasmid, such as pBR322 and a selective marker (usually one of the mentioned above). What is important is they do not contain yeast origin of replication, therefore, they cannot replicate in any other way than through the process of integration with chromosomal DNA. In terms of transformation frequency, YIps come at the very last place with the number significantly lower than 1000 transformed cells per μg and usually only one copy per cell. On the other hand, their recombinants are very stable, usually making them the top pick for the experimental purposes.

Yeast replicative plasmids (YRps) are another type of yeast cloning vectors. They contain a backbone from *E. coli* vector, the yeast origin of replication in close proximity to the selective marker. Such structure suggests that YRps can replicate independently with relatively high transformation frequency ranging from 1000 to 10000 transformed cells per μg and a number of copies between 5 and 100 per cell. YRps recombinants are as unstable as the YEps ones, making them one of the last choice for laboratory work.

Along with the time, there was a growing demand for much larger pieces of DNA to be manipulated through the techniques of genetic engineering. It was at this point that the last type of yeast cloning vectors was developed, namely the yeast artificial chromosomes (YACs). The general idea behind those constructs was that yeast chromosomes usually carry several hundred kilobases of genetic material, so why not imitate the native DNA? YACs were thought to contain the three key elements of a chromosome:

- Centromeres required for the proper chromosome positioning during the cell division.
- Origins of replication, which are the places on chromosome where the replication of genetic material starts.
- Telomers as the defenders of the chromosomes against exonucleases.

Several types of YACs have been developed over the years, they usually consist mostly of a *E. coli* pBR322 plasmid with some yeast genes, such as selective markers (usually at least two located oppositely), origin of replication, gene CEN4 coding for centromere region, the two telomeres fragments TEL, and an additional selective marker with a restriction enzyme site, such as SUP4 gene that compensates for a mutation-causing accumulation of a red pigment. YACs are equipped with a restriction enzyme site between two TEL sites, so that upon cleavage, an 'artificial' linear chromosome is created, subsequently SUP4 is cut, and the chromosome is divided into two arms, between which a DNA fragment to be cloned is ligated, thus recreating the single-line chromosome structure. The next step is to transform double auxotrophic mutant organisms that will not be able to survive in the minimum medium without the properly received YAC. Additional experimental control can be then tested by simple optical inspection—colonies with disrupted SUP4 gene will appear as white, meaning they are transformed, any other color means that the colony has not been properly transformed.

Yeast Expression System

Recombinant proteins are the biomolecules of great importance, because among other things, they are able to mimic the functions of native proteins, hence, they are extensively studied in biotechnology and biopharmaceutical research. The critical point of target protein production is the choice of efficient expression system which enables obtaining functional product with high yield.

Yeast expression system constitutes a good alternative for widely used bacterial and higher eukaryote expression systems. They are genetically well defined and are known to perform many post-translational modifications, including proper protein folding, disulphide bond formation, and glycosylation. The culturing of yeast is also easy, rapid, and cheap, which is their big advantage over the insect or mammalian cells. They easily undergo genetical manipulation and adapt to fermentation processes, therefore, using yeasts as a cell factory is convenient and enables to obtain a fair amount of target protein. In contrast to bacteria, recombinant proteins obtained in yeast expression systems are free of endotoxins that make this system safer, especially in terms of medical and food application.

In fact, about 20% of all biopharmaceuticals are produced by *S. cerevisiae*. Among them, the most dominant are insulin, human serum albumin, hepatitis vaccines, and virus-like particles used for vaccination against human papillomavirus.

However, yeast cells are limited in the production of human-like glycoproteins by their inability to produce complex *N*-linked glycans. In addition, *S. cerevisiae* produce the hypermannosylated *N*-linked glycans with the mannose residues being attached to the chitobiose core (a dimmer of β-1,4-linked glucosamine units). Hypermannosylation also results in a short half life *in vivo* and thereby compromises the efficacy of most therapeutic glycoproteins. To overcome this issue, great deal of effort has been put into altering the glycosylation pathways in *P. pastorsis* to produce strains possessing human-like *N*-glycosilation patterns. This achievement has contributed to the increased usefulness of yeast in industry for the production of stable and recombinant glycoproteins.

There is no ultimate procedure for yeast expression system that could work equally well for the production of all kinds of proteins. Optimisation of whole process is the critical step to obtain sufficient amounts of pure, properly folded and secreted protein of interest. While small and simple in structure proteins are easy to obtain, the big and multi-domain protein could require certain chaperones to facilitate the folding process. The advantage of yeast expression system is that it allows extracellular secretion of produced protein when proper signalling sequence has been attached to the structural gene. It significantly facilitates the recombinant protein purification process from the culturing medium and allows to optimise the culturing conditions. In order to increase protein secretion level, a few strategies have been developed. One of them is protein engineering of a desired product, for instance by modifying protein coding sequences and signalling sequences. Since this methodology is highly specific against each protein, the conditions optimised for one protein do not always work for another. Different approach is to engineer the host strains and tune-up folding and secretory machinery by over expression or deletion genes that are critical for the protein secretion. Additionally, it has been shown that expression in low temperatures enhances the level of secretion. There are numerous varieties of expression vectors available for producing heterologous proteins in yeast, and these are the derivatives of YIp, YEp, and YRp plasmids described previously. The DNA coding for the protein of interest is inserted into the vector. The type of selective marker and promoter strength are key factors that determine the plasmid copy number and the mRNA level of the recombinant protein. Varieties of inducible and constitutive promotors have been applied for gene expression in yeasts in the past.

The first of these allow the controllable gene expression. Most of inducible promoters are responsive to catabolite repression or react to other environmental conditions, like in-cell iron concentration, stress, or lack of essential amino acids. GAL promoter, which is induced by adding galactose, provides a straightforward system for expression regulation of the cloned foreign gene. Another good example can be CUP1 promoter, which is induced by copper or heat shock factor promoter, induced by heat stress at 39°C. There are also some other groups of promoters that initiate strong and constitutive expression. TEF1 promoter, as an example of *S. cerevisiae*, is a widely used representative of this group, as it can drive high gene expression in both high and low glucose conditions. Selection of a suitable promoter depends on specific process requirements and the properties of the target protein to be produced. Additionally, yeasts are recognised as a generally recognised as safe (GRAS) organism, which only strengthens its position as the most frequently used microbial eukaryote for recombinant protein synthesis.

Yeast's Two Hybrid System

Ever since the Field and Song discovery described in 1989, a new approach toward the examination of protein-protein interactions emerged, it was named as the yeast two-hybrid system. It allows to detect the interaction of two proteins in the yeast cell, and it can be used to select an interacting partner of a known protein. This technique takes the advantage of the fact that majority of eukaryotic transcriptional factors, such as Gal4p, consist of two independent, functional DNA domains: binding domain (BD) and transcription activation domain (AD). While the two domains are normally on the same polypeptide chain, the transcription factor also functions when these two domains are brought together by noncovalent protein-protein interactions. In yeast hybrid system, each of these domains is connected to the one from the studied protein. As a result, two fusion proteins are created: one combined to DNA-binding domain (BD) and the other joined to activation domain (AD). The genes coding for both fusion proteins are carried by different plasmids, but each plasmid undergoes expression in the same yeast cell. If the interaction between studied proteins occurs, BD and AD domains are close enough to activate transcription of a reporter gene that is regulated by the transcription factors.

General idea of the yeast two-hybrid system can be represented by an example of transcriptional factors and a gene coding for β-galactosidase, wherein the interaction between studied proteins may potentially lead to the expression of the reporter gene coding for β-galactosidase in *E. coli* lacZ reporter gene. The presence of this enzyme can later be verified by simple reaction with X-gal, which yields a blueish insoluble product, thereby confirming or denying the association of studied proteins. Since 1989, yeast two-hybrid system has been studied extensively and further developed to find countless new applications, some of which are summarised and generally described in Ref.

Discussed unique yeasts features, which are fundamental for their versatile applications are still examined and after finishing the 'Synthetic Yeast Genome Project (Sc2.0)' the new perspectives of the applying them will be opened as well as in the molecular biology and in the industrial applications.

Chapter 25

Biotechnology of Brewer's Yeast

INTRODUCTION

Yeasts are eukaryotic micro-organisms classified as members of the fungus kingdom with 1500 species currently identified and are estimated to constitute 1% of all described fungal species. Yeasts are unicellular, although some species may also develop multicellular characteristics by forming strings of connected budding cells known as pseudohyphae or false hyphae. Yeast sizes vary greatly, depending on species and environment, typically measuring 3–4 μm in diameter, although some yeasts can grow to 40 μm in size. Most yeasts reproduce asexually by mitosis, and many do so by the asymmetric division process known as budding.

Yeasts do not form a single taxonomic or phylogenetic grouping. The term 'yeast' is often taken as a synonym for *Saccharomyces cerevisiae*, but the phylogenetic diversity of yeasts is shown by their placement in two separate phyla: the *Ascomycota* and the *Basidiomycota*. The budding yeasts ('true yeasts') are classified in the order *Saccharomycetales*.

By fermentation, the yeast species *Saccharomyces cerevisiae* converts carbohydrates to carbon dioxide and alcohols–for thousands of years the carbon dioxide has been used in baking and the alcohol in alcoholic beverages. Other species of yeasts, such as Candida albicans, are opportunistic pathogens and can cause infections in humans.

BIOTECHNOLOGY OF BREWER'S YEAST

The progress of chemistry, physiology and microbiology during the 19th Century, allowed a scientific approach to brewing that caused a tremendous advancement on the production of beer. Louis Pasteur demonstrated that alcoholic fermentation is a process caused by living yeast cells. His conclusion was that fermentation is a physiological phenomenon by which sugars are converted in ethanol as a consequence of yeast metabolism.

GENETIC CONSTITUTION OF BREWER'S YEAST

Saccharomyces cerevisiae is one of the best genetically characterised yeast as its genome is fully sequenced and analysed exhaustively. Being a eukaryotic, the key of its success lies in the selection of

337

a model strain with a perfect heterothallic life cycle. In contrast, brewer's yeast is refractory to the genetic procedures used with laboratory strains. The main reason is its low sexual fertility. Like most other industrial yeast, brewing strains do not sporulate or do so with low efficiency. Even in those cases that they show a suitable sporulation frequency, most spores are not viable.

Strain Types

There are basically two kinds of yeast used in brewing that correspond to the ale and lager types of beer. Ale beer is produced by a top-fermenting yeast that works at about room temperature, ferments quickly, and produces beer with a characteristic fruity aroma. The bottom-fermenting lager yeast works at lower temperatures, about 10–14°C, ferments more slowly and produces beer with a distinct taste. The vast majority of beer production worldwide is lager. It is difficult to make generalisations concerning the yeast strains used for the industrial production of beer, since they are generally ill characterised and very few comparative studies have been reported. Bottom fermenting, lager strains are usually labelled *Saccharomyces carlsbergensis*. Although strains from different sources show differences regarding cell size, morphology and frequency of spore formation, it is unlikely that these differences reflect a significant genetic divergence. Only one strain, Carlsberg production strain 244, has been extensively analysed and most of the studies described in the following sections have been conducted with this strain.

Genetic Crosses

Early attempts to carry out conventional genetic analysis with brewer's yeast faced the problems of poor sporulation and low viability. To overcome this difficulty, several researchers hybridised brewing strains with laboratory strains of *S. cerevisiae*. Notwithstanding the poor performance of brewing strains, viable spores were recovered from them. Some of the spores had mating capability and could be crossed with *S. cerevisiae* to generate hybrids easier to manipulate. The meiotic offspring of the hybrids was repeatedly backcrossed with laboratory strains of *S. cerevisiae* to bring particular traits of the brewing strain into an organism amenable to analysis. This procedure was followed to study flocculence, an important character in brewing. Gjermansen and Sigsgaard carried out a detailed analysis of the meiotic offspring of *S. carlsbergensis* strain 244. They obtained viable spore clones of both mating types. Celllines with opposite mating type were crossed pairwise to generate a number of hybrids that were tested for brewing performance. One of them was as good as the original strain. Additionally, the clones derived from strain 244 with mating capability served as starting material for further genetic analysis which are described in the following section.

The kar mutations have been particularly useful tools to investigate cytoplasmic inheritance. Additionally, the kar mutations supplied new genetic techniques. For instance, the chromosome number of virtually any *Saccharomyces* strain can be duplicated upon mating with a *kar*2 partner. These new tools and techniques opened a new way for the characterisation of the brewer's yeast. Since the brewing strain does not mate normally, the strain used in *kar* crosses was a meiotic derivative of strain 244 with mating capability. When disomic strains for chromosome III (also referred to as chromosome addition strains) were crossed to haploid *S. cerevisiae* strains, normal spore viability was obtained, allowing tetrad analysis. In this process, one of the two copies of chromosome III can be lost. If the original *S. cerevisiae* copy is lost, the result is a 'chromosome substitution strain' carrying a complete *S. cerevisiae* chromosome set, except chromosome III, which comes from *S. carlsbergensis*. Meiotic analysis of crosses between chromosome III addition strains and laboratory strains of *S. cerevisiae* revealed two important facts: (i) the functional equivalence of chromosome III for the brewing strain and *S. cerevisiae*, since ascospore viability and chromosome segregation were normal, and (ii) in grite of the functional equivalence, the two copies of chromosome III were different since the overall frequency of recombination between them was much lower than that expected for perfect homologues. The new procedure allowed the analysis of entire chromosomes from the brewing strain, placed into a laboratory yeast that could easily be manipulated genetically.

Molecular analysis

A clear picture of the genetic composition of *S. carlsbergensis* emerged from Southern hybridisation experiments and from the first gene sequences from this yeast. Nilsson-Tillgren reported the transfer of a chromosome III from the brewing strain to *S. cerevisiae*. Determination of the nucleotide sequence of a number of *S. carlsbergensis* genes provided a precise characterisation of the difference between the two types of homologous alleles present in the brewing yeast.

Ploidy

Finding a sound answer for the long-standing question of how many chromosomes are contained in brewer's yeast, has taken a long time. The relative DNA content of *S. carlsbergensis* 244 has been recently determined by flow cytometry. Results obtained show that the genetic constitution of this strain must be close to tetraploidy.

Origin of brewing strains

The hybrid nature of the brewing yeast explains its poor sexual performance. Divergence between homologous sequences impairs chromosome pairing and recombination, which are requisites for a proper meiotic function. Sexual reproduction appears in Evolution as a mechanism that recombines the genetic material of organisms to generate variability. It offers adaptive advantages to a changing environment through the random generation of new genotypes. On the contrary, abolition of sex is advantageous when the purpose is to keep unchanged a given property. The maintenance over the centuries of a brewing procedure to produce beer with particular organoleptic properties likely caused the selection of a particular type of yeast. The hybrid, vegetative vigor of this yeast assured a good fermentative capability, whereas its sexual infertility would keep fixed the genetic constitution responsible for the 'good beer' phenotype. Sequence analysis shows that one of the two parental species that generated *S. carisbergensis* was *S. cerevisiae*, but the precise identification of the other contributor is less clear. Thus, lager strains of different origin, labelled *S. carlsbergensis*, could be independently generated hybrids of slightly different genetic constitution.

Genetic Manipulation

Yeast and barley play an active, primary role in the brewing process. The other two beer ingredients, water and hops, have secondary roles. Yeast is the fermenting agent, which transforms the carbohydrates stored in the grain of barley into ethanol. It produces a battery of compounds that ultimately result in the aroma and flavour of the beer. Barley is not solely a source of fermentable sugars. During the process of malting, cells in the germinating barley seeds secrete enzymes that are required to digest the starch into simpler sugars, mainly maltose and glucose, which can be assimilated by the yeast. Many properties of barley, in particular those affecting its carbohydrate content and composition, but also other characteristics, are very important for the quality of beer. Genetic engineering can be used to modify the properties of yeast and barley in ways that improve their performance in brewing. Different experimental approaches directed to the modification of the brewer's yeast, to produce beer with better properties or new characteristics. In most cases, technical advances allow the construction of new strains of yeast with the desired properties. Currently however, public concern about the use of genetically modified food poses a barrier to the industrial use of these strains.

Accelerated maturation of beer

The production of lager beer comprises two separate fermentation stages. The main fermentation, in which the fermentable sugars are converted in ethanol, is followed by a secondary fermentation, referred to as maturation or lagering. The most important function of maturation is the removal of diacetyl, a compound that causes an unwanted buttery flavour in beer. Diacetyl is formed by the spontaneous (non-enzymatic) oxidative decarboxylation of α-acetolactate, an intermediate in the biosynthesis of valine. In yeast, as in other organisms, the two branched-chain amino acids, isoleucine and valine, are synthesised in an unusual pathway in which a set of enzymes, acting in parallel reactions, lead to the formation of different end products. Like diacetyl is formed as a by-product of valine biosynthesis, a related compound, 2-3-pentanedione, is formed by decarboxylation of α-aceto-α-hydroxybutirate in the isoleucine biosynthesis. Both compounds, diacetyl and α-aceto-α-hydroxy-butirate produce a similar undesirable effect in beer, although much more pronounced in the case of diacetyl.

Together, they are referred to as vicinal diketones. Diacetyl is converted to acetoin by the action of diacetyl reductase, an enzyme from the yeast. The maturation period, which lasts several weeks, assures the conversion of the available a-acetolactate into diacetyl and the subsequent transformation of diacetyl into acetoin. The amounts formed of this last compound do not have a significant influence on beer flavour. Preventing diacetyl formation would reduce or even make unnecessary the lagering period. This would represent a considerable benefit for the brewing industry.

Different approaches have been devised to eliminate diacetyl. A first one requires the manipulation of the isoleucine-valine biosynthetic pathway, either by blocking the formation of the diacetyl precursor α-acetolactate, or by increasing the flux of the pathway at a later stage, channelling the available α-acetolactate into valine before it is converted into diacetyl. Masschelein and collaborators were first to suggest that a deleterious mutation of the brewer's yeast *ILV2* gene would solve the diacetyl problem. This gene encodes the enzyme acetohydroxyacid synthase, which catalyses the synthesis of α-acetolactate, from which diacetyl is formed.

This or any alternative action on the valine pathway requires the manipulation of specific genes encoding enzymes of the pathway. These genes have been cloned from *S. cerevisiae* and characterised. *S. carlsbergensis*-specific alleles of the *ILV* genes from the brewer's strain have also been cloned. Because of the genetic complexity of the brewing strain (a hybrid with about four copies of each gene, two from each parent), the abolition of the *ILV2* function requires the very laborious task of eliminating each of the four copies of the gene present in the yeast.

An alternative could be to boost the activity of the enzymes that direct the following steps in the conversion of α-acetolactate into valine: the reductoisomerase, encoded by *ILV5* and possibly the dehydrase, encoded by *ILV3*. To achieve the desired effect, it could be sufficient to manipulate only one of the four copies of the *ILV* genes present in the brewer's yeast. A clever procedure to inhibit the *ILV2* function, by using an antisense RNA of the gene, has been reported. However, a later note from the same laboratory stated that the reported results were incorrect. Another approach makes use of an enzyme, α-acetolactate decarboxylase, which catalyses the direct conversion of acetolactate into acetoin, bypassing the formation of dyacetyl. This enzyme is produced by different micro-organisms.

Beer Attenuation and the Production of Light Beer

Conversion of barley into wort that can be fermented requires two previous processes: Malting and mashing. During malting, the barley grain is subjected to partial germination, achieved by moistening, and subsequent drying. Germination induces the synthesis of amylase and other enzymes that allow the seed to mobilise its reserves. The dried malt is milled and the resulting powder is mixed with water and allowed to steep at warm temperatures. During mashing, amylases digest the seed's starch, liberating simpler sugars, chiefiy maltose.

This process is critical, since the brewing yeast is unable to hydrolyse starch. The enzymatic action of barley's amylases on starch yields fermentable sugars, but also oligosaccharides (dextrins) which remain unfermented during brewing. Dextrins represent an important fraction of the caloric content of beer. In current brewing practice, it is quite common to add exogenous enzymes. Thus glucoamylase can be added to the mash to improve the digestion of the starch. If the enzymatic treatment is carried out exhaustively, the dextrins are completely hydrolysed, and the result is a light beer with substantially lower caloric content, for which there is a significant market demand in some parts of the world. A convenient alternative to the addition of exogenous glucoamylase is to endow the brewer's yeast with the genetic capability of synthesising ibis enzyme. A variety of *S. cerevisiae*, formerly classified as a separate species (*S. diastaticus*), produces glucoamylase. Because of its close phylogenetic relationship with the brewing yeast, *S. diastaticus* is an obvious source of the glucoamylase gene.

The percentage of the sugar in the wort that is converted into ethanol and CO_2 by the yeast is called attenuation. Microbial contamination of beer is often associated with a pronounced increase in the attenuation value, which is known as superattenuation. This effect is due to the fermentation of dextrins, which are hydrolysed by amylases produced by the contaminant micro-organisms. *S. diastaticus* was characterised as a wild yeast that caused superattenuation. Similarly to the synthesis of invertase or maltase by *Saccharomyces*, the synthesis of glucoamylase is controlled by a set of at least three polymeric genes, designated *STA1*, *STA2* and *STA3*. This genetic system is complicated by the existence in normal *S. cerevisiae* strains of a gene, designated *STA10*, which inhibits the expression of the other *STA* genes. Recently, the *STA10* gene has been identified with the absence of *FLO8p*, a transcriptional regulator of both glucoamylase and flocculation genes. The sequence of the *STA1* gene was first determined by Yamashita and others. Different species of filamentous fungi, in particular some of the genus *Aspergillus*, produce powerful glucoamylases. The gene that encodes the enzyme of *A. awamori* has been expressed in *S. cerevisiae*. Available information about the genetic control of glucoamylase production by *Saccharomyces* and current technology makes the construction of brewing strains with this capability relatively easy.

Beer filterability and the action of β-glucanases

Brewing with certain types or batches of barley, or using certain malting or brewing practices, can yield wort and beer with high viscosity, very difficult to filtrate. When this problem arises, the beer may also present hazes and gelatinous precipitates. Scott pointed out that this problem was caused by a deficiency in β-glucanase activity. The substrate of this enzyme, β-glucan, is a major component of the endosperm cell walls of barley and other cereals. During the germination of the grain, β-glucanase degrades the endosperm cell walls, allowing the access of other hydrolytic enzymes to the starch and protein reserves of the seed. Insufficient β-glucanse activity during malting gives rise to an excess of β-glucan in the wort, which causes the problems. The addition of bacterial or fungal β-glucanases to the mash, or directly to the beer during the fermentation, is a common remedy. The construction of a brewing yeast with appropriate β-glucanase activity would make unnecessary the treatment with exogenous enzymes. Suitable organisms to be used as sources of the β-glucanase gene are *Bacillus subtilis* and *Thricoderma reesei*, from which the commercial enzyme preparations used in brewing are prepared. The genes from both have been characterised and brewer's yeast expressing β-glucanase activity have been constructed. An alternative is to make use of the gene encoding barley β-glucanase, the enzyme that naturally acts in malting. This gene has been characterised and expressed in *S. cerevisiae*. However, the barley enzyme has lower thermal resistance than the microbial enzymes, which is a limitation for its use against the β-glucans present in wort. Consequently, the enzyme has been engineered to increase its thermal stability.

Control of sulphite production in brewer's yeast

Sulphite has an important, dual function in beer. It acts as an antioxidant and a stabilising agent of flavour. Sulphite is formed by the yeast in the assimilation of inorganic sulphate, as an intermediate of the biosynthesis of sulphur-containing amino acids, but its physiological concentration is low. Hansen and Kielland-Brandt have engineered a brewing strain to enhance sulphite level to a concentration that increases flavour stability. The formation of sulphite from sulphate is carried out in three consecutive enzymatic steps catalysed by ATP sulphurylase, adenylsulphate kinase and phosphoadenyl sulphate reductase. In *S. cerevisiae*, these enzymes are encoded by *MET3*, *MET14* and *MET16*. In turn, sulphite is converted firstly into sulphide, by sulphite reductase, and then into homocysteine by homocysteine synthetase. This last compound leads to the synthesis of cysteine, methionine and *S-adenosylmethionine*. It has been proposed that *S-adenosylmethionine* plays a key regulatory role by repressing the genes of the pathway. However, more recent evidence assigns this function to cysteine. Anyhow, because of the regulation of the pathway, yeast growing in the presence of methionine contains very little sulphite. To increase its production in the brewing yeast, Hansen and Kielland-Brandt planned to abolish sulphite reductase activity. This would increase sulphite concentration, as it cannot be reduced. At the same time, the disruption of the methionine pathway prevents the formation of cysteine and keeps free from repression the genes involved in sulphite formation.

Sulphite reductase is a tetramer with an $\alpha_2 \beta_2$ structure. The α and β subunits are encoded by the *MET10* and *MET5* genes, respectively. Hansen and Kielland-Brandt undertook the construction of a brewing strain without *MET10* gene function. The allotetraploid constitution of S. carlsbergensis made it extremely difficult to perform the disruption of the four functional copies of the yeast. Therefore, they used allodiploid strains, obtained as meiotic derivatives of the brewer's yeast. These allodiploids contains two homologous alleles of the *MET10* gene, one similar to the version normally found in *S. cerevisiae* and another which is *S. carlsbergensis*-specific. It is known that some allodiploids can be mated to each other to regenerate tetraploid strains with good brewing performance. The functional *MET10* alleles present in the allodiploids were replaced by deletion-harbouring, non-functional copies, by two successive steps of homologous recombination. New allotetraploid strains with reduced or abolished *MET10* activity were then generated by crossing the manipulated allodiploids. The brewing performance of one of these strains, in which the *MET10* function was totally abolished, met the expectations. Hansen and Kielland-Brandt have used another strategy to increase the production of sulphite which relies in the inactivation of the *MET2* gene function. The *MET2* gene encodes *O*-acetyl transferase. This enzyme catalyses the biosynthesis of *O*-acetyl homoserine, which binds hydrogen sulphide to form homocysteine. Similarly to the inactivation of *MET10*, inactivation of *MET2* impedes the formation of cysteine, depressing the genes required for sulphite biosynthesis.

Yeast flocculation

As beer fermentation proceeds, yeast cells start to flocculate. The flocs grow in size, and when they reach a certain mass start to settle. Eventually, the great majority of the yeast biomass sediments. This phenomenon is of great importance to the brewing process because it allows separation of the yeast biomass from the beer, once the primary fermentation is over the small fraction of the yeast that is left in the green beer is sufficient to carry out the subsequent step, the lagering. Flocculation is a cell adhesion process mediated by the interaction between a lectin protein and mannose. Stratford and Assinder carried out an analysis of 42 flocculent strains of Saccharomyces and defined two different phenotypes. One was the known pattern observed in laboratory strains that carried the *FLO1* gene. They found, in some ale brewing strains, a new flocculation pattern characterised by being inhibited by the presence in the medium of a variety of sugars, including mannose, maltose, sucrose and glucose, whereas the *FLO1* type was sensitive only to mannose. The genetic analysis of flocculation has revealed the existence of a polymeric gene family analogous to the *SUC*, *MAL*, *STA* and *MEL* families. The *FLO1* gene has been extensively characterised, which encodes a large, cell wall protein of 1,537 amino acids. The protein is highly glycosylated. It has a central domain harbouring direct repeats rich in serine and threonine (putative sites for glycosylation). Kobayashi and others have isolated a flocculation gene homolog to *FLO1* that corresponds to the new pattern described by Stratford and Assinder. This result is consistent with the hybrid nature of the brewing yeast. In addition to the structural genes encoding flocculins, other *FLO* genes play a regulatory role. For instance, the *FLO8* gene (alias *STA10*) encodes a transcriptional activator that in addition to flocculation regulates glucoamylase production, filamentous growth and mating.

Beer spoilage caused by micro-organisms

Microbial contamination of beer, caused by bacteria or wild yeast is a serious problem in brewing. To overcome the contamination, commonly sulphur dioxide and other chemicals are added, but this practice faces restrictive legal regulation and consumer rejection. An attractive alternative is to endow the brewing yeast with the capability of producing anti-microbial compounds. A specific example is the expression in *S. cerevisiae* of the genes required for the biosynthesis of pediocin, an antibacterial peptide from *Pediococcus acidilactici*. Another example is the transfer to brewing strains of the killer character, conferred by the production of a toxin active against other yeasts.

Enhanced synthesis of organoleptic compounds

The yeast metabolism during beer fermentation gives rise to the formation of higher alcohol, esters and other compounds which make an important contribution to the aroma and taste of beer. A first group of compounds important to beer flavour are isoamyl and isobutyl alcohol and their acetate esters. These compounds derive from the metabolism of valine and leucine. Two genes, *ATF1* and *LEU4*, encoding enzymes involved in the formation of these compounds, have been successfully manipulated to increase theirs synthesis. *ATF1* encodes alcohol acetyl transferase. It has been shown that its over-expression causes increased production of isoamyl acetate. *LEU4* encodes α-isopropylmalate synthase, an enzyme that controls a key step in the formation of isoamyl alcohol from leucine. This enzyme is inhibited by leucine. Mutant strains resistant to a toxic analog of leucine are insensitive to leucine inhibition. Mutants of this type, obtained from a lager strain, produce increased amounts of isoamyl alcohol and its ester.

Chapter 26

Biotechnology of Beer and Wine Industry

INTRODUCTION

Beer is an undistilled beverage produced from fermentation of barley malt by yeast especially *Sacccharomyces crevisiae* and *S. carsibergenesis*. Generally materials rich in starch like rice, wheat and maize are also added to increase the amount of fermentable sugars. They are called as adjuncts. The brewing of beer is a complex process that draws on a diversity of sciences and technology, of which chemistry is but one. This chapter focuses on the chemistry of the brewing process and of the finished product. It examines each of the main classes of molecule found in beer, considers their contribution to quality and their origins in the brewing process. The study of beer and its production provides an excellent illustrative example for teaching how raw materials and the manner by which they are processed determine the acceptability of a product. Beer, whilst 90% + water, contains a wide range of chemical species which establish its properties.

Apart from ethanol (the common denominator amongst all alcoholic beverages), beer contains substances that determine its flavor, foam, and colour. The flavorsome components of beer include the bitter iso-a-acids and aromatic essential oils from hops, along with esters, acids, sulphur-containing compounds and vicinal diketones from yeast. The foaminess of beer depends on the presence of carbon dioxide but also of surface-active materials like amphipathic polypeptides from malt and the bitter substances from hops.

The colour is due to Maillard reaction products generated largely during the kilning of malt. The malting and brewing processes (which are briefly described) are designed to maximise the extraction and digestion of barley starch and protein, yielding highly fermentable wort. The processes are also designed to eliminate materials that can have an adverse effect on beer quality, such as the haze-forming polyphenol from barley and hops and the lipids and oxygen that, together, can cause beer to stale.

The word beer comes from the Latin word *Bibere* (to drink). The basic ingredients for most beers are malted barley, water, hops and yeast, indeed, the 500-year-old Bavarian purity law (the *Reinheitsgebot*) restricts brewers to these ingredients for beer to be brewed in Germany. Most other brewers worldwide have much greater flexibility in their production process opportunities, yet the largest companies are ever mindful of the importance of tradition.

Compared to most other alcoholic beverages, beer is relatively low in alcohol. The highest average strength of beer [alcohol by volume (ABV) indicates the millilitres of ethanol per 100 ml of beer] in any country worldwide is 5.1 per cent and the lowest is 3.9 per cent. By contrast, the ABV of wines is typically in the range 11–15 per cent.

OVERVIEW OF MALTING AND BREWING

Brewer's yeast *Saccharomyces* can grow on sugar anaerobically by fermenting it to ethanol:

$$C_6H_{12}O_6 \rightarrow 2C_2H_5OH + 2CO_2$$

while malt and yeast contribute substantially to the character of beers, the quality of beer is at least as much a function of the water and, especially, of the hops used in its production.

Barley starch supplies most of the sugars from which the alcohol is derived in the majority of the world's beers. Historically, this is because, unlike other cereals, barley retains its husk on threshing and this husk traditionally forms the filter bed through which the liquid extract of sugars is separated in the brewery. Even so, some beers are made largely from wheat while others are from sorghum.

The starch in barley is enclosed in a cell wall and proteins and these wrappings are stripped away in the malting process (essentially a limited germination of the barley grains), leaving the starch largely preserved. Removal of the wall framework softens the grain and makes it more readily milled. Not only that, unpleasant grainy and astringent characters are removed during malting.

In the brewery, the malted grain must first be milled to produce relatively fine particles, which are for the most part starch. The particles are then intimately mixed with hot water in a process called mashing. The water must possess the right mix of salts. For example, fine ales are produced from waters with high levels of calcium while famous pilsners are from waters with low levels of calcium. Typically mashes have a thickness of three parts water to one part malt and contain a stand at around 65°C, at which temperature the granules of starch are converted by gelatinisation from an indigestible granular state into a 'melted' form that is much more susceptible to enzymatic digestion. The enzymes that break down the starch are called the amylases. They are developed during the malting process, but only start to act once the gelatinisation of the starch has occurred in the mash tun.

Some brewers will have added starch from other sources, such as maize (corn) or rice, to supplement that from malt. These other sources are called adjuncts. After perhaps an hour of mashing, the liquid portion of the mash, known as wort, is recovered, either by straining through the residual spent grains or by filtering through plates. The wort is run to the kettle (sometimes known as the copper, even though they are now-a-days fabricated from stainless steel) where it is boiled, usually for around 1 hr. Boiling serves various functions, including sterilisation of wort, precipitation of proteins (which would otherwise come out of solution in the finished beer and cause cloudiness), and the driving away of unpleasant grainy characters originating in the barley. Many brewers also add some adjunct sugars at this stage, at which most brewers introduce at least a proportion of their hops.

The hops have two principal components: resins and essential oils. The resins (so-called α-acids) are changed ('isomerised') during boiling to yield iso-α-acids, which provide the bitterness to beer. This process is rather inefficient. Now-a-days, hops are often extracted with liquefied carbon dioxide and the extract is either added to the kettle or extensively isomerised outside the brewery for addition to the finished beer (thereby avoiding losses due to the tendency of the bitter substances to stick on to yeast). The oils are responsible for the 'hoppy nose' on beer. They are very volatile and if the hops are all added at the start of the boil, then all of the aroma will be blown up the chimney (stack). In traditional lager brewing, a proportion of the hops is held back and only added towards the end of boiling, which allows the oils to remain in the wort. For obvious reasons, this process is called late hopping. In traditional ale production, a handful of hops is added to the cask at the end of the process, enabling a complex mixture of oils to give a distinctive character to such products. This is called dry hopping. Liquid carbon dioxide can be used to extract oils as well as resins and these extracts can also be added late in the process to make modifications to beer flavour.

After the removal of the precipitate produced during boiling ('hot break', 'trub'), the hopped wort is cooled and pitched with yeast. There are many strains of brewing yeast and brewers carefully look after their own strains because of their importance in determining brand identity. Fundamentally brewing yeast can be divided into ale and lager strains, the former type collecting at the surface of the fermenting wort and the latter settling at the bottom of a fermentation (although this differentiation is becoming blurred with modern fermenters).

Both types need a little oxygen to trigger off their metabolism, but otherwise the alcoholic fermentation is anaerobic. Ale fermentations are usually complete within a few days at temperatures as high as 20°C, whereas lager fermentations at temperatures as low as 6°C can take several weeks. Fermentation is complete when the desired alcohol content has been reached and when an unpleasant butterscotch flavour, which develops during all fermentations, has been mopped up by yeast. The yeast is harvested for use in the next fermentation.

In traditional ale brewing, the beer is now mixed with hops, some priming sugars and with isinglass finings from the swim bladders of certain fish, which settle out the solids in the cask.

In traditional lager brewing, the 'green beer' is matured by several weeks of cold storage, prior to filtering. Now-a-days, the majority of beers, both ales and lagers, receive a relatively short conditioning period after fermentation and before filtration. This conditioning is ideally performed at –1°C or lower (but not so low as to freeze the beer) for a minimum of 3 days, under which conditions more proteins drop out of the solution, making the beer less likely to cloud in the package or glass. The filtered beer is adjusted to the required carbonation before packaging into cans, kegs or glass or plastic bottles.

Barley

Although it is possible to make beer using raw barley and added enzymes (so-called barley brewing), this is extremely unusual. Unmalted barley alone is unsuitable for brewing beer because: (i) it is hard and difficult to mill, (ii) it lacks most of the enzymes needed to produce fermentable components in wort, (iii) it contains complex viscous materials that slow down solid-liquid separation processes in the brewery, which may cause clarity problems in beer, and (iv) it contains unpleasant raw and grainy characters and is devoid of pleasant flavours associated with malt.

The first stage in malting is to expose the grain to water, which enters an undamaged grain solely through the micropyle and progressively hydrates the embryo and the endosperm. This switches on the metabolism of the embryo, which sends hormonal signals to the aleurone layer, triggers that switch on the synthesis of enzymes responsible for digesting the components of the starchy endosperm. The digestion products migrate to the embryo and sustain its growth.

The aim is controlled germination, to soften the grain, remove troublesome materials and expose starch without promoting excessive growth of the embryo that would be wasteful (malting loss). The three stages of commercial malting are:

1. Steeping, which brings the moisture content of the grain to a level sufficient to allow metabolism to be triggered in the grain.

2. Germination, during which the contents of the starchy endosperm are substantially degraded ('modification') resulting in a softening of the grain.

3. Kilning, in which the moisture is reduced to a level low enough to arrest modification.

Mashing: The Production of Sweet Wort

Sweet wort is the sugary liquid that is extracted from malt (and other solid adjuncts used at this stage) through the processes of milling, mashing and wort separation. Larger breweries will have raw materials delivered in bulk (rail or road) with increasingly sophisticated unloading and transfer facilities as the size increases. Smaller breweries will have malt, etc. delivered by sack.

The conscientious brewer will check the delivery and the vehicle it came in for cleanliness and will representatively sample the bulk. The resultant sample will be inspected visually and smelled before unloading is permitted. Most breweries will spotcheck malt deliveries for key analytical parameters to enable them to monitor the quality of a supplier's material against the agreed contractual specification. Grist materials are stored in silos sized according to brewhouse throughput.

Milling

Before malt or other grains can be extracted, they must be milled. Fundamentally the more extensive the milling, the greater the potential there is to extract materials from the grain. However, in most systems for separating wort from spent grains after mashing, the husk is important as a filter medium. The more intact the husk, the better the filtration. Therefore, milling must be a compromise between thoroughly grinding the endosperm while leaving the husk as intact as possible. There are fundamentally two types of milling: dry milling and wet milling. In the former, mills may be either roll, disk or hammer. If wort separation is by a lauter tun (discussed later), then a roll mill is used. If a mash filter is installed, then a hammer (or disk) mill may be employed because the husk is much less important for wort separation by a mash filter. Wet milling, which was adopted from the corn starch process, was introduced into some brewing operations as an opportunity to minimise damage to the husk on milling. By making the husk 'soggy', it is rendered less likely to shatter than would a dry husk.

Mashing

Mashing is the process of mixing milled grist with heated water in order to digest the key components of the malt and generate wort containing all the necessary ingredients for the desired fermentation and aspects of beer quality. Most importantly it is the primary stage for the breakdown of starch. The starch in the granules is very highly ordered, which tends to make the granules difficult to digest. When granules are heated (in the case of barley starch beyond 55–65°C), the molecular order in the granules is disrupted in a process called gelatinisation. Now that the interactions (even to the point of crystallinity) within the starch have been broken down, the starch molecules become susceptible to enzymic digestion. It is for the purpose of gelatinisation and subsequent enzymic digestion that the mashing process in brewing involves heating. Although 80–90 per cent of the granules in barley are small, they only account for 10–15 per cent of the total weight of starch. The small granules are substantially degraded during the malting process, whereas degradation of the large granules is restricted to a degree of surface pitting. (This is important, as it is not in the interests of the brewer (or maltster) to have excessive loss of starch, which is needed as the source of sugar for fermentation.) The starch in barley (as in other plants) is in two molecular forms: amylose, which has very long linear chains of glucose units, and amylopectin, which comprises shorter chains of glucose units that are linked through side chains.

Several enzymes are required for the complete conversion of starch to glucose. α-Amylase, which is an endo enzyme, hydrolyses the $\alpha 1$–4 bonds within amylose and amylopectin. β-Amylase, an *exo* enzyme, also hydrolyses $\alpha 1$–4 bonds, but it approaches the substrate (either intact starch or the lower molecular weight 'dextrins' produced by α-amylase) from the non-reducing end, chopping off units of two glucoses (i.e. molecules of maltose). Limit dextrinase is the third key activity, attacking the $\alpha 1$–6 side chains in amylopectin.

α-Amylase develops during the germination phase of malting. It is extremely heat resistant, and also present in very high activity, therefore, it is capable of extensive attack, not only on the starch from malt but also on that from adjuncts added in quantities of 50 per cent or more.

Furthermore, it is developed much later than the other two enzymes, and germination must be prolonged if high levels of this enzyme are to be developed. It is present in several forms (free and bound): the bound form is both synthesised and released during germination. Like the proteinases, there are endogenous inhibitors of limit dextrinase in grain, and this is probably the main factor which determines that some 20 per cent of the starch in most brews is left in the wort as non-fermentable dextrins. Although it is possible to contrive operations that will allow greater conversion of starch to fermentable sugar, in practice, many brewers seeking a fully fermentable wort add a heat-resistant glucoamylase (e.g. from *Aspergillus*) to the mash (or fermenter). This enzyme has an *exo* action like β-amylase, but it chops off individual glucose units.

There are several types of mashing which can broadly be classified as infusion mashing, decoction mashing and temperature-programmed mashing. Whichever type of mashing is employed, the vessels these days are almost exclusively fabricated from stainless steel (once they were copper). What stainless steel loses in heat transfer properties is made up for in its toughness and ability to be cleaned thoroughly by caustic and acidic detergents. Irrespective of the mashing system, most mashing systems (apart from wet milling operations) incorporate a device for mixing the milled grist with water (which some brewers call 'liquor'). This device, the 'pre-masher', can be of various designs, the classic one being the Steel's masher, which was developed for the traditional infusion mash tun.

Infusion mashing is relatively uncommon, but still championed by traditional brewers of ales. It was designed in England to deal with well-modified ale malts that did not require a low temperature start to mashing in order to deal with residual cell-wall material (β-glucans). Grist is mixed with water (a typical ratio would be three parts solid to one part water) in a Steel's masher en route to the preheated mash tun, with a single holding temperature, typically 65°C, being employed. This temperature facilitates gelatinisation of starch and subsequent amylolytic action. At the completion of this 'conversion', wort is separated from the spent grains in the same vessel, which incorporates a false bottom and facility to regulate the hydrostatic pressure across the grains bed. The grist is sparged to enable leaching of as much extract as possible from the bed.

Decoction mashing was designed on the mainland continent of Europe to deal with lager malts which were less well-modified than ale malts. Essentially it provides the facility to start mashing at a relatively low temperature, thereby allowing hydrolysis of the β-glucans present in the malt, followed by raising the temperature to a level sufficient to allow gelatinisation of starch and its subsequent enzymic hydrolysis. The manner by which the temperature increase was achieved was by transferring a portion of the initial mash to a separate vessel where it was taken to boiling and then returned to the main mash, leading to an increase in temperature. This is a rather simplified version of the process, which traditionally involved several steps of progressive temperature increase.

Temperature-programmed mashing: Although there are some adherents to the decoction-mashing protocol, most brewers now-a-days employ the related but simpler temperature-programmed mashing. Again, the mashing is commenced at a relatively low temperature, but subsequent increases in temperature are effected in a single vessel by employing steam-heated jackets around the vessel to raise the temperature of the contents, which are thoroughly mixed to ensure even heat transfer. Mashing may commence at 45–50°C, followed by a temperature rise of $1°C.min^{-1}$ until the conversion temperature (63–68°C) is reached. The mash will be held for perhaps 50 minutes to 1 hr, before raising the temperature again to the sparging temperature (76–78°C). High temperatures are employed at the end of the process to arrest enzymic activity, to facilitate solubilisation of materials and to reduce viscosity, thereby allowing more rapid liquid-solid separation.

Adjuncts

A key aspect of solid adjuncts is the gelatinisation temperature of the starch. A higher gelatinisation temperature for corn, rice and sorghum means that these cereals need treatment at higher temperatures than do barley, oats, rye or wheat. If the cereal is in the form of grits (produced by the dry milling of cereal in order to remove outer layers and the oil-rich germ), then it needs to be 'cooked' in the brewhouse. Alternatively, the cereal can be preprocessed by intense heat treatment in a micronisation or flaking operation. In the former process, the whole grain is passed by conveyor under an intense heat source (260°C), resulting in a 'popping' of the kernels (puffed breakfast cereals). In flaking, grits are gelatinised by steam and then rolled between steam-heated rollers. Flakes are not required to be milled in the brewhouse, but micronised cereals are.

Cereal cookers employed for dealing with grits are made of stainless steel and incorporate an agitator and steam jackets. The adjunct is delivered from a hopper and the adjunct will be mixed with water at a rate of perhaps 15 kg per hl of water. The adjunct will be mixed with 10–20 per cent of malt as a source of enzymes. The precise temperature employed in the cooker will depend on the adjunct and the preferences of the brewer.

Following cooking, the adjunct mash is likely to be taken to boiling and then mixed with the main mash (at its mashing-in temperature), with the resultant effect being the temperature rise to conversion for the malt starch (decoction mashing). This is sometimes called 'double mashing'.

Wort separation

Traditionally, recovering wort from the residual grains in the brewery is perhaps the most skilled part of brewing. Not only is the aim to produce a wort with as much extract as possible, but many brewers prefer to do this such that the wort is 'bright', that is, not containing many insoluble particles which may present difficulties later. All this needs to take place within a time window, for the mashing vessel must be emptied in readiness for the next brew.

Mash filters: Increasingly, modern breweries use mash filters. These operate by using plates of polypropylene to filter the liquid wort from the residual grains. Accordingly, the grains serve no purpose as a filter medium and their particle sizes are irrelevant. The high pressures that can be used in the squeezing of the plates together overcome the reduced permeability due to smaller particle sizes (the sand versus clay analogy used earlier). Furthermore, the grains bed depth is particularly shallow (2–3 inches), being nothing more than the distance between the adjacent plates.

Water

Since water represents at least 90 per cent of the composition of most beers, it will clearly have a major direct impact on the product, particularly in terms of flavour and clarity. The nature of the water, however, exerts its influence much earlier in the process, through the impact of the salts it contains on enzymic and chemical processes, through the determination of pH, etc. Water in breweries comes either from wells owned by the brewer or from municipal supplies, especially in the latter instance, the water will be subjected to clean-up procedures, such as charcoal filtration, to eliminate undesirable taints and colours. The water in some places is very hard, while in some places it is soft.

The water composition can be adjusted, either by adding or by removing ions. Thus, calcium levels may be increased in order to promote the precipitation of oxalic acid as oxalate, to lower the pH by reaction with phosphate ions ($3\ Ca^{2+} + 2HPO_4^{2-} \rightarrow Ca_3(PO_4)_2 + 2H^+$) and to promote amylase action. (The optimum pH for mashing is between 5.2 and 5.4.) The alkalinity of water used for sparging (alkalinity is largely determined by the content of carbonate and bicarbonate) may be reduced to less than 50 ppm in order to limit the extraction of tannins.

Ions such as iron and copper must be as low as possible to preclude oxidation. Furthermore, water may need to be of different standards for different purposes. The microbiological status of water used for slurrying yeast or for use downstream generally is important. Water used for diluting high-gravity streams must be of low oxygen content, and its ionic composition will be critical. When ions need to be removed, the likeliest approach is ion-exchange resin technology.

Hops

The hop, *Humulus lupulus*, is rich in resins and oils, the former being the source of bitterness, the latter the source of aroma. The hop is remarkable amongst agricultural crops in that essentially its sole outlet is for brewing. Hops are grown in all temperate regions of the world, with approximately one-third coming from Germany.

Hops are hardy, climbing herbaceous perennial plants grown in gardens using characteristic string frameworks to support them. It is only the female plant that is cultivated, as it is the one that develops the hop cone. Hops are generally classified into two categories: aroma hops and bittering hops. All hops are capable of providing both bitterness and aroma.

The use of whole cone hops is comparatively uncommon now-a-days. Many brewers use hops that have been hammer-milled and then compressed into pellets. In this form they are more stable, more efficiently utilised and do not present the brewer with the problem of separating out the vegetative parts of the hop plant. Some use hop extracts that are derived by dissolving the resins in liquid carbon dioxide, followed by a chemical isomerisation if the bitterness is to be added to the finished beer rather than in the boiling stage. Recent years have been marked by an enormous increase in the use of such pre-isomerised extracts after they have been modified by reduction.

Wort Boiling and Clarification

The boiling of wort serves various functions, primary amongst which are the isomerisation of the hop resins (α-acids) to the more soluble and bitter iso-α-acids, sterilisation, the driving off of unwanted volatile materials, the precipitation of protein/polyphenol complexes (as 'hot break' or 'trub') and concentration of the wort. The extent of wort boiling is normally described in terms of percentage evaporation. After boiling, wort is transferred to a clarification device.

The system employed for removing insoluble material after boiling depends on the way in which the hopping was carried out. If whole hop cones are used, clarification is through a hop jack (hop back), which is analogous to a lauter tun, but in this case the bed of residual hops constitutes the filter medium. If hop pellets or extracts are used, then the device of choice is the whirlpool, a cylindrical vessel, into which hot wort is transferred tangentially through an opening 0.5–1 metre above the base. The wort is set into a rotational flux, which forces trub to a pile in the middle of the vessel.

Wort Cooling

Almost all cooling systems these days are of the stainless steel plate heat exchanger type, sometimes called 'paraflows'. Heat is transferred from the wort to a coolant, either water or glycol depending on how low the temperature needs to be taken. At this stage, it is likely that more material will precipitate from solution ('cold break'). Brewers are divided on whether they feel this to be good or bad for fermentation and beer quality. The presence of this break certainly accelerates fermentation and, therefore, it will directly influence yeast metabolism. As in so much of brewing, the aim should be consistency: either consistently 'bright worts' or ones containing a relatively consistent level of trub.

Yeast

Brewing yeast is *Saccharomyces cerevisiae* (ale yeast) or *Saccharomyces pastorianus* (lager yeast). There are many separate strains of brewing yeast, each of which is distinguishable phenotypically [e.g. in the extent to which it will ferment different sugars, or in the amount of oxygen it needs to prompt its growth, or in the amounts of its metabolic products (i.e. flavour spectrum of the resultant beer), or its behaviour in suspension (top versus bottom fermenting, flocculent or non-flocculent)] and genotypically, in terms of its DNA fingerprint.

The fundamental differentiation between ale and lager strains is based on the ability or otherwise to ferment the sugar melibiose: ale strains cannot whereas lager strains can because they produce the enzyme (α-galactosidase) necessary to convert melibiose into glucose and galactose. Ale yeasts also move to the top of open fermentation vessels and are called top-fermenting yeasts. Lager yeasts drop to the bottom of fermenters and are termed bottom-fermenting yeasts. Now-a-days it is frequently difficult to make this differentiation, when beers are widely fermented in similar types of vessel (deep cylindro-conical tanks) which tend to equalise the way in which yeast behave in suspension.

Oxygen is needed by the yeast to synthesise the unsaturated fatty acids and sterols it needs for its membranes. This oxygen is introduced at the wort cooling stage in the quantities that the yeast requires — but no more, because excessive aeration or oxygenation promotes excessive yeast growth, and the more yeast is produced in a fermentation, the less alcohol will be produced. Different yeasts need different amounts of oxygen.

Yeast uses its stored reserves of carbohydrate in order to fuel the early stages of metabolism when it is pitched into wort, for example, the synthesis of sterols. There are two principal reserves: glycogen and trehalose. Glycogen is similar in structure to the amylopectin fraction of barley starch. Trehalose is a disaccharide comprising two glucoses linked with an α-1,1 bond between their reducing carbons. The glycogen reserves of yeast build up during fermentation and it is important that they are conserved in the yeast during storage between fermentations. Trehalose may feature as more of a protection against the stress of starvation. It certainly seems to help the survival of yeast under dehydration conditions employed for the storage and shipping of dried yeast.

Now-a-days brewers maintain their own pure yeast strains. While it is still a fact that some brewers simply use the yeast grown in one fermentation to 'pitch' the following fermentation, and that they have done this for many tens of years, it is much more usual for yeast to be repropagated from a pure culture every 4–6 generations. (When brewers talk of 'generations', they mean successive fermentations, strictly speaking, yeast advances a generation every time it buds, and therefore there are several generations during any individual fermentation.)

The majority of brewing yeasts are resistant to acid (pH 2.0–2.2) and so the addition of phosphoric acid to attain this pH is very effective in killing bacteria with which yeast may become progressively contaminated from fermentation to fermentation. Many brewers use such an acid washing of yeast between fermentations.

There are two key indices of yeast health: viability and vitality. Both should be high if a successful fermentation is to be achieved. Viability is a measure of whether a yeast culture is alive or dead. While microscopic inspection of a yeast sample is useful as a gross indicator of that culture (e.g. presence of substantial infection), quantitative evaluation of viability needs a staining test. The most common is the use of methylene blue: viable yeast decolourises it, dead cells do not. Although a yeast cell may be living, it does not necessarily mean that it is healthy. Vitality is a measure of how healthy a yeast cell is. Many techniques have been advanced as an index of vitality, but none has been accepted as definitive.

Preferably yeast is stored in a readily sanitised room that can be cleaned efficiently and which is supplied with a filtered air supply and possesses a pressure higher than the surroundings in order to impose an outwards vector of contaminants. Ideally it should be at or around 0°C. Even if storage is not in such a room, the tanks must be rigorously cleaned, chilled to 0–4°C and have the facility for gently rousing (mixing) to avoid hot spots.

Yeast is stored in slurries ('barms') of 5–15 per cent, solids under 6 inches of beer, water or potassium phosphate solution. An alternative procedure is to press the yeast and store it at 4°C in a cake form (20–30 per cent dry solids). Pressed yeast may be held for about 10 days, water slurried and beer slurried for 3–4 weeks and slurries in 2 per cent phosphate, pH 5 for 5 weeks.

Apart from the importance of pitching yeast of good condition, it is also important that the amount pitched is in the correct quantity. The higher the pitching rate, the more rapid the fermentation. As the pitching rate increases, initially so too does the amount of new biomass synthesised, until at a certain rate, the amount of new yeast synthesised declines. The rate of attenuation and the amount of growth directly impacts the metabolism of yeast and the levels of its metabolic products (i.e. beer flavour) hence the need for control.

Yeast can be quantified by weight or cell number. Typically some 10^7 cells per ml will be pitched for wort of 12°Plato (1.5–2.5 g pressed weight per l). At such a pitching rate, lager yeast will divide 4–5 times in fermentation. Yeast numbers can be measured using a haemocytometer, which is a counting chamber loaded onto a microscope slide. It is possible to weigh yeast or to centrifuge it down in pots which are calibrated to relate volume to mass, but in these cases it must be remembered that there are usually other solid materials present, for example, trub.

Another procedure that has come into vogue is the use of capacitance probes that can be inserted in-line. An intact and living yeast cell acts as a capacitor and gives a signal whereas dead ones (or insoluble materials) do not. The device is calibrated against a cell number (or weight) technique and therefore allows the direct read-out of the amount of viable yeast in a slurry. Other in-line devices quantify yeast on the basis of light scatter.

Brewery Fermentations

Primary fermentation is the fermentation stage proper in which yeast, through controlled growth, is allowed to ferment wort to generate alcohol and the desired spectrum of flavours. Increasingly brewery fermentations are conducted in cylindro-conical vessels. The fermentation is regulated by control of several parameters, notably the starting strength of the wort (°Plato, which approximates to percentage sugar by weight, or Brix), the amount of viable yeast ('pitching rate'), the quantity of oxygen introduced and the temperature. Fermentation is monitored by measuring the decrease in specific gravity (alcohol has a much lower specific gravity than sugar).

Ales are generally fermented at a higher temperature (15–20°C) than lagers (6–13°C) and therefore attenuation (the achievement of the finished specific gravity) is achieved more rapidly. Thus, an ale fermenting at 20°C may achieve attenuation gravity in 2 days, whereas a lager fermented at 8.5°C may take 10 days. The temperature has a substantial effect on the metabolism of yeast, and the levels of a flavour substance like iso-butanol will be 16.5 and 7 mg l^{-1}, respectively, for the ale and the lager. Some brewers add zinc (e.g. 0.2 ppm) to promote yeast action—it is a cofactor for the enzyme alcohol dehydrogenase. During fermentation, the pH falls because yeast secretes organic acids and protons.

Surplus yeast will be removed at the end of fermentation, either by a process such as 'skimming' for a traditional square fermenter employing top fermenting yeast, or from the base of a cone in a cylindro-conical vessel. This is not only to preserve the viability and vitality of the yeast, but also to circumvent the autolysis and secretory tendencies of yeast that will be to the detriment of flavour and foam. There will still be sufficient yeast in the beer to effect the secondary fermentation.

The green beer produced by primary fermentation needs to be 'conditioned', in respect of establishment of a desired carbon dioxide content and refinement of the flavour. This is called secondary fermentation. Above all at this stage, there needs to be the removal of an undesirable butter-scotch flavour character due to substances called vicinal diketones (VDKs, discussed later). Traditionally it is the lager beers fermented at lower temperatures that have needed the more prolonged maturation (storage: 'lagering') in order to refine their flavour and develop carbonation. The latter depends on the presence of sugars, either those (perhaps 10 per cent) which the brewer ensures are residual from the primary fermentation or those introduced in the 'krausening' process, in which a proportion of freshly fermenting wort is added to the maturing beer. Many brewers are unconvinced by the need for prolonged storage periods (other than for its strong marketing appeal) and they tend to combine the primary and secondary fermentation stages. Once the target attenuation has been reached, the temperature is allowed to rise (perhaps by 4°C), which permits the yeast to deal more rapidly with the VDKs. Carbonation will be achieved downstream by the direct introduction of gas.

Once the secondary fermentation stage is complete (and the length of this varies considerably between brewers), then the temperature is dropped, ideally to –1° or –2°C to enable precipitation and sedimentation of materials which would otherwise cause a haze in the beer. The sedimentation of yeast is also promoted in this 'cold conditioning' stage, perhaps with the aid of isinglass finings. These are solutions of collagen derived from the swim bladders of certain species of fish from the South China Seas. Collagen has a net positive charge at the pH of beer, whereas yeast and other particulates have a net negative charge. Opposite charges attracting, the isinglass forms a complex with these particles and the resultant large agglomerates sediment readily because of an increase in particle size. Sometimes, the isinglass finings are used alongside 'auxiliary finings' based on silicate, the combination being more effective than is in glass alone. Some brewers centrifuge to aid clarification.

For the most part, fermenters these days are fabricated from stainless steel and will be lagged and feature jackets that allow coolant to be circulated (the heat generated during fermentation is sufficient to effect any necessary warming—so the temperature is regulated by balancing metabolic heat with cooling afforded by the coolant in the jacket, which may be water, glycol or ammonia depending on how much refrigeration is demanded). Modern vessels tend to be enclosed, for microbiological reasons. However, across the world there remain a great many open tanks. Cylindro-conical vessels can have a capacity of up to 13000 hl and are readily cleaned using CIP operations.

Filtration

After a period of typically 3 days minimum in 'cold conditioning', the beer is generally filtered. Diverse types of filter are available, perhaps the most common being the plate-and-frame filter which consists of a series of plates in sequence, over each of which a cloth is hung. The beer is mixed with a filter-aid porous particles which both trap particles and prevent the system from clogging. Two major kinds of filter aid are in regular use: kieselguhr and perlite. The former comprises fossils or skeletons of primitive organisms called diatoms. These can be mined and classified to provide grades that differ in their permeability characteristics. Particles of kieselguhr contain pores into which other particles (such as those found in beer) can pass, depending on their size. Perlites are derived from volcanic glasses crushed to form microscopic flat particles. They are better to handle than kieselguhr, but are not as efficient as filter aids. Filtration starts when a pre-coat of filter aid is applied to the filter by cycling a slurry of filter aid through the plates. This pre-coat is generally of quite a coarse grade, whereas the filter-aid (the body feed) which is dosed into the beer during the filtration proper tends to be a finer grade. It is selected according to the particles within the beer that need to be removed. If a beer contains a lot of yeast, but relatively few small particles, then a relatively coarse grade is best. If the converse applies, then a fine grade with smaller pores will be used.

Stabilisation of beer

Apart from filtration, various other treatments may be applied to beer downstream, all with the aim of enhancing the shelf-life of the product. A haze in beer can be due to various materials, but principally it is due to the cross-linking of certain proteins and certain polyphenols. Therefore, if one or both of these materials is removed, then the shelf-life is extended. Brewhouse operations are in part designed to precipitate protein-polyphenol complexes. Thus, if these operations are performed efficiently, then much of the job of stabilisation is achieved. Good, vigorous, 'rolling' boils, for instance, will ensure precipitation. Before that, avoidance of the last runnings in the lautering operation will prevent excessive levels of polyphenol entering the wort. The cold conditioning stage also has a major role to play, by chilling out protein-polyphenol complexes, enabling them to be taken out on the filter. Control over oxygen and oxidation is important because it is particularly the oxidised polyphenols that tend to cross-link with proteins. For really long shelf-lives, though, and certainly if the beer is being shipped to extremes of climate, additional stabilisation treatments will be necessary. Polyphenols can be removed with PVPP. Protein can be precipitated by adding tannic acid, hydrolysed using papain (the same enzyme from paw paw that is used as meat tenderiser) or, and most commonly, adsorbed on silica hydrogels and silica xerogels.

Gas Control

Final adjustment will now be made to the level of gases in the beer. As we have seen, it is important that the oxygen level in the bright beer is as low as possible. Unfortunately, whenever beer is moved around and processed in a brewery, there is always the risk of oxygen pick-up. For example, oxygen can enter through leaky pumps. A check on oxygen content will be made once the bright beer tank (filtered beer is bright beer) is filled and, if the level is above specification (which most brewers will set at 0.1–0.3 ppm), oxygen will have to be removed. This is achieved by purging the tank with an inert gas, usually nitrogen, from a sinter in the base of the vessel. The level of carbon dioxide in a beer may either need to be increased or decreased. The majority of beers contain between two and three volumes of CO_2, whereas most brewery fermentations generate 'naturally' no more than 1.2–1.7 volumes of the gas. The simplest and most usual procedure by which CO_2 is introduced is by injection as a flow of bubbles as beer is transferred from the filter to the bright beer tank. If the CO_2 content needs to be dropped, this is a more formidable challenge. It may be necessary for beers that are supposed to have a relatively low carbonation and, as for oxygen, this can be achieved by purging. However, concerns about 'bit' production have stimulated the development of gentle membrane-based systems for gas control. Beer is flowed past membranes, made from polypropylene or polytetrafluoroethylene, that are water-hating and therefore do not 'wet-out'. Gases, but not liquids, will pass freely across such membranes, the rate of flux being proportional to the concentration of each individual gas and dependent also on the rate at which the beer flows past the membrane.

Packaging

The packaging operation is the most expensive stage in the brewery, in terms of raw materials and labour. Beer will be brought into specification in the Bright Beer Tank (sometimes called the Fine Ale Tank or the package release tank). The carbonation level may be higher (e.g. by 0.2 volume) than that specified for the beer in package, to allow for losses during filling.

Although beer is relatively resistant to spoilage, it is by no means entirely incapable of supporting the growth of micro-organisms. For this reason, most beers are treated to eliminate any residual brewing yeast or infecting wild yeasts and bacteria before or during packaging. This can be achieved in one of two ways: pasteurisation or sterile filtration. Pasteurisation can take one of two forms in the brewery: flash pasteurisation for beer pre-package and tunnel pasteurisation for beer in can or bottle. The principle in either case, of course, is that heat kills micro-organisms. One PU is defined as exposure for 1 minute at 60°C. The higher the temperature, the more rapidly the micro-organisms are destroyed. A 7°C rise in temperature leads to a ten-fold increase in the rate of cell death. The pasteurisation time required to kill organisms at different temperatures can be read off from a plot. Typically, a brewer might use 5–20 PU but higher 'doses' may be used for some beers, for example, low alcohol beers which are more susceptible to infection. In flash pasteurisation, the beer flows through a heat exchanger (essentially like a wort cooler acting in reverse), which raises the temperature typically to 72°C. Residence times of between 30 and 60 seconds at this temperature are sufficient to kill off virtually all microbes. Ideally there will not be many of these to remove: good brewers will ensure low loadings of micro-organisms by attention to hygiene throughout the process and ensuring that the prior filtration operation is efficient. An increasingly popular mechanism for removing micro-organisms is to filter them out by passing the beer through a fine mesh filter. The rationale for selecting this procedure rather than pasteurisation is as much for marketing reasons as for any technical advantage it presents: many brands of beer these days are being sold on a claim of not being heat-treated, and therefore free from any 'cooking'. In fact, provided the oxygen level is very low, modest heating of beer does not have a major impact on the flavour of many beers, although those products with relatively subtle, lighter flavour will obviously display 'cooked' notes more readily than will beers that have a more complex flavour character. The sterile filter must be located downstream from the filter that is used to separate solids from the beer. Sterile filters may be of several types, a common variant incorporating a membrane formed from polypropylene or polytetrafluoroethylene and with pores of between 0.45 and 0.8 μm.

Filling bottles and cans

Bottles entering the brewery's packaging hall are first washed and, if they are returnable bottles (i.e. they have been used previously to hold beer), they will need a much more robust cleaning and sterilisation, inside and out, involving soaking and jetting with hot caustic detergent and thorough rinsing with water. The beer coming from the bright beer tanks is transferred to a bowl at the heart of the filling machine. Bottle fillers are machines based on a rotary carousel principle. They have a series of filling heads: the more the heads, the greater the capacity of the filler. The bottles enter on a conveyor and, sequentially, each is raised into position beneath the next vacant filler head, each of which comprises a filler tube. An air-tight seal is made and, in modern fillers, a specific air evacuation stage starts the filling sequence. The bottle is counter-pressured with carbon dioxide, before the beer is allowed to flow into the bottle by gravity from the bowl. The machine will have been adjusted so that the correct volume of beer is introduced into the vessel. Once filled, the 'top' pressure on the bottle is relieved, and the bottle is released from its filling head. It passes rapidly to the machine that will crimp on the crown cork but, en route, the bottle will have been either tapped or its contents 'jetted' with a minuscule amount of sterile water in order to fob the contents and drive off any air from the space in the bottle between the surface of the beer and the neck (the 'headspace'). Next stop is the tunnel pasteuriser if the beer is to be pasteurised after filling, but if sterile filtration is used, the filler and capper are likely to be enclosed in a sterile room. The bottles now head off for labelling, secondary packaging and warehousing.

Putting beer into cans has much in common with bottling. It is the container, of course, that is very different—and definitely one trip. Cans may be of aluminium or stainless steel, which will have an internal lacquer to protect the beer from the metal surface and *vice versa*. Cans arrive in the canning hall on vast trays, all pre-printed and instantly recognisable. They are inverted, washed and sprayed, prior to filling in a manner very similar to the bottles. Once filled, the lid is fitted to the can basically by folding the two pieces of metal together to make a secure seam past which neither beer nor gas can pass.

Quality of Beer

Flavour

The flavour of beer can be split into three separate components: taste, smell (aroma) and texture (mouthfeel). There are only four proper tastes: sweet, sour, salt and bitter. They are detected on the tongue. A related sense is the tingle associated with high levels of carbonation in a drink: this is due to the triggering of the trigeminal nerve by carbon dioxide. This nerve responds to mild irritants, such as carbonation and capsaicin (a substance largely responsible for the 'pain delivery' of spices and peppers).

Carbon dioxide is also relevant insofar as its level influences the extent to which volatile molecules will be delivered via the foam and into the headspace above the beer in a glass. The sweetness of a beer is due, of course, to its level of sugars, either those that have survived fermentation or those introduced as primings. The principal contributors to sourness in beer are the organic acids that are produced by yeast during fermentation. These lower the pH: it is the H^+ ion imparted by acidic solutions that causes the sour character to be perceived on the palate. Most beers have a pH between 3.9 and 4.6.

Saltiness in beer is afforded by sodium and potassium, while of the anions present in beer, chloride and sulphate are of particular importance. Chloride is said to contribute a mellowing and fullness to a palate, while sulphate is felt to elevate the dryness of beer. Perhaps the most important taste in beer is bitterness, primarily imparted by the iso-α-acids derived from the hop resins.

Many people believe that they can taste other notes on a beer. In fact they are detecting them with the nose, the confusion arising because there is a continuum between the back of the throat and the nasal passages. The smell (or aroma) of a beer is a complex distillation of the contribution of a great many individual molecules. No beer is so simple as to have its 'nose' determined by one or even a very few substances. The perceived character is a balance between positive and negative flavour notes, each of which may be a consequence of one or a combination of many compounds of different chemical classes. The 'flavour threshold' is the lowest concentration of a substance which is detectable in beer.

The substances that contribute to the aroma of beer are diverse. They are derived from malt and hops and by yeast activity (leaving aside for the moment the contribution of contaminating microbes). In turn there are interactions between these sources, insofar as yeast converts one flavour constituent from malt or hops into a different one, for example.

Various alcohols influence the flavour of beer, by far the most important of which is ethanol, which is present in most beers at levels at least 350-fold higher than any other alcohol. Ethanol contributes directly to the flavour of beer, registering a warming character. It also influences the flavour contribution of other volatile substances in beer. Because it is quantitatively third only to water and carbon dioxide as the main component of beer, it is not surprising that it moderates the flavour impact of other substances. It does this by affecting the vapour pressure of other molecules (i.e. their relative tendency to remain in beer or to migrate to the headspace of the beer). The higher alcohols in beer are important as the immediate precursors of the esters, which are proportionately more flavour active. And so it is important to be able to regulate the levels of the higher alcohols produced by yeast if ester levels are also to be controlled.

Foam

A point of difference between beer and other alcoholic beverages is its possession of stable foam. This is due to the presence of hydrophobic (amphipathic) polypeptides, derived from cereal, that cross-link with the bitter iso-α-acids in the bubble walls to counter the forces of surface tension that tend to lead to foam collapse.

Gushing

Foaming can be taken to excess, in which case the problem which manifests itself in small pack is 'gushing', that is, the spontaneous generation of foam on opening a package of beer. This is due to the presence of nucleation sites in beer that cause the dramatic discharging of carbon dioxide from solution. These nucleation sites may be particles of materials like oxalate or filter aid, but most commonly gushing is caused by intensely hydrophobic peptides that are produced from Fusarium that can infect barley unless precautions are taken.

Spoilage of Beer

Compared with most other foods and beverages beer is relatively resistant to infection. There are several reasons for this, namely the presence of ethanol, a low pH, the relative shortage of nutrients (sugars, amino acids), the anaerobic conditions and the presence of antimicrobial agents, notably the iso-α-acids.

The most problematic gram-positive bacteria are lactic acid bacteria belonging to the genera *Lactobacillus* and *Pediococcus*. At least ten species of lactobacillus spoil beer. They tolerate the acidic conditions. Some species (e.g. *Lactobacillus brevis* and *Lactobacillus plantarum*) grow quickly during fermentation, conditioning and storage, while others (e.g. *Lactobacillus lindner*) grow relatively slowly. Spoilage with lactobacilli is especially problematic during the conditioning of beer and after packaging, resulting in a silky turbidity and off-flavours. Pediococci are homofermentative. Six species have been identified, the most important being *Pediococcus damnosus*. Such infection generates lactic acid and diacetyl. The production of polysaccharide capsules can cause ropiness in beer.

Many gram-positive bacteria are killed by iso-α-acids. These agents probably disrupt nutrient transport across the membrane of the bacteria, but only when they are present in their protonated forms (i.e. at low pH). This is one of the reasons why a beer at pH 4 will be more resistant to infection than one at pH 4.5. Some Gram positives are resistant to iso-α-acids and most gram negatives are.

Important gram-negative bacteria include the acetic acid bacteria (*Acetobacter*, *Gluconobacter*), Enterobacteriaceae (*Escherichia*, *Aerobacter*, *Klebsiella*, *Citrobacter*, *Obesumbacterium*), *Zymomonas*, *Pectinatus* and *Megasphaera*. Acetic acid bacteria produce a vinegary flavour in beer and a ropy slime. It is most often found in draft beer, where there is a relatively aerobic environment close to the beer, for example, in partly emptied containers. Enterobacteriaceae are aerobic and cannot grow in the presence of ethanol. They are a threat in wort and early in fermentation and they produce cabbagy/vegetable/eggy aromas. *Zymomonas* is a problem with primed beers (it uses invert sugar or glucose, but cannot use maltose). Although it has a metabolism reminiscent of *Saccharomyces* (it's actually used to produce alcoholic beverages in some countries), it does tend to produce large amounts of acetaldehyde.

ENZYMES IN BREWING

Enzymes are proteins with a special structure capable of accelerating the breakdown of different substrates. They act as catalysts to increase the speed of a chemical reaction without themselves undergoing any permanent chemical change. They are not used up in the reaction or appear as reaction products.

Enzymes bind temporarily to the substrate, thereby lowering the amount of activation energy, enabling the reaction to occur faster and at lower temperatures. Like most chemical reactions, the rate of an enzyme-catalysed reaction increases as the temperature is raised. A 10°C rise in temperature will increase the activity of most enzymes by 50 to 100 per cent. Variations in reaction temperature as small as 1–2°C may introduce changes of 10–20 per cent in the results. In the case of enzymatic reactions, this is complicated by the fact that high temperatures adversely affect many enzymes. The reaction rate increases with temperature to a maximum level, then abruptly falls off with further increase of temperature. Many enzymes start to become denatured at temperatures above 40°C. Over a period of time, enzymes will be deactivated, at even moderate temperatures. Storage of enzymes at 5°C or below is generally the most suitable.

Enzymes are equally affected by changes in pH. The optimum pH value will vary greatly from one enzyme to another. Most of the brewing enzymes have an optimum pH in the range 4.5 to 6.0, which is the operating range of most brewing processes. Concentration of enzyme used, with reference to the substrate concentration, plays an important role. With an excess concentration of substrate, such as starch in a brewers' wort, there is a linear effect of increasing the enzyme concentration upon the reaction rate. Hence, if all other factors are kept constant, malts with higher enzymatic power will break down starch faster. It has been shown that if the amount of the enzyme is kept constant and the substrate concentration is gradually increased, the reaction velocity will increase until it reaches a maximum. After this point, increases in substrate concentration will not increase the velocity.

In addition to temperature and pH, there are other factors, such as ionic strength, which can affect the enzymatic reaction. Each of these physical and chemical parameters must be considered and optimised, in order for an enzymatic reaction to be accurate and reproducible.

Biochemical Changes During Brewing

Enzymes are essential in catalysing the biochemical changes, which occur in the brewing process. There are two principal processes of interest to the brewer:

1. The breakdown of carbohydrate, principally starch, in malted barley, to sugars.
2. The fermentation of sugars and other nutrients by yeast, under anaerobic conditions, to release energy and produce ethanol, as a metabolic by-product.

These biological reactions are catalysed by enzymes from the barley and yeast respectively. Barley is able to produce all the enzymes needed to degrade starch, β-glucan, pentosans, lipids and proteins, which are the major compounds of concern to the brewer. During malting, the barley corn is allowed to germinate and produce enzymes for breaking down the cell walls in the corn and release energy stored as starch in the endosperm. The starch is present as concentric granules surrounded by a protein matrix, and has to be broken down during mashing before the enzyme amylases can gain access.

Three principal enzymic reactions occur in malt during the mashing process. The principal reaction is the hydrolysis of starch to sugars by α and β-amylase. Before enzyme hydrolysis can occur, it is necessary to exceed the starch gelatinisation temperature of malt. Therefore, it is necessary to select the optimum conditions for the saccharifying enzymes to operate. This is achieved by stabilising the enzymes by:

1. Optimising pH at mashing (usually between pH5 and 6).
2. Adding calcium ions to stabilise the enzyme.
3. Using thick mash (high concentration substrate to insulate the enzymes against denaturing).
4. Optimising temperature to favour the activity of both the α and β-amylase.

The amylase enzymes break the α-1,4 links in amylose and amylopectin to give a mixture of glucose, maltose, maltotriose and higher sugars called dextrins, which are unfermentable, to give a wort (malt derived sugar solution), which is fermentable up to 70 per cent.

α-amylase produces random hydrolyses of starch to dextrins, while β-amylase attacks the starch and dextrins from the reducing end, stripping off pairs of sugar molecules (maltose). By varying the mashing temperature, it is possible to preferentially favour one enzyme reaction over the other and hence influence the fermentability of the wort, with the lower temperatures giving higher fermentable worts.

Apart from starch, barley contains a number of non-starch polysaccharides, principal being β-glucan (~75 per cent of the cell wall). The molecule has a distinctive linear structure containing 70 per cent β-1,4 linkages and 30 per cent β-1,3 linkages. Mostly it is water soluble, but a proportion is bound covalently to cell wall proteins. If there is insufficient degradation of the cell walls, then enzymic access to the protein and starch will be restricted, and the extract from the malt is reduced.

Although much of the necessary β-glucanase activity occurs during malting, there is inevitably some survival of cell wall material (even in the most fully modified malt). This will be exacerbated if adjuncts such as barley and wheat are also used. Consequently, it is necessary to ensure the continued activity of β-glucanase during mashing. If the large viscous β-glucan molecules are not broken down during malting or mashing, other process problems can also occur. These include:

1. Reduced extract recovery.
2. High wort viscosity.
3. Poor run off performance.
4. Beer filtration problems.
5. Beer haze problems.

Most brewers are very careful in selecting malt with low β-glucan levels, and it is common practice in many breweries to add exogenous β-glucanase to decrease wort and beer viscosity and to improve filterability. Advantage of addition of β-glucanase can be shown by an increasing in filter flow rate and decrease in wort viscosity.

Hydrolysis of proteins and polypeptides

While about 95 per cent of the starch from malt is solubilised by the end of mashing, only about 35–40 per cent of the malt protein (total nitrogen) is solubilised. This is referred to as the TSN (total soluble nitrogen) in an unboiled wort. The principal enzymes involved in the breakdown of malt proteins are:

1. Endoproteases, which break the large protein molecules into polypeptide chains.
2. Exopeptidases, which attack polypeptides, stripping off small polypeptides to produce amino acids.

Endopeptidases have a low optimum temperature and hence with high temperature mashing most of the protein breakdown will have taken place during the malting process. They randomly attack the protein. Exopeptidases are able to withstand higher temperatures and release the amino acids from the polypeptide chains.

There are two principal groups of exopeptidase enzymes:

1. Carboxypeptidase which attacks the proteins from the carbonyl end. This enzyme is not present in raw barley, but is rapidly produced during steeping and is active at normal mash pH. Optimum conditions for its use are: pH 3.9–5.5, temperature: 45°–50°C, and inactivation temperature: 70°C.
2. Aminopeptidase, which attacks the proteins from the amino end, is much less active at mash pH and does not play a significant role in protein breakdown during mashing. Optimum conditions for its use are: pH: 4.8–5.2, temperature: 50°C, inactivation temperature: >70°C.

Most of the proteolysis occurs during malting. It is impossible to completely compensate for a nitrogen deficiency in malt by introducing a prolonged mash stand at <50°C without adding exogenous enzymes. Nitrogenous materials account for 5–6 per cent of wort solids, which is equivalent to around 30–40 per cent of the total nitrogen in malt. Good yeast growth and rapid fermentation requires 160 mg/l of free amino nitrogen (at 12°P wort) depending on the yeast strain.

Carboxypeptidases can release amino acids in mashing, provided the endopeptidase has broken down the protein substrate during the malting process. The optimum temperature to produce free amino nitrogen production is 50°C. Proteins in the mash dissolve at these low temperatures and then precipitate at 65°C, which can inhibit lautering. Excessive proteolysis in malting and mashing will reduce foam stability and the pH of a normal mash is not optimal for proteolysis.

Proteins Found in Wort

Much of the surplus protein is left behind in the spent grains, but when oxidised can form a protein 'scum', which causes run off problems. Some of the soluble proteins play an essential role as enzymes, catalysing the reactions described above.

Polypeptides

These are long-chain sequences of relatively high molecular weight amino acids. There are two important groups in brewing: Hydrophobic polypeptides, which make up beer foam, and acidic polypeptides, which can combine with polyphenols to produce hot and cold break, and if not removed, these contribute to colloidal instability in beer. This group of compounds is also probably important in contributing to the texture and mouth-feel of the beer.

Peptides

These are short chain sequences of amino acids, usually 2 to 10 units long, and probably have a minor effect on body and mouth-feel.

Amino acids

These make up 10–15 per cent of the TSN and are an essential source of nutrient for yeast growth. The usual concentration of soluble free amino nitrogen (FAN) in wort is required to be above 160-mg/l, lower levels can lead to a defective fermentation. In addition to their role in yeast growth, amino acids are also involved in a number of metabolic pathways, producing significant flavour active compounds, which contribute to the final flavour of the beer. The activity of proteolytic enzymes is effected by temperature of mashing, which, in turn, will effect the total nitrogen, amino nitrogen, head retention and shelf life stability.

Fermentation

Yeasts are able to respire anaerobically, but under anaerobic conditions they can only partially break down sugar molecules to ethanol to release energy in the form of ATP (adenosine triphosphate).

The role of yeast in the fermentation is that of a living catalyst, effecting the reaction without becoming part of the finished product. During the course of the fermentation, the yeast cells grow and replicate up to five times. Although the yeast gains its energy from the sugar, which it converts to alcohol it can only utilise simple sugars. The sugars are taken up in a specific order, with the monosaccharides, glucose and fructose used first, together with sucrose.

Although the latter is a disaccharide, it behaves like a monosaccharide, since it is broken down to glucose and fructose outside the cell, through the action of the yeast enzyme invertase. Once the wort glucose level falls, the yeast starts to use the disaccharide, maltose, which is usually the most abundant sugar in brewers, wort. Maltose has to be transported into the cell, where it is broken down to glucose. Lastly most yeast strains can utilise the trisaccharide, maltotriose, but only slowly. Brewing strains of yeast cannot generally ferment the longer chained or branched sugars (called dextrins), which persist in to the finished beer as unfermentable extract to give the beer body and mouth-feel. As well as sugars, yeast requires nitrogen, which in wort comes from the malt, in the form of soluble amino nitrogen.

A healthy fermentation yeast requires more than 160 mg/l of soluble nitrogen. If there is insufficient soluble nitrogen, for example when high cereal or sugar adjunct are used, then additional nitrogen may be required in the form of simple ammonium salts.

Syrup manufacture

A number of brewers use brewing syrups, which are manufactured from hydrolysed starch solution. Since the starch is not malted, microbial exogenous enzymes have to be used and by selecting different enzyme combinations the syrup producer can control the composition and fermentability of the syrup.

Brewing

As already discussed, beer is produced by mixing crushed barley malt and hot water in a large copper mash. Besides malt, adjuncts such as maize, sorghum, rice and barley or pure starch itself, are added to the mash. The mash is filtered in a lautertun. The resulting liquid, sweet wort, is run off to the copper where it is boiled with hops. The hopped wort is cooled and transferred to the fermentation vessels where yeast is added. After the fermentation, the so-called green beer, is matured before the final filtration and bottling. This is a much-simplified account of making beer.

The traditional source of enzymes used for the conversion of cereals into beer is barley malt, one of the key ingredients in brewing. If too little enzyme activity is present in the mash, there will be several undesirable consequences, such as:

1. Too low yield of extract.
2. Longer time for wort separation.
3. Slower rate of fermentation.
4. Lower alcohol content.
5. Inferior flavour.
6. Lower stability.

Enzymes are used to supplement malt's own enzymes in order to prevent these problems. Rossari biotech has developed a compounded formulation — Fermross BR — combining benefits of various enzymes to ensure:

1. Optimal and efficient liquefaction of adjuncts.
2. Accelerate rate of fermentation.
3. Shorten the beer maturation time.
4. Allow cheaper raw materials for beer manufacture.

Brewing with Adjuncts

Malt is the traditional source of α-amylase for the liquefaction of adjuncts. Nevertheless, it needs to be supplemented with heat stable α-amylase to ensure:

1. More predictable and simpler liquefaction of adjuncts, shorter process time and increased productivity.
2. The malt enzymes are preserved for subsequent saccharification process, resulting in a better wort and beer quality.
3. Eliminating the malt from the adjunct cooker, rendering more freedom in balancing volumes and temperatures in the mashing programme, while using high adjunct ratio.

Brewing with Barley

Traditionally, the use of barley has been limited to 10–20 per cent of the grist when using high-quality malts. Low-quality malts need to be supplemented with extra enzyme activity, to maintain adequate brewing performance. Brewers may add Fermross BR blend of thermo stable α-amylase, β-glucanase and protease at the mashing-in stage or later, separately as required.

Wort separation and beer filtration are two common bottlenecks in brewing. Poor lautering reduces production capacity and lowers extract yields. A thorough breakdown of β-glucans and pentosans during mashing is essential for fast wort separation. Non-degraded β-glucans and pentosans carried over into the fermenter reduce the beer filtration capacity. Fermross BR improves wort separation thereby increasing rate of filtration.

Enzymes in Improving Fermentation

Improvements in fermentability for various substrates can be achieved by adding Fermross BR which contains α-amylase, at the start of fermentation, together with a glucoamylase at mashing-in. Fungal α-amylases produce maltose and dextrins, whereas glucoamylase produces glucose from both linear and branched dextrins. The alcohol content is an important parameter that brewers must control. The amount of alcohol in a beer is controlled by the level of fermentable sugar in the extract, which in turn, is controlled by the amount of amylases in the mash, and the saccharifying enzymes used during fermentation. Yeast needs proteins in order to multiply. If the level of available amino nitrogen is less, the fermentation is poor, leading to inferior beer quality. Fermross BR contains a neutral protease, when added at mashing-in, raise the level of free amino nitrogen. This is beneficial when working with poor malts or with high adjunct ratios.

Diacetyl Control

Beer is said to be mature for racking when the diacetyl level drops below 0.07 ppm. Diacetyl gives beer an off-flavour like buttermilk. During 'maturing' diacetyl content drops to a level where it can't be tasted. Diacetyl is formed by oxidative decarboxylation of α-acetolactate during primary fermentation, which is removed again by the yeast, during beer maturation by conversion to acetoin, which is almost tasteless.

By adding Fermross BR, which contains an enzyme α-acetolactate decarboxylase at the beginning of the primary fermentation, it is possible to convert α-acetolactate directly into acetoin. Most of the α-acetolactate is degraded before it has a chance to oxidise and thus less diacetyl is formed. Thus, use of Fermross BR shortens or completely eliminates the maturation period. The brewery enjoys greater fermentation and maturation capacity without investing in new equipment.

WINE

Winemaking, or vinification, is the production of wine, starting with selection of the grapes or other produce and ending with bottling the finished wine. Although most wine is made from grapes, it may also be made from other fruit or nontoxic plant material.

Process of Manufacture

After the harvest, the grapes are taken into a winery and prepared for primary ferment, at this stage red wine making diverges from white wine making. Red wine is made from the must (pulp) of red or black grapes that undergo fermentation together with the grape skins. White wine is made by fermenting juice which is made by pressing crushed grapes to extract a juice, the skins are removed and play no further role. Occasionally white wine is made from red grapes, this is done by extracting their juice with minimal contact with the grapes' skins. Rosé wines are made from red grapes where the juice is allowed to stay in contact with the dark skins long enough to pick up a pinkish colour, but little of the tannins contained in the skins.

To start primary fermentation yeast is added to the must for red wine or juice for white wine. During this fermentation, which often takes between one and two weeks, the yeast converts most of the sugars in the grape juice into ethanol (alcohol) and carbon dioxide. The carbon dioxide is lost to the atmosphere. After the primary fermentation of red grapes the free run wine is pumped off into tanks and the skins are pressed to extract the remaining juice and wine, the press wine blended with the free run wine at the wine makers discretion. The wine is kept warm and the remaining sugars are converted into alcohol and carbon dioxide. The next process in the making of red wine is secondary fermentation. This is a bacterial fermentation which converts malic acid to lactic acid. This process decreases the acid in the wine and softens the taste of the wine.

Red wine is sometimes transferred to oak barrels to mature for a period of weeks or months, this practice imparts oak aromas to the wine. The wine must be settled or clarified and adjustments made prior to filtration and bottling (Fig 26.1).

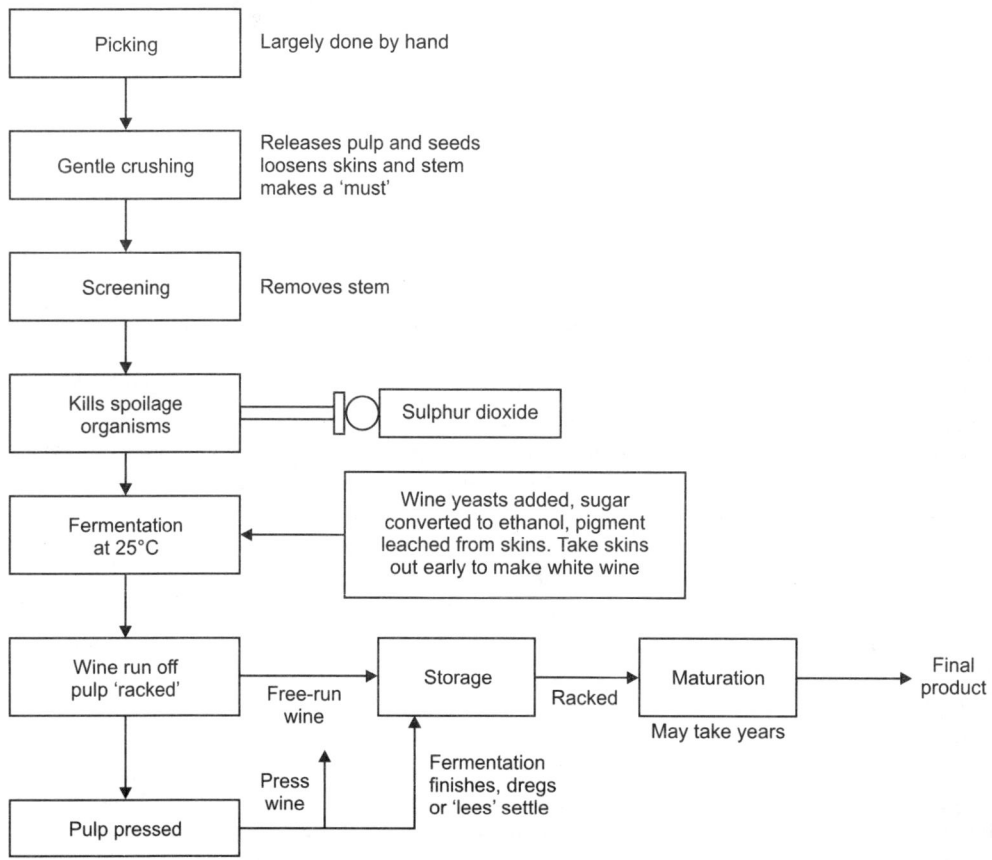

Fig. 26.1: Wine production.

The time from harvest to drinking can vary from a few months for *Beaujolais nouveau* wines to over twenty years for top wines. However, only about 10 per cent of all red and 5 per cent of white wine will taste better after five years than it will after just one year. Depending on the quality of grape and the target wine style, some of these steps may be combined or omitted to achieve the particular goals of the winemaker. Many wines of comparable quality are produced using similar but distinctly different approaches to their production, quality is dictated by the attributes of the starting material and not necessarily the steps taken during vinification.

Variations on the above procedure exist. With sparkling wines such as Champagne, an additional fermentation takes place inside the bottle, trapping carbon dioxide and creating the characteristic bubbles. Sweet wines are made by ensuring that some residual sugar remains after fermentation is completed. This can be done by harvesting late (late harvest wine), freezing the grapes to concentrate the sugar (ice wine), or adding a substance to kill the remaining yeast before fermentation is completed, for example, high proof brandy is added when making port wine. In other cases the winemaker may choose to hold back some of the sweet grape juice and add it to the wine after the fermentation is done, a technique known as süssreserve. The process produces waste-water, pomace, and lees that require collection, treatment, and disposal or beneficial use. The quality of the grapes determines the quality of the wine more than any other factor. Grape quality is affected by variety as well as weather during the growing season, soil minerals and acidity, time of harvest, and pruning method. The combination of these effects is often referred to as the grape's terroir.

Harvest is the picking of the grapes and in many ways the first step in wine production. Grapes are either harvested mechanically or by hand. The decision to harvest grapes is typically made by the winemaker and informed by the level of sugar (called °Brix), acid (TA or Titratable Acidity as expressed by tartaric acid equivalents) and pH of the grapes. Other considerations include phenological ripeness, berry flavour, tannin development (seed colour and taste). Overall disposition of the grapevine and weather forecasts are taken into account. Manual harvesting is the hand-picking of grape clusters from the grapevines. Destemming is the process of separating stems from the grapes. Depending on the winemaking procedure, this process may be undertaken before crushing with the purpose of lowering the development of tannins and vegetal flavours in the resulting wine.

Crushing and primary fermentation

Crushing is the process of gently squeezing the berries and breaking the skins to start to liberate the contents of the berries. Desteming is the process of removing the grapes from the rachis (the stem which holds the grapes). In traditional and smaller-scale wine making, the harvested grapes are sometimes crushed by trampling them barefoot or by the use of inexpensive small scale crushers. These can also destem at the same time. However, in larger wineries, a mechanical crusher/destemmer is used. The decision about desteming is different for red and white wine making. Generally when making white wine the fruit is only crushed, the stems are then placed in the press with the berries. The presence of stems in the mix facilitates pressing by allowing juice to flow past flattened skins. These accumulate at the edge of the press.

For red winemaking, stems of the grapes are usually removed before fermentation since the stems have a relatively high tannin content, in addition to tannin they can also give the wine a vegetal aroma (due to extraction of 2-methoxy-3-isopropylpyrazine which has an aroma reminiscent of green bell peppers). On occasion, the winemaker may decide to leave them in if the grapes themselves contain less tannin than desired. This is more acceptable if the stems have 'ripened' and started to turn brown. If increased skin extraction is desired, a winemaker might choose to crush the grapes after destemming. Removal of stems first means no stem tannin can be extracted. In these cases the grapes pass between two rollers which squeeze the grapes enough to separate the skin and pulp, but not so much as to cause excessive shearing or tearing of the skin tissues.

Most red wines derive their colour from grape skins (the exception being varieties or hybrids of non-vinifera vines which contain juice pigmented with the dark Malvidin 3,5-diglucoside anthocyanin) and therefore contact between the juice and skins is essential for colour extraction. Red wines are produced by destemming and crushing the grapes into a tank and leaving the skins in contact with the juice throughout the fermentation (maceration). It is possible to produce white (colourless) wines from red grapes by the fastidious pressing of uncrushed fruit. This minimises contact between grape juice and skins (as in the making of *Blanc de noirs* sparkling wine, which is derived from Pinot noir, a red vinifera grape).

Most white wines are processed without destemming or crushing and are transferred from picking bins directly to the press. This is to avoid any extraction of tannin from either the skins or grapeseeds, as well as maintaining proper juice flow through a matrix of grape clusters rather than loose berries. In some circumstances winemakers choose to crush white grapes for a short period of skin contact, usually for three to 24 hr. This serves to extract flavour and tannin from the skins (the tannin being extracted to encourage protein precipitation without excessive Bentonite addition) as well as potassium ions, which participate in bitartrate precipitation (cream of tartar). It also results in an increase in the pH of the juice which may be desirable for overly acidic grapes.

In the case of rosé wines, the fruit is crushed and the dark skins are left in contact with the juice just long enough to extract the colour that the winemaker desires. The must is then pressed, and fermentation continues as if the wine maker was making a white wine. Yeast is normally already present on the grapes, often visible as a powdery appearance of the grapes. The fermentation can be done with this natural yeast, but since this can give unpredictable results depending on the exact types of yeast that are present, cultured yeast is often added to the must. One of the main problems with the use of wild ferments is the failure for the fermentation to go to completion, that is some sugar remains unfermented. This can make the wine sweet when a dry wine is desired. Frequently wild ferments lead to the production of unpleasant acetic acid (vinegar) production as a by-product. During the primary fermentation, the yeast cells feed on the sugars in the must and multiply, producing carbon dioxide gas and alcohol.

The temperature during the fermentation affects both the taste of the end product, as well as the speed of the fermentation. For red wines, the temperature is typically 22° to 25°C, and for white wines 15° to 18°C. For every gram of sugar that is converted, about half a gram of alcohol is produced, so to achieve a 12 per cent alcohol concentration, the must should contain about 24 per cent sugars. The sugar percentage of the must is calculated from the measured density, the must weight, with the help of a saccharometer. If the sugar content of the grapes is too low to obtain the desired alcohol percentage, sugar can be added (chaptalisation). In commercial winemaking, chaptalisation is subject to local regulations.

During or after the alcoholic fermentation, malolactic fermentation can also take place, during which specific strains of bacteria convert malic acid into the milder lactic acid. This fermentation is often initiated by inoculation with desired bacteria.

Pressing: Pressing is the act of applying pressure to grapes or pomace in order to separate juice or wine from grapes and grape skins. Pressing is not always a necessary act in winemaking, if grapes are crushed there is a considerable amount of juice immediately liberated (called free-run juice) that can be used for vinification. Typically this free-run juice is of a higher quality than the press juice. However, most wineries do use presses in order to increase their production (gallons) per ton, as pressed juice can represent between 15–30 per cent of the total juice volume from the grape.

Presses act by positioning the grape skins or whole grape clusters between a rigid surface and a moveable surface and slowly decrease the volume between the two surfaces. Modern presses dictate the duration and pressure at each press cycle, usually ramping from 0 to 2.0 Bar. Sometimes winemakers choose pressures which separate the streams of pressed juice, called making 'press cuts'. As the pressure increases the amount of tannin extracted from the skins into the juice increases, often rendering the pressed juice excessively tannic or harsh. Because of the location of grape juice constituents in the berry (water and acid are found primarily in the mesocarp or pulp, whereas tannins are found primarily in the pericarp, or skin, and seeds), pressed juice or wine tends to be lower in acidity with a higher pH than the free-run juice.

With red wines, the must is pressed after primary fermentation, which separates the skins and other solid matter from the liquid. With white wine, the liquid is separated from the must before fermentation. With rose, the skins may be kept in contact for a shorter period to give colour to the wine, in that case the must may be pressed as well. After a period in which the wine stands or ages, the wine is separated from the dead yeast and any solids that remained (called lees), and transferred to a new container where any additional fermentation may take place.

Pigeage: Pigeage is a French winemaking term for the traditional stomping of grapes in open fermentation tanks. To make certain types of wine, grapes are put through a crusher and then poured into open fermentation tanks. Once fermentation begins, the grape skins are pushed to the surface by carbon dioxide gases released in the fermentation process. This layer of skins and other solids is known as the cap. As the skins are the source of the tannins, the cap needs to be mixed through the liquid each day or punched, which traditionally is done by stomping through the vat.

Cold and heat stabilisation

Cold stabilisation is a process used in winemaking to reduce tartrate crystals (generally potassium bitartrate) in wine. These tartrate crystals look like grains of clear sand, and are also known as 'wine crystals' or 'wine diamonds'. They are formed by the union of tartaric acid and potassium, and may appear to be sediment in the wine, though they are not. During the cold stabilising process after fermentation, the temperature of the wine is dropped to close to freezing for 1–2 weeks. This will cause the crystals to separate from the wine and stick to the sides of the holding vessel. When the wine is drained from the vessels, the tartrates are left behind. They may also form in wine bottles that have been stored under very cold conditions. During 'heat stabilisation', unstable proteins are removed by adsorption onto bentonite, preventing them from precipitating in the bottled wine.

Secondary fermentation and bulk ageing

During the secondary fermentation and ageing process, which takes three to six months, the fermentation continues very slowly. The wine is kept under an airlock to protect the wine from oxidation. Proteins from the grape are broken down and the remaining yeast cells and other fine particles from the grapes are allowed to settle. Potassium bitartrate will also precipitate, a process which can be enhanced by cold stabilisation to prevent the appearance of (harmless) tartrate crystals after bottling. The result of these processes is that the originally cloudy wine becomes clear. The wine can be racked during this process to remove the lees.

The secondary fermentation usually takes place in either large stainless steel vessels with a volume of several cubic meters, or oak barrels, depending on the goals of the winemakers. Unoaked wine is fermented in a barrel made of stainless steel or other material having no influence in the final taste of the wine. Depending on the desired taste, it could be fermented mainly in stainless steel to be briefly put in oak, or have the complete fermentation done in stainless steel. Oak could be added as chips used with a non-wooden barrel instead of a fully wooden barrel. This process is mainly used in cheaper wine.

Malolactic fermentation

Malolactic fermentation occurs when lactic acid bacteria metabolise malic acid and produce lactic acid and carbon dioxide. This is carried out either as an intentional procedure in which specially cultivated strains of such bacteria are introduced into the maturing wine, or it can happen by chance if uncultivated lactic acid bacteria are present. Malolactic fermentation can improve the taste of wine that has high levels of malic acid, because malic acid in higher concentration generally causes an often unpleasant harsh and bitter taste sensation, whereas lactic acid is perceived as more gentle and less sour. The process is used in most red wines and is discretionary for white wines.

Tests for wine

Whether the wine is ageing in tanks or barrels, tests are run periodically in a laboratory to check the status of the wine. Common tests include °Brix, pH, titratable acidity, residual sugar, free or available sulphur, total sulphur, volatile acidity and percent alcohol. Additional tests include those for the crystallisation of cream of tartar (potassium hydrogen tratrate) and the precipation of heat unstable protein, this last test is limited to white wines. These tests are often performed throughout the making of the wine as well as prior to bottling. In response to the results of these tests, a winemaker can then decide on appropriate remedial action, for example the addition of more sulphur dioxide. Sensory tests will also be performed and again in response to these a wine maker may take remedial action such as the addition of a protein to soften the taste of the wine.

°Brix is one measure of the soluble solids in the grape juice and represents not only the sugars but also includes many other soluble substances such as salts, acids and tannins, sometimes called Total Soluble Solids (TSS). However, sugar is by far the compound in greatest quantity and so for all practical purposes these units are a measure of sugar level. The level of sugar in the grapes is important not only because it will determine the final alcohol content of the wine, but also because it is an indirect index of grape maturity. Brix is measured in grams per hundred grams of solution, so 20 Bx means that 100 grams of juice contains 20 gm of dissolved compounds.

A Brix test can be run either in the lab or in the field for a quick reference number to see what the sugar content is. Brix is usually measured with a refractometer while the other methods use a hydrometer. Generally, hydrometers are a cheaper alternative. For more accurate use of sugar measurement it should be remembered that all measurements are affected by the temperature at which the reading is made. Suppliers of equipment generally will supply correction charts.

Volatile acidity test verifies if there is any steam distillable acids in the wine. Mainly present is acetic acid but lactic, butyric, propionic and formic acids can also be found. Usually the test checks for these acids in a cash still, but there are new methods available such as HPLC, gas chromatography and enzymatic methods. The amount of volatile acidity found in sound grapes is negligible, since it is a by-product of microbial metabolism. It's important to remember that acetic acid bacteria require oxygen to grow. Eliminating any air in wine containers as well as a sulphur dioxide addition will limit their growth. Rejecting moldy grapes will also prevent possible problems associated with acetic acid bacteria.

Blending and fining

Different batches of wine can be mixed before bottling in order to achieve the desired taste. The winemaker can correct perceived inadequacies by mixing wines from different grapes and batches that were produced under different conditions. These adjustments can be as simple as adjusting acid or tannin levels, to as complex as blending different varieties or vintages to achieve a consistent taste.

Fining agents are used during winemaking to remove tannins, reduce astringency and remove microscopic particles that could cloud the wines. The winemakers decide on which fining agents are used and these may vary from product to product and even batch to batch (usually depending on the grapes of that particular year).

Gelatine has been used in winemaking for centuries and is recognised as a traditional method for wine fining, or clarifying. It is also the most commonly used agent to reduce the tannin content. Generally no gelatine remains in the wine because it reacts with the wine components, as it clarifies, and forms a sediment which is removed by filtration prior to bottling.

Besides gelatine, other fining agents for wine are often derived from animal and fish products, such as micronised potassium casseinate (casein is milk protein), egg whites, egg albumin, bone char, bull's blood, isinglass, lysozyme, and skim milk powder. Some aromatised wines contain honey or egg-yolk extract. Nonanimal-based filtering agents are also often used, such as bentonite (a volcanic clay-based filter), diatomaceous earth, cellulose pads, paper filters and membrane filters (thin films of plastic polymer material having uniformly sized holes).

Preservatives

The most common preservative used in winemaking is sulphur dioxide. Another useful preservative is potassium sorbate.

Sulphur dioxide has two primary actions, firstly it is an anti microbial agent and secondly an anti oxidant. In the making of white wine it can be added prior to fermentation and immediately after alcoholic fermentation is complete. If added after alcoholic ferment it will have the effect of preventing or stopping malolactic fermentation, bacterial spoilage and help protect against the damaging effects of oxygen. Additions of up to 100 mg per litre (of sulphur dioxide) can be added, but the available or free sulphur dioxide should be measured by the aspiration method and adjusted to 30 mg per litre. Available sulphur dioxide should be maintained at this level until bottling. For rose wines smaller additions should be made and the available level should be no more than 30 mg per litre.

In the making of red wine sulphur dioxide may be used at high levels (100 mg per litre) prior to ferment to assist stabilise colour otherwise it is used at the end of malolactic ferment and performs the same functions as in white wine. However, small additions (say 20 mg per litre) should be used to avoid bleaching red pigments and the maintenance level should be about 20 mg per litre. Furthermore, small additions (say 20 mg per litre) may be made to red wine after alcoholic ferment and before malolactic ferment to overcome minor oxidation and prevent the growth of acetic acid bacteria.

Without the use of sulphur dioxide, wines can readily suffer bacterial spoilage no matter how hygienic the winemaking practice.

Potassium sorbate is effective for the control of fungal growth, including yeast, especially for sweet wines in bottle. However, one potential hazard is the metabolism of sorbate to geraniol a potent and very unpleasant by-product. To avoid this, either the wine must be sterile bottled or contain enough sulphur dioxide to inhibit the growth of bacteria. Sterile bottling includes the use of filtration.

Filtration

Filtration in winemaking is used to accomplish two objectives, clarification and microbial stabilisation. In clarification, large particles that affect the visual appearance of the wine are removed. In microbial stabilisation, organisms that affect the stability of the wine are removed therefore reducing the likelihood of re-fermentation or spoilage.

The process of clarification is concerned with the removal of particles, those larger than 5–10 micrometers for coarse polishing, particles larger than 1–4 micrometers for clarifying or polishing. Microbial stabilisation requires a filtration of at least 0.65 micrometers. However, filtration at this level may lighten a wines colour and body. Microbial stabilisation does not imply sterility. It simply means that a significant amount of yeast and bacteria have been removed.

Bottling

A final dose of sulphite is added to help preserve the wine and prevent unwanted fermentation in the bottle. The wine bottles then are traditionally sealed with a cork, although alternative wine closures such as synthetic corks and screwcaps, which are less subject to cork taint, are becoming increasingly popular. The final step is adding a capsule to the top of the bottle which is then heated for a tight seal.

BREWERY WASTE TREATMENT METHODS

After all cleaner production options have been implemented, it will still be necessary to manage two waste streams: waste-water and solid waste. The same concept of the eco-efficiency should be applied to the treatment of both types of waste. Brewery waste-water is generally high in organic material. This section mainly describes alternative treatment to reduce the effect of this organic material on receiving waters. Possibilities for combined treatment of municipal and brewery waste-waters are also addressed.

Characteristics of Brewery Waste-water

Waste-water from breweries is divided into three types:
1. Process waste-water (PWW) from production.
2. Sanitary waste-water (SWW) from toilets and kitchens.
3. Rainwater.

This section focuses on PWW. The brewery's SWW will contribute only a small loading, whether measured as organic material or as flow, but it will require attention in regard to the clogging of pumps and screens. Rainwater should be discharge to a separate drainage system, as it can interfere with the operation of a waste-water treatment plant (WWTP).

The amount of PWW from a brewery will depend on the extent of production and the efficiency of water usage. The waste-water-to-beer ratio will normally be 1.3–2 hl/hl beer less than the ratio of brewing water to beer. The difference is due to water entering the product or being disposed with spent grains and surplus yeast, as well as to evaporation in cooling towers and during wort boiling.

Peak PWW flow will be on the order of 2.5–3.5 times the average flow, depending on how close to the production area the measurement is made. The period of peak flows is normally short. Peak flow occurs in the brewhouse and beer processing area in connection with cleaning operations. In the packaging area peak flows occur when the line is closed down, as bottle washers and tunnel pasteurisers are emptied. Peaks can also occur in the water treatment area during the backwash of filters.

The concentration of organic material will depend on the waste-water-to-beer ratio and the discharge of organic material as waste-water. The concentration of organic material is usually measured in COD (chemical oxygen demand) or BOD (biological oxygen demand). If not otherwise indicated, BOD is measured for a five-day period. The typical discharge of organic material from a brewery varies, but is normally in the range of 0.6–1.8 kg BOD/hl beer. Larger discharges can occur, but they may be attributable to the discharge of surplus yeast, trub or other concentrated wastes, which could be disposed of in better ways. Production of non-alcoholic beer can result in very high discharges if the condensed alcohol ends up in waste-water. Normally PWW is low in non-biodegradable components. Brewery waste-water generally has a COD/BOD ratio of 1.5–1.7, indicating that it is easily degradable. Bioassays, such as the activated sludge oxygen consumption inhibition test, usually show an increased respiration rate compared with that of ordinary municipal waste-water. Organic material in PWW will be present in the form of soluble material, as demonstrated by a typical COD (filtered)/COD (total) ratio of 0.9. The discharge of suspended solids is the range of 0.2–0.4 kg SS/hl beer. Discharge of mash, yeast, kieselguhr and paper pulp will increase this ratio significantly.

Nitrogen concentration will often be in the range of 30–100 g N/m^3. Nitrogen comes from malt and adjuncts. Nitric acid used for cleaning may contribute to the total nitrogen content. Often there is not enough nitrogen for aerobic treatment of PWW and it has to be added. However, the concentration will depend on the water ratio, amount of yeast discharged, and the cleaning agents used.

Phosphorus can also come from cleaning agents. Concentrations vary, but are usually in the range of 30–100 g P/m^3. As with nitrogen, the actual phosphorous concentration will depend on the water ratio and the cleaning agents used. The concentration of heavy metals is normally very low. Wear on the machines, especially conveyors in the packaging lines, can be a source of nickel and chrome. A summary of the characteristics of brewery waste-water is given in Table 26.1.

Table 26.1: Characteristics of brewery waste-water.

Characteristics	*Amount*
Water-to-beer ratio	4–10 hl water/hl beer
Waste-water-to-beer ratio	1.3–2 hl/hl lower than water-to-beer ratio
BOD	0.6–1.8 kg BOD/hl beer
Suspended solids	0.2–0.4 kg SS/hl beer
COD/BOD	1.5–1.7
Nitrogen	30–100 g/m^3 waste-water
Phosphorus	30–100 g/m^3 waste-water
Heavy metal concentration	Very low

Pretreatment

Process waste-water is pretreated in order to reduce the risk of harm to the sewer system, and to avoid occupational health risks to sewer workers. It is also pretreated for operational reasons. Pretreatment takes place either at the brewery or at a municipal WWTP.

A common form of pretreatment is neutralisation of the PWW, which can be carried out:

1. In production areas.
2. In central neutralisation tanks with acid/caustic.
3. Through CO_2 neutralisation.
4. Through biological neutralisation.

In production areas, spent detergent can be neutralised by dosing acids or caustics into the tank before the waste is discharged. The acid or caustic used for neutralisation could be a residue from cleaning. Surplus CO_2 can be used for the neutralisation of caustic in CIP plants or of overflow from bottle washers. The central neutralisation of PWW requires a tank with a hydraulic retention time of approximately 20 minutes. The mixing capacity should be sufficient to keep the tank completely mixed. Sodium hydroxide, hydrogen chloride and sulphuric acids are normally used as neutralisation agents.

Other caustics and acids could be used, but they are normally too expensive. The dosing capacity of the neutralisation plant will depend on the operation of the brewery and the requirements to be met. Since both caustic and acidic cleaning agents are used in breweries, reduction in chemical usage for neutralisation can be obtained by increasing the hydraulic retention time in the neutralisation tank. Neutralisation tanks are often also used as equalisation tanks, with a hydraulic retention time of three of six hours.

It is possible to neutralise caustic PWW using flue gas from the boiler plant. The neutralisation plant can be in the form of a scrubber or a simpler system with venting of the gas to a sump. The efficiency of the system will depend on the flue gas-to-PWW contact. Neutralisation with flue gas is inexpensive since the chemical costs are low. The system can be difficult to operate, and will normally require that PWW is pumped to the scrubber system.

Partial neutralisation through biological conversion will normally occur in PWW. It has been observed that the pH in equalisation tanks can drop without the addition of acids. The effect is difficult to control, but will reduce the dosing requirements of acid to caustic PWW. In order for biological acidification to occur, the hydraulic retention time should be three to four hours.

Anaerobic treatment

The treatment of PWW in an anaerobic reactor converts organic material to methane (CH_4) and CO_2. Compared to aerobic treatment, anaerobic treatment has the advantages/disadvantages shown in Table 26.2. Anaerobic treatment is feasible if the influent concentration is approximately 100 g BOD/m^3. This is possible at many breweries, but will depend on the organic material discharge and the water-to-beer ratio. If the influent concentration is too low, separation of the waste-water from the brewhouse and beer processing areas can be an option. Discharge of trub and yeast will increase the organic concentration, but it will normally be advantageous to dispose of these by-products in other ways.

Table 26.2: Advantages/Disadvantages of anaerobic treatment.

Advantages	*Disadvantages*
High loading of reactor volume	Sensitive process of temperatures, pH and loading
Reduced structural and area requirement	Effluent will normally require further treatment before discharge to recipient
Less energy consumption	Long start-up period
Process produces biogas	Potential for odour problems
Smaller sludge generation	Requires higher influent concentration
Reduced disposal cost	
Limited nutrient addition	
Less biological CO_2 generation (if used for fuel)	

Anaerobic treatment of brewery PWW is developing rapidly, as required influent and obtained effluent concentration are lowered. Loading of anaerobic treatment plants is normally in the range of 5–10 kg COD/m^3/d. Other more highly loaded anaerobic treatment plants are also available. Sludge generation is typically on the order of 0.04–0.08 kg SS/kg COD removed. The effluent concentration for a plant in stable operation is in the range of 100–500 g COD/m^3. It is not normally possible to discharge effluent with these concentration to receiving waters and further treatment is therefore necessary. Before the anaerobic reactor, an equalising/hydrolysing tank with a retention time of four to eight hours is installed. To filter out suspended material, the incoming PWW is passed through a microscreen. Different types of anaerobic reactors are available. The most common type for the treatment of brewery waste-water is the upflow anaerobic sludge blanket (UASB). PWW enters the UASB at the bottom and flows upwards through a bed which consists of granular sludge. In the top sludge, water and biogas and separated. Several versions of this reactor are available. The biogas can be used in the boiler plant, substituting for fuel oil.

Aerobic treatment

Organic material is converted into CO_2 and sludge (biomass) through aerobic treatment. The conversion is done with O_2 supplied to the reactor tank either mechanically or by diffusion from the atmosphere. There are many different types of aerobic WWTP available, from very simple lagoons to complex deep shaft plants. Common types of aerobic WWTP systems for the treatment of PWW are:

1. Activated sludge
2. Biofilters
3. Lagoon

Normally activated sludge systems are the most common and applicable for treating PWW. There are different types of treatment plants. The main difference lies in the loading of the oxidation tanks, ranging from high loading to extended aeration. The efficiency of an aerobic WWTP will depend on several environmental conditions, such as the content of easily degradable organic compounds, temperature, pH, oxygen concentration, and nutrient concentration.

PWW will often be most efficient in the temperature range 25–35°C. In tropical countries attention should be given to possible variations in the influent temperature, as efficiency will decrease rapidly when the temperature exceeds 38–40°C. For biofilters and other systems with plug-flow, a large equalisation tank might be required to prevent temperature shock to the biomass. Activated sludge treatment plants will normally not require an equalisation tank. At 25–35°C, an activated sludge WWTP can be loaded with 1.2–1.8 kg COD/m³/d and obtain a effluent quality of 15–25 g BOD/m³. The degradation of the organic material will result in generation of sludge. Typical generation will be on the order of 0.45–0.55 kg SS/kg BOD removed.

A frequent problem with activated sludge plants is the build-up of filamentous biomass, which can result in excess suspended solids in the effluents and, in the worst case, a sludge washout. The formation of filamentous sludge can be reduced by introducing selector tanks, which benefits biomass with good settlement characteristics. A typical aerobic waste-water treatment plant is shown in Fig. 26.2.

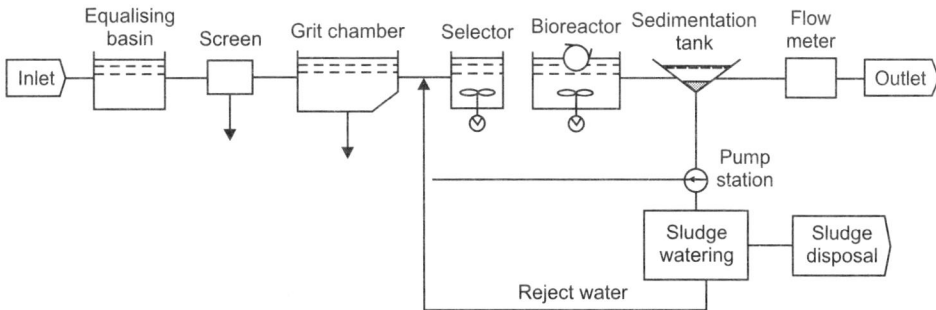

Fig. 26.2: Aerobic waste-water treatment plant for a brewery.

The separation of sludge and treated PWW (effluent) is performed in a sedimentation tank. Normal loading of the tank is in the range of 0.5–1.0 m^3/hr. Loading will depend on the sedimentation characteristics of the sludge. An efficient sedimentation tank can secure a concentration of suspended solids in the effluent of 20–30 g SS/m^3.

For both anaerobic and aerobic sludge, the typical dewatering methods are sludge concentration in a tank, followed by mechanical dewatering in a belt filter press or centrifuge.

If anaerobic treatment is used, the sludge generated is of minimal concern with respect to cost or environment effects and can often be sold. In many cases, excess sludge can be applied to land since the heavy metals content is low. Expert evaluation is needed to determine the most appropriate effluent treatment process for a particular brewery, operating in a particular cost and regulatory environment. The amount of sludge from a WWTP can vary considerably from 0.1 kg/hl to 0.8 kg/hl dry matter.

If effluent requirements are more stringent than 15 g/m^3 BOD and 20–30 g/m^3 SS, further treatment will usually be necessary. It will often comprise the installation of sand filters to remove the suspended solids, as the soluble BOD is very low after extended aerobic treatment.

Application to land

The treatment of PWW by application to land allows the water and nutrients to be used in agriculture. Compared to aerobic treatment and anaerobic treatment, application to land has the following advantages/disadvantages (Table 26.3).

Table 26.3: Advantages/Disadvantages of application to land.

Advantages	*Disadvantages*
Less energy consumption	Requires large amount of land
No nutrient addition/valuable for crop growth	Requires intensive agricultural management
No sludge disposal	Large waste-water storage requirement depending on climate
Recycles water and nutrient	Increased monitoring requirements (especially for metals, etc.)
Net producers of carbon	
No polymer or chemical requirements	

A typical range of loadings for land application is as follows:

1. Hydraulic   5–30 mm/d.
2. BOD   25–100 kg/ha/d.
3. TSS   12–37 kg/ha/d.
4. Nitrogen   220–617 kg/ha/d.

In areas where climatic and soil conditions are favourable, land application of waste-water is typically the low-cost and low environmental impact alternative. The typical crop is grass, harvested either for animal feed or landscaping. Crop use must follow regulatory requirements.

Combined treatment

Many breweries located in urban areas have their waste-water treated by public WWTP. This is in part due to historical reasons or the lack of a feasible water course near the brewery. Combined treatment of municipal and brewery waste-water can in many cases be advantageous for both parties and for the environment.

Some examples of the positive effects of combined treatment in biological WWTPs are:

1. The easily biodegradable organic compounds in brewery waste-water are favourable to removal of nitrogen through the denitrification process.

2. In cold climates, biological processes are affected by the temperature in the WWTP. A temperature increase from warmer brewery waste-water may improve the treatment processes.

3. In biological processes, nutrients (N and P) are needed for the growth of biomass. With an aerobic process in a brewery, waste-water often lacks nitrogen. The need for nutrient addition can be avoided through combined treatment. The increased assimilation can also lead to reduced discharge of nitrogen, as the nitrogen assimilated is removed through the excess sludge.

4. There can also be economic advantages to shared costs.

The combined treatment of municipal and brewery waste-water should be carefully considered in a case by case basis. In some cases some form of pretreatment of the brewery waste-water will be needed. Since the capital investments in WWTP are large, long-term agreements between the parties are encouraged.

Solid Waste Handling

The most important consideration in handling solid waste is that it can be separated into different categories. In breweries, most solid waste can be reused or recycled. At breweries using returnable bottles, a large category will be broken glass. It is possible to recycle this glass in the production of new glass, but it should be relatively free of other materials such as metals and paper. As an alternative, the glass can be disposed at landfills.

Kieselguhr constitutes a very large category. Various recycling methods are under development, but at present they are not capable of totally replacing new kieselguhr with recycled. The handling of kieselguhr as waste is therefore often limited to pressing it and disposing it at sanitary landfills. In some areas disposal on farmland can be an alternative, but warnings concerning possible runoff to nearby water courses must be issued. Kieselguhr may also be used in the production of cement and as a component in brick making.

Label pulp from washing of returnable bottles can be disposed by composting or recycling. Otherwise, the pulp must be disposed at sanitary lanfills. It may contain caustic liquor and heavy metals from the ink. Breweries using one-way packaging materials such as cans and one-way bottles usually generate particularly large amounts of plastic and cardboard waste. Both the plastic and cardboard should be source-separated for recycling. The best applicable handling method for the rest of the solid waste is source separation of materials such as paper, cardboard, metals and wood. Most of this waste can be recycled or burned.

Waste oil and chemicals are more problematic categories. Both types of waste should be separated and sent to facilities outside the brewery. Caution should be taken in storing them at the brewery. Provision of chemical waste handling facilities is normally the responsibility of local public authorities. Other wastes include solvents and paints, a certain amount of which may be recovered and reused. The remaining of these materials which cannot be reused should be incinerated.

GENETICALLY MODIFIED ORGANISMS IN THE WINE INDUSTRY

The process used in creating the transgenic grape was originally developed by McKersie to genetically engineer cold tolerance into alfalfa. The grape was created by inserting naturally occurring genes for cold tolerance into single grape plant cells. The gene introduced into the grape was isolated from a wild cousin of broccoli called *Arabidopsis thaliana*. This gene produces the enzyme superoxide dismutase, which detoxifies the toxic metabolites of oxygen called oxygen-free radicals, which cause many of the damaging reactions associated with stress. During freezing and drought, these molecules attack plant cells and cause decay. Cold tolerance depends on how well a plants proteins can detoxify the oxygen molecules.

The gene was inserted into single grape plant cells via a common soil bacteria, *Agrobacterium tumefaciens*, which was genetically programmed to deliver the coldtolerance genes. The transgenic grape cells grew in an incubator, a growth chamber and the Guelph Transgenic Plant Research Complex before being planted at Château des Charmes. This new transgenic technique could eventually be applied to almost all perennial fruit crops and winter annual plants growing in a cold climate.

Wine is arguably the oldest biotechnological endeavor, with humans having been involved in wine production for at least 7000 years. Despite the artisan nature of its production, work by pioneering scientists such as Antoine-Laurent de Lavoisier and Louis Pasteur placed wine research in a prominent position for the application of cutting-edge biological and chemical sciences, a position it still holds to this day. Technologies such as whole-genome sequencing and systems biology are now revolutionising wine making by combining the ability to engineer phenotypes rationally, with a precise understanding of the genetic makeup and key phenotypic drivers of the key organisms that contribute to this age-old industry.

Vitis vinifera is thought to have originated in Europe and consists of approximately 5000 cultivars used in wine, table and dried grape industries around the world. Initial improvement of grapewines was reliant on random selection of natural mutations which led to an improvement in cultivation and/or some aspect of the fruit quality.

Grapewine improvement has remained relatively untouched however by classical breeding programs, and few new cultivars, e.g. Pinotage, have become commercial successes. In comparison, classical breeding has played a major role in the development of rootstock varieties which are resistant to soil-borne pests and pathogens. Plant genetic engineering has been proposed to have potential for grapewine/wine improvement in the wine industry.

Genetically Modified Grapewines

Vitis vinifera consists of 38–40 chromosomes, the entire DNA sequence of which has recently been determined. Worldwide, all major viticultural research centres are carrying out some form of grapewine biotechnology, including Australia, Chile, France, Germany, Italy, South Africa, Spain and the USA. Initial studies of plant genetic engineering proved difficult since *Vitis vinifera* is a woody perennial, and hence difficult to genetically manipulate (recalcitrant). Only a few limited successes have been reported and there is still no universal protocol for genetically modifying all grapewine cultivars. The first significant progress was made in 1989 using embryonic cell suspensions derived from the plant's anthers as the target tissue for grapewine transformations. Both the techniques of *Agrobacterium*-mediated transformation and particle bombardment have been used for such experiments.

Trials are also being carried out in other countries such as Canada, South Africa and Australia. In South Africa, genetically modified grapewines have recently been planted at Welgevallen, an experimental farm at the University of Stellenbosch.

In South Africa, Winetech (Wine Industry Network of Expertise and Technology) has also been established, which is an organisation performing research into various aspects of genetically improving organisms for the wine industry. Their program involves 'Improving grapewine, wine yeast and bacteria for a quality focused, market directed wine industry'. The mission of Winetech is 'To provide the South African wine industry with a sustainable basis of forefront technology and human resources in order to strengthen both local and international competitiveness and profitability'.

Fungal resistance

Since 1999 there have been field trials with gene-modified fungal-resistant grape wines in two areas in Germany, the Pfalz and Franken. The trials were planned to last for at least 10 years, and examined mainly the varieties Riesling and Chardonnay, for which it had not been possible previously to breed fungus-resistant wines (e.g. against grey mould, powdery and downy mildew). The trials however, were suspended at the beginning of 2005. Recently however, new trials have been initiated in Germany where the wines have been modified with genes from barley, to protect the grapes from fungal infections. The Institute of Wine Cultivations said there would be no noticeable difference in the wine's taste. The first wine made from these altered grapes will most likely only be available in several decades.

In South Africa, a field trial for fungal resistance is being carried out on the Welgevallen Experimental Farm in Stellenbosch. One hundred non-transgenic and virus-free 238 US Vit8-7 rootstocks have been planted and onto these, transgenic Chardonnay or Sultana has been grafted. The grape wines have been genetically modified with a grape gene (Vvpgip) which protects against fungal pathogens, and a grape gene (VvNCED) which protects against water stress.

The wines will be monitored for at least 5 seasons as to their performance. The Institute of Wine Biotechnology (IWBT) is largely focusing on fungal resistance to grey rot (*Botrytis cinerea*), as well as powdery and downy mildew.

Researchers believe the production of fungal disease resistant plants using transgenic technology is an attractive alternative to chemical treatments and should encourage environmentally friendly practices in the wineyard. The IWBT is focusing on the use of chitinases, polygalacturonase-inhibiting proteins.

Virus resistance

Grapewines with genetically engineered resistance to Fanleaf degeneration caused by a virus are being tested in field trials in Colmar (Alsace, France). The virus is transmitted by nematodes and is the most widespread nepovirus involved in grapewine degeneration.

In South Africa, Winetech established a 'Grapewine Virus research program' in 2000 which 'aims to alleviate the serious virus disease problems in the South African wine industry by thorough characterisation of grapewine viruses, to develop strategies to manage the diseases they cause, to prevent further spread of these diseases, and to use the latest technologies for the establishment of genetic virus resistance in wine grape cultivars in the longer term'.

Resistance/tolerance to abiotic stress

Apart from biotic threats such as viruses, fungi, bacteria and insects, grapewines are also threatened by abiotic stresses such as drought, heat, cold, water logging and salt. Grapewines can be affected significantly by water loss in the form of factors such as canopy growth, bunch quality and photosynthetic efficiency. The IWBT is researching into the development of grapewines with enhanced capabilities to grow under adverse conditions including water stress, and are studying the carotenoid biosynthetic pathway of grapewines in this regard. This pathway has been found to produce compounds involved in environmental stress responses.

Examples of genes which have been isolated in the IWBT biotechnology program include:

- Carotenoid biosynthetic pathway: *VvPSY, VvPDS, VvZDS, VvCiso, VvLECY , VvECH, VvLBCY, VvBCH, VvZEP , VvVDE , VvNSY.*
- Abscisic acid biosynthetic pathway: *VvNCED, VvCCD.*

These genes have been used in:

1. Isolation and characterisation of carotenoid pathway genes and promoters from *Vitis vinifera* as resources towards stress-tolerant grapes with superior quality.
2. Functional analysis of central metabolic pathways with regards to roles in stress-tolerance, colour development or sugar metabolism.
3. Metabolic engineering of grapewine towards enhanced abiotic stress resistance and improved quality parameters.

Quality traits and plant development

Australia currently has 4 GM field trials underway. They have field-tested grapes which have had their composition modified of components such as sugar content, colour and grape size. It is believed that modifying flower and fruit development may allow an increase in the harvest.

In South Africa, IWBT are using their research into the carotenoid biosynthetic pathway to also investigate quality parameters of grapes. These metabolic pathways have been found to also be involved in quality aspects and are thus being investigated as to their ability to improve flavour and aroma compounds in grapes. Carotenoids serve as important precursors for apocarotenoids, which are involved in a wide range of functions in plants including pigments, flavours and aromas.

Genetically Modified Yeasts

Yeasts are single-celled organisms that are classified as fungi. According to American Tartaric Products, the first genetically modified wine yeast, ML01, was released in 2005 by Springer Oenologie (a division of Lesaffre Yeast Corporation). This yeast was first available only in North America where GMOs were unregulated, but has since spread to other countries. The yeast was developed by Dr. HJJ van Vuuren from the Wine Research Centre, University of British Columbia, Vancouver, Canada, and was modified by inserting two foreign genes, one from the pombe yeast, a yeast found in Africa and used to make beer, and one from the bacteria *O. oeni*, so that the alcoholic and malolactic fermentations, could occur at the same time. This may thus be a convenience to winemakers, especially those producing large quantities of wine. In 2003, the FDA designated the yeast as GRAS (generally recognised as safe), however, Professor Joseph Cummins, genetics professor at the University of Western Ontario, has stated that wine yeasts are unstable, and that by genetically altering them, it could lead to unexpected toxicity in the final product. There has also been concern that should certain wineries choose to use ML01, the GM wine yeast could contaminate native and traditional wine yeasts through the air, surface waste and water runoff. Many wineries in the Napa Valley are very particular about their choice of wine yeast, and contamination of these other yeast strains would be undesirable to them.

There are now two commercially available GM yeasts. One such yeast has been genetically manipulated to better degrade urea during the wine making process (*Saccharomyces cerevisiae* strain ECMo01). The benefit of such a characteristic is that the wine contains less ethyl carbamate, a chemical considered by some regulatory bodies to be a human health risk. In January 2006, this yeast strain was declared GRAS by the FDA.

The second GM yeast (ML01, been designed to allow malolactic fermentation to proceed more efficiently, thereby producing fewer biogenic amines, such as histamines, which cause headaches and asthmatic-type reactions in some people. Other experiments to genetically modify yeasts include improvement in culture maintenance and the viability of cells, improved yield of fermentable sugars and an improved assimilation of nitrogen. They have also involved an improved tolerance to anti-microbial substances, a reduction in the formation of foam, improved flocculation ability, and an improvement in wine processing with the use of polysaccharide degrading yeast strains which are capable of degrading natural wine polymers (glucans, xylans, pectins). In addition, some yeasts have been genetically modified in order to correct wine acidity.

Genetically Modified Wine Bacteria

Lactic acid bacteria have historically been associated with food and beverage fermentations as they occur naturally in the starting materials used. Lactic acid bacteria occur in must and wine and perform the secondary fermentation known as malolactic fermentation. They are also considered beneficial to the wine's sensory qualities due to flavour modification, and have been found to control spoilage bacteria. *Oenococcus oeni* is a lactic acid bacteria used commercially in the wine industry. Internationally, rapid progress has been made in the last 20 years in the development of tools for the genetic modification of lactic acid bacteria.

Countries involved in the improvement of lactic acid bacteria strains for wine making include Australia, France, Germany, Italy, South Africa and the USA. Major targets of the IWBT are to select for strains that are better adapted as starter cultures for malolactic fermentation. They have also been investigating specific enzymes that are involved in the production of wine aroma compounds, and bacteriocins that can be used as an alternative to chemical preservatives. The genome sequences of several lactic acid bacteria have recently become available contributing to the study of genes of these bacteria.

SECTION IX

Environmental and Ecological Aspects of Food Biotechnology

Chapter 27

Carbon Footprint of Food Industry

INTRODUCTION

The carbon footprint is a measure of the amount of greenhouse gases (GHG) produced by our activities in relation to carbon dioxide (CO_2) or carbon. All activities caused by mankind from building our homes, using our cars to flying on holiday can be the subject of carbon footprinting. The carbon footprint on food is an estimate of all the emissions caused by the production (e.g. farming), manufacture and delivery to the consumer and the disposal of packaging. For example a 1.4L petrol car emits about 160g CO_2 per kilometer. So a carbon footprint of 80g CO_2 for a standard packet of crisps is about the same as driving a typical petrol powered car half of a kilometer.

Thus, the carbon footprint is a measure of the exclusive total amount of carbon dioxide emissions that is directly and indirectly caused by an activity or is accumulated over the life stages of a product. This includes activities of individuals, populations, governments, companies, organisations, processes, industry sectors, etc. Products include goods and services. In any case, all direct (on-site, internal) and indirect emissions (off-site, external, embodied, upstream, downstream) need to be taken into account.

ECOLOGICAL FOOTPRINT OF THE GLOBAL FOOD SYSTEM

After air and water, food is the most essential resource people require to sustain themselves. These resources are provided by the layer of interconnected life that covers our planet: the biosphere. Yet the way the food system provides food often severely damages the health of the biosphere through soil and aquifer depletion, deforestation, aggressive use of agrochemicals, fishery collapses, and the loss of biodiversity in crops, livestock, and wild species.

The global food system has become such a dominant force shaping the surface of this planet and its ecosystems that we can no longer achieve sustainability without revamping the food system. At the same time sustainable food systems provide great hope for building a sustainable future—a future in which all can lead satisfying lives within the means of the biosphere.

The section discusses the Ecological footprint analysis to document the current food systems demand on the biosphere. Ecological footprint accounts track the area of biologically productive land and water needed to produce the resources consumed by a given population and to absorb its waste.

In order to create a sustainable food system, we must break down the food Footprint into its primary components: the cropland Footprint, the pasture Footprint, the fisheries Footprint, and the energy Footprint. By understanding consumption patterns by sector, it becomes easier to target specific areas of consumption.

Cropland Footprint

The world's cropland produces human food, animal feed, fiber, and other non-food crops, and makes up 53 per cent of the global food Footprint. Footprint accounts analyse the consumption of 75 primary crop products and 15 secondary products.

The cropland Footprint has steadily increased with global population. The intensification of farming, using agrochemicals, irrigation, and monoculture cropping, has slowed the expansion of cropland: the cropland Footprint grew by less than 10 percent over the last 40 years, while the world population doubled. But these gains have ecological costs: a swollen energy Footprint and increased demands on neighbouring ecosystems to cope with nutrient loading, soil erosion, toxicity, and water shortages. The concentration of global food production under the control of a few transnational corporations, bolstered by free trade agreements, structural adjustment policies, and subsidies for the overproduction of crop commodities, has created North-South food trade imbalances and import dependencies that underlie a growing food insecurity in many countries. Production of cash crop exports in exchange for food imports can undermine food self-sufficiency and threaten local ecosystems, adding to the global Footprint.

Agribusiness consolidation and large-scale, monoculture cash-cropping also leads to the loss of crop and livestock diversity. Wheat, rice, and corn are now the three most abundant plants on Earth, providing 60 percent of human food. At the same time, industrial agriculture threatens crop diversity through the replacement of native varieties with hybrid strains and the contamination of crop and wild species from the introduction of genetically modified organisms. As the global food supply relies on a diminishing variety of crops, it becomes vulnerable to pest outbreaks, the breeding of superbugs, and climate disruptions, all of which could further expand the human Footprint even as it must shrink.

Pasture Footprint

The world's grazing lands provide us with meat, milk, wool, and hides and represent 13 per cent of the global food Footprint. Footprint accounts analyse eight pasturedependent categories and show a growing pasture Footprint as the world consumes more animal products. While the pasture and grassland Footprint has not grown as rapidly as the consumption of animal products, this is due to the increased use of fertilised pastures for grazing, breeding and managing livestock to boost production efficiency, and feeding livestock from cropland production. In many countries, livestock are at least partially, sometimes exclusively, fed from corn, soyabeans, other crops and crop residues, and fishmeal.

Fisheries Footprint

Industrial scale fisheries are rapidly changing the ecology of ocean ecosystems, while an increasing number of studies document the damaging effects of aquaculture and farmed fish.

Footprint accounts analyse 22 fish and aquaculture categories, incorporating 40 species groups. The global fisheries Footprint has risen more dramatically than other food categories, as the world craves more and bigger fish—that is, fish higher on the food chain. Overall, the world's fisheries are losing productivity. There are fewer and smaller fish. While the catch tonnage remains constant, the quality of fish is declining, as measured by their average trophic level—their status on the food chain. If trends continue, and fish populations from higher trophic levels continue to be overfished or collapse, we may be moving toward oceans of jellyfish or other sea life low on the food chain and with little economic value.

Energy Footprint

The global food system is responsible for a sizeable portion of the world's fossil fuel consumption and corresponding carbon dioxide emissions. Estimates vary depending on how the food system is defined or bounded—we use 10 per cent as a conservative placeholder for calculating the global food energy This 10 per cent includes the energy used in food production, for inputs like fertilisers, pesticides, and irrigation, and in post-produ :tion.

Post-production, which accounts for 80–90 per cent of the food system's fossil fuel use, includes processing, packaging, transportation, storage, and retail. An increasingly globalised food supply means a hefty transport Footprint.

The food system's thirst for fossil fuel energy leads to stunning imbalances: the energy required to produce, process, package, and distribute a can of corn is six times the food energy contained in that corn. The packaging alone uses more than twice the energy of production, driving the corn home from the store and preparing it also uses more energy than production.

Other Food System Impacts—Beyond the Ecological Footprint

As we have seen, the global food Footprint represents a significant portion of the Earth's total biomass production, yet even this is a conservative underestimate of the true area required for food production. Several other factors could be included, as described below.

Unsustainable yields: Footprint accounts currently do not reflect the environmental damage associated with industrial yields, such as soil degradation from intensive agricultural practices, water eutrophication, salinisation from irrigation, or pesticide toxicity.

Climate change: Besides the CO_2 from its fossil fuel use, agricultural production adds to the atmospheric carbon stock through forest clearing and the release of soil carbon through cultivation. Food production also contributes to global warming through the release of methane from livestock, rice cultivation, and the burning of agricultural residues. Yet agriculture has the potential to act beneficially as a carbon sink, through farming practices like conservation tillage that build up organic carbon in soil rather than release it to the atmosphere.

Fresh water: The shortage of fresh water is one of the most immediate and potentially devastating environmental challenges facing humanity. Agriculture depletes water stocks and compromises water quality through increasing loads of organic and inorganic pollutants. Despite its significance, current Ecological Footprint accounts leave out the consumption of water due to a lack of adequate data documenting the impact of a given unit of water, which varies widely depending on soil composition, watershed hydrology, seasonal availability, withdrawal methods, and water quality. Footprint accounts also do not incorporate human activities that cause irreversible damage to the environment, such as aquifer depletion or the bioaccumulation of persistent toxins from pesticides.

EATING LESS BEEF WILL REDUCE CARBON FOOTPRINT MORE THAN CARS

Beef's environmental impact dwarfs that of other meat including chicken and pork. New research reveals, that eating less red meat would be a better way for people to cut carbon emissions than giving up their cars. The heavy impact on the environment of meat production was known but the research shows a new scale and scope of damage, particularly for beef. The popular red meat requires 28 times more land to produce than pork or chicken, 11 times more water and results in five times more climate-warming emissions. When compared to staples like potatoes, wheat, and rice, the impact of beef per calorie is even more extreme, requiring 160 times more land and producing 11 times more greenhouse gases.

Agriculture is a significant driver of global warming and causes 15% of all emissions, half of which are from livestock. Furthermore, the huge amounts of grain and water needed to raise cattle is a concern to experts worried about feeding an extra 2 billion people by 2050. But previous calls for people to eat less meat in order to help the environment, or preserve grain stocks, have been highly controversial.

Researchers analysed how much land, water and nitrogen fertiliser was needed to raise beef and compared this with poultry, pork, eggs and dairy produce. Beef had a far greater impact than all the others because as ruminants, cattle make far less efficient use of their feed. 'Only a minute fraction of the food consumed by cattle goes into the bloodstream, so the bulk of the energy is lost.' Feeding cattle on grain rather than grass exacerbates this inefficiency, although Eshel noted that even grass-fed cattle still have greater environmental footprints than other animal produce.

'The biggest intervention people could make towards reducing their carbon footprints would not be to abandon cars, but to eat significantly less red meat,' 'Another recent study implies the single biggest intervention to free up calories that could be used to feed people would be not to use grains for beef production.' The study of British people's diets was conducted by University of Oxford scientists and found that meat-rich diets defined as more than 100 g per day resulted in 7.2 kg of carbon dioxide emissions. In contrast, both vegetarian and fish-eating diets caused about 3.8 kg of CO_2 per day, while vegan diets produced only 2.9 kg. The research analysed the food eaten by 30000 meat eaters, 16000 vegetarians, 8000 fish eaters and 2000 vegans.

SHRINK YOUR FOOD FOOTPRINT

A person's food footprint (foodprint) is all the emissions that result from the production, transportation and storage of the food supplied to meet their consumption needs. We chose to focus on food supply, rather than only food consumption, because a large proportion of food is lost at retail and consumer level. Although emissions also occur when people transport, store and cook food, these emissions are omitted from our calculations as they are captured in travel and housing footprints.

The biggest sources of emissions are from the beef (0.8 T), dairy (0.5 T) and chicken (0.3 T) food groups. This isn't because these food groups dominate intake in the US diet, but rather that they are relatively carbon intensive to produce, in particular beef and dairy. The remaining food footprint is reasonably well distributed across other groups like cereals, vegetables, fruit, oils, sugars and drinks. In terms of shares, the footprint breaks down like this.

Food Waste

The simplest and most cost-effective way to reduce your food footprint is to minimise food waste. Although not all food waste is within your control, your purchasing and cooking habits can play a large part in reducing food losses. For each of our food groups we divide total food supply into three groups: retail losses, consumer losses and consumption. Each are expressed as a share of total food supplied. Retail losses are food that is supplied to stores but never sold, due to spoilage and processing. Consumer losses are food purchased but not eaten due to a combination of spoilage, the non-edible share of food, cooking waste and plate waste. Consumption is the portion of total food supplied that is actually eaten. Even though not all food supplied can be eaten (e.g. bones, cores, skins), the scale of food loss is still surprising. While small amount of fresh food can be lost during repackaging within stores, the vast majority of retail losses occur because shops are unable to sell food before it goes out of date. In fruit, vegetables, chicken and beef this is more than a quarter of all food supplied.

Consumer losses are between 10% and 25% of total food supplied for each group, meaning that as a share of purchased food they are actually much higher. Take the beef example, 23% of total supply is consumer loss, which corresponds to a loss of 33% of purchased food. Consumer losses are made up of food that is non-edible, allowed to go out of date, wasted during cooking or wasted from the plate.

With the exception of the non-edible share of food you can generally make large reductions in your food footprint by simply ensuring that you eat everything you buy. Doing so just takes some common sense and is action we would naturally take if food was more expensive or scarce. By being careful about the portions you cook, eating what you cook or storing leftovers, you can easily reduce plate waste. By being careful about what you buy, eating perishable things on time and being less squeamish about used by dates where sensible, you can also minimise food spoilage in your home.

A large proportion of retail losses is food that is perfectly fine to eat but is discarded due to supermarket caution and food aesthetics. Although retail losses are largely out of our hands we can still affect them to a degree. Buying food which is close to its 'best before' date often avoids such waste. This makes sense if you know you are going to eat it soon after purchase and such food will often be subject to discounts. By reducing food waste you may be able to shrink your food footprint by as much as a quarter. Although it can be hard to quantify this improvement in your calculations, limiting food waste should result both in a reduction in your kitchen waste volume and food costs. As such reducing food waste is the natural place to start shrinking your food footprint.

Carbon Intensity of Food

The great variation in how foods are produced, processed and transported means their footprints are very different. The vast majority of emissions, typically around 80%, occur during food production. This means how your food is produced is the most important factor in your food footprint. Unlike in other sectors of your personal footprint, which are typically dominated by carbon dioxide emissions, both nitrous oxide and methane play an important role in food production. Nitrous oxide emissions are significant in most food groups due to the widespread use of nitrogen based fertilisers in agriculture. In the US nitrous oxide emissions makes up around a third of total food emissions. Methane emissions occur mostly due to enteric fermentation in animals like cows, sheep and goats so they are largely limited to the beef and dairy food groups. Despite this, methane emissions account for around a quarter of total food emissions in the US.

By using weighted averages for the production of foods within each group we can calculate an average carbon intensity for each. Although such averages aren't accurate for specific foods, they are very good for assessing complete diets. Having converted our intensities to food energy we also want to account for the emissions from retail and consumer losses. By dividing our intensities by one minus the percentage loss for both consumer and retail losses, we produce intensities in terms of grams of carbon dioxide equivalents per kilocalorie (g CO_2e/kcal) of food eaten. This captures all the emissions from the food the food typically supplied for each kilocalorie consumed. After being adjusted for energy content and losses, our results look similar but there are subtle changes. Because of their high energy content snacks, oils and cereals are the least carbon intensive ways to supply food energy. Beef at 14 g CO_2e/kcal remains the most intensive while fruit, dairy and chicken are also relatively high. Again, these are weighted averages so within each group there may be wide variation depending on production methods, food losses and energy content. Using food energy rather than food weight to assess footprints is very useful, because it gives us an idea of the most and least carbon intensive ways we can supply our energy needs. Although daily kilocalorie intake is not the only important factor in diet, it is a simple and clear way to think about different diets while ensuring estimated consumption is in a reasonable range. And the better you understand the carbon intensity of the food you eat, the more effectively you will be able to reduce your footprint.

Different Diets

Because the majority of food supply emissions occur during production, changing your diet is the most effective way to shrink your footprint. Reducing your consumption of beef, lamb and dairy products will have the largest effect, due to their high carbon intensity. Additional reductions may also be found by limiting consumption of other meats and certain types of fruits. Choosing to switch food consumption from high carbon intensity foods (eg beef, cheese) to low-carbon intensity foods (bread, potatoes, grains) can reduce emissions by as much as 90% for that food consumption.

Food Miles

Despite often being highlighted in the media the significant distance that much of our food travels is not the major source of food emissions. This is confirmed by bottom up studies of individual foods and well as top down studies of entire food supply markets. Although this varies from country to country, a rough guide of food supply emissions is something like 80% production, 10% transport and 10% storage at wholesale and retail level. Transport emissions are in fact dominated by the upstream emissions of moving food stuffs to production facilities, with the remainder of transport emissions being from delivering final products to wholesalers and retailers. Reducing these transport emissions can help reduce total emissions, but only if it makes sense in the context of the foods entire footprint. Foods for which transport is a significant factor are often those which are shipped by air due to their perishing quickly and having high value per unit weight.

Cooking and Storage

After its sale, food is generally transported home, stored and often cooked before being eaten. Because emissions from these processes are accounted for in our travel and housing footprints, we do not calculate them as part of our food footprint. This is not to say they are unimportant, it merely avoids double counting. Because fridges and freezers are on all day every day they are often a major use of electricity. In the US refrigeration accounts for 12% of domestic electricity use, which has a footprint of around 400 kg CO_2e a year per person. To a limited degree you reduce the electricity your fridge and freezer consume by not setting them too cold, insuring they have proper seals, are well defrosted and located in the coolest area possible. Much more important however is the choice you make when replacing an old fridge and freezer.

In most countries there are efficiency standards, like the A-G energy labelling in Europe and Energy Star in the US, which you can use to inform your choice. While these tell you how efficient the fridge or freezer is for its class they are only so helpful, because they don't account for size. If you buy the highest class possible, and also keep your fridge-freezer as small as possible given you needs, you will limit how many kilowatt-hours a year your fridge uses. And thus reduce its carbon footprint. Cooking can also be a significant source of emissions. In the US it is around 200 kg CO_2e a year per person per year. In general cooking with gas creates fewer emissions than cooking with electricity, unless you have access to low-carbon electricity.

Land Use Emissions

As we stressed earlier in this guide, emissions from land use, land-use change and forestry (LULUCF) are a major part of global emissions. Although these emissions are generally not calculated as part of the food supply footprint, the expansion of agriculture is a major driver of deforestation, which dominates LULUCF emissions. More than 60% of these land use emissions occurred in Brazil and Indonesia, largely as the result of deforestation. The major driver of this deforestation was conversion to agricultural land. In Brazil as much as 80% of deforestation occurs due to the reclaiming of land for cattle rearing or growing feeds like soya for cattle. In Indonesia deforestation in recent years has been largely driven by clearing land for palm oil plantations, which as one of the cheapest vegetable oils is used widely in food production. By helping to drive deforestation the production of beef, soya and palm oil, along with other crops in tropical regions, has a very large indirect footprint. Though these emissions are generally not quantified as part of the food supply footprint, they are hugely important and an additional reason to consider switching away from particular foods.

FOOD TRANSPORTATION ISSUES AND REDUCING CARBON FOOTPRINT

Transportation is the largest end-use contributor toward global warming in the United States and many other developed countries. Transportation has a significant impact within the food and beverage sector because food is often shipped long distances and not infrequently via air.

Although the impact of transportation is important, full life cycle analyses indicate that for most foods transportation does not have the largest environmental impact. Some analysts, such as Weber and Matthews, estimate that given the typical household food basket, aggregate transportation accounts for just 11% of total carbon emissions associated with food production.

Therefore, it is still worthwhile to consider improving the food distribution system. There are often many options for delivering food to consumers, and these supply chain configurations can result in vastly differing energy and emissions profiles. This chapter provides the background and tools for analysing the energy intensity and resultant emissions of a food distribution system, evaluating tradeoffs and identifying opportunities for significant improvement.

Supply Chain Basics

Before we can further investigate transportation impacts, we must first introduce the concept of the supply chain: the sequenced network of facilities and activities that support the production and delivery of a good or service. A supply chain starts with basic suppliers and extends all the way to consumers via stages. These stages may include such facilities as suppliers, factories, warehouses and other storage facilities, distribution centers, and retail outlets. Figure 27.1 shows a sample supply chain, where the arrows denote the flow of a product toward the consumer. This figure depicts both inbound logistics (the delivery of raw materials and packaging to the manufacturer) as well as outbound logistics (the transportation and storage of the finished good to the end consumer). This section focuses on outbound logistics, colloquially known as 'gate-to-kitchen' and 'farm-to-fork' in the food and beverage industry. The emissions associated with outbound logistics vary by origin and type of food. Weber and Matthews estimate that food transportation may account for 50% of total carbon emissions for many fruits and vegetables, but less than 10% for red meat products. Although inbound logistics can require substantial energy use, it is considered part of the production process. Although the interrelationships between supply chain stages may be quite complex, all supply chains have one aspect in common—they end with a consumer.

Supplier   Manufacturer   Distributor   Retailer   Consumer

Fig. 27.1: A simple supply chain.

Supply chains for different products may be interlinked, one supply chain's end consumer may represent an intermediate node for another supply chain. Examples include a firm that buys components and assembles them into consumer items, and a soft drink producer that buys cylinders of compressed CO_2 to carbonate its products.

Much supply chain complexity results from the fact that few supply chains are completely controlled by one firm or vertically integrated. For example, producers and retailers are not typically owned by the same firm. Companies may outsource supply chain activities, especially transport and storage activities, which are handled more effectively by third party logistic (3PL) providers. Outside firms that form a part of a company's supply chain are channel partners. These partnerships require collaboration across organisations. The supply chain management (SCM) can be defined as the coordination of business functions within an organisation and between the organisation and its channel partners. SCM strives to provide goods and services that fulfill customer demand responsively, efficiently, and sustainably.

SCM includes such functions as demand forecasting, purchasing (also known as sourcing), customer relationship management (CRM), and logistics. Logistics concerns the movement and storage of goods, services, and information. It is an umbrella term for such important functions as transportation, inventory management, packaging, and returns/reverse logistics. Some terminology will be helpful to understand who is doing what. The shipper initiates the movement of the product forward into the supply chain, the carrier is the party that does the actual moving of the product, and the consignee receives the product.

Transport Modes

Within the developed world there are four basic transport modes for shipping large quantities of packaged products: water, rail, truck, and air. Trucking dominates, comprising more than 75% of the total US. Freight transit bill. Trucking variables include truck type, ownership model (such as 3PL or company-owned fleet), and loading option (less-than-truckload or full-truckload). The dominant transport mode has shifted over time. Short sea shipping, using ocean-going vessels for delivering cargo domestically, is popular in Europe and also holds promise for replacing many truck deliveries in the United States. To compare transport modes with regard to energy usage and resultant emissions, we define a ton-km as the movement of 1 metric ton of cargo over 1 km. Table 27.1 shows that these modes have very different energy and emissions profiles. The water and rail transport modes are contingent upon the availability of navigable water and established railroad tracks. An additional consideration is the potential need for supply chain responsiveness: air freight may be the only viable option for long-distance transport when customer orders require immediate fulfillment.

Intermodal transport

Before we choose one mode over another, we should consider intermodal transport. Defined as using more than one transportation mode to move a shipment between two points, an intermodal route might involve shipping cargo by water, then by rail, then by truck. Intermodal transport became practical with the advent of containerisation, where products stay in the same container throughout their entire journey.

Table 27.1: Energy and emissions per T-km.

	Mega Joules per T-km	*kg CO₂e per T-km*
International water-container	0.2	0.14
Inland water	0.3	0.21
Rail[a]	0.3	0.18[a]
Truck[b]	2.7	1.8
Air[c]	10	6.8

Note that utilisation and backhaul rates will affect all figures:
[a] May depend on whether diesel or electric power is used
[b] Depends on size and type of truck, power source
[c] Includes effects from radiative forcing

Containerisation was made possible through global standardisation of container size and features, which dramatically reduced intermodal transfer times and significantly increased cost efficiency. From a sustainability viewpoint, the advantage of intermodal transport is that we can utilise more efficient modes for major transport corridors, and then shift to trucks for transport to remote destinations. Shippers can also use a 3PL provider to oversee the entire shipping process. One disadvantage of intermodal transport is its inherent complexity of coordination and the information technology support required to address that complexity. Another issue is the movement and repositioning of empty containers.

Utilisation and backhaul

Many carbon analysers base calculations on only transport mode and shipping distance. In analysing, we will take into account additional factors, including vehicle utilisation (how full the vehicle is) and backhaul (whether or not the vehicle carries freight on its return journey). Although fully laden vehicles use more fuel than nearly empty ones, most of the energy expended during a trip is used to move the vehicle and not its cargo. Underutilised vehicles waste energy, as do vehicles that return empty. Also, weight and volume limits must be respected, and all but the lightest and bulkiest cargo loads tend to 'weigh out' rather than 'cube out.'

It can be difficult to determine utilisation fractions and backhaul percentages, as these are likely to vary with each trip. Such information is even more challenging to obtain when transportation functions have been outsourced. However, some assumptions can be made. For example, vehicles chartered by 3PL providers are likely to have higher utilisation fractions because they often carry cargo from multiple companies. Third party logistic providers are also likely to have higher backhaul rates, because they have more opportunities for obtaining return freight owing to their broader customer base.

Warehousing

Logistics involves not only the movement of goods, but their storage. Unless a product is custom ordered by an onsite client, it is likely that the product will enter storage at some point in its journey to the consumer. Such storage can occur at any supply chain stage: at the producer, distributor warehouse, and/or retailer stockroom. Intermediate supply chain stages range from pure storage centers to dedicated cross-dock facilities, in which cargo from upstream supplier trucks/railcars is transferred directly to outbound trucks/railcars destined for downstream stages. In addition to storage, warehouses can provide additional services: pick and pack (repackaging palletised products to smaller quantities destined for either retailers or end consumers), customs clearance, or even house product-finishing functions such as customising goods to the local marketplace.

Packaging

Packaging decisions are inherently linked to the supply chain. Goods are frequently shipped in bulk and broken into consumer-sized quantities at a warehouse or other facility, and individual commodities are sometimes bundled into larger end-items, such as multipacks, and palletised. Packaging materials (pallets, boxes, totes, slipsheets, etc.) for both finished goods and intermediate support functions may be designed to be recyclable, compostable, or reusable. Non-landfilled packaging is highly desirable, but creates other challenges, such as the impact of reusable packaging in the reverse supply chain.

Packaging can often be reengineered to reduce package weight or bulk, which can translate into savings in raw materials, landfill impacts, and transport/storage energy use, but extra costs may be incurred elsewhere.

Complexity of food supply chains

The supply chains can be long and complex. Food supply chains are some of the most difficult to manage as they must often address time constraints to avoid spoilage, as well as concerns about contamination, high weight-to-value ratios, fragility, unique packaging requirements, and the potential impact of food being wasted rather than consumed. We show here how these considerations affect outbound logistics.

One challenge relates to food production being inherently dependent on nature. Not only is the cultivation of many foods restricted geographically, but also temporally. Fruits, vegetables, and grains typically have fixed growing cycles with short and specific annual harvest periods. There are three options for supplying fresh produce that is out of season locally: sourcing from distant growing areas, using long-term storage, or cultivating in a protected environment such as a greenhouse. Importing produce often results in lower overall emissions than harvesting and storing local produce for several months. Indeed, energy needed for long-term cold storage can dominate a product's overall emissions profile. Carlsson-Kanyama shows, for example, that storage accounts for 60% of the carbon emissions associated with carrots. Higher emissions can result not only from the energy needed for climate control, but also from the inherent yield losses that occur during storage. Protected cultivation is even more energy intensive. Carlsson-Kanyama and others show that tomatoes produced locally in Swedish greenhouses require ten times the energy as field-grown tomatoes imported from Southern Europe. Thus, long-distance supply chains, even though they are energy intensive, may yield the lowest overall footprint for providing out-of-season product to consumers.

A second challenge is related to situations where similar food commodities are produced locally as well as imported from distant locations, the emissions intensity of the production methods must be considered in any comparison of overall supply chain emissions. For example, Saunders and Barber find that milk solids produced locally in the United Kingdom generate 34% more emissions than the same product imported from New Zealand, even with transport included. This result reflects the more energy-intensive dairy production system in the United Kingdom.

A third challenge is that highly perishable foods require special handling to avoid yield loss and potential health issues. These foods often require cooling, refrigeration, or freezing during transport and/or storage. It may also be necessary to control other conditions, such as humidity, exposure to air, or contact with other items.

These requirements increase energy usage and emissions. A fourth challenge is that the location of facilities within a food supply chain can also affect emissions. For example, Sim and others find that overall carbon emissions can be significantly reduced by locating processing and storage facilities in countries where more electricity is generated from renewable fuels or cleaner energy.

Fifth, when time is of the essence, as in the transport of highly perishable produce such as berries, air freight may be the only viable transport option. Air freighting may also be necessary in regions such as Africa where no other viable alternative exists for transporting produce to market. As previously shown, air freighting is highly energy intensive. Scholz and others report that fresh salmon air freighted from overseas has about twice the environmental impact as frozen salmon transported by container ships over the same distance. The difference owing to transport modes is far more significant in this case than production choices such as wild versus farmed or organic versus conventional.

A sixth challenge is that safe food storage not only requires climate controls, but also a high degree of sanitation. In most developed countries, warehouses must be built and maintained to stringent guidelines to be certified as 'food grade.' In the United States, wood pallets may not be reused and may soon be phased out as unsanitary.

The process of packaging food is yet another challenge. Twede and others emphasise that packaging beverage products is a high-speed automated process involving expensive equipment. Such capital investment and the need for a controlled environment favours centralising packaging at the point of production, even if it might be more energy efficient to ship product in bulk. Food and beverage products typically require extensive packaging, which adds both weight and volume to the product. Additional energy and materials are required to create the packaging and transport it to the production site. Smith performs a life cycle assessment of the Nova Scotia wine industry and finds that the largest contribution to emissions is owing to the production and transport of wine bottles.

Measuring transportation-related carbon emissions

This section presents the basics of performing a carbon audit and concludes with some examples from practice. Although other gases such as methane and nitrous oxide may contribute to global warming, aggregate greenhouse gas measures are typically reported in CO_2 equivalents (CO_2e), which is kgs of CO_2 emitted per kg of product.

Carbon dioxide dominates, comprising 95% of total greenhouse gas emissions by volume. Emissions are colloquially called 'carbon emissions,' which can lead to confusion as some older studies only weigh the carbon component of the gas, which is 30% of the total mass of CO_2. It is also now standard practice to report all significant greenhouse gas emissions as a single carbon emissions.

The scope of the analysis depends on the purpose of the study. Scope 1 includes only direct emissions, whereas Scope 2 also includes indirect emissions from any consumption of purchased electricity, heat, or steam. Scope 3 is the broadest, including all other indirect emissions, such as the extraction and production of purchased materials and fuels, all outsourced activities, and waste disposal. Scope 3 can also include the substantial impact of radiative forcing from the contrails in tallying airplane emissions.

Scope must be carefully considered because incomplete framing (inappropriate scope) may lead to incorrect conclusions. For example, food miles are defined as the distance between the production source and the retail store, or 'farm-to-fork.'

In addition to providing consumers with information, carbon audits can provide useful insights for companies evaluating their operations. However, the figures from carbon audits should be viewed as guidelines rather than as precise and absolute truths.

To sum up, the transportation-related carbon footprint varies from a few per cent to more than half of the total carbon footprint associated with food production, distribution, and storage. Supply chains are complex and varied, and food supply chains are especially challenging because of seasonality, freshness, spoilage, and sanitary considerations. Measuring transportation-related carbon footprint involves careful choice of the scope of the analysis, and there is much uncertainty in the results. Caution is warranted regarding the absolute numbers from carbon assessments, so it may be best to focus primarily on relative comparisons.

Supply chain planners must carefully consider the trade-off between transportation-related energy cost and carbon footprint and storage-related energy cost and carbon footprint. Also, the frequent small deliveries called for by lean manufacturing practices, although optimising efficiency within a facility, can increase overall carbon footprint. Packaging is another important consideration, and the use of plastics rather than glass tends to lower carbon footprint. Benefits for the environment and health are further accrued by plastic recycling.

To reduce carbon footprint, suppliers are consolidating their operations, increasing their use of rail and water transit, and increasing transport efficiency by filling trucks and considering backhaul opportunities. Food waste is another potentially significant contributor to carbon emissions, which could potentially be reduced via alternative packaging options.

CARBON FOOTPRINT RANKING OF FOOD

Table 27.2 shows the greenhouse gas emissions produced by one kilo of each food. It includes all the emissions produced on the farm, in the factory, on the road, in the shop and in your home. It also shows how many miles you need to drive to produce that many greenhouse gases. For example, you need to drive 63 miles to produce the same emissions as eating one kilogram of beef.

Table 27.2: Greenhouse gas emission produce by 1 kg of each food.

Rank	Food	CO_2 Kilos equivalent	Car miles equivalent
1	Lamb	39.2	91
2	Beef	27.0	63
3	Cheese	13.5	31
4	Pork	12.1	28
5	Turkey	10.9	25
6	Chicken	6.9	16
7	Tuna	6.1	14
8	Eggs	4.8	11
9	Potatoes	2.9	7
10	Rice	2.7	6
11	Nuts	2.3	5
12	Beans/tofu	2.0	4.5
13	Vegetables	2.0	4.5
14	Milk	1.9	4
15	Fruit	1.1	2.5
16	Lentils	0.9	2

Meat, cheese and eggs have the highest carbon footprint. Fruit, vegetables, beans and nuts have much lower carbon footprints. If you move towards a mainly vegetarian diet, you can have a large impact on your personal carbon footprint.

Tips for Reducing Carbon Footprints

Tips for reducing carbon footprints are given below:

- Eat vegetarian
- Bring back home-cooking
- Cooking smartly
- Eat organic
- Save water
- Shop wisely
- Shop local
- Reuse and recycle
- Grow your own food

Chapter 28

Utilisation of Food Wastes for Sustainable Development

INTRODUCTION

While it is true that the principle of waste prevention is universally accepted, the practice has lagged far behind. Food industry will also have to concentrate on waste avoidance as well as utilisation of process wastes. Application of clean technologies enhances the safety and quality of the product as well as reducing the energy requirements and environmental impact of the food industry. The main environmental impacts of the food sector are aquatic, atmospheric and solid waste emissions. By choosing proper separation technology, waste-water treatment is usually carried out and is implemented in process installations. The atmospheric emissions are mainly caused by extensive energy use. The food industry consumes a great deal of energy for heating buildings, processes, and process water, for refrigeration and for the transportation of raw materials and products. The increased share of renewable energy sources could slowly reduce the amount of conventional fossil fuel utilisation. Solid by-products and wastes are also generated in high amounts in the food industry. The main treatment method of solid wastes is, at present, composting. Recovery and reuse of by-products and wastes as raw materials is another option. However, microbiological quality and safety is always of major concern.

ANALYSIS OF FRUIT PROCESSING AND EVALUATION OF WASTE MINIMISATION POTENTIAL

From an environmental point of view, processing of berries produces large amounts of effluents and solid waste. In fruit juice processing large amounts of water are used, mainly for cleaning purposes. Due to hygienic and food safety considerations, most of the utilised water is drinking water quality and the amount of water effluent can be up to 10 m^3/T of raw material. The water is used for raw material washing, plant and equipment cleaning, and other industrial utilisation. The resultant waste-water has a high organic content, containing parts of the fruits, cleaning agents, salts and suspended solids.

As the amount and quality of the effluent greatly influences the economic feasibility of a company, efforts should be made to minimise the use of water and therefore to:

1. Use dry methods such as vibration or air jets to clean raw fruit.

2. Separate and recirculate process waste-waters.

3. Minimise the use of water for cleaning purposes.

4. Remove solid wastes without the use of water.

5. Use counter-current systems where washing is necessary.

Solid wastes usually originate from pretreatment—washing and sorting, and they consist of damaged fruits, stems and stalks. A major source of solid waste generation is the pressing process, in which peels, seeds, pulps are separated from the fruit juice. There is a large unused potential in the juice processing wastes, as they contain a sizeable amount of healthy substances, such as flavonoids, colours and pectins.

Another perceived innovative technology development need is called upon for the utilisation of solid wastes from the juice pressing operation. The remaining waste after the berry pressing and separation process steps, i.e. peels and seeds, do contain valuable compounds such as flavonoids or aromatic oils. When applying proper extraction technology, i.e. supercritical fluid extraction, these healthy compounds can be recovered and applied either by the food industry or the cosmetic or pharmaceutical industries.

Application of Membrane Processes for Waste Minimisation

Membrane technology is based on a thin physical barrier through which materials can either pass (the permeate) or be rejected and retained (the retentate) due to a driving force that can be pressure difference, concentration gradient, temperature gradient, and/or electrical potential difference. Appropriately-used membrane separation can provide financial savings and conserve resources. Maximum benefits are obtained when one or both the output streams from the membrane system are recycled or reused, thereby reducing process materials requirement and minimising waste disposal costs. Compared with conventional processing, membrane technology has many advantages. By implementing membranes, the separated substances are often recoverable in a chemically unchanged form and are therefore easily reused. Membrane separation units are compact and their modular construction means that they can be scaled up or down easily.

Membrane filtration processes offer new ways of food processing to fulfil the consumer demand for healthy food rich in valuable components and preserved without chemical additives. With the application, the process becomes simpler, shorter and takes place at lower temperatures, therefore the valuable and heat-sensitive compounds are not lost to any great extent. State-of-the-art membrane technology methods offer the possibility to enhance food safety, and reduce the energy consumption and the environmental impact of food processing. Membrane separations are applied for:

1. Concentration (removal of a diluting solvent such as water).

2. Purification (separation of contaminants).

3. Fractionation (resolution into two or more component substances).

The pressure driven membrane processes are divided based on the membrane pore size to:

1. Micro-filtration (0.1–10 µm).

2. Ultra-filtration (0.01–0.1 µm).

3. Nano-filtration (1–10 nm).

4. Reverse osmosis (0.1–1 nm)

Typical food industrial applications of micro-filtration are:

1. Cold sterilisation of beverages.

2. Clarification of fruit juices, beers and wines.

3. Continuous fermentation.

4. Separation of oil-water emulsions.

5. Waste-water treatment.

Applications of ultra-filtration are:

1. Concentration of milk.

2. Recovery of whey proteins.

3. Recovery of potato starch and proteins.

4. Concentration of egg.

5. Clarification of fruit juices and alcoholic beverages.

Main application of nano-filtration:

1. Removal of micro-pollutants.

2. Water softening.

3. Waste-water treatment.

Typically reverse osmosis is used in:

1. Desalination.

2. Concentration of food juice and sugars.

3. Concentration of milk.

When operating a membrane system, optimal conditions should be found. The measurement of the process efficiency can be the selectivity and the permeate flux. Flux is typically expressed as volume or mass per unit membrane area per unit time, for example litres/m^2/hr. Temperature can affect flux significantly. Operating at high flux levels means that less membrane area is required and economies can be made in terms of capital, operating and membrane replacement costs.

JUICE CONCENTRATION WITH MEMBRANES

Preservation of fruit juices also contributes to waste minimisation by the means of avoiding the spoilage of the product. The traditional preservation methods are based on the addition of chemicals or physical methods such as pasteurisation, evaporation. Comparing to the evaporation, which is widely used for fruit juice concentrate production, energy efficiency is of great importance. The end-product is clean and of good quality, while the by-product of the final concentration step is a clear water that can be re-used in a process, e.g. for the first rinse of the berry fruits or for floor washing purposes.

A complex method based on pressure-driven membrane technologies was carried out for grape juice processing (Fig. 28.1). A two-stage process resulted in a fruit juice concentrate, while the valuable compounds were retained in the juice. As a first step, micro-filtration was applied as a pretreatment for clarification and sterilisation of juice samples. Microbiological experiments and sensory analysis was carried out to prove the efficiency of the process step. Based on the microbiological experiments on YEPD sterile medium that allows the growth of all types of micro-organisms, a six order of magnitude decrease in the total cell number was achieved.

Independent analysers performed profile analysis of micro-filtered juice samples. They analysed the taste and smell of grape juice samples. The analysis took place in a special laboratory, where the analysers were seated in separate cubicles and performed the evaluation aided by a computer program. The following effects were observed on the juice properties after clarification:

1. Loss of the original taste and smell to a certain extent.

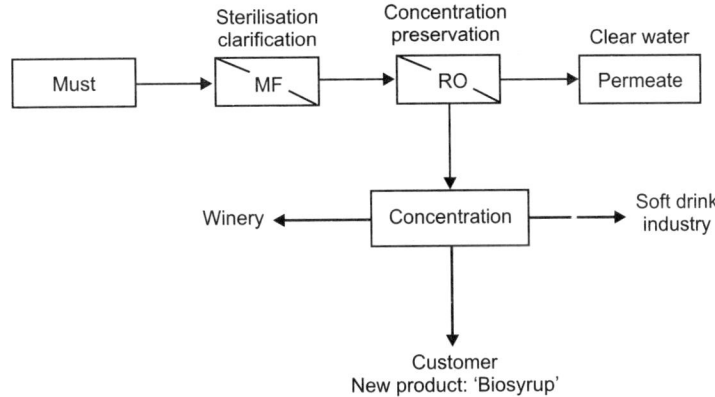

Fig. 28.1: Complex method for fruit juice processing.

2. The appearance of the filtered musts was enhanced
3. Due to lower side taste intensity, the taste of the treated samples was preferred.

As a next processing step, reverse osmosis was applied for concentration, i.e. preservation of the fruit juice. During the filtration procedure, water is removed from the juice as permeate, while the sugar content is enhanced in the retentate, therefore preservation occurs. The measurement of total solids by refractometer was carried out continuously and antocyanin analysis by spectrophotometer was done. Based on these results, concentration has been achieved and the antocyanins were retained in the juice by 99.4 per cent. The conclusions of the work were that application of membrane technologies in fruit juice processing can help to make the procedure cost effective, and environmentally sound. In the future, the application of this process will be tested on Finnish berry juices. However, because of the higher pectin content and higher acidity of berry juices, the application is challenging.

The high pectin content can cause fouling of the membranes due to their molecular weight, which may mean that the concentration procedure might fail. Therefore, an enzymatic pectin breakdown will be carried out before the membrane filtration procedure.

Studies have shown that berry juices have high organic acid content, and low fermentable sugar content. Due to this, the development of berry products is limited. Berry wine production is especially challenging, given that fermentation necessitates sugar addition. However, this results in aroma weakening of the berry wine. To improve the application of berry juices, malolactic fermentation was tested and found to be a promising way for acidity reduction without the loss of natural sugar content.

SFE FOR RECOVERY OF VALUABLE COMPOUNDS FROM SOLID WASTE

Promising technology was found in supercritical fluid extraction (SFE) with natural CO_2 for the recovery of valuable compounds and can be considered as an environmentally friendly solvent-free extraction method that results in minimal oxidative and thermal stress. With SFE, high-value oils as well as aromas can be fully recovered in their natural composition. For these high-value compounds, SFE is not only the most favourable but also the least expensive method of production.

After the quantitative and qualitative analysis of Finnish berry processing by-products, laboratory tests for the recovery of aroma compounds, flavonoids from peels, as well as seed oil recovery by SFE method will be performed.

RECYCLING OF FRUIT AND VEGETABLES

Recycling of fruit and vegetable waste is one of the most important means of utilising it in a number of innovative ways yielding new products and meeting the requirements of essential products required in human, animal and plant nutrition as well as in the pharmaceutical industry. Microbial technology is available for recycling and processing of fruit and vegetable waste and following products can be made out of the different processes.

Fermented Edible Products

A number of beverages such as cider, beer, wine and brandy, and vinegar can be obtained from the fermentation of fruit wastes. Apple pomace has been utilised for the production of cider. Best quality of cider can be made by carbonating it. Good quality apple cider and brandy can also be produced by fermenting milled apple pulp. The possibility of making brandy from dried culled and surplus apples, grapes, oranges and other fruits have also been explored. Vinegar can also be prepared from fruit wastes. The fruit waste is initially subjected to alcoholic fermentation by acetic acid fermentation by Acedobacter bacteria, which produce acetic acid. Vinegar production by fermenting waste from pineapple juice has been reported. Vinegar production by fermenting orange peel juice has also been attempted successfully. Apple pomace extract can also be mixed with molasses in the ratio of 2:1 for producing vinegar.

Single Cell Proteins

Single cell proteins can be produced from dried and pectin extracted apple pomace by using *Trichoderma viride* and *Aspergillus niger*. The grape waste and pressed apple pulp have also been employed as a substrate for Aspergillus niger to generate crude protein and cellulose. Pineapple waste for single cell protein production has also been utilised. Using Fusarium has also used citrus peel juice to generate single cell protein. The waste from brewery and distilleries can also be used for the production of single cell proteins. Potato peels supplemented with ammonium chloride have also been used for the production of protein by using a non-toxic fungi pleurotus ostreatus. Similarly, waste from orange, sugarcane and grape processing industry have also been utilised for the production of single cell protein.

Animal Feed

The waste obtained from processing of fruits and vegetables is rich in fibre, which includes cellulose, hemi-cellulose, lignin and silica with poor quality of protein. Fermented potato waste has been successfully tried as animal feed. Apple pojmace after fermentation with different species yeast, followed by drying, makes the feed enriched with proteins, vitamins, minerals and fats and which can be used for feeding animals. Waste from wineries, breweries and distilleries can be used for feeding livestock. Animal feed can also be obtained from grape pomace after fermentation. Dry brewer's grains after addition of molasses become a very good cattle feed.

Ethanol

The waste from fruits and vegetable processing industries being polysaccharides (cellulose, hemi-cellulose and lignin) can be subjected to solid-state fermentation for the production of ethanol, which has several uses. It can be used as a liquid fuel supplement and as a solvent in many industries. Process for production of ethanol from apple has been developed. Pear and cherry waste have also been utilised for production of ethanol. Orange peel after enzymatic hydrolysis was found suitable for the production of ethanol by use of *Saccharomyces cerevisiae*.

Biogas Production

Biomass consisting of agricultural, forest, crop residues, solid and liquid wastes from industries, sewage and sludge can be utilised for production of biogas through microbial technology. Similarly, the waste from fruit and vegetable processing industries has been used for production of biogas. Biogas is produced by anaerobic digestion of fruit and vegetable wastes. Methanotropic bacteria like *Methanobacterium* and *Methanococus* spp. can utilise CO_2 from waste materials to produce methane. During this process, the complex polymers are first hydrolysed into simple substances by acid forming bacteria and finally these are digested anaerobically by methanotropic bacteria and methane gas is liberated.

Thus, the waste from fruit and vegetable processing in real sense is not a waste as every thing can be profitably recycled, bioconverted and utilised in one or the other form as food, feed or fodder. However, most of the technologies for the waste utilisation are developed at the laboratory scale, so these technologies needed to be standardised for commercial exploitation by the industry. Since the waste is a source of pollution, it has to be treated before discharging into the environment. The regulatory agencies can act as catalyst in developing different processes for the utilisation and management of waste arising out of processing industries and industries engaged in food processing should invest a part of their investment on research and development for waste utilisation and standardisation of the various processes which are commercially viable. Furthermore, the wastage in fruit and vegetable can be effectively managed by the use of biotechnology, by maintaining efficient food distribution system and by promoting domestic and international trade. Thus, proper waste utilisation will add to the wealth of the nation and will benefit all involved in the process.

UTILISATION OF FOOD WASTES FOR SUSTAINABLE DEVELOPMENT

This section deals with the alternative means of handling food wastes including the conversion of wastes from gari industry, cassava peels, shaft from processed cassava, yam, plaintain, banana agro-waste, corncobs, citrus, pastures and sugar cane, forage, organic wastes, etc. into useful products for human and animal consumption. The biconversion of waste to usable energy is also a part of utilisation of waste, as by burning solid fuel for heat, by fermenting plant matter to produce fuel, as ethanol or by bacterial decomposition of organic waste to produce methanol. Alternative means of handling food wastes focus more on utilisation rather than disposal. Thus, the possibility of producing a useful product from wastes will greatly ehnance and ensure sustainable economic development in the world at large. More recently the problem of effluent from processing operation and their disposal has gained public recognition. In many areas of the world, especially the developing countries, the environmental issues are the same. Human beings produce large quantities of wastes as we go about our daily lives. From our homes come wastes from food preparation, washing machines, baths, toilets, newspapers, junk mail, packaging, hobbies, auto and home maintenance projects, and the landscape. In addition, wastes are generated in producing the goods and services we utilise.

Waste is defined as any material, which has not yet been fully utilised, i.e. the leftovers from production and consumption. However, waste is an expensive and sometimes unavoidable result of human activity. It includes plant materials, agricultural, industrial, and municipal wastes and residues. Waste also refers to liquid or solid discharged from residences, business premises, small scale industries, and institutions. In general, waste can be characterised based on its bulk or organic contents, physical characteristics, and specific contaminants. According to Okonko, each waste contains its unique quality and characteristics, which then suggests the type of treatment required. The two divisions of waste—domestic and industrial effluent have different make-ups and often require various treatment processes. Though, waste treatment is generally classified into four levels: primary, secondary, tertiary, and quaternary treatment with each treatment level aimed at removing a more specific class of contaminants.

Sustainable Development

Sustainable development is a process in which the exploitation of resources, the direction of investments, the orientation of technological development and the institutional changes are all made consistent with future as well as the present needs. Sustainable development helps achieve the necessary balance between the resolution of social and economic problems and the protection of the environment, the provision of desirable living conditions for the present generations and measures taken to preserve these conditions for future generations.

Sustainable development is also a key phrase used by politicians, economists and environmentalists. Sustainable advancement and development in relation to a nation is the process of making living, that area of land and/or water more useful or profitable for mankind. The life sickness affects over 30 per cent of global socio-economic and sustainable development turnover by way of healthcare, food and energy, agriculture and forestry. This percentage impact will grow with biotechnological developments which are increasingly improving the efficiencies of production processes in all spheres of life. This therefore implies that biotechnology occupy a very strategic position in the socio-economic advancement and sustainable development of the nation in particular and the world at large. Scientific advances through the years have relied on the development of new tools to improve socio-economy such as health care, agricultural production, and environmental protection.

Sustainable development is a policy which aims to ensure that development meets the needs of our present society without compromising the ability of future generations to meet their own needs. Sustainable development aims to ensure that the development needs of the present do not compromise the needs of future generations. A way of addressing this sensitive issue could be through the sustainable development of waste as an alternative and renewable energy resource.

Waste Utilisation

Wastes are produced by virtually all types of industries, although many cleanup and disposal options exist, no single process can be applied to all types of waste steams. The trend in the world today is to convert waste into useful products through the manipulation of micro-organisms and to recycle waste product as much as possible and the role of micro-organisms in waste utilisation has been studied extensively by several authors. Some workers have thus explored ways of minimising the environmental hazard posed by the gari industry effluent, not just by getting rid of it but by converting it to useful products. Waste utilisation is another approach in waste management practice. Waste utilisation is an ecologically safe and economically efficient method of waste management since, the waste is not treated spending money or disposed off in the landfill causing pollution. Waste utilisation could be brought about by the following methods.

Bioconversion

Biological processes for the conversion of wastes to fuels include ethanol fermentation by yeast or bacteria, and methane production by microbial consortia under anaerobic conditions. Bioconversion is referred to as the enzyme-mediated conversion of organic substrates, such as cellulose, to other more valuable substances, such as protein, by other organisms. The conversion of biomass to usable energy, as by burning solid fuel for heat, by fermenting plant matter to produce fuel, as ethanol or by bacterial decomposition of organic waste to produce methanol is also referred to as bioconversion.

Bioremediation

One of the promising methods for toxic waste cleanup problems is bioremediation. Bioremediation is a environmental biotechnology process that use either naturally occurring or deliberately introduced micro-organisms, to consume and breakdown environmental pollutants into harmless by-products such as water, CO_2 and salts, in order to cleanup a polluted site. Naturally occurring bacteria or fungi that degrade specific substances are isolated, cloned, and manufactured in large quantities and introduced as combinations of micro-organisms into a hazardous waste site to eliminate specific contaminants. Under carefully controlled conditions, it is a practical and cost effective method to remove pollutants from contaminated surfaces and subsurfaces.

Biotechnological processes

In the industry the production processes are now being modified using biotechnology for reduction in pollution caused by the conventional methods. The biotechnological processes also prove to be very economical and also they provide products, which are better or at least equal in quality to the conventional methods. But in these processes the cost of pollution eradication is also saved as these processes generally give out very little or nil pollution and are more efficient than the conventional processes. Biotechnology serves as a solution to many problems in various fields ranging from fuels to many other cleaner and innovative clean up technologies. Some examples are:

1. Biotechnological production of biosurfactants.
2. Biochemical conversion of lignocelluloses substrates to cellulose, liquid glucose, and value added chemicals.

Biotechnological, bioremediation or bioconversion process is often successful and the most inexpensive method, it is only one of many techniques for dealing with hazardous wastes. This biological waste treatment or bioconversion is desirable because it is inexpensive, can be done at the site of pollution, and causes minimal physical disturbance to the surrounding area compared to other methods.

Biocatalysis

To develop biocatalytic methods for the conversion of crop derived carbohydrates to high value polysaccharides or oligosaccharides. The project composed of two major objectives. Develop biocatalytic methods for the conversion of starch, corn coproducts, beet sugar or cane sugar to value-added oligosaccharides.

Biofilm reactors

Nicolella reported that biofilm reactors are in operation at industrial scale throughout the world. Use of biofilm reactors is anticipated to be economical for the production of these industrial chemicals. It has been reported that the best biofilms were obtained with *Pseudomonas fragi*, *Streptomyces viridosporus*, and *Thermoactinomyces vulgaris* when used in combination with polypropylene composite chips.

Sources and Nature of Food Waste for By-product Developments

Waste contains three primary constituents: Cellulose, hemicellulose and lignin, and can contain other compounds (e.g. extractives). Cellulose and hemicellulose are carbohydrates that can be broken down by enzymes, acids or other compounds to simple sugars, and then fermented to produce ethanol renewable electricity, fuels, and biomass-based products. When the amount of organic agricultural waste, such as corn stalks, leaves and wheat straw from wheat-processing facilities, sawdust and other residues from wood mills, is also considered, this component of solid waste could be a principal resource for biodevelopment. Materials of organic origin are known as biomass (a term that describes energy materials that emanate from biological sources) and are of major importance to sustainable development because they are renewable as opposed to non-organic materials and fossil carbohydrates.

Common farm organic wastes such as maize cobs, banana peels, pawpaw fruit peels, maize chaff, stumps of palm tree, palm tree inflorescence, maize stem, rice straw and spinch weeds were earthworm (*Eudrilus eugeniae*) garden snail (*Limicolaria aurora*) and palm grub (*Oryctes rhinoceros*). It was demonstrated by Omoyinmi that animal protein production varied from 0.91 g/kg to 1.41 g/kg of waste in earthworm, from 1.15 g/kg to 1.40 g/kg of waste in garden snail and from 0.90 g/kg to 1.60 g/kg of waste in palm grub. It was also shown that the short life-cycle and production of large number of offsprings could be harnessed for the raising of feed for fish/livestock and in some cases human consumption. This culture of invertebrates offered economic benefits to the farmer and it improved on the environmental quality by transforming wastes into beneficial products.

A surveys on the potential for biomass waste to alleviate energy problems in Tanzania through utilisation of agro-industrial residues for anaerobic conversion into biogas and biodiesel, sisal industry, the largest producer of agro-industrial residues, has a potential to produce energy that could greatly supplement the current shortfall of hydropower generation.

In 2008, Kareem and Akpan reported that the use of agricultural by-products as substrate for enzyme production was cheap and could facilitate large scale production of industrial enzymes in the tropics. Eight isolates of *Rhizopus* sp. was obtained from the environment and were grown on solid media for the production of pectinase enzymes. Three media formulated from agricultural materials were the following: Medium A (ricebran + cassava starch, 10:2 w/w), medium B (cassava starch + soyabean, 1:2 w/w), medium C (ricebran + soyabean + casein hydrolysate, 10:20.5 w/w). The result obtained by Kareem and Akpan showed that medium A gave the highest pectinase activity of 1533.33 µ/ml followed by medium A and C with 1366.66 and 1066.00 µ/ml respectively after 72 hr fermentation. The three solid media supported profuse mycelial growth of *Rhizopus* species and enhanced its pectinase producing potential. A comparative study of the performance of cow dung and poultry manure as alternative nutrient sources in a bioremediation process was described by Obire and Akinde and Chukwura. Obire and Akinde also reported that amelioration of oil polluted soil with cow dung and poultry manure facilitates the disappearance of crude oil in the soil thereby increasing the rate of soil recovery. Poultry manure performed better than cow dung which will greatly enhanced food productivity at such a time like this when the world at large is facing food crisis.

Coffee-husk and pulp

Coffee husk and coffee pulp are coffee processing by-products. Some of the husk is used as organic fertiliser while coffee pulp has its application and utilisation in Swine feeding. The presence of tannins and caffeine diminishes acceptability and palatability of husk by animals.

Caffeine

Caffeine is also a component of several cola drinks. The addition of caffeine in cola drinks is responsible for almost 70 per cent of the world's pure caffeine trading. Asano reported a successful microbial production of theobromine from caffeine while Braham and Bressani and Bressani and Braham have reported the potential uses of coffee berry by-products and the composition, technology, and utilisation of coffee pulp in other species as well as its antiphysiological factors. The popularity of coffee beverage is also based on the stimulant effect of caffeine, because of this pharmacological effect, caffeine has long been added to medical formulations to compensate the depressive effects of other drugs.

Citrus pulp

According to Wing and Wing, the Florida Citrus Exchange established a fellowship for research into uses of citrus waste and thus launched an area of investigation which remains strongly productive, involved primarily is citrus pulp, consisting mainly of the rag, peel, and seeds of oranges with minor amounts from other fruits. This waste collects on concrete slabs or in open pits at canneries. Cattle eat citrus pulp in the fresh state, but it accumulates too fast for current consumption, and it ferments and spoils too rapidly to save as it is produced. The feeding value and nutritive properties of citrus by-products proved that the digestible nutrients of dried grapefruit refuse were good for growing heifers.

Citrus molasses

Citrus molasses also serves as a substrate for fermentation in the beverage-alcohol industry. The remaining distillery waste can be condensed to a very acceptable feedstuff high in pentose sugars and, because of yeast used for fermentation, high in good quality protein. Large and increasing amounts of citrus molasses are used for production of beverage alcohol. The remaining sugars, which are pentoses, cannot be used by the beverage industry, but they are an excellent source of energy for cattle.

Cassava wastes

Disposal of agricultural by-products such as cassava wastes from processing activities is becoming a concern in Nigeria due to its foul odour. Conversion of these low-value cassava wastes into biosorbent that can remove toxic and valuable metals from industrial waste-water would increase their market value and ultimately benefit the millions of cassava starch, garri and fufu producers. Cassava is a major staple food in Nigeria and therefore produces large volumes of wastes, which has been creating environmental nuisance in the region.

Cassava (*Manihot esculenta* Crantz), a major staple food in many tropical countries like Nigeria and therefore produces large volumes of wastes, which has been creating environmental nuisance in the region. Disposal of agricultural by-products such as cassava wastes from processing activities is becoming a concern in Nigeria due to its foul odour. Conversion of these low-value cassava wastes into biosorbent that can remove toxic and valuable metals from industrial waste-water would increase their market value and ultimately benefit the millions of cassava starch, garri and fufu producers.

Cassava also has to be processed in various ways to reduce its cyanide content to safe levels for human consumption. One of the traditional forms into which the tuber is processed is a granular starchy product called 'gari'. In the processing of gari, the cassava roots are peeled, washed, grated and dewatered with presses. It is sometimes allowed to ferment in the press bag for 1–4 days for flavour development. The dewatered meal is sieved to remove large fibres and roasted in large shallow open pots to get garri. The grating and dewatering are the most important unit processes vital for the detoxification of the product, though the roasting step also serves to volatise remnant HCN. Large volumes of starchy effluent are generated during the dewatering or pressing step as waste. In gari-producing communities, huge amounts of this effluent are generated and constitute an environmental menace. In Nigeria, high levels of gari effluent are produced daily and carelessly drained onto roads, streets, rivers and agricultural lands, thereby, constituting a serious environmental hazard. Not only does the effluent result in a strong stench, it could also result in loss of aquatic life because it is toxic. During the grating of cassava, cyanohydrins are produced at maceration by action of endogenous linamarase on the cyanogenic glucosides of cassava, linamarin and lotastrain. Being unstable above pH4 (effluent pH = 4.3), these would degrade non-enzymatically to give free cyanide.

There is thus likely to be higher percentage of HCN than cyanohydrin in the effluent. It is expected that boiling to gelatinise the effluent starch would greatly reduce the HCN content by volatilisation and during fermentation, more HCN would be lost as the broth is agitated by CO_2 released. Also during distillation, if the distillate is collected only at the boiling point of alcohol, most of the residual HCN would be lost before ethanol collection. In this way, a distilled alcoholic beverage with acceptable cyanide content can be prepared from the effluent. The following therefore is a test of hypothesis and is a report of an attempt to prepare an alcoholic beverage fit for human consumption from gari industry effluent.

Smith carried out an investigation into the possibility of producing a useful product from gari industry effluent, which is an environmental menace, using a fermentation process. The effluent starch with an initial HCN concentration of 9 mg/ml was gelatinsed by cooking, hydrolysed using commercial alpha and beta amylases and fermented with *Saccharomyces cerevisiae* from palm wine. The fermented broth was distilled in the presence of dry orange peel. HCN concentration in the final product was reduced by 99.9 per cent, to a level that is safe for human consumption. Ethanol content was 50.1 per cent (v/v). According to Smith and others, sensory evaluation showed that the product was moderately acceptable, though the ethanol yield was low.

This showed that maximum starch conversion to fermentable sugars by the commercial alpha and beta amylases occurred at the 90th minute with sugar concentration increasing from the initial 3 to 23 per cent suggesting that any residual cyanide did not inhibit the amylses. They concluded that the environmentally undesirable gari effluent can be given economic value by being converted to portable spirit which can be employed for the preparation of various beverages and foods as well as pharmaceuticals for internal consumption. The process also detoxifies the final effluent, making it safe for discharge into streams and rivers.

Pineapple peels

According to Tran selection of a strain of *Aspergillus* for the production of citric acid from pineapple waste in solid-state fermentation proved valuable.

Corn cobs/local plant tubers

Corncobs is waste produced from an agricultural products, it was converted into fermentable sugars by the pretreatment processes of dilution with distilled water and the action of concentrated hydrochloric acid respectively. According to Ashiru, the presence of these fermentable sugars, i.e. hexose sugars and additional nutrients added to this substrate, were then utilised each by two yeasts, *Candida albicans*, *Saccharomyces cerevisiae* and a mould, *Neurospora crassa* and converted it into ethanol after incubation for 96 hours at a temperature of 37°C. Growth culture media has also been composed from corn cobs and local plant turbers for microbial isolation as reported in a study by Bankole. All media prepared from local plant tubers supported the growth fungi. Corncob was also acid hydrolysed by Ashiru to obtained glucose and pentose sugars. At the end of acid hydrolysis, corncob demonstrated to give high yield in hexose sugars with a brix degree of 26°.

Sugarcane bagasse

Sugarcane bagasse is also waste product generated in large quantities in Nigeria and is classified as lignocelluloses. It was acid hydrolysed by Ashiru to obtained glucose and pentose sugars. At the end of acid hydrolysis, detoxification was carried out for all substrates using potassium hydroxide and this was followed by the addition of other nutrients to increase yield and facilitate better fermentation. The highest yield and productivity was recorded with the fungus *Neurospora crassa* using sugarcane bagasse with values being $77.88 \text{ g} 1^{-1}$.

Sugar cane molasses

There has been continuous production of citric acid from sugar cane molasses using a combination of submerged immobilised and surface stabilised cultures of *Aspergillus niger*, KCU520 as reported by Smith. Molasses are wastes products produced from sugar refineries and as an agricultural waste respectively, it was converted into fermentable sugars by the pretreatment processes of dilution with distilled water and the action of concentrated hydrochloric acid respectively.

The presence of these fermentable sugars, i.e. hexose sugars and additional nutrients added to these substrates, were then utilised each by two yeasts, *Candida albicans*, *Saccharomyces cerevisiae* and a mould, *Neurospora crassa* and converted into ethanol after incubation for 96 hr at a temperature of 37°C. *Neurospora crassa* produced the highest percentage yield of ethanol from the substrates, with 33.65 per cent yield. *Candida albicans* produced the lowest percentage yield with 14.46 per cent from molasses. *Saccharomyces cerevisiae* produced a percentage 23.59 per cent ethanol from molasses.

Yam tubers waste

Two species each of Yam (*Dioscorea rotundata*) and (*Dioscorea alata*) which are local common plant tuber in Nigeria was sourced and used for the production of amala flour which is being consumed as staple food in Southwestern, Nigeria.

Different species of yam has also been sourced and compared with Irish potato for the growth support capability for fungi (*Aspergillus niger, Rhizopus nigicans* and *Saccharomyces cerevisiae*) in a study by Bankole and Aina in 2007. The raw material was reported to have supported the growth of inoculated fungi with cooked yam by the 3rd day of growth.

Cocoyam tubers

Cocoyam (*Colocesia esculenta*) is an edible root crop belonging to the family Aracea. It makes significant contribution both as root crops and vegetables in the diet of people, particularly Nigerians and Africans at large. The percentage of starch (72 per cent) in cocoyam was exploited in the production of ethanol and vinegar by Braide. This was achieved by two distinct biochemical processes brought about by the action of micro-organisms under controlled environmental conditions.

The first process called alcoholic fermentation was brought about by yeast (*Saccharomyces carlsbergensisi*) and the second process acetic acid fermentation (acetification) was brought about by an acetic acid bacteria (*Acetobacter* sp.). Saccharification of gelatinised mash undergo two stages of enzyme hydrolysis, namely bacteria alpha-amylase (Amylic-TS) and fungal alpha-amylase (AGM) to produce fermentable sugars. Statistically, there exist a significant difference in the aroma and colour, but not in taste between cocoyam vinegar compared with commercial cider and white vinegar at 95 per cent confidence limit.

Soya whey/soyabean curd residue

Agarose-entraped *Aspergillus niger* cells has been used for the production of citric acid from soya whey. Soyabean curd residue supplement has been found to be very significant for enhancement of methane production from pretreated woody waste.

Mango peels

The use of fermented and unfermented mango peels (*Mangifera indica*-R) as animal feeds was reported by Ojokoh. Ripe mango peels (*Mangifera indica*-R) was naturally fermented for 96 hours at room temperature (30°C). The quality of the unfermented and fermented mango peels were accessed by determining the proximate composition, mineral contents, anti-nutritional content as well as the microbiological quality. The result of the proximate analysis revealed that there was an increase in the protein content of the ripe mango peels fermented with value of 8.64 per cent contents. Ojokoh used the fermented and unfermented samples to feed albino rats and found that there was an increase in the daily weight of the albino rat feds with these mango peels.

Banana agro-waste

Banana is major cash crop of this region generating vast agricultural waste after harvest. The agro-waste including dried leaves and psuedostem after harvest was used as substrate for the release of sugars. Thus, under these conditions the agro-waste left behind for natural degradation can be utilised affectively to yield fermentable sugars which can be converted into other substances like alcohol.

Processed food waste

Food waste can be defined as any edible material or by-product that is generated in the production, processing, transportation, distribution or consumption of food. The primary waste products fed to swine are plate and kitchen waste, bakery waste, and food products from grocery stores and this has proved very valuable as reported by many authors and dehydrated restaurant food waste products are used as feedstuffs for finishing pigs.

Biosolids

Biosolids are nutrient-rich, predominantly organic materials. Although classified as a waste material, biosolids can be a beneficial agricultural or horticultural resource because they contain many essential plant nutrients and organic matter. Following proper treatment and processing, biosolids can be recycled as fertilisers or soil amendments to improve and maintain productive soils and stimulate plant growth, with negligible human health or environmental impacts.

Lignocellulose as a resource for bioproducts

Lignocellulose consists of lignin, hemicellulose and cellulose, Sun and Cheng shows the typical compositions of the three components in various lignocellulosic materials. Hydrolysis methods have also been used to degrade lignocellulose to alkaline and acid. However, many processes enzymes are preferred to acid or alkaline processes since they are specific biocatalysts, can operate under much milder reaction conditions, do not produce undesirable products and are environmentally friendly. The future of fermentation technology will be greatly enhanced by lignocellulose conversion. The findings of Okeke and Obi on the lignocellulose and sugar compositions of some agro wastes materials showed that saccharification of agro wastes materials by fungal cellulases and hemicellulases yielded high sugar content. To develop the carbohydrate potential of biowaste materials, its cellulose content has to be converted into sugars such as glucose that can be used as starting compounds in the biosynthesis of many bioproducts. This conversion process could either be acid or enzyme catalysed, the concentrated acid process for producing sugars was reported as early as 1883 and there has been a simultaneous saccharification and fermentation of cellulose into different useful products. This treatment disrupts the hydrogen bonding between cellulose chains and once it has been decrystallised it becomes extremely susceptible to hydrolysis. Diluted acid was introduced in the process as a pretreatment agent to remove the hemicellulose content of biomass before decrystallisation and hydrolysis of the cellulose fraction while Castellonos reported the comparative evaluation of hydrolytic efficiency toward microcrystalline cellulose of *Penicillium* and *Trichoderma* cellulases. Improvements in acid–sugar separation and recovery have opened the door for commercial applications and the current economics of this opportunity are driven by the availability of a cheap feedstock that usually poses a disposal or waste problem.

Potential Biobased Products and Applications

Wastes from biomass can also provide raw materials for a diversity of biobased products. For e.g. plastics from biomass are being produced using polylactic acid from corn.

Production of biofuel

The demand for ethanol has the most significant market where ethanol is either used as a chemical feedstock or as an octane enhancer or petrol additive Brazil produces ethanol from the fermentation of cane juice whereas in the US corn is used. In the US, fuel ethanol has been used in gasohol or oxygenated fuels since the 1980s. These gasoline fuels contain up to 10 per cent ethanol by volume. The production of ethanol from sugars or starch impacts negatively on the economics of the process, thus making ethanol more expensive compared with fossil fuels. However, the huge amounts of residual plant biomass considered as waste can potentially be converted into various different value-added products including biofuels, chemicals, and cheap energy sources for fermentation, improved animal feeds and human nutrients. High energy liquid fuels are also derived from plants.

Production of biogas

Biogas is a renewable fuel and electricity produced from it can be used to attract renewable energy subsidies in some parts of the world. Depending on where it is produced, biogas can also be called swamp, marsh, landfill or digester gas. A biogas plant is the name often given to an anaerobic digester that treats farm wastes or energy crops. Biogas can be produced utilising anaerobic digesters. These plants can be fed with energy crops such as maize silage or biodegradable wastes including sewage sludge and food waste. The prospects for biogas cannot be underestimated. The composition of biogas varies depending upon the origin of the anaerobic digestion process. Landfill gas typically has methane concentrations around 50 per cent. Advanced waste treatment technologies can produce biogas with 55–75 per cent CH_4.

Biogas can be utilised for electricity production, space heating, water heating and process heating. If compressed, it can replace compressed natural gas for use in vehicles, where it can fuel an internal combustion engine or fuel cells. Methane within biogas can be concentrated to the same standards as natural gas, when it is, it is called biomethane. If concentrated and compressed it can also be used in vehicle transportation.

Production of biorefinery products

The most profitable way to operate a biomass-to-ethanol plant is as a refinery producing a variety of products from processing all the chemical components (hemicelluloses, cellulose, lignin, and extractives) of cellulosic feedstock. The plant could make use of extractives by converting them to resin acids or pharmaceuticals (taxols from specific conifers, for example). Cellulose derivatives can be processed into a variety of products including higher value animal feeds.

Production of biodiesel

Biodiesel production is a completely renewable resource. Biodiesel product is made from soya and canola, which is a self-sustaining fuel. Best of all it provides a market for excess soyabean oil production. Biodiesel is a substitute for fuels that produce a lot of soot and carbons. These poisonous elements, which are, associated with regular diesel fuel emissions (especially buses). However, biodiesel has been around for decades as a supplement that is added to conventional diesel fuel to improve the lubricity of diesel engines. A biodiesel fuel consists of methyl esters of soyabean oil. Many car manufacturers are seeing the wisdom of creating vehicles that can accommodate a biodiesel product by creating a diesel car that is friendly to the use of vegetable oil blended with diesel fuel. In addition to displacing North America's reliance on imported petroleum, the use of biodiesel product has been shown to reduce air pollution and greenhouse gases.

Production of Enzymes

Production of cellulases and hemicellulases

Cellulases and hemicellulases have numerous applications and biotechnological potential for various industries including chemicals, fuel, food, brewery and wine, animal feed, textile and laundry, pulp and paper and agriculture. Hemicellulases are used for pulping and bleaching in the pulp and paper industry where they are used to modify the structure of xylan and glucomannan in pulp fibres to enhance chemical delignification. Goyal expounded more on the characteristics of fungal cellulases while Kim studied the factorial optimisation of a six cellulase mixture.

Production of xylanases

In the baking industry xylanases are used for improving desirable texture, loaf volume and shelf-life of bread. A xylanase has shown excellent performance in the wheat separation process.

Production of new enzyme systems

Valuable substances produced by organisms can be growing on food processing wastes. Diacetyl reductase, for example, has been isolated, purified, and characterised by some researchers. This enzyme is of industrial interest for the production of speciality chemicals from fats and oils. β-glucosidase was produced by *Aspergillus niger* grown on fruit pomace. This enzyme is now under investigation for enzymatic release of natural flavours from food processing residues.

Production of Other High-value Bioproducts

Production of xanthan

Xanthan are produced by fermentation using glucose as the base substrate but theoretically these same products could be manufactured from 'lignocellulose waste'.

Production of xylitol

Xylitol used instead of sucrose in food as a sweetner, has odontological applications such as teeth hardening, remineralisation, and as an antimicrobial agent, it is used in chewing gum and toothpaste formulations. Various bioconversion methods, therefore, have been explored for the production of xylitol from hemicellulose using micro-organisms or their enzymes.

Production of Industrial Chemicals

In recent times, many developed countries have returned to carbohydrate-based industries for industrial products. Renewable fuel production includes fuels such as ethanol, methanol, and hydrogen, biodiesel, and Fischer-Tropsch (FT) liquids (fuels derived using the Fischer-Tropsch conversion process). The FT process can produce diesel, naptha, and other fuels that can be used as substitutes for gasoline. Landfills produce a methane rich biogas that is most commonly used for power generation as previously discussed by many authors. Bioconversion of wastes could make a significant contribution to the production of organic chemicals. Other examples of production of industrial chemicals produced in biofilm reactors include acetic acid or vinegar, lactic acid, succinic acid, and fumaric acid.

Production of ethanol

There have been increased industrial uses of agricultural commodities to produce different products. The production of ethanol from biowaste can improve energy security and decrease pollution. Akpan produced ethanol from gari effluents using *Aspergillus niger* in a one-step fermentation process. Rajagopal prepared a beer using cassava. The same can be done using garri effluent to produce different types of alcoholic beverages. Ethanol has been produced continuously in an attached biofilm expanded bed bioreactor of *Zymomonas mobilis* and *Saccharomyces cerevisiae* in biofilm reactors while Krug and Daugulis used *Zymomonas mobilis* immobilised on an ion exchange resin for ethanol production. Adsorbed cells of *Saccharomyces cerevisiae* have also been used in a packed bed continuous bioreactor to produce ethanol from molasses. Lynd investigated ethanol yield and tolerance in continuous culture during thermophilic ethanol production.

Ethanol is an excellent transportation fuel and blends of it have benefits, such as reduced gasoline use, thus lowering the need for fossil fuels. It also improves the performance of an ethanol–gasoline blend and ethanol provides oxygen for the fuel resulting in a more complete combustion with a low atmospheric photochemical reactivity. Even though CO_2 is released during the fermentation of sugars to form ethanol and the burning of ethanol as a fuel, the CO_2 is reutilised to grow new biomass replacing that harvested for ethanol production.

Production of butanol/2,3-butanediol

Butanol is an important industrial chemical that can be produced from a number of carbohydrates using a number of microbial cultures. Butanol can be used as a fuel and has higher/greater energy content than ethanol. Continuous production of 2,3-butanediol from whey permeate using cells of *Klebsiella pneumoniae* immobilised on to bonechar was reported by Maddox.

Production of acetic acid/vinegar

Commercial production of acetic acid or vinegar using biofilm reactors as a bioconversion technology has been exercised for many years. Production of these chemicals has been reported by Crueger and Crueger. The percentage of starch (72 per cent) in cocoyam was exploited in the production of ethanol and vinegar by Braide. Statistically, there exist a significant difference in the aroma and colour, but not in taste between cocoyam vinegar compared with commercial cider and white vinegar at 95 per cent confidence limit. The acetic acid is produced by one of the bacteria grouped in the two genera, *Gluconobacter* and *Acetobacter*. The species that are used commercially include *Acetobacter aceti*, *A. pasteurianus*, and *Gluconobacter oxydans*.

Production of fumaric acid

Biofilm reactors as a bioconversion technology have also been used successfully for the production of fumaric acid from glucose and mineral ore treatment.

Production of citric acid

Citric acid (CA) is a carboxylic organic acid that is soluble in water with a pleasant taste. It is the most important acid used in the food industries. Until about 1920, all commercial CA was produced from lemon and lime juices. Rohr reported that CA can be produced by fermentation process using species of micro-organisms namely *Aspergillus niger*, a fungus which was used commercially for the first time in 1923 and in solid-substrate fermentation. Sugar source has effect on the citric acid production by *Aspergillus niger*.

Production of lactic acid

Production of lactic acid in biofilm reactors is another example of industrial chemical production in such reactors as a bioconversion technology. Demirci evaluated a number of supports for biofilm formation using lactic acid producing cultures. Lactic acid was produced in repeated batch cultures in a biofilm reactor.

Production of succinic acid

Succinic acid is a chemical that has been produced in biofilm reactors during bioconversion processes. The industrial potential for succinic acid fermentation was recognised as early as the late 1970s. Succinic acid ($HOOCCH_2CH_2COOH$) is a dicarboxylic acid, which can be used as a feedstock chemical for the production of high value products such as 1,4-butanediol, tetrahydrofuran, adipic acid, γ-butyrolactone, and n-methylpyrrolidone for applications in agriculture, food, medicine, plastics, cosmetics, and textiles.

Production of forage

Manures have been used as fertiliser for centuries, but crop fertilisation with manure has received renewed attention in recent years as concern for water pollution potential from excess manure has increased. However, the content will vary depending on the type of animal, the feed that was consumed, the way the manure was handled, and other factors. Large, concentrated animal operations such as dairies and layer houses generally import more nutrients onto the farm in feed than goes out in milk, eggs, and animals.

Forage producers can make good use of many materials considered wastes. Layer and broiler manure, dairy manure, septage and biosolids, composted urban plant debris, phosphogypsum, and waste lime from municipal water treatment plants are examples of materials that can be used for fertilising and liming pastures and fields. Often these materials can be obtained at little or no cost to the farmer or rancher. In some circumstances the landowner may be paid for taking the waste, thus collecting a disposal fee while the soil and crops benefit from the materials added. Forage crops such as corn silage and hay are excellent candidates for such application since they remove nutrients in significant quantities.

Application Potentials of Bioproducts

The question now will be do food wastes pollute the soil? Not if the materials are chosen carefully and applied properly. A waste in one situation can be a resource in another. Animal manures have been returned to the land as fertiliser and soil conditioner for centuries. Yard wastes have much the same composition as crop residues, which are considered an asset in good soil management. The Centre for Biomass Utilisation focuses on developing the following technologies: Cofiring biomass with coal using agricultural wastes and food-processing wastes to produce transportation fuels and chemical feedstocks. A wide variety of biomass resources are available on our planet for conversion into bioproducts. These may include whole plants, plant parts (e.g. seeds, stalks), plant constituents (e.g. starch, lipids, protein and fibre), processing by-products (distiller's grains, corn solubles), materials of marine origin and animal by-products, municipal and industrial wastes. These resources can be used to create new biomaterials and this will require an intimate understanding of the composition of the raw material whether it is whole plant or constituents, so that the desired functional elements can be obtained for bioproduct production.

Application of biotechnology to the processing of food (including beverages) produced from agriculture has proved highly valuable. Indeed, the combination of bio-based feedstock, bioprocesses and new products offers the potential to revolutionise chemical industry structures. In less than 10 years, integrated biorefineries will play a role comparable to today's oil and gas crackers. They will make use of row crops, energy crops, agricultural waste and food waste as inputs to extract oil and starch for food, protein for feed, lignin for combustion, cellulose for conversion into fermentable sugars, as well as other by-products.

Sugar will be the key feedstock of the future, as it can be used to ferment ethanol for transportation fuel, but also for a whole set of new, basic building blocks. Molecules such as lactic acids, succinic acid, propylene glycol or 3-hydroxy propionic acid produced at 20 cents per pound can catalyse the innovation of new chemical product families, similar to the innovation boost based on the cracker chemicals in the middle of this 21st century. Indeed, the combination of bio-based feedstock, bio-processes and new products offers the potential to revolutionise chemical industry structures.

Biogas is used extensively throughout rural China and where waste-water treatment and industry coincide. The Biogas Support Program in Nepal has installed over 100000 biogas plants in rural areas. Vietnam's Biogas Program for Animal Husbandry Sector has led to the installation of over 20000 plants throughout that country. Biogas is also in use in rural Costa Rica. In Colombia experiments with diesel engines-generator sets partially fuelled by biogas demonstrated that biogas could be used for power generation, reducing elecricity costs by 40 per cent compared with purchase from the regional utility. Owing to simplicity in implementation and use of cheap raw materials in villages, it is one of the most environmentally sound energy sources for rural needs.

Fossil oil and natural gas are being replaced by carbohydrates from renewable resources as low-cost, renewable feedstock. In particular, the development of technology to convert cellulosic biomass from agricultural waste, food waste or energy crops into fermentable sugars offers the perspective of producing ethanol and other bulk organic chemicals at low cost. The cost of biomass-based ethanol produced on a commercial scale, for example, is expected to undercut the cost of gasoline with oil at $30 per barrel. A first semi-commercial ethanol plant using straw is being operated by Iogen and Shell in Canada. More companies have started their own endeavours in this field, including Dupont, John Deere, Genencor, Novozymes, and Abengoa.

Entirely new bio-based products are competing against conventional products on the basis of a superior cost/performance ratio or are even fulfilling unmet market needs. Enzymes, for example, are fast-growing bioproducts that make washing powder more effective, allow softer processing of textiles and pulp and paper, and reduce nitrogen emissions from animal farming. Other examples are biopolymers, cargill dow's pla (polylactide) offers a green alternative to pla at similar cost and performance—two large-scale plants for further new biopolymers are under construction—Dupont's Sorona and Metabolix's phas (polyhydroxyalkanoate).

Benefits of Sustainable Utilisation of Food Waste

Compared to burning fossil fuel, there may be benefits associated with greenhouse gas reduction with the creation of markets for agricultural waste where its disposal has a negative effect on the environment. For example, the use of biomass as feedstock reduces net carbon emissions to the atmosphere and provides reductions in methane emissions from natural decay processes. The ecologically responsible removal of slash from logged areas benefits the environment. The use of other agricultural residues also reduces emissions of volatile organic compounds, odours, dust, and nuisances associated with agricultural operations such as dairies and animal feeding operations. Improved management of animal manure and solid wastes also reduces groundwater contamination. There are economic benefits from the biomass based industrial activities such as electricity generation, erosion control, and for the production of fuels, animal feed, and green or renewable chemicals (such as solvents and lubricants, polymers and plastics) and increased food productivity. Total economic benefits derived from these activities depend on the mixture of biomass based products generated from it, as well as the state of maturity of these industries.

Considerations for Sustainable Utilisation of Food Waste

1. The effect of utilisation of food waste on the water budget should be considered, particularly where a shallow groundwater table is present or in areas prone to runoff.
2. Minimise the impact of odours of land-applied food wastes by making application at times when temperatures are cool and when wind direction is away from neighbours.
3. Food and agricultural wastes contain pathogens. Wastes should be utilised in a manner that minimises the disease potential to humans or animals.
4. Flushing of animal wastes offers several advantages for the livestock producer including labour efficiency and a more pleasant if not actually healthier environment in animal housing areas.

Challenges of Sustainable Waste Utilisation

Pollution of surface waters

Pollution of surface waters may result from runoff of applied food waste materials. To prevent this, one should not apply waste to frozen or saturated soils. Incorporating surface-applied food waste into soil will limit overland movement of nutrients and organic matter detrimental to water quality. Check 'hydraulic loading' of waste application to ensure that the volume of liquid applied does not exceed the soil's percolation and water holding potential.

High production costs

The cost of processing cellulosic materials to produce ethanol is high. The development of new technologies has significantly decreased the cost of producing ethanol, particularly corn-derived ethanol. However, for some, technologies for cellulosic ethanol production are currently in the experimental state, while others believe that these technologies are already sufficiently mature but have not been widely applied due to lack of capital, which is difficult to attract for the implementation of new technologies.

WASTE TO ENZYMES THROUGH SOLID-STATE FERMENTATION

Waste disposal is a major problem, as the costs associated with waste management are skyrocketing. Annually around 1200 million tons of agricultural wastes are being produced globally. Only 15 per cent is used as animal feed, particularly as cattle supplement, plant fertiliser or soil conditioner, while the rest are dumped in landfills, which leads to adverse environmental impacts including global warming. In order to minimise the waste, one option is to convert the waste to value-added products. brewery spent grain (BSG), sugarcane baggase (SCB) and spent mushroom compost (SMC) are the agro-industrial waste disposed abundantly into landfills in Malaysia. These lignin and cellulose rich agro-industrial wastes have a big potential in enzyme production where it can supply and become economical source of raw material. It can act as an inducer in ligninolytic activities. High sugar content in the wastes made it economically feasible for enzyme production via solid-state fermentation (SSF).

The objective of this section is to examine the prospects of using agro-industrial wastes for the production of value-added product like as enzymes. The study involved the utilisation of three types of agroindustrial wastes such as BSG, SCB and SMC as substrate in SSF trials. Two species of fungi namely *Aspergillus niger* and *Schizophyllum commune* were used as the fermenting organisms. From the fermented extracts, the activities of laccase, a hydrolytic enzyme was assayed. Results indicated that extracts from SCB fermented by *A. niger* showed highest enzyme activity. The laccase activity in SCB was 608 per cent higher than the control. The chemical property, such as cellulose content of the substrate contributes to higher laccase production. Extracts from other substrates (BSG and SMC), fermented by *A. niger* showed 30–145 per cent higher laccase activity compared to control. Laccase activity in all substrates, fermented by *S. commune* shows lower activity than the substrates fermented by *A. niger*. Among the substrates tested, sugar cane baggase seems to be more suitable for extracellular enzyme production by fungal solid state fermentation.

Cultivating fungus using an inert substrate without any free-flowing water is known as solid-state fermentation (SSF). SSF utilises agro-industrial wastes as the substrates in the enzyme production. Substrates (agro-industrial wastes) act as either inert or non-inert material, supporting the fermentation process. Inert substrates only act as an attachment place for the fungal growth, while non-inert substrate also supply nutrients for the fungal growth.

Solid-state fermentation is a suitable process for enzyme production using filamentous fungi since, SSF is well adapted to the metabolism of the fungus and it gives the natural habitat for fungal growth. Enzymes are protein biomolecules that serve as catalysts to speed up or slow down the chemical reactions. Laccases (EC 1.10.3.2) are copper-containing oxidase enzymes that are found in many plants, fungi, and micro-organisms. Laccase has a variety of application in detoxification of environmental pollutants, wine stabilisation, paper processing, and enzymatic conversion of chemical intermediates and production of useful chemicals from lignin. In Malaysia, enzymes have been used in many industries.

Globally approximately 1200 million tons of agricultural wastes are being produced annually. Agricultural wastes are either sold as cattle feed, composted, burned or disposed into landfill, which leads to adverse environmental impacts. Using agro-industrial wastes for SSF is economically feasible and eventually solves the environmental problems caused by their disposal. Agricultural wastes such as brewery spent grain (BSG), sugarcane bagasse (SCB) and spent mushroom compost (SMC) are abundantly found in Malaysia, and these are considered for valourisation. Brewery spent grain is the solid residue left after the separation during the brewing process. It contains of grain husks and other residual compounds not converted to fermentable sugars by the mashing process. In a typical brewery producing about 700000 hL of beer annually, 12125 T of spent grain are generated in one year. Spent compost mushroom (SMC) material is a waste from the production process of edible mushroom, which is composed mainly of saw dust.

The generation rate of SMC in UK alone reaches 200000 T/annum. Sugarcane bagasse is the fibrous (lingo-cellulosic) waste that remains after crushing the sugarcane stalk and extraction of its juice. Every 100 tons of raw sugarcane generate 4 tons of SCB. In Malaysia, approximately 800000 metric tons of sugarcane was produced, resulting in 32000 metric tons of bagasse during 2006.

Materials and Method

Fungal selection

Two different types of fungi, *Schizophyllum commune* and *Aspergillus niger* were screened for their potential in enzyme production via solid-state fermentation.

Substrate for SSF: Sugar cane bagasse (SCB), spent mushroom compost (SMC) and brewery spent grains (BSG) were used as the substrate for solid-state fermentation process. BSG were collected from Carlsberg Brewery Malaysia, SMC from Ganofarm, Tanjung Sepat, Selangor and SCB from SS2, Petaling Jaya. The chemical properties and moisture content of the substrates were studied. Twenty five gm of fresh samples were sterilised at 121°C for 20 minutes in 250 ml flasks and left to cool.

Composition analysis: The moisture content and the chemical properties of the substrates were measured.

Solid-state fermentation

Schizophyllum commune and *Aspergillus niger* were grown on Potato Dextrose Agar (PDA) at 30°C for seven days. Four pieces of seven days old mycelial disk, measuring 1 cm × 1 cm, were inoculated into 250 ml conical flask containing sterile substrate. Negative blanks were prepared without any inocula. All flaks were incubated at 30°C for seven days.

Extraction: Two hundred ml of distilled water were added to each flask and mixed using an incubator shaker at 180 rpm for 45 minutes (Series 25 Incubator Shaker).

Enzyme activity assay: The extracts were tested for laccase activity according to Saito method. A volume of 0.1 ml of the crude extract was mixed with 2.9 ml of Syringaldazine (20 μm in 0.1 m phosphate buffer). The absorbance was read at 525 nm. One unit of laccase activity was defined as one unit of enzyme producing absorbance chance/minute/g of substrate.

To sum up, the current trend in the world today is to utilise and convert waste into useful products and to recycle waste product as means of achieving sustainable development. Among the conversion processes and product developments discussed thus far, methane and ethanol production from various food wastes and energy food crops is economically feasible within the restraints of scale and location. Although biological processes for the production of gaseous and liquid fuels have been well demonstrated with cultured microalgal biomass as reported by most authors, these processes must still be integrated into a system capable of meeting basic requirements for overall efficiency of converting solar energy into biofuels. Furthermore, a model system must at least in principle, be capable of easy scale-up and not be limited by either engineering or economic factors.

Bioproduct development would influence the activities of the food, pharmaceutical, cosmetic and petroleum sectors more in the future as the pressure on waste management and biodevelopment increases. As with any development, the sustainable reuse of food waste resources would not be without difficulties, but it would open up the opportunity for biotechnological developments. The economic environment would need to foster the type of conditions in which the emergent food industry can thrive.

Although, developing countries are still grappling with socio-economic issues including meeting the massive energy shortage demands, food security and developing biotechnological solutions in the agriculture, agro-processing and other related manufacturing sectors. Therefore to manage food wastes, the general pathways of industrial-food waste generation should be reduced, recycled or reused and what is left must be treated and disposed of in an environmentally acceptable way. If a process is not environmentally friendly, it should be redesigned such that it becomes so and where a process cannot be redesigned, then it is necessary to reconsider whether it should be undertaken at all.

Chapter 29

Environmental and Ecological Aspects of Genetically Modified Crops

INTRODUCTION

Risk is defined as the probability or potential for harm from an activity. Environmental risk assessment (ERA) is a structured, reasoned, science-based approach for considering the chance of environmental harm from a particular activity, in this case, the widespread cultivation of a GE plant. The goal of the risk assessment is to identify, characterise and evaluate risks to the health and safety of the environment from the cultivation of the GE plant that resulted from the genetic engineering process. The risk assessment identifies risks by considering a wide range of potential pathways through which harm might occur. Risks are then characterised by considering how serious the harm could be (consequences) and how likely it is that a particular harm could occur.

The risk is then evaluated by integrating the consequences and the likelihood. It is important to remember that all agricultural practices, including traditional and organic agriculture, pose the risk of adverse environmental impacts. Therefore, ERA for the cultivation of GE plants takes a comparative approach: the assessment evaluates any risks posed by the GE plant in comparison to the risks posed by the non-GE plant. For example, several species of the genus *Brassica* are grown worldwide as valuable crops, such as mustard and canola, however many of these species can become weeds. Therefore, in the assessment of environmental risks posed by a GE variety of mustard (*Brassica juncea*), the risk assessment process must evaluate the weediness potential of the GE variety in comparison to the known weediness of the species. A comparative analysis can help risk assessors identify unintended environmental effects resulting from the genetic engineering process. If there are no biologically significant differences, the GE plant is considered to be 'substantially equivalent' to the non-GE variety. If there are significant differences between the GE and non-GE varieties, the analysis focuses on the impact of these differences. A difference does not automatically indicate the potential for harm, many differences will have no adverse environmental effects or even beneficial effects. However, if the risk assessors determine that the difference poses environmental harm, either because an existing risk has increased or because a new risk has been identified, the impact of these differences on the environment should be further assessed.

RISK ASSESSMENT PROCESS

Just as risk assessments tend to reflect the same fundamental principles, they tend to share the same basic organisational framework. That is not surprising, because if every risk assessment was performed in a unique way, there would be no basis for decision makers or the public to compare the results of one risk assessment with another. All risks are relative and the evaluation of a particular risk, e.g. the use of new pesticide, is meaningless unless it was performed using the same process as the assessments of existing pesticides already on the market. Similarly, risk assessments of GE plants should be performed using the same basic process each time, so that valid, robust comparisons can be made between multiple risk assessments. This should be true even if the assessments were performed at different times, by different risk assessors.

Risk assessment, including the assessment of risks from GE plants, can be described as a four-step process.

1. Risk identification (What could go wrong?) Regulators consider a broad range of scenarios in which the release of a GE plant, for purposes of cultivation, could possibly cause harm to people or the environment. In each scenario there must be a causal link between the cultivation of the GE plant and the harm. Risk identification should be comprehensive and rigorous, however, care should be taken to avoid over-emphasising insubstantial risk scenarios. Risks that warrant detailed consequence and likelihood assessments to determine the level of risk they pose to human health and safety or to the environment are generally identified by considering the following questions.

   • Is the potential harm attributable to the genetic engineering process? Any harm not posed by or resulting from the use of gene technology should not be considered.

   • Is there a plausible and observable pathway linking the proposed cultivation of the GE plant to the potential harm? In cases where no plausible or observable pathways link the proposed cultivation to the potential harm, the risk scenario should not be considered further.

   • Is the risk substantive? After an initial consideration of the chance and seriousness of harm, does the risk scenario warrant more detailed consideration?

2. Risk characterisation consequence assessment (How serious could the harm be?) Once a risk has been identified, regulators assess the severity of the potential harm. The seriousness of harm is dependent on the scale at which impacts are considered. Harm to humans may be considered significant at the level of an individual, whereas harm to the environment is usually considered significant at the level of species, communities or ecosystems. Assessing the seriousness of harm to people or to the environment may include consideration of the following questions.

   • What is the magnitude of each potential adverse impact: does it cause a large change over baseline conditions?

   • What is the spatial extent or scale of the potential adverse impact?

   • What is the temporal component of the impact, namely, the duration and frequency? Does it cause a rapid rate of change? Is it likely to occur in the short or long term? What is the duration (day, year, decade) over which an impact may be discernible? Will the nature of the impact change over time? Is it intermittent or repetitive? Will the impact disappear at some point?

3. Risk characterisation: likelihood assessment (How likely is the harm to occur?) Regulators examine the causal link between the cultivation of the GE plant and a particular harm and determine how likely it is that the harm will occur. In the chance of harm is close to zero, then risk is considered

minimal and needs no further analysis. However, care needs to be exercised when considering the remote possibility of risks that may have extreme adverse impacts.

4. Risk evaluation (What is the level of concern?) Once regulators have assessed the severity of the harm and the likelihood of its occurrence, they evaluate whether the risk is negligible, low, moderate or high. Risk is evaluated against the objective of protecting the health and safety of people and the environment to determine the level of concern and subsequently, the need for controls to mitigate or reduce risk. Risk evaluation may also aid consideration of whether the proposed cultivation should be authorised, whether further assessment is necessary or whether additional data must be collected. Risk evaluation combines the findings from the consequence (hazard) and likelihood (exposure) assessments, using a matrix to determine the level of risk and whether risk mitigation is needed to reduce the level of risk. To help inform the regulatory decision making process and make the process more transparent, it is useful to define discrete levels of risk. The risk assessment process is frequently iterative in nature: regulators may analyse the data they have collected relative to a particular risk hypothesis and determine that they need to return to Problem Formulation to collect more data or to restate the risk hypothesis. This iteration is common in all fields of risk assessment and generally results in a better outcome from the assessment process.

PROBLEM FORMULATION FOR ENVIRONMENTAL RISK ASSESSMENT

Problem formulation is a multi-step framework that provides the means to organise an environmental risk assessment so that the assessment is done in a logical and transparent way. Typically the problem formulation process is represented as a series of five steps:

1. Identify the protection goal.
2. Derive the operational goal.
3. Determine the assessment endpoint.
4. Formulate the risk hypothesis.
5. Determine the measurement endpoints.

This stepwise process helps risk assessors decide what questions the assessment will address and what data are most relevant to those questions. In the end, problem formulation facilitates both the decision-making processes in risk assessment and clarifies to stakeholders on how the decisions are made. It is a five-step process, presented below. In this example, the risk assessors are assessing the potential risks of growing an insect-resistant GE cotton variety, the cotton plant produces a protein (*Bt* protein) that is toxic to certain insects.

Identify the protection goal: The purpose of an environmental risk assessment for the commercial release of a GE plant is to determine whether the plant can be released while protecting valued environmental resources. A Protection Goal is a broad statement of national policy focused on the protection of a key environmental resource of recognised value, such as water quality, human health or agricultural productivity. The purpose of a risk assessment for the commercial release of a GM plant is to determine whether the plant can be released while protecting these valued environmental resources.

Example: 'Protect biodiversity': In other words, the assessment addresses the question of whether the commercial release of insect-resistant cotton will impair India's ability to meet one of its protection goals, in this example, the protection of biodiversity.

Derive the operational goal: Generally, protection goals are articulated using very broad language, sometimes including legal or technical terms, however, the risk assessment process is case-specific, grounded in science and based on testable hypotheses. So before the risk assessment process can begin, assessors must derive one or more specific operational goals from the protection goal, which suggests the types of questions the assessors must address and the data they must consider. For example, from a broad protection goal, such as 'protect biodiversity,' the risk assessors could derive a more specific operational goal that relates to the context of crop production.

Example: 'Protect agriculturally important pollinators.' This goal suggests the types of questions the risk assessors must address and it begins to narrow the scope of the data the assessors must consider to assess potential risks from the commercial release of a GM crop plant. In this case, it is data regarding risks to agriculturally important pollinators.

Determine the assessment endpoint: Next, the risk assessors must determine one or more Assessment Endpoints appropriate to the Operational Goal. An Assessment Endpoint specifies the nature of the protection given to the environmental resource, i.e. what specifically will be protected, how much protection will be given and for how long.

Example: 'Cultivation of insect-resistant cotton will not threaten long-term sustainability of honeybee populations, compared to cultivation of the non-GE cotton.' For example, this assessment endpoint identifies honeybees as a valued environmental resource that India intends to protect, it defines sustainable bee populations as the nature of the protection, and it states that protection will be provided to bees for a long period of time.

Formulate the risk hypothesis: The Assessment Endpoint is then reformulated into a Risk Hypothesis, which is a statement that can be tested and found to be either true or false, using specific scientific data.

Example: 'Cultivation of insect-resistant cotton will adversely affect honeybee populations, compared to cultivation of non-GE cotton.' In this example, the risk hypothesis has been phrased as a positive statement, but it is also acceptable to phrase the hypothesis as a negative statement, i.e. cultivation of the GE plant will not adversely affect. In either case, the job of the risk assessors is the same: to determine, using scientific data, whether the hypothesis is true or false.

Once the risk hypothesis has been formulated, the assessors return to the original causal pathway they developed during the risk identification step. The assessors may have hypothesised a causal link between *Bt* cotton and honeybees in which the bees could be adversely affected, simply because bees are known to pollinate cotton. However, this type of preliminary causal relationship is insufficient for the purposes of risk assessment. Identifying all the intervening steps in a causal pathway leading to harm is crucial to determining the likelihood that a particular harm may occur. A causal pathway leading to increased harm may involve many steps, all of which must occur. If some of the steps have only a small chance of occurring, then the overall pathway has an extremely limited chance of occurring due to the combination of several steps with low probablility. Alternatively, one step may have almost no chance of occurring, resulting in a very low overall probability even if all other steps have a reasonable chance of occurring. Some pathways can be complex, but identifying each of the steps makes the analysis simpler to do and easier for stakeholders to understand.

Using the example of a GE cotton variety that has been genetically engineered to produce bacterial toxins that kill insect pests that feed on cotton, regulators may be concerned that the commercial cultivation of this GE cotton variety may cause harm to honeybees. To assess the risk, they develop a detailed causal pathway (also called a 'pathway to harm') to help evaluate the probability that bees will be harmed. This pathway starts with the risk hypothesis and breaks it into discrete steps, all of which must happen for the harm to occur.

Determine the Measurement Endpoints: Once the Risk Hypothesis has been formulated and the detailed pathway to harm has been prepared, it is straightforward for the assessors to determine the specific types of data, whether qualitative or quantitative, that will enable them to test the Risk Hypothesis. These data are called Measurement Endpoints.

Example: Data regarding honeybee mortality when exposed to GE and non-GE cotton plants.

The goal is to identify specific ways, including both intentional changes and unintended ones, in which the GE plant is significantly different from the non-GE version and how those differences could impact the long-term sustainability of honeybee populations.

The Risk Hypothesis is based on a comparison between the GE plant and the non-GE version of the plant and so the data collection process must first collect sufficient information to fully characterise the biology of the non-GE version. Then data must be collected that might identify and characterise significant differences between the GE and non-GE versions of the plant that might adversely impact honeybees. The 'pathway to harm' method helps with the determination of Measurement Endpoints in two important ways. First, each step suggests the specific data that the risk assessors will need to collect and analyse. This data will probably be provided by the applicant or it may be available in the published scientific literature. Second, this method helps developers of new GE crops identify and understand which data they will need to collect for the risk assessment process. A third advantage to this technique is that it describes a very logical process that is easily understood by stakeholders.

For example, given the hypothesis regarding impacts to honeybee populations, the risk assessors must fully understand how the non-GM version of the plant interacts directly or indirectly with honeybees and evaluate the repercussions of those interactions, in terms of negative impacts on the bees. Then the potential adverse impacts of the GE plants on bees are compared with those posed by the non-GE version of the plant. Ultimately, the risk assessors will determine whether the differences between the GM and non-GM plants are likely to result in significantly different impacts on bees.

Returning to the pathway to harm provided above, it is obvious that one step is crucial: whether the bacterial toxin produced by the cotton plant is toxic to honeybees. If there is data in the application relevant to this question, possibly supplemented with data from the scientific literature, the risk assessors should be able to answer this key question. For example, it is well known that the commonly used bacterial toxins used in insect-resistant crops are not toxic to honeybees. Returning to the causal pathway, it is clear that if honeybees are not affected by the toxin, then it is not possible for honeybees to be harmed from the commercial cultivation of insect-resistant cotton in a manner that is different from non-GE cotton.

The process outlined above demonstrates how problem formulation should be used to correctly frame each environmental risk assessment in a structured, transparent and efficient way. Problem formulation focuses attention on key questions, the answers to which determine whether a particular course of action, i.e. the commercial release of a GE plant, will adversely affect India's capacity to meet its designated Protection Goals. Problem formulation also helps risk assessors determine their data needs to answer these questions and provides them with tools to determine whether data are relevant and sufficient to adequately test plausible and relevant risk hypotheses.

ECOLOGICAL BENEFITS OF GM CROP CULTIVATION

GM-crops have been adopted for commercial cultivation by farmers in a number of regions over the last decade, as farmers expected potential benefits compared to conventional crops. In order to evaluate the benefits, GM crop cultivation needs to be compared with risks and benefits of the conventional management practices. Conventional crop protection methods relying on chemical pesticides have damaged agricultural land and the environment. In addition, soil cultivation practices such as tilling have largely contributed to soil degradation by increasing erosion, nutrient loss and degradation of biological processes. The introduction of GM crops has helped to reduce some environmental impacts of conventional crops and farming methods. This section concentrates on actual observed benefits following a decade of commercial cultivation of GM crops distinguishing between direct benefits and those, which are indirectly caused by changes in pest and weed management when adopting GM crops.

PESTICIDE REDUCTIONS DUE TO INSECT RESISTANT *BT*-CROPS

The adoption of *Bt*-maize expressing the insecticidal protein *Cry*1Ab has resulted in only modest reductions in insecticide applications due to the small area of conventional maize treated with insecticides. The results of the large-scale studies performed during the last few years provides evidence that *Bt*-maize provides more specific insect control and has fewer side effects on non-target arthropods than most insecticides currently used.

The commercial cultivation of *Bt*-cotton, in contrast, has proven to have resulted both in a significant reduction in the quantity and in the number of insecticide applications. Cotton is highly susceptible to several serious insect pests belonging to the budworm-bollworm complex (tobacco budworm, cotton and pink bollworm). These insects constitute a major problem in most cotton-growing areas, because they can cause considerable damage. Conventional cotton cultivation therefore heavily relies on repeated insecticide applications throughout the growing season. Although estimates on pesticide use vary because pesticide use is depending on regional pest pressures, management practices and yearly variations, it appears that the adoption of *Bt*-cotton has significantly reduced the numbers of pesticide applications in every country where *Bt*-cotton has been grown. Moreover, most studies estimate a reduction in the amounts of pesticides used. Direct environmental benefits of reduced insecticide applications in *Bt*-cotton resulted in fewer non-target effects and in reduced pesticide inputs in water. In China, for example, the number of pesticide applications against lepidopteran pests in cotton has considerably dropped from nine in 2001 to four applications in 2016 following the adoption of *Bt*-cotton.

Concerns have been raised that environmental benefits may be lowered by additional spraying against secondary pests that were formerly controlled by the broad spectrum pesticides. There is no published evidence, however, that *Bt*-cotton has resulted in a general change in the pest spectrum leading to an overall increase of pesticide applications.

A new *Bt*-cotton variety (Bollgard II) containing two *Bt*-genes (*Cry*1Ac and *Cry*2Ab2) was commercially cultivated for the first time in 2003 in Australia and in the United States. Besides its improved performance against the cotton bollworm, Bollgard II confers an additional resistance against secondary pests including soyabean and cabbage looper, saltmarsh caterpillar as well as beet and fall armyworm. Data from two years of cultivation in Australia indicate that this new variety has even higher environmental benefits due to the reduction of insecticide use by 92%. The two-gene varieties, which are expected to replace single *Bt*-gene varieties in the future, can provide better efficacy, reduce or eliminate the necessity for additional chemical pesticides and lower the rate of resistance development.

Pesticide reductions related to the adoption of *Bt*-cotton have also shown to have reduced many immediate as well as longer-term risks to human health. The use of pesticides is yearly causing considerable poisoning incidences in many, often developing countries. From 1992–1996, for example, there has been an annual average of 54000 pesticide poisonings of farmers or farm workers in China causing approximately 490 deaths. As a result of less chemical pesticide spraying in *Bt*-cotton, demonstrable health benefits for farm workers have been documented in China and South Africa. Similarly, the adoption of *Bt*-rice in China would prove to have positive impacts on productivity and farmers health. Especially small and poor household farmers would thereby benefit from an 80% reduction in the use of pesticides compared to conventional varieties.

New Weed Control Strategies Offered by GM Herbicide Tolerant Crops

The adoption of GMHT crops has resulted in several weed management changes compared to conventionally managed crops. GMHT crops allow the use of a single broad spectrum herbicide that has a wider spectrum of activity and that may reduce the need for herbicide combinations or chemicals that require multiple applications. The herbicides used in GMHT crops (glyphosate or glufosinate) are foliar-applied, post-emergence herbicides, which usually allow using herbicides in a more targeted manner. They can be applied after weeds have emerged, i.e. areas with high weed densities can be identified and treated, while areas with low weed pressure can be treated with reduced herbicide amounts. Post-emergence herbicides are thus generally applied at lower rates than soil applied, pre-emergence herbicides, also because absorption by soil colloids and degradation are reduced. Glyphosate and glufosinate are generally considered toxicologically more benign, being less toxic to human health and the environment than many of the herbicides they replace. In addition, glyphosate and glufosinate have relatively short soil half-lives and they persist almost half as long in the environment compared to the replaced herbicides. Neither of these moves readily to ground water, which results in fewer losses of chemicals by leaching and run-off from the field.

Perhaps the most important environmental benefit of the adoption of GMHT crops is the role they have played in facilitating conservation tillage agriculture. Prior to the introduction of transgenic varieties, most growers used tillage to prepare the soil for planting. Excessive tillage, however, is known to cause soil structure changes, increase the susceptibility to soil erosion and reduce soil moisture. Loss of top soil due to tillage therefore causes environmental damage that can last for centuries. Since the early 1990's, growers have been reducing their tillage operations for soil conservation benefits. The possibility offered by GMHT crops to use broad spectrum herbicides has further encouraged growers to adopt conservation tillage strategies. According to USDA survey data, about 60% of the area planted with GMHT soyabean was under conservation tillage in 1997, compared with only about 40% for conventional soyabean. Because weed control can be done during the postemergence phase, farmers can use direct-seeding techniques since there is no need for pre-seeding tillage. Conservation tillage leaves a layer of plant residues on the soil surface, preventing soil erosion, reducing evaporation and increasing the ability of the soil to absorb moisture. A richer soil biota develops that can improve nutrient recycling and this may also help combat crop pests and diseases. Earthworm populations are generally higher in no-till fields than in conventionally tilled fields. In addition to a reduction in soil erosion and degradation, less frequent soil cultivation also results in a decrease in the emission of greenhouse gases, partly arising from a reduction in fuel use. Gianessi cites a survey by the American Soyabean Association, indicating that US soyabean growers reported making fewer tillage passes through their fields since 1995 when GMHT soyabean was first introduced. There is also evidence that conservation tillage can provide a wide range of benefits to farmland biodiversity by improving agricultural land as habitat for wildlife. The greater availability of crop residues and weed seeds can improve food supplies for insects, birds, and small mammals.

Agricultural production systems are complex and diverse. As with the adoption of any new technology, the use of agricultural biotechnology might include positive and possibly less favourable environmental impacts. GM crops systems can help to reduce some environmental risks associated with conventional agriculture, but they will also introduce new challenges that must be addressed. In order to valuate the environmental impacts of GM crop systems, their risks should always be weighed considering their potential benefits and current agricultural practice.

SCIENTIFIC DEBATES ON RISKS OF GM CROPS

The interpretation of collected scientific data is debated controversially by different stakeholders involved in the debate on potential risks of GM crops on biodiversity. Although some groups argue that experiences and solid scientific knowledge are still lacking, the ongoing debate is not primarily due to a lack in scientific data, but more to a lack in clear definitions on how to put a value on effects of GM crops on biodiversity in the context of current agriculture. The interpretation of study results is thereby often challenged by the absence of a baseline for the comparison of effects of GM crops on biodiversity. Consequently, some consider any effect related to GM crops as being undesired, while others correlate it to effects caused by modern agricultural practices recognising that a multitude of factors involved cause environmental effects. The interpretation of study results is often challenged by knowledge gaps on the natural variation occurring in any biological system, which is caused by a multitude of factors. Rather than the GM crop alone being the influencing factor, environmental effects are caused by agricultural production systems where the GM crop is one factor among others. Although science can help to assess these natural variations, it will most probably not be possible to elucidate all ecological interactions taking place in such systems.

In addition, not every environmental effect is automatically of ecological significance and leading to relevant impacts on biodiversity. In practice, decision-making will thus have to be not purely based on scientific criteria, but will also be strongly influenced by political, social, economical and ethical factors. Ecologically significant effects are only judged unacceptable (i.e. representing a damage) by the society if they are perceived as being linked to a deterioration in quality of a particular entity (e.g. biodiversity). Valuation of scientific data is thus influenced by the individual and subjective perceptions of the terms safety, risk and uncertainty by the society and particularly by the persons involved in decision-making. By performing a risk/benefit assessment comparing positive and negative effects of the GM crop system with current agricultural practice, it is in the end the society's decision whether genetic engineering is considered being safe enough.

Effects of GM Crops on Non-target Organisms

- There is scientific controversy on the baseline that should be applied when assessing potential effects of insect-resistant GM crops. It is discussed whether this should be the most common agricultural practice used (e.g. pesticides) or a practice that is only used by a few farmers (such as organic farming, which accounts for less than 3% of arable crop production in Switzerland).
- There is a debate to what extent indirect toxic effects, i.e. effects on natural enemies that largely depend on the target pest, should be valuated considering that such effects are common for all pest control methods and not restricted to the use of insect-resistant GM crops.
- It is unclear, which changes in population size and community structures of natural enemies could have an impact on functions and ecosystem services for natural pest regulation.

Impacts of GM Crops on Soil Ecosystems

- A common definition for soil quality has not been found yet.

- Population sizes and community structure of soil micro-organism are subject to high variation, and the baseline comparison for ecological implication is still not clear. Standard indicator species have not been defined. Different studies use a range of different parameters and techniques.

- Should influences of plant characteristics (higher/lower lignin content) associated with a particular *Bt*-variety, yet, unrelated to the inserted transgene, be compared to influences caused by plant characteristics of conventional cultivars?

- Is the presence of low percentages of transgenic *Bt*-toxins from *Bt*-crops in soils a reason for concern, considering that *Bt*-toxins are naturally occurring in soils due to the soil bacteria *Bacillus thuringiensis*, and due to *Bt*-spray formulations that are commonly used for insect control in agriculture and forestry?

Gene Flow from GM Crops to Wild Relatives

- In most agricultural landscapes, there is usually a gradual transition from peri-agricultural to semi-natural habitats. Although 'wild plants' can usually be distinguished from 'agricultural weeds', a clear definition of what plant species are considered being truly wild plants is lacking.

- Should effects occurring within agricultural or peri-agricultural environments be given the same importance as those effects, which could occur in natural habitats?

- Should gene flow from GM crops to wild relatives be valuated in a different way than gene flow from conventional crops?

Invasiveness of GM Crops into Natural Habitats

Is the presence of volunteer GMHT oilseed rape in habitats such as field borders or road verges an unwanted environmental effect, considering that non-transgenic oilseed rape is regularly occurring in such habitats and that HT is not considered to confer a selective advantage in natural habitats?

Impacts of GM Crops on Pest and Weed Management and their Ecological Consequences

- Is it better to have a high biodiversity in-crop (i.e. to have weedy crops), or to enhance off-crop biodiversity (e.g. separate buffer strips outside the fields) providing food for insects and birds?

- There is a need to define criteria for what is considered an agronomic problem and what is regarded being only an agronomic nuisance that could emerge from the cultivation of GMHT crops.

- Should herbicide-resistant weeds that have been caused by GMHT crops be valuated differently than herbicide-resistant weeds that have been caused by unsustainable (non-transgenic) weed management?

SECTION X

Bioethics and Intellectual Property Rights

Chapter 30

Safety and Bioethics of Food Biotechnology

INTRODUCTION

Biotechnology is at the intersection of science and ethics. Technological developments are shaped by an ethical vision, which in turn is shaped by available technology. Much in biotechnology can be celebrated for how it benefits humanity. But technology can have a darker side. Biotechnology can produce unanticipated consequences that cause harm or dehumanise people. The ethical implications of proposed developments must be carefully examined. The ethical assessment of new technologies, including biotechnology, requires a different approach to ethics. Changes are necessary because new technology can have a more profound impact on the world, because of limitations with a rights-based approach to ethics, because of the importance and difficulty of predicting consequences, and because biotechnology now manipulates humans themselves. The ethical questions raised by biotechnology are of a very different nature. Given the potential to profoundly change the future course of humanity, such questions require careful consideration. Rather than focussing on rights and freedoms, wisdom is needed to articulate our responsibilities towards nature and others, including future generations. The power and potential of biotechnology demands caution to ensure ethical progress.

Biotechnology, at its core, is about understanding life and using this knowledge to benefit people. Many see biotechnology as a significant force in improving the quality of people's lives in the 21st century. Obviously, biotechnology is intimately tied to science and scientific knowledge. At the very least, biotechnology promotes a certain vision of life, one in which some things are viewed as good and to be encouraged or pursued, and other things are bad and should be avoided or eliminated. That vision influences people's choices and what is viewed as ethically appropriate. A two-way flow exists in which ethics influences biotechnology even while the science impacts ethics.

At times, the relationship between biotechnology and ethics is portrayed as one of conflict. Sometimes the impression is conveyed that ethics is needed only when someone wants to tell others that what they are doing is wrong. To a degree, this is understandable since controversy, debate and argument are usually integral to ethics discussions.

But ethics is just as important when there is consensus that a direction is good and right. The role of ethics is often invisible at this stage. There wasn't an ethical debate over whether to search for a cure for cancer. But the decision to pursue such research was motivated by a common vision that curing cancer was the ethical thing to do. Ethical examination of issues is important not only as a form of critique but also to identify and celebrate the right things people do.

The effort, resources and creativity focussed on developing better treatments are ethically laudable. As such, there is much to celebrate about biotechnology. Society and individuals have benefited in many ways from technology. Many technological developments protect people from illnesses and natural disasters, giving some people 'liberation from the tyranny of nature'. In some parts of the world, people have higher living standards. Travel and communication have developed in unprecedented ways. Many of these changes can be welcomed as ethical developments.

Yet at the same time, other ethical considerations must be considered. At what price are some of these developments realised? Some developments seem motivated by a desire to find treatment at any price. Assisted human reproduction is a particularly controversial area where biotechnological treatment of infertility leads to many ethical dilemmas. Even with less controversial conditions like heart disease or cancer, developments have left people with high expectations that cures should exist. Some are concerned that technological developments lead to dehumanisation or in healthcare lead to less emphasis on caring. Ethical concerns exist about justice, and how fairly these technological benefits are distributed—both within society and around the world. With all the options now available for some, concerns are raised about whether too much choice is bad for us.

Overall, though, technology has a strong ethical foundation. The appropriate response to misgivings and concerns is not to reject technology. 'By turning our backs on technological change, we would be expressing our satisfaction with current levels of hunger, disease, and privation.' The benefits of technology, realised and potential, point to a technological mandate: biotechnology should strive to benefit people's lives. Many of the concerns about technology can be traced to the technological imperative the idea that something should be developed because we can, or we think we can.

GOALS OF BIOTECHNOLOGY

Ethics includes assessment of the rights and wrongs of specific technologies and applications (like cloning or genetic diagnosis). Another important pursuit within ethics is examining the broader goals and aims of enterprises like biotechnology. The relief of sickness is one goal, but there are others that can be more ethically controversial. Developing the necessary biotechnology for engineered negligible senescence assumes that indefinite life extension is good for humanity. Even if accepted as an ethical goal, it would be one goal among many.

Taking the time to reflect on these aspects of scientific developments can be difficult, especially with the pace and focus within biotechnology. The pressures of competing for funding, making breakthroughs, securing intellectual property, and obtaining market share all push against calls for caution or time-consuming reflection. Technological development can seem like a motorway, everyone on the fast track to success. Ethics, even when well intentioned, can seem like a diversion or a road-block that prevents biotechnology reaching its destination, or delays it inexcusably. However, there is a growing realisation that ethics must be a part of the planning process within biotechnology. In many areas of research, ethics does impact the design of scientific experiments. Any research involving human or animal participants will be scrutinised by ethics committees. The methodology must conform with ethical codes and guidelines. An argument can be made that publicly funded research should be conducted in ways that conform with society's values. 'When the nation decides an activity is worth its public money, it declares that the activity is valued, desired, and favoured'. Therefore it is important to ensure that what is publicly funded is ethically acceptable in society. The goal of relieving suffering is widely accepted, yet it must be balanced against other societal goals. The ethics of proposed biotechnological developments must be scrutinised carefully.

Darker Side

Even such a laudable goal as relieving human suffering cannot be taken as condoning any and all biotechnology. Humanity's creativeness and resourcefulness have long been recognised and praised. But human activity can have a darker side. The ancient Greek philosopher Sophocles reflected on these two sides of technological development. On the one hand he noted many human accomplishments in transport, agriculture and medicine. But he also pointed to problems with this same inventiveness.

Biotechnology is a particularly fitting example of technology with such fundamentally different characteristics that it requires a careful re-examination of how its ethical dimensions are evaluated. Biotechnology 'raises moral questions that are not simply difficult in the familiar sense but are of an altogether different kind'.

Challenging characteristics of biotechnology

The capacity for new technology to have global impact shows that ethics needs to broaden its focus. Environmental problems and the existence of nuclear technology demonstrate the importance of ethical examination of more than just human–human interactions. New technology also highlights the vulnerability of nature. Previous technological developments appeared to assume that natural resources were in endless supply and that nature could rebound from any human impact. Environmental changes show these assumptions were problematic. Ethical evaluations of biotechnology need to take the vulnerability of nature into account. These issues also point to limitations in previous ethical approaches that focussed only on humans. At the same time, a concern for these broader issues can lead to new technological challenges and exciting research opportunities, such as has occurred with research into renewal energy sources stemming from ethical concern for the environment.

Limitations with rights

Rights-based approaches to ethics have made important contributions to human welfare. They provide a means by which vulnerable humans can argue for more ethical treatment. However, such approaches have their limitations. A rights-based approach to ethics must include some method of identifying those who bear rights. Those who have rights place duties on others to uphold those rights. It has proved very difficult to find consensus on how rights are to be ascribed. One approach is that all humans are inherently entitled to all human rights. This raises questions about when a human is given these rights (at fertilisation or birth or some other point). It also leaves no guidance on how to treat the non-human world. Biotechnology requires answers to these questions to address ethical concerns about non-human species and nature as a whole.

This has led to an approach where rights are granted based on particular abilities and attributes. There is little consensus over what abilities entitle an organism to rights. Philosophically, it is also difficult to justify why any particular attribute should lead to the granting of rights. The whole approach is criticised as being motivated by a desire to treat unethically those not given rights. This is particularly relevant to research on human embryos, especially embryonic stem cell research.

Developments in biotechnology point to serious limitations with a rights-based approach to ethics. Rather than providing insurmountable problems for ethics, these point to the need for a different approach to ethics. Jonas and others point out that rather than focussing exclusively on human rights and entitlements, the new technological era requires a greater focus on human responsibility.

Future consequences: Earlier technology impacted humans and their lives, but did not have the potential to change human nature but biotechnology does. With that comes the potential for broader and long-range consequences. Predictions about these consequences can be difficult and unreliable. This is particularly cogent with genetic technology. The consequences of our ability to manipulate the human genome could impact many, if not all, future generations. The way genes interact with one another means that manipulating one gene could have unintended effects on other genes or their expressed proteins. This is especially important given the recent realisation that the human genome contains fewer genes than originally presumed.

The medicalisation of patients and the instrumentalisation of people are consequences of technology's successes. This can have a dehumanising effect on human life, which makes it easier to treat some humans as less than fully human. This is a way in which technology can take on a life of its own and have much more profound ethical consequences.

Biotechnology has the added capacity to produce products that literally do take on life. The technology humans developed in the past was inanimate and could be left unused if found to be ethically problematic—as difficult as that might have been.

Impact on human nature and personhood: No area of biotechnology more clearly brings to focus the need for careful ethical reflection than its potential to impact human nature. Previous technology has provided new tools that impacted human activities and society. Humans were the makers of technology. Some aspects of biotechnology now make humans the objects of technology. Humans have turned upon themselves and are ready 'to make over the maker of all the rest'.

Recent developments with stem cell research and cloning have been the lightning rod for debate over human personhood. These discussions point to the gulf between proponents on the different sides. Some have viewed embryos as 'featureless bundles of cells'. From this perspective the human embryo is a human non-person that can be used and destroyed in research. Others disagree and maintain that the human embryo should be treated as a person, making it unethical to treat it merely as a means to others' ends. Personhood can be viewed as an inherent attribute of all humans. This confers all humans with certain rights and determines how persons should be treated ethically. This approach protects humans, especially the vulnerable, from unethical treatment. The other approach makes personhood conditional on reaching some stage of development or possessing certain abilities. Only humans with those capacities are then entitled to protection. A fundamental problem with this approach is that it always arises to justify killing those declared to be human non-persons. How will it affect us to treat human lives as commodities to be manipulated and destroyed at will? When we justify doing so with embryos, will it become easier to do so at later stages of development?

This points to the difficulty of determining public policy when sections of society have irreconcilable positions on matters of fundamental importance. We must also examine how biotechnology itself impacts our view of human nature. Leon Kass asks how will it affect us 'to look upon nascent human life as a natural resource to be mined, exploited, commodified. The little embryos are merely destroyed, but we— their users—are at risk of corruption'. This is much more than a debate over rights. This is about human dignity, including what it means for humans to act with dignity. This changes the focus from ascribing rights to determining responsibilities.

Central place of responsibility: The enormity of the potential impact of biotechnology on human nature should cause us to proceed cautiously. Biotechnology has the potential to do great good. But it also has the potential to cause much harm. This could arise in the physical realm through unexpected consequences of the technology itself. But other harms could arise through the non-physical impacts of biotechnology. Cars and computers have affected many aspects of human life and society. Biotechnology could change what it means to be human.

A rights approach to ethics makes clear where people have rights. Each right carries a corollary duty or responsibility. If people have a right to healthcare, someone has the responsibility to provide healthcare resources. Much energy has been expended identifying and defending human rights. We now need a similar emphasis on human responsibilities.

Responsibility is also a corollary of power. Biotechnology brings new powers to humanity. These powers should remind us of our responsibility to nature and the environment, to all of life, to the future, and to human nature and personhood. To understand these responsibilities entails the development of wisdom. That wisdom requires ethical reflection before developing specific forms of biotechnology. Taking the time for that reflection can go against the pace of biotechnological developments and hubris over human wisdom.

Researchers warned that new technology was propelling us towards a utopian future. These developments have the potential for much good, but also risk changing, harming or even destroying some species, including ourselves. To make the right ethical decisions 'requires supreme wisdom—an impossible situation for man in general, because he does not possess that wisdom, and in particular for contemporary man, because he denies the very existence of its object, objective value and truth. We need wisdom most when we believe in it least'.

In contrast, ethics searches for better answers to ethical questions. It acknowledges the limitations in current wisdom, and strives to improve our understanding. The way forward is muddied by our inability to accurately predict the consequences of proposed biotechnological developments. Some argue that we should push ahead and deal with problems as they arise. But given the scale of disaster that biotechnological mistakes could trigger, Jonas' guiding principle contains much wisdom. He argued that 'ignorance of the ultimate implications becomes itself a reason for responsible restraint—as the second best to the possession of wisdom itself'.

Time and resources must be committed to examining the ethical implications of proposed biotechnological developments. The potential impact on all aspects of nature must be considered. The social, emotional and spiritual implications of developments in biotechnology must also be examined. When humans themselves are the objects of biotechnology, great caution is necessary least we promote a view of ourselves and our neighbours as nothing more than living bits of technology.

ETHICAL ASPECTS OF FOOD AND AGRICULTURAL BIOTECHNOLOGY

Many of the expressed concerns about food and agricultural biotechnology are described as ethical. Decision leaders should interpret the expression of ethical concerns as a demand for competing visions of nature and the public good to be expressed in public dialogue about food and agricultural biotechnology, for those who feel that their values have been neglected to have an adequate opportunity to express their concerns in their own words, and for their voices to be heard.

Those who call for attention to ethical issues appeal to many diverse values. Their concerns can be classified into two broad categories. On the one hand, some see the very act of using genetic technology to raise ethical issues that would not apply to other applications of food and agricultural technology. On the other hand, some believe that specific applications of biotechnology raise ethical issues that are not being adequately addressed, even if these issues may be raised in connection to other, more conventional types of agricultural technology, as well.

Special Arguments Pertaining to the use of rDNA Technology

There are several types of concern noted by those who question whether the use of biotechnology may be intrinsically questionable.

- Genes and essences: Longstanding religious and cultural traditions associate the idea of a particular Aessence with different species of living organisms, and specify an obligation for human beings to respect these essences. Some may associate the modern notion of genes with this traditional notion of essence.
- Species boundaries and natural kinds: The idea that there is a specified order of nature may involve the belief that the species of plants and animals we find around us represent natural kinds. Some may fear that biotechnology disturbs this order and thereby violates absolute limits on what human beings are ethically permitted to do.
- Religious arguments: Many religious traditions prohibit acts that involve transpecies reproduction, or ban the consumption of some species groups for food, and the mixing of foods from different groups. Biotechnology may be interpreted as contrary to some of these religious traditions.
- Emotional repugnance: Cultural traditions dictate that some potentially consumable substances (e.g. species such as cat and dog, or particular parts of plants and animals) are not suitable for use as food. Western food systems currently respect the repugnance that people feel toward these substances as a sufficient ground for policies that help people avoid consuming them. Some individuals may feel a similar repugnance toward bioengineered foods.

General technological ethics

There are a number of ethical questions that can be raised with respect to virtually any new food or agricultural technology. As they are raised in connection with biotechnology, these questions suggest the following types of ethical concern:

- Environmental ethics: Technology raises environmental issues when there are environmental exposures that pose risk to humans, wildlife or to ecosystem integrity. It has been alleged that agricultural biotechnology may pose risks to wildlife in or near farm fields. There are also issues associated with the question of whether agricultural ecosystems can themselves to exhibit features of ecological integrity.
- Food safety: Many of the issues associated with the safety of eating bioengineered foods are technical, but the question of whether regulators should make this decision based on an assessment of the risks, or whether individual consumers should be placed in a position to make the choice themselves is an ethical one.
- Moral status of animals: If genetic engineering of livestock would compromise animal welfare, there are ethical questions that can be raised. There are also ethical questions about whether it would be ethical to use biotechnology to make animals more tolerant of production settings that are currently regarded as inimical to animal welfare.
- Impact on farming communities: Some critics of agricultural biotechnology have alleged that it will contribute to farm bankruptcies and the depletion of farming population in rural communities.

There has been a longstanding ethical debate as to whether technology or policy that has these effects on farming communities can be ethically justified in virtue of offsetting benefits in the form of efficient production and lower food prices. The concern is particularly relevant to the impact of biotechnology in developing regions where many farm at the subsistence level.

- Shifting power relations: Related to the concern on farming communities, some have argued that biotechnology will help a few well-capitalised firms control decision making in agriculture (including future research), and limit farmers ability to choose from an array of production possibilities. This concern is related to a general ethical concern with the distribution of economic power and wealth in democratic societies.

Responses to these issues

This section discusses several approaches that have been discussed as a possible response to these various ethical issues:

- Uncertainty and the precautionary principle: Many of these ethical issues involve uncertainty about the risks or outcomes associated with biotechnology. The Precautionary Principle has been suggested as the appropriate decision rule to utilise in response to such situations. It suggests that decision makers should not permit technological innovations to go forward simply because alleged harms have not been proven to exist. However, it is not clear how the Precautionary Principle should be applied in the case of food and agricultural biotechnology.

- Consent, labels and consumer choice: Various proposals for labelling products of biotechnology have been discussed. On the one hand, these proposals are supported by an informed-consent approach to issues in food safety, and may be the most satisfactory response to concerns based on religious values, emotional repugnance and other intrinsic objections to biotechnology. Labels might give individuals who have these concerns an opportunity of exit, to opt out of a food system that causes them anxiety or concern. On the other hand, labels may stigmatise bioengineered foods, and may not provide information that would be useful for consumers trying to make choices on the basis of nutrition and food safety.

- Methods in applied ethics: How do methods in ethics suggest a response to these concerns. One approach suggests that common ethical principles can be applied to provide definitive answers to the questions raised above. A more promising approach suggests that only open public discussion of these issues can produce an adequate basis for responding to the questions that critics of biotechnology raise.

Trust and public confidence

As debates over food and agricultural biotechnology become politicised, with activist organisations opposing both industry and governmental spokespersons, there is a growing tendency for public discourse on biotechnology to reflect the strategic interests of industry and activists. There is a grave risk that as science becomes deployed in these debates, scientists themselves will become so tainted by the strategic character of debate that the public will begin to lose confidence in the objectivity and judgment of scientists. Scientific spokes persons thus have an ethical responsibility to develop a capacity to participate in ethically-charged public discussions of biotechnology without either denigrating the values of others by characterising them as irrational, or presuming uncritically that their science-based perspectives are the ethically proper approach to take.

In the light of above mentioned facts the Ethical issues associated with food and agricultural biotechnology must be regarded as open-ended and in great need of more structured and serious dialogue. Both specialists and members of the public should be encouraged to articulate their concerns, and to respond to the views of others in a considered and respectful manner.

VARIETY OF CHANGES SHAPING BIOETHICS TODAY

The following development reflects a variety of changes that are shaping bioethics today:

- The rapid pace of development in science and technology: ethical guidelines cannot be developed quickly enough to keep up.
- The lack of public trust in scientists, corporations and regulators, partly due to the failure to safeguard the public against infected blood and, more recently, against 'mad cow' disease in Europe.
- The poor ability of institutions to educate the public on complex scientific issues without obfuscating or patronising the public.
- Increasing democratisation and demands for accountability and transparency.
- Weakening of political control.
- Rapid increase in availability of information, especially through the Internet.
- Development of technology for which there is little precedent in its own terms and in terms of its ethical use.
- The lack of coherence within the bioethical establishment itself in terms of having strong foundational values accepted by all.
- The technical complexity of the issues.
- Globalisation of economies, in which policy decisions in one country have implications for many other countries.
- Vocal, well-organised, protest groups.
- An increasing public recognition of the impact of our actions on the environment.

These factors have resulted in two primary outcomes a public that has become empowered to demand consultation, and policymakers who are reluctant to make risky decisions–including those that may have been acceptable in the past–without public consultation. For bureaucrats, seeking to involve the public has been a way of coping with their diminished credibility.

In addition to public consultation, there is a need for dialogue among different constituencies. Academics, industries, the public, the media and applied ethicists must collectively identify issues, plan an agenda for vigorous in-depth studies of those issues, engage in transparent discussion of the results of the research in a truth-seeking, nonconfrontational way, and reach consensus in the development of guidelines. If this is done well, the guidelines will be clear, understood by all and, more importantly, supported by all stakeholders. The current confrontational methods have not served the public well. Attempts by industry public relations officers to provide more and more information to a distrusting public have also been ineffective. In this context, bioethicists can have an important function: to bring their specialised knowledge and analytical skills to clarify and facilitate, rather than simplistically to preach or propound on what is right or wrong in their opinion.

An interesting recent phenomenon, likely to become more common as biotechnology issues become more complex, is that bioethicists are being held legally responsible for their advice. The parents of a teenager who died as a result of a gene therapy experiment have recently sued a prominent bioethicist from a well-known US university. The bioethicist had advised the scientist-clinicians who performed the procedure.

SAFETY OF FOOD BIOTECHNOLOGY

The Food and Drug Administration (FDA), along with the Environmental Protection Agency (EPA) and the United States Department of Agriculture (USDA) combine to regulate genetically engineered foods. The FDA ensures that foods made from genetically engineered plants are safe for humans and animals to consume, the USDA makes sure the plants are safe to grow, and the EPA ensures that pesticides introduced into these plants are safe for both human and animal consumption and for the environment. Foods produced through either biotechnology or conventional methods must all meet the same high safety standards.

Labelling Requirements

The FDA has decided that this new technique for changing the genetic make-up of plants does not differ significantly from traditional plant breeding techniques. Therefore, no special labelling is required. However, common food allergy proteins would require labelling. For example, if genetic material from a peanut is put into a tomato, the tomato would require labelling. The special labelling requirement would let people with an allergy to peanuts know that the tomato may contain peanut proteins which could cause an allergic reaction. Also, if the nutritional content or the composition of the food changes substantially, additional labelling is required.

MODERN FOOD BIOTECHNOLOGY: DEFINITION AND OVERVIEW OF POTENTIAL BENEFITS AND RISKS

The application of modern biotechnology to food production presents new opportunities and challenges for human health and development. Recombinant gene technology, the most well-known modern biotechnology, enables plants, animals and micro-organisms to be genetically modified (GM) with novel traits beyond what is possible through traditional breeding and selection technologies. It is recognised that techniques such as cloning, tissue culture and marker-assisted breeding are often regarded as modern biotechnologies, in addition to genetic modification.

The inclusion of novel traits potentially offers increased agricultural productivity or improved quality and nutritional and processing characteristics, which can contribute directly to enhancing human health and development. From a health perspective, there may also be indirect benefits, such as reduction in agricultural chemical usage, and enhanced farm income, crop sustainability and food security, particularly in developing countries.

The novel traits in genetically modified organisms (GMOs) may also, however, carry potential direct risks to human health and development. Many, but not all, genes and traits used in agricultural GMOs are novel and have no history of safe food use. Several countries have instituted guidelines or legislation for mandatory premarket risk assessment of GM food. At the international level, agreements and standards are available to address these concerns.

GMOs may also affect human health indirectly through detrimental impacts on the environment or through unfavourable impacts on economic (including trade), social and ethical factors.

These impacts need to be assessed in relation to the benefits and risks that may also arise from foods that have not been genetically modified. For example, new, conventionally bred varieties of a crop plant may also have impacts—both positive and negative—on human health and the environment.

Recent International Controversies and Study Initiative

Conflicting assessments and incomplete substantiation of the benefits, risks and limitations of GM food organisms by various scientific, commercial, consumer and public organisations have resulted in national and international controversy regarding their safe use as food and safe release into the environment. An example is the debate on food aid that contained GM material offered to countries in southern Africa in 2002, after 13 million people faced famine following failed harvests. This international debate highlighted several important issues, such as health, safety, development, ownership and international trade in GMOs.

Such controversies have not only highlighted the wide range of opinions within and between Member States, but also the existing diversity in regulatory frameworks and principles for assessing benefits and risks of GMOs.

CURRENT USE, RESEARCH AND IMPENDING DEVELOPMENT OF FOODS PRODUCED THROUGH MODERN BIOTECHNOLOGY

Foods produced through modern biotechnology can be categorised as follows:

1. Foods consisting of or containing living/viable organisms, e.g. maize.
2. Foods derived from or containing ingredients derived from GMOs, e.g. flour, food protein products or oil from GM soyabeans.
3. Foods containing single ingredients or additives produced by GM micro-organisms (GMMs), e.g. colours, vitamins and essential amino acids.
4. Foods containing ingredients processed by enzymes produced through GMMs, e.g. high-fructose corn syrup produced from starch, using the enzyme glucose isomerase (product of a GMM).

Crops

Crop breeding and the introduction of GM crops for food production

Conventional breeding, especially of crops, livestock and fish, focuses principally on increased productivity, increased resistance to diseases and pests, and enhanced quality with respect to nutrition and food processing.

Various transformation methods are used to transfer recombinant DNA into recipient species to produce a GMO. For plants, these include transformation mediated by *Agrobacterium tumefaciens* (a common soil bacterium that contains genetic elements for infection of plants) and biolistics — shooting recombinant DNA placed on microparticles into recipient cells. The methods used in the transformation of various animal species include microinjection, electroporation and germ-line cells. The success rate of transformations in animals tends to be lower than in plants, and to vary from species to species, thus requiring the use of many animals.

Genetic modification is often faster than conventional breeding techniques, as stable expression of a trait is achieved using far fewer breeding generations. It also allows a more precise alteration of an organism than conventional methods of breeding, as it enables the selection and transfer of a specific gene of interest. However, with the present technology, in many cases it leads to random insertion in the host genome, and consequently may have unintended developmental or physiological effects. However, such effects can also occur in conventional breeding and the selection process used in modern biotechnology aims to eliminate such unintended effects to establish a stable and beneficial trait. It should be noted that conventional breeding programs directed by the molecular analysis of genetic markers are also of critical importance to modern plant and animal breeding. However, human and environmental health consequences of these techniques are not considered here.

GM crops currently in commercial production

At present, only a few GM crops are permitted for food use and traded on the international food and feed markets. These include herbicide and insect-resistant maize (*Bt* maize), herbicide-resistant soyabean, rape (canola) oilseed, and insect and herbicide-resistant cotton (primarily a fibre crop, though refined cottonseed oil is used as food). In addition, several government authorities have approved varieties of papaya, potato, rice, squash, sugar beet and tomato for food use and environmental release. The latter crops, however, are currently grown and traded only in a limited number of countries, mainly for domestic consumption.

Future trends in GM crops

The commercial introduction of transgenic crop plants with agronomic traits is often referred to as the first generation of transgenic plants. Further development of GM crops with agronomic traits is continuing, and production of a range of GM crops with enhanced nutritional profiles is also underway. Various novel traits are currently being tested in laboratories and field tests in a number of countries. Many of these second-generation GM crops are still in the development stage and are unlikely to enter the market for several years.

Agronomic traits

Pest and disease resistance: In the short-term, most newly commercialised GM crops will continue to concentrate on agronomic traits, especially herbicide resistance and insect resistance and, indirectly, yield potential.

Virus resistance: Virus resistance could be extremely important to improving agricultural productivity. Field tests of the following virus-resistant crops are currently being conducted in various parts of the world: sweet potato (feathery mottle virus), maize (maize streak virus), and African cassava (mosaic virus).

Altered nutrition and composition

Vitamin-A-enhanced rice: The best-known example of a GM crop conferring enhanced nutritional properties is rice containing a high level of β-carotene — a vitamin A precursor (so-called 'golden rice'). Vitamin A is essential for increasing resistance to disease, protecting against visual impairment and blindness, and improving the chances of growth and development. Vitamin A deficiency is a public health problem that contributes to severe illness and childhood mortality. This preventable condition increases the burden of disease on the health systems of developing countries. A number of strategies have been suggested for combating vitamin A deficiency, including dietary approaches (e.g. fortification of foods) and supplementation via pills.

'High iron' rice: Prevalence of iron deficiency is very high in those parts of the world in which rice is the daily food staple. This is because rice has a very low iron content. Transgenic rice seeds with the iron-carrier protein ferritin from soya were found to contain twice as much iron as seeds of non-transformed rice. Rice has been transformed with three genes which increase iron storage in rice kernels and iron absorption from the digestive tract.

Improved protein content: Researchers are also investigating methods that could improve the protein content of staple vegetables, such as cassava, plantain and potato. Results from greenhouse trials show that these tubers have 35–45 per cent more protein, and enhanced levels of essential amino acids.

Removing allergens and antinutrients: Cassava roots naturally contain high levels of cyanide. As they are a staple food in tropical Africa, this has led to high blood-cyanide levels which have harmful effects. Application of modern biotechnology to decrease the levels of this toxic chemical in cassava would reduce its preparation time. In potatoes, insertion of an invertase gene from yeast reduces the natural levels of glycoalkaloid toxin.

Altered starch and fatty acid profile: In the quest to provide healthier foods, there is an effort to increase the starch content of potatoes so that they absorb less fat during frying. To create healthier fats, the fatty-acid composition of soya and canola has been altered to produce oils with reduced levels of saturated fats. R&D is currently focusing on GM soyabean, oilseed rape and oil palm.

Increased antioxidant content: The lycopene and lutein contents of tomatoes have been increased as have isoflavones in soya. These phytonutrients are known to improve health or prevent disease.

Livestock and Fish

In terms of food production, the application of modern biotechnology to livestock falls into two main areas: animal production and human nutrition.

Fish

The projected increasing demand for fish suggests that GM fish may become important in both developed and developing countries. Enhanced-growth Atlantic salmon containing a growth hormone gene from Chinook salmon is likely to be the first GM animal on the food market. These fish grow 3–5 times faster than their non-transgenic counterparts, to reduce production time and increase food availability. At least eight other farmed fish species have been genetically modified for growth enhancement. Other fish in which genes for growth hormones have been experimentally introduced include grass carp, rainbow trout, tilapia and catfish. In all cases, the growth-hormone genes are of fish origin.

Livestock and poultry

Foods derived from GM livestock and poultry are far from commercial use. Several growth-enhancing novel genes have been introduced into pigs that have also affected the quality of the meat, i.e. the meat is more lean and tender. This research was initiated over a decade ago, but owing to some morphological and physiological effects developed by the pigs, these have not been commercialised.

Micro-organisms

Micro-organisms as foods

Currently, there are no known commercial products containing live genetically modified micro-organisms (GMMs) on the market. In the United Kingdom, GM yeast for beer production has been approved since 1993, but the product was never intended to be commercialised. Other micro-organisms used in foods (which are in the R&D phase) include starter fermentation cultures for various foods (bakery and brewing), and lactic acid bacteria in cheese. R&D is also aimed at minimising infections by pathogenic micro-organisms and improving nutritional value and flavour. Attempts have been made to genetically modify ruminant micro-organisms for protecting livestock from poisonous feed components. Micro-organisms improved by modern biotechnology are also under development in the field of probiotics, which are live micro-organisms that, when consumed in adequate amounts as part of food, confer a health benefit on the host.

Food ingredients, processing aids, dietary supplements and veterinary chemicals derived from GM micro-organisms

Many enzymes used as processing aids in food and feed production are derived through the use of GMMs. This means that the GM micro-organisms are inactive, degraded or removed from the final product. GM yeasts, fungi and bacteria have been in commercial use for this purpose for over a decade. Examples include: alpha-amylase for bread-making, glucose isomerase for fructose production, and chymosin for cheese-making. Most of the micro-organisms modified for food processing are derivatives of micro-organisms used in conventional food biotechnology.

GMMs are also permitted in a number of countries for the production of micronutrients, such as vitamins and amino acids used for food or dietary supplement purposes. An example is the production of carotenoids (used as food additives, colourants or dietary supplements) in GM bacterial systems. In the future, complete metabolic pathways could be integrated in GM micro-organisms, enabling them to produce new compounds. For animal husbandry, veterinary products such as bovine somatropin, used for increasing milk production, have been developed using genetic engineering. Bovine somatropin has been on the market in several countries for over a decade.

RISK OF GMOS AND GM FOODS TO HUMAN HEALTH AND THE ENVIRONMENT

Introduction of a transgene into a recipient organism is not a precisely controlled process, and can result in a variety of outcomes with regard to integration, expression and stability of the transgene in the host.

When new foods (crop varieties, animal breeds or micro-organisms) are developed by traditional breeding methods, they are usually not subject to specific pre- or post-market risk or safety assessment by national authorities or through international standards. This is in contrast to requirements introduced for GMOs and GM foods.

The discovery of recombinant DNA had raised concerns among researchers regarding the potential creation of recombinant viruses whose escape would threaten public health. Fourteen months after a voluntary moratorium on research involving recombinant DNA techniques, guidelines for the physical and biological containment of riskier experiments were drafted and agreed.

To provide international consistency in risk analysis of GMOs and GM foods which incorporates risk assessment, management and communication components, a number of international regulatory and standard-setting bodies have introduced uniform standards. These include standards for human-health and environmental-safety assessment of GMOs and GM foods, and notification of their movement across national borders. The objective of uniform global standards for risk assessment would be challenging as countries are bound to reach different decisions on the scope of the assessment, particularly the resolution of whether or not to include social or economic aspects.

International regulatory systems covering GM food safety (Codex Principles) and environmental safety (Cartagena Protocol on Biosafety) came into force in 2003.

The concept that allows for the comparison of a final product with one having an acceptable standard of safety is an important element of a GM food safety assessment. This principle was elaborated by FAO, WHO and OECD in the early 1990s and referred to as 'substantial equivalence'. The principle suggests that GM foods can be considered as safe as conventional foods when key toxicological and nutritional components of the GM food are comparable to the conventional food (within naturally occurring variability), and when the genetic modification itself is considered safe. However, the concept has been criticised by some researchers.

Assessment of the Impact of GM Foods on Human Health

Principles for the safety assessment of GM foods

The Codex safety assessment principles for GM foods require investigation of:

1. Direct health effects (toxicity).
2. Tendency to provoke allergic reactions (allergenicity).
3. Specific components thought to have nutritional or toxic properties.
4. Stability of the inserted gene.
5. Nutritional effects associated with the specific genetic modification.
6. Any unintended effects which could result from the gene insertion.

The 2003 Expert consultation on the safety assessment of foods derived from GM animals, including fish formed the opinion that to further develop the risk-assessment process with current scientific knowledge, integrated toxicological and nutritional evaluations should be conducted in order to identify food-safety issues that may need further investigation (Fig. 30.1). Both evaluations combine data from the hazard identification and characterisation, and food intake assessment steps. It should be noted that such newly suggested further developments of the risk-assessment process have not yet been considered by Codex, and that the international principles and guidelines for risk analysis and safety assessment of foods derived from biotechnology are as accepted by Codex in 2003.

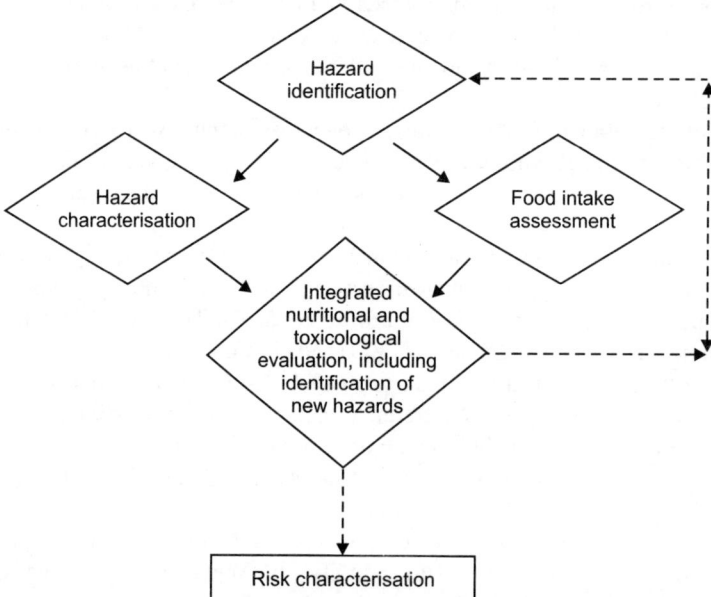

Fig. 30.1: Schematic overview of a suggested further development of the risk assessment process.

Potential direct effects on human health

The potential direct health effects of GM foods are generally comparable to the known risks associated with conventional foods, and include, for example, the potential for allergenicity and toxicity of components present, and the nutritional quality and microbiological safety of the food.

As mentioned above, many of these issues have not traditionally been specifically assessed for conventional food, but in one area—toxicity of food components—there is ample experience related to the use of animal experiments to test potential toxicity of targeted chemical components. However, the intrinsic difficulty in testing whole foods, as opposed to specific components, in animal feeding experiments have resulted in the development of alternative approaches for the safety assessment of GM foods.

The safety assessment of GM food follows a stepwise process aided by a series of structured questions. Factors taken into account in the safety assessment include:

1. Identity of gene of interest, including sequence analysis of flanking regions and copy number.
2. Source of gene of interest.
3. Composition of GMO.
4. Protein expression product of the novel DNA.
5. Potential toxicity.
6. Potential allergenicity.
7. Possible secondary effects from gene expression or the disruption of the host DNA or metabolic pathways, including composition of critical macronutrients, micronutrients, antinutrients, endogenous toxicants, allergens and physiologically active substances.

Potential unintended effects of GM foods on human health

Unintended effects, such as elevated levels of antinutritional or toxic constituents in food, have on occasion been characterised in conventional breeding methods, e.g. glycoalkaloid levels in potatoes. Organisms derived from conventional breeding methods, including tissue cultures, may have a somewhat enhanced possibility for genetic (and epigenetic—environmentally induced changes that affect the expression of a gene without changing the DNA sequence) instabilities, such as the activity of mobile elements and gene-silencing effects. These effects could increase the probability of unintended pleiotropic effects (affecting more than one phenotypic trait), e.g. increased or decreased expression of constituents or possibly modifications in expressed proteins, as well as epistasis (the interaction of the inserted gene with other genes). It has been argued that random insertion of genes in GMOs may cause genetic and phenotypic instabilities but, as yet, no clear scientific evidence for such effects is available. A better understanding of the impact of natural transposable elements on the eukaryotic genome may shed some light on the random insertion of sequences. Gene expression in conventional and GM crops is subject to environmental influences. Environmental conditions such as drought or heat can stimulate some genes, turning the expression up or down. The assessment of potential synergistic effects is necessary in the risk assessment of organisms derived from gene stacking, i.e. breeding of GMOs containing genetic constructs with multiple traits. Internationally agreed procedures for the assessment of such organisms are desirable. Unintended effects can be classified as insertional effects, i.e. related to the position of insertion of the gene of interest or as secondary effects, associated with the interaction between the expressed products of the introduced gene and endogenous proteins and metabolites. There is common agreement that targeted approaches, i.e. the measuring of single compounds, is very useful and adequate to detect such effects, as has been done with conventionally bred products. To enhance and improve the identification and analyses of these unintended effects, profiling methods have been suggested. This untargeted approach allows detection of unintended effects at the mRNA (microarray), protein (proteomic) and metabolite (metabolomic) level.

Potential human-health effects from horizontal gene transfer

Natural genetic transformation has been found to occur in different environments, e.g. in food. In addition, it has been shown that ingested DNA from food is not completely degraded by digestion, and that small fragments of DNA from GM foods can be found in different parts of the gastrointestinal tract. As the consequences of horizontal gene transfer (HGT) may be significant in some human-health conditions, the potential for HGT needs to be part of the risk assessment of GM food.

WHO consultations have also discussed the potential risks of gene transfer from GM foods to mammalian cells or gut bacteria. These panels have suggested that it may be prudent in a food-safety assessment to assume that DNA fragments survive in the human gastrointestinal tract and can be absorbed by either the gut microflora or somatic cells lining the intestinal tract. It was agreed that the assessment needs to take into account a number of factors including, but not limited to, the specific characteristics encoded by the DNA sequences, the characteristics of the receiving organism, and the selective conditions of the local environment of the receiving organisms. Some scientists have pointed to the present methodological limitations of a comprehensive scientific evaluation of this problem (mainly because of estimations that only approximately 1 per cent of naturally existing bacteria can be cultured, and therefore analysed). Discussion also addresses the consequences of a rare probability of a transfer event against the high numbers of bacteria and genes available for transfer.

The DNA construct used to change the genetic composition of a recipient organism should be considered within an assessment, especially if the gene or its promoter (e.g. cytomegalovirus promoter) has been derived from a viral source. Sequences unrelated to the target gene could be introduced as part of the construct. Inadvertent introduction of such sequences into the germ-line of a GM animal not only has the potential for creating unintended genetic damage, but can also contribute by recombination to the generation of novel infectious viruses. A well-known example is the generation of a replication-competent murine leukaemia virus during the development of a vector containing a globin gene.

The horizontal transfer of recombinant genetic material to micro-organisms has demonstrated an enhanced stability of DNA under certain conditions. Natural transformation of DNA to bacteria involves the active uptake of extracellular DNA by bacteria in a status of competence or in rare, illegitimate recombination events. The probability of such an event occurring appears to be extremely low, and very much related to the genes, constructs and organisms in question. WHO expert panels concluded that horizontal gene transfer is a rare event that cannot be completely discounted, and that the consequences of such transfer should be considered in a safety assessment. The panels encouraged the use of recombinant DNA without antibiotic-resistance genes (particularly those that could interfere with human or animal therapies) or any other sequences which could stimulate transfer. The panels also discouraged the use of any unnecessary DNA sequences, including marker genes in the genetic construct. The safety assessment of a genetic construct should also examine the included marker genes. Commonly used marker genes code for antibiotic resistance. Risk assessment of these selectable genes should focus on gene transfer to micro-organisms residing in the gastrointestinal tract of humans or animals. As the potential of this gene transfer cannot be completely ruled out, the safety assessment should also consider information on the role of the antibiotic in human and veterinary medical uses.

Potential immune responses and allergenicity induced by GM foods

Food allergies or hypersensitivities are adverse reactions to foods triggered by the immune system. Within the different types of reactions involved, non-immunological intolerances to food and reactions involving components of the immune system need to be differentiated. The former may invoke reactions such as bloating or other unpleasant reactions, but are thought not to involve the immune system and called 'food intolerances'.

Allergic reactions to traditional foods are well known. The major food allergens are proteins in and derived from eggs, fish, milk, peanuts, shellfish, including crustaceans and molluscs (e.g. clams, mussels and oysters), soya, tree nuts (e.g. almonds, Brazil nuts, cashews, hazelnuts/filberts, macadamia nuts, pecans, pine nuts, pistachios and walnuts) and wheat. Whereas the groups of main allergens are well known and advanced testing methods have been elaborated, traditionally developed foods are not generally tested for allergens before market introduction.

The application of modern biotechnology to crops has the potential to make food less safe if the newly added protein proves to cause an allergic reaction once in the food supply. A well-known case is the transfer of a gene encoding a known allergen, the 2S-Albumin gene from the Brazil nut, to a previously safe soyabean variety. When the allergenic properties of the transgenic soyabean were tested, sera from patients allergic to Brazil nuts cross-reacted with the transgenic soyabean. For this reason, a commercial product was never pursued. On the other hand, the introduction of an entirely new protein that has not been previously found in the food chain represents a different case.

In the first case, guidelines for assessing foods with known allergens are clear. The second case is more difficult to assess because there is no definitive test to determine the potential allergenicity of a novel protein. Instead, several risk factors provide a rough guide as to the likelihood of allergenicity.

Risk-assessment protocols for food allergy examine four elements: (i) allergenicity assessment (is the food or elements in the food a potential allergen), (ii) dose response assessment (is there a safe concentration of the allergen), (iii) exposure assessment (how likely is it that people will encounter the allergen), and (iv) susceptible subpopulations (how do those prone to allergy react to this new food).

WHO expert panels have established protocols for evaluating the allergenicity of GM foods on the basis of the weight of evidence. The strategy adopted is applicable to foods containing a gene derived from either a source known to be allergenic or a source not known to be allergenic. The panels have, however, discouraged the transfer of genes from known allergenic foods unless it can be demonstrated that the protein product of the transferred gene is not allergenic. These principles have been applied by many regulatory agencies assessing the safety of GM foods and have provided the basis for Codex guidelines for the safety assessment of foods derived from biotechnology.

Safety aspects of food derived from GM animals

Genetically modified animals have mainly been produced for biomedical research purposes. To date, no GM food animals have been introduced onto international markets. But GM food animals such as fish can be expected in the near future. In principle, the assessment of food and feed safety for GM animals follows the general principles of the assessment of GMOs outlined above. However, the specificities of the introduction of transgenes into animals, often using viral constructs for introduction into the germ-line, need distinct consideration.

The risk assessment of foods derived from GM animals needs to be undertaken, as for other GM foods, on a case-by-case basis. This includes an assessment of potential recombination of viral vectors used for transformation with wild-type viruses, especially in poultry, where potential incomplete digestion could lead to intestinal uptake of orally administered proteins, and an assessment of peptide expression that may have hormonal activity (e.g. in fish).

WHO expert consultation on the Safety assessment of foods derived from GM animals, including fish held in 2003 addressed the key issues for food safety and evaluated the extent of scientific knowledge with regard to hazard identification and characterisation unique to transgenic animals.

Phenotypic analysis: Because of their size, and limitations in the generation process, there will in general be few initial founders for screening of GM animals, meaning that information on the variation range between animals with the same genetic modification will be rather limited. This will make interpretation of differences difficult. Furthermore, a selection of the edible tissues and products to be analysed has to be made for the different animal species. In specific cases, phenotypic analysis may also be advisable after processing or, for fish, during the various stages of spoilage. For example, adverse biogenic amines can be formed during spoilage in salmon, tuna, herring and other fish species. Similarly, formaldehyde may be produced in spoiled shrimp, cod, hake and many other species.

Compositional analysis: Background data on the natural variation for individual constituents in different tissues need to be generated. Data in existing databases must be evaluated for their quality and value for use in comparative compositional analysis.

Safety aspects of foods derived from or produced with GMMs

The production of food additives or processing aids using GMMs, where the micro-organism is not a part of the food, has become an important and generally well-accepted technology, with a significant number of such products on the market. Experience with the purification of proteins in the biomedical field suggests that well-standardised purification protocols are of central importance for the safety of these products.

Potential effect of GMOs on human health mediated through environmental impact

Work on environmental-health indicators suggests that various agricultural practices have direct and indirect effects on human health and development. Hazards can take many forms—wholly natural in origin or derived from human activities and interventions. The need to assess indirect effects of the use of GMOs in food production has been emphasised by many countries. For example, the production of chemicals or enzymes from contained GM micro-organisms (e.g. chemicals, pharmaceuticals or food additives) have contributed significantly to decreases in the amount of energy use, toxic and solid wastes in the environment, thereby significantly enhancing human health and development.

GMOs and Environmental Safety

Principles of environmental risk assessment

In many national regulations, the elements of environmental risk assessment (ERA) for GM organisms include the biological and molecular characterisations of the genetic insert, the nature and environmental context of the recipient organism, the significance of new traits of the GMO for the environment, and information on the geographical and ecological characteristics of the environment in which the introduction will take place. The risk assessment focuses especially on potential consequences for the stability and diversity of ecosystems, including putative invasiveness, vertical or horizontal gene flow, other ecological impacts, effects on biodiversity and the impact of the presence of GM material in other products. Different approaches in the ERA regulations of different countries have often resulted in different conclusions on the environmental safety of certain GMOs, especially where the ERA focuses not only on the direct effects of GMOs, but also addresses indirect or long-term effects on ecosystems, e.g. impact of agricultural practices on ecosystems.

Internationally, the concept of 'familiarity' was developed also in the concept of environmental safety of transgenic plants. The concept facilitates risk/safety assessments, because to be familiar means having enough information to be able to make a judgement of safety or risk. Familiarity can also be used to indicate appropriate management practices, including whether standard agricultural practices are adequate or whether other management practices are needed to manage the risk. Familiarity allows the risk assessor to draw on previous knowledge and experience with the introduction of plants and micro-organisms into the environment and informs appropriate management practices. As familiarity depends also on knowledge of the environment and its interaction with introduced organisms, the risk/safety assessment in one country may not be applicable in another country.

Potential unintended effects of GMOs on nontarget organisms, ecosystems and biodiversity

Potential risks for the environment include unintended effects on nontarget organisms, ecosystems and biodiversity. Insect-resistant GM crops have been developed by expression of a variety of insecticidal toxins from the bacterium *Bacillus thuringiensis* (*Bt*). Detrimental effects on beneficial insects or a faster induction of resistant insects (depending on the specific characteristics of the *Bt* proteins, expression in pollen and areas of cultivation), have been considered in the ERA of a number of insect-protected GM crops. Studies on the toxicity of *Bt* maize on the monarch butterfly in the USA indicate that for most commercially available hybrids, the *Bt* expression in pollen is low, and laboratory and field studies show that no acute toxic effects at any pollen density would be encountered in the field. These questions are considered an issue for monitoring strategies and improved pest-resistance management.

Outcrossing: Outcrossing of transgenes has been reported from fields of commercially grown GM plants, including oilseed rape and sugar beet, and has been demonstrated in experimental releases for a number of crops, including rice and maize. Outcrossing could result in an undesired transfer of genes such as herbicide-resistance genes to nontarget crops or weeds, creating new weed-management problems.

GM animals: Genetically engineered insects, shellfish, fish and other animals that can easily escape, are highly mobile and that form feral populations easily, are of concern, especially if they are more successful at reproduction than their natural counterparts. For example, it is possible that if released into the natural environment, transgenic salmon, with genes engineered to accelerate growth could compete more successfully for food and mates than wild salmon, thus endangering wild populations.

GMMs: Gene transfer from bacteria to bacteria in the soil has been demonstrated in some systems, e.g. for antibiotic-resistance genes, and only a limited number of releases of GMMs (e.g. *Pseudomonas* and *Rhizobium*) has been permitted, mainly to explore the spread and the fate of micro-organisms in nature. Risk assessment in this field is impeded by a number of factors, such as the limited knowledge of micro-organisms in the environment (only approximately 1 per cent of soil bacteria are currently described), the existence of natural transfer mechanisms between micro-organisms, and the difficulties in controlling their spread.

DEVELOPING REGULATORY AND SAFETY SYSTEMS FOR MODERN FOOD BIOTECHNOLOGY: A ROLE FOR CAPACITY BUILDING

It is widely accepted that the application of modern biotechnology could be important to economic development, but may also involve inherent risks. Thus, all countries, be they developers or net importers of products derived from modern biotechnology, should introduce measures that safeguard human health and environmental safety. In fact, many governments are in the process of developing legal instruments/ regulatory systems that address human health and environmental safety. The effectiveness of such measures will be determined by a country's capacity (both in terms of human resources and infrastructure) to expeditiously handle the evaluation, management and risk communication of each new product of modern biotechnology. While evaluation and risk management may be done on a case-by-case basis, risk communication activities undertaken by governments should address the process according to which decisions are taken. Safety issues with regard to protecting the environment and human health are different and require different expertise. Biosafety tends to be the responsibility of the department of environment or agriculture, whereas the authority for food safety often lies with the department of health. Hence, the legal instruments for regulation may differ.

At the international level, 15 legally binding instruments and nonbinding codes of practice address some aspect of GMOs, but none of these on its own integrates the regulation of biotechnology across all sectors. Such sector-based regulations and powers increase the already overstretched capacity needs of developing countries, and present challenges to developing a fully coherent policy and regulatory framework for modern biotechnology. The challenge for developing countries is to achieve coherence in national legislation for crops, livestock, fish, forest trees and micro-organisms, while meeting international obligations and ensuring harmonisation.

The shortcomings of most capacity-building programs lie in the simplistic notion that assumes a 'one size fits all' development path. Donors often prescribe programs that are largely based on the experiences of developed countries, on the assumption that these will work equally well for developing countries. Unfortunately, this is rarely the case and can result in limited or disappointing outcomes.

Capacity-building initiatives must be sustained beyond the life of the activity as an integral part of a development program and not be a once-off activity. In turn, developing countries must participate in and take ownership of an activity and be encouraged to take charge of their own development. Demand-driven knowledge development is more likely to be absorbed if it reflects local circumstances, and more likely to be applied by society.

Capacity Needs

Food safety is attracting increased attention because of its implications for public health. In general, food control systems in developing countries are poorly developed, and less organised than in most developed countries. Their overall capacity needs in terms of food safety can be summarised as follows: (i) basic infrastructure, (ii) national food control strategy, (iii) food legislation and regulatory framework, (iv) food inspection services, (v) food control laboratories and equipment, and (vi) implementation of food quality and safety assurance systems.

The work in food safety is multidimensional, and there are frequently several food laws under the authority of different agencies. In many countries, effective food control is undermined by the existence of fragmented legislation, multiple jurisdictions and weaknesses in surveillance, monitoring and enforcement mechanisms. Food-safety legislation developed specifically for the safety of GM foods should be integrated within the existing food laws, taking into account the special risk-management requirements.

In order to make informed decisions on the safety of GMOs and GM foods, governments need substantial human and institutional resources in the disciplines required for assessing the risks for the environment and for human food presented by GMOs. Developing countries have limited expertise in the required fields of science, as biotechnologists in these countries are generally engaged in research and therefore mostly unavailable to the regulatory bodies and as policy-makers. In most developing countries, those same scientists sit on national biosafety committees, and are involved in both risk assessment and policy-making.

There are three vulnerabilities in this scenario: (i) when developers are also risk assessors, the potential for conflict of interest is magnified, (ii) because most members of the national biosafety committee are recruited on a voluntary basis, they do not devote too much time to this responsibility, and (iii) because membership of the national biosafety committee generally rotates, there is no continuity in the capacity gained through experience.

A Potential Role for Modern Biotechnology

The convention on biological diversity dictates the use and application of relevant technologies as a means of achieving the objectives of conservation and sustainable use with specific reference to biotechnology. Modern biotechnology is purported, from a technical perspective, to have a number of products for addressing certain food-security problems of developing countries. It offers the possibility of an agricultural system that is more reliant on biological processes rather than chemical applications. The potential uses of modern biotechnology in agriculture include: increasing yields while reducing inputs of fertilisers, herbicides and insecticides, conferring drought or salt tolerance on crop plants, increasing shelf-life, reducing postharvest losses, increasing the nutrient content of produce, and delivering vaccines. The availability of such products could not only have an important role in reducing hunger and increasing food security, but also have the potential to address some of the health problems of the developing world. Achieving the improvements in crop yields expected in developing countries can help to alleviate poverty: directly by increasing the household incomes of small farmers who adopt these technologies, and indirectly, through spill-overs, as evidenced in the price slumps of herbicides and insecticides. Indirect benefits as a whole tend to have an impact on both technology adopters and non-adopters, the rural and urban poor.

Indeed, some developing countries have identified priority areas such as tolerances to alkaline earth metals, drought and soil salinity, disease resistance, crop yields and nutritionally enhanced crops. The adoption of technologies designed to prolong shelf-life could be valuable in helping to reduce postharvest losses in regionally important crops. Prime candidates in terms of crops of choice for development are the so-called 'orphan crops', such as cassava, sweet potato, millet, sorghum and yam. Multinationals have found no incentive to develop these crops and have instead invested in marketable crops with high profit returns. This strategy is intended to target wealthier farmers in temperate-zone countries with the financial capacity and tradition of supporting new seed products. However, here is a potential for multinational companies to develop crops grown largely in developing countries. The investment costs are low and the potential markets considerably large.

While some public-sector research institutes in developing countries are forging ahead with the application of modern biotechnology, a small number are supported by government policy and therefore follow a defined agenda. Still other governments believe that the risks (safety, environmental and/or economic) associated with modern biotechnology outweigh the benefits.

Currently, the many promises of modern biotechnology that could have an impact on food security have not been realised in most developing countries. In fact, the uptake of modern biotechnology has been remarkably low owing to the number of factors that underpin food security issues. In part, this could be because the first generation of commercially available crops using modern biotechnology were modified with single genes to impart agronomic properties with traits for pest and weed control, and not complex characteristics that would modify the growth of crops in harsh conditions. Secondly, the technologies are developed by companies in industrialised countries with little or no direct investment in, and which derive little economic benefit from, developing countries.

Thirdly, many developing countries do not have the necessary biosafety frameworks to regulate the products of modern biotechnology. For example, it took over two years for the Kenyan authorities to approve the field-testing of a virus-resistant sweet-potato variety because the scientific capacity for evaluating the product was unavailable. It should be noted, however, that such delays in the approval process have also been seen in developed countries, especially during the initiation of national regulatory evaluation. Developing countries with limited financial and human resources need to find the right balance for investing in conventional and modern biotechnology research programs. While alliances with the private sector may contribute to the search for new technologies, the public sector needs to focus on crops and traits in which the former may be unwilling or unable to invest. The extent to which priority is given to modern biotechnology over other research methods should be linked to a country's agricultural priorities and objectives as well as to its environmental concerns.

Ultimately, investment in interventions that support good governance, the development of rural infrastructure and market access is required before any of the promises of modern biotechnology can be realised. In general, policies that stimulate economic growth and target poverty reduction may have significant bearing on the health and well-being of the population.

SOCIAL AND ETHICAL CONCERNS ABOUT GM FOODS

Cultural Variability and Public Perception

Across the world, food is a part of cultural identity and societal life, and has religious significance to people. Therefore, any technological modification, including changes to the genetic basis of crops or animals used for food, may be met with social resistance. In many countries, people's interaction with nature, often correlated with religious perspectives, causes social and ethical resistance to modifications that interfere with genes. Whereas the objectives of food safety in its limited sense are more clearly realised and harmonised internationally, the objectives of nature protection, environmental safety and sustainable agriculture are much more complex, unclear and variable in different regions of the world. Investigations of public perception in areas of the world with relatively high resistance to GM foods indicate that lack of information is not the primary reason.

The public is not for or against GMOs *per se*—people discuss arguments both for and against GMOs, and are aware of contradictions within these arguments. Also, people do not demand zero risk.

They are quite aware that their lives are full of risks that need to be balanced against each other and against the potential benefits. People may also discriminate in their perception of different technologies where a general positive perception can be observed for applications with a clear benefit for society, e.g. for modern medicines. A key finding is that people do not react so much to genetic modification as a specific technology, but rather to the context in which GMOs are developed and the purported benefits they are to produce.

Nevertheless, the techniques of genetic engineering are often described as 'pushing nature beyond its limits'. Many of the concerns expressed about GMOs, including those about 'unnaturalness', have also been expressed in relation to other agricultural innovations, such as the use of pesticides, animal-derived animal feed and antibiotics in animal feed. Organic agriculture is perceived as reversing or opposing these developments, whereas GMOs are perceived as the ultimate manifestation of this trend. GM-free areas are, therefore, seen as a way of preserving nature.

The opposition to GM crops and foods has as much to do with social and political values as with concerns about health and safety. Consumers' growing awareness of their rights and farmers' increasing fear of dependence on multinational companies are symptoms of a deeper concern about values and priorities, the type of environment people want, the role of biodiversity, tolerance of risk and the price that people are prepared to pay for regulation. Some people are concerned about the level of control exercised by a few chemical companies on seed markets. GMOs are emblematic of the powerful economic fears that globalisation inspires. In certain regions, hostility to GMOs is symbolic of a broader opposition to the encroachment of market forces. These are perceived to be creating a world in which money rules with little consideration for historical traditions, cultural identities and social needs.

Labelling of GM Foods and Consumer Choice

In establishing GM food labelling policies to ensure that consumers receive meaningful information, regulatory authorities have had to grapple with a complex array of issues related to GMOs. These have included scientific, health, environmental, political, cultural and economic issues, as well as the appropriate compliance and enforcement requirements.

Chapter 31

Food Security and Intellectual Property Rights in Developing Countries

INTRODUCTION

Food insecurity is a major problem throughout the World. It is a concern at all levels, from individuals to states. At a basic level, food security is about fulfilling each individual's human right to food. Within the broad question of the human right to food, food security also relates more specifically to issues of agricultural policy, economic development and trade.

Food security is defined as physical and economic access to sufficient, safe and nutritious food by all people to meet their dietary needs and food preferences for an active and healthy life. Meeting food security objectives implies improving access to food which is itself linked to poverty eradication. Undernourishment is linked to inadequate access to means of production such as 'land, water, inputs, improved seeds and plants, appropriate technologies and farm credit' which in turn implies an incapacity to produce or purchase sufficient food. The Plan of Action also notes the significance of environmental threats to food security which can come in the form of drought, land degradation or loss of biodiversity and negatively impact on food production.

Another definition of food security, though widely accepted, has been criticised from different standpoints. Some actors tend to use a more restrictive definition which focuses more on the question of global increases in food production than on the issue of household access to food. Other actors have criticised the definition because it does not go far enough insofar as it does not include a rights dimension. Notwithstanding disagreements on the exact definition of food security, the fulfilment of food needs constitutes a generally accepted goal.

This chapter discusses the specific link between food security and intellectual property rights (IPRs), one – but only one – of the important perspectives from which food security must be analysed.

IPRs have become increasingly important in the past couple of decades in a number of fields. This includes, for instance, agricultural biotechnology where IPRs provide a basic incentive for the development of the private sector in this area. The extension of IPRs to agriculture is of special significance because agriculture and food security are closely interlinked. In other words, the introduction of IPRs in agriculture is directly linked to the realisation of basic food needs. The introduction and strengthening of IPRs in the agricultural sector of developing countries has been and remains contentious. On the whole, food security constitutes the central concern of all relevant actors. The introduction of IPRs in plant varieties is justified by the need to foster food security in the long-term. Similarly, arguments in favour of an open system where private IPRs are not enforced are also based on the premise that this will contribute to food security. At present, IPRs in agriculture have been and are being introduced in developing countries that are members of the World Trade Organisation (WTO). This is taking place in a context where food insecurity remains a central concern for a majority of developing countries where a large proportion of the population does not have access to sufficient good quality food. A host of conceptual and practical issues need to be addressed in the context of the paradigmatic shift from a system seeking to foster food security on the basis of the free exchange of knowledge to a system seeking to achieve the same goal on the basis of the private appropriation of knowledge. This is not only due to the fact that IPRs provide different kinds of incentives for inventiveness than a system based on the free sharing of knowledge but also because some of the new plant varieties are the product of genetic engineering. The latter bring in other environmental and socio-economic dimensions to the subject considered.

This chapter discusses the issue of food security from the narrow perspective of intellectual property. The first section provides a general introduction to the issues and challenges in this field. The second section goes on to introduce the relevant international legal framework for food security and intellectual property. The third section examines some of the implications of recent developments in international law for developing countries and looks in more detail at the way in which India has been implementing its international obligations in this field. Finally, the fourth section, building on the analysis provided in the previous sections provides recommendations for the implementation of existing international legal obligations and the further development of the legal regime in this field.

INTELLECTUAL PROPERTY RIGHTS IN DEVELOPING COUNTRIES

Food security remains an overwhelming concern for developing countries even though some countries classified as developing countries have virtually eradicated hunger.

Food Security

Food security can be understood at different levels, from the household to the international level. While the overall availability of food at a global level is not a major concern at present, food availability in specific regions of the world and access to food by specific individuals remains a major concern in most parts of the world. Further, population growth in countries where undernourishment is already a problem and diminishing arable land availability make food insecurity one of the most important policy challenges of coming years.

Food security is not only dependent on the availability of food but also on effective access and appropriate distribution of existing foodstuffs. Unavailability of foodstuffs is not a major concern at present a worldwide level since the world produces enough food for its present population. Availability is a concern at present in the case of countries suffering from armed conflicts, in situations where sufficient arable land is not available or in the case of persistent drought. Food availability will also be an increasing concern in the future if food production does not keep pace with population growth. At present, however, the problem of under nourishment is often more linked to the problem of lack of access to food and maldistribution of foodstuffs than the problem of unavailability. In countries like

India, overall food availability has been more than sufficient for a number of years but the numbers of undernourished keep rising. This indicates that food security must be analysed at different levels at the same time. The availability of sufficient food within the country does not indicate that each and every household and every individual has access to sufficient food, the latter being the ultimate measure of food security. Food security at an individual level implies that people must either have a sufficient income to purchase food or the capacity to feed themselves directly by growing their own food. There is therefore a direct link between poverty and food security. More specifically, food security is influenced by individuals capacity to work, individual and household access to land and their control over the land and other productive assets, including seeds. Further, food security is also influenced by policies concerning the management of the environment in general and agricultural biodiversity specifically. Diversity constitutes from an environmental point of view one of the ways in which resilience of agricultural systems can be ensured while from a socio-economic point of view, agro-biodiversity constitutes to a large extent one of the basic productive assets of poor farmers.

One of the major debates with regard to food security today is the contribution that agro-biotechnology can make to meeting the food needs of the world's population. This happens in a context where it is expected that most of the increase in food production will continue to come from further intensification of crop production where part of this increase will come in the form of higher yields and part in the increase of multiple cropping and reduced fallow periods. It is hoped that transgenic plant varieties can contribute to at least part of this food production increase. In practice, the impacts of transgenic plant varieties on agricultural management are partly similar to the impacts of Green Revolution varieties. The main differences are concerns over environmental safety on the one hand and the impacts of the close link between agro-biotechnology and IPRs. At present, the potential of modern biotechnology for food security in developing countries remains an open question. Firstly, it appears that plant biotechnology research is only likely. This may have significant implications for local and national food security in a context where it is expected that the development of agro-biotechnology may lead to further market concentration and where access to genetically modified seeds may be hampered by their higher cost to benefit poor farmers if it is applied to 'well defined social or economic objectives'. To date, commercialised genetically modified crops have generally not focused on the needs of developing country agriculture. In fact, it is uncertain whether the large life-science companies that are responsible for most of the applied agro-biotechnology research. Thanks to the incentives provided by IPRs can ever be expected to focus their research efforts on plant varieties of specific interest to poor farmers and consumers in developing countries. Secondly, the scale of overall benefits derived from the introduction of transgenic plant varieties remains a matter of debate when agricultural and other factors, such as environmental and socio-economic factors are taken into account. Thirdly, according to projections showing an increase in agricultural trade in coming years, it is possible that further specialisation will occur whereby some developing countries may be led to increase the production of non-food cash crops at the expense of basic food crops.

The policy challenges concerning food security are immense. Guaranteeing access to food for each individual around the world today and in the future requires measures to create wealth in poor communities, measures to enhance poor farmers' control over their land and productive assets, measures to conserve the natural resource base while increasing either agricultural productivity or arable land availability and measures to ensure effective distribution of existing food supplies. In addition to the dimensions highlighted, the question of food security can also be looked at from a rights perspective. The human right to food provides, for instance, that freedom from hunger requires steps to improve methods of production, conservation and distribution of food. Further, states have to proactively engage in activities to strengthen people's access to and utilisation of resources and means to ensure their livelihood and food security. This includes measures such as land reform, ensuring physical and economic access to credit, natural resources, new technologies, rural infrastructure, irrigation, and provision of explicit farmers rights through legislation. Building on the human rights approach the concept of 'food sovereignty' is also noteworthy. Food sovereignty implies the recognition of the freedom and capacity of people and their communities to exercise and realise their right to food, right to produce food and the assurance of access to productive resources. It is a valuable addition to the food security discourse insofar as it is a concept which applies from individual level to the level of nation states.

Intellectual Property Rights and Food Security

There are a number of links between IPRs and food security. In general, IPRs such as patents or plant breeders' rights seek to give incentives, mainly to private sector actors, to develop seeds that either produce higher yields or have specific characteristics which will improve food security and agro-biodiversity management. IPRs were for a long time underdeveloped in the context of agriculture. Firstly, in many countries and at the international level, agricultural management was premised on the basis of the free exchange of germplasm and knowledge, a system wherein IPRs did not fit well. Secondly, it was generally recognised that agriculture was substantially different from other fields of technology because farmers were often used to save seeds from previous crops and because the link between the fulfilment of basic food needs and agriculture made it undesirable to foster commercialisation in this field. IPRs have progressively been introduced in agriculture in two main phases.

Firstly, a number of developed countries adopted over time a form of intellectual property protection for plant varieties – plant breeders' rights – which is derived from the patent model. Secondly, in the context of the development of genetic engineering, the progressive introduction of patents over life forms has constituted a major incentive for the overall growth of agro-biotechnology. At present, the Agreement on Trade-Related Aspects of Intellectual Property Rights (TRIPS Agreement) provides a number of specific minimum levels of protection that all WTO member states must respect. This includes, for instance, the patentability of micro-organisms and a form of intellectual property protection for plant varieties. Beyond these minimums, there is no uniformity around the world insofar as some countries like the United States have gone further than the TRIPS minimums and accept, for instance, the patentability of plant varieties.

A number of justifications can be offered for the introduction of IPRs with a view to foster food security in developing countries. In general, the legal protection offered by IPRs is one of the most important incentives for private sector involvement in agro-biotechnology. IPRs are thus primordial in ensuring the participation of the private sector in the development of improved plant varieties. Improvements that can be brought about by agro-biotechnology include plant varieties that produce higher yields by enhancing the capacity of the plant to absorb more photosynthetic energy into grain rather than stem or leaf, varieties that have the capacity to combat pests and varieties modified to grow faster through enhanced efficiency in the use of inputs such as fertilisers, pesticides and water. From a food security point of view, another potentially interesting feature of agro-biotechnology is the possibility to modify varieties to improve their nutritional value, such as in the case of the pro-vitamin A rice. Other arguments include the potential of the introduction of IPRs in developing countries to increase foreign direct investment, increase technology transfers and R&D by foreign companies while at the same time giving domestic actors incentives to be more innovative.

Policy Considerations for Food Security in the Context of Intellectual Property Rights

IPRs have the potential to enhance agricultural production. However, in the context of developing countries, this contribution must be analysed in a broader perspective which takes into account a number of other variables. The introduction of IPRs in agriculture has important links with other forms of property rights directly relevant in agriculture, such as land rights and rights over biological resources. In fact, the question of access to biological and genetic resources for food and agriculture has been at the centre of significant debates at the international level for a number of years. Control by individual farmers, private companies and states over the genetic and biological resources they hold and related knowledge has become increasingly contentious with the progressive introduction of IPRs over certain types of plant varieties for instance. While the sharing of resources and knowledge was emphasised until the 1980s, the new system which promotes individual appropriation has led to the formulation of a new set of rules concerning control over knowledge and resources. At the international level, while private individual appropriation of inventions through IPRs has been condoned, state control over primary resources has at least in principle been reinforced. At the national level, the role of farmers in conserving and enhancing agro-biodiversity has generally been recognised but this is not necessarily translated into specific claims over resources or knowledge.

The introduction of IPRs in agriculture raises specific concerns with regard to farmers' control over their resources and knowledge. In general, IPRs tend to facilitate control over seeds and related knowledge by agri-businesses at the expense of small and subsistence farmers. This is linked in part to the royalties that farmers must pay to acquire protected seeds together with the associated restrictions on saving, replanting and selling saved seeds. In principle, it appears essential that farmers should retain control over plant varieties so that they may continue to innovate, improve and adapt varieties to suit changing needs and conditions. At present it is unlikely that IPRs holders will be able to control farmers' ability to save and replant seeds as much as in countries like the United States where IPRs protection is often enhanced with contractual obligations. However, the introduction of genetic use restriction technologies would constitute a specific challenge in this context since this would provide a tool for patent holders to ensure that farmers fully respect patent rights. The challenge that the progressive introduction and strengthening of IPRs in agriculture imposes on relevant actors is, for instance, quite severe for the Consultative Group on International Agricultural Research (CGIAR). Faced with the complete overhaul of the international agricultural system which is taking place, the International Agricultural Research Centres (IARCs) have specifically indicated that 'there is some concern that even the Right to Food, as defined by various governments, could be compromised by certain interpretations of intellectual property and other agreements'. From a broader perspective, the impacts of IPRs can be compared to the broader impacts of globalisation in food in agriculture of which they are one segment. As noted by the FAO, globalisation can have a number of positive impacts but at the same time may contribute to the disempowerment of certain communities and countries. In other words, the potential of transgenic plant varieties to foster food security is partly linked to the development of mechanisms to foster their transfer and ways to ensure that they are affordable for poor farmers.

The introduction and strengthening of IPRs in agriculture fosters two kinds of concerns linked to R&D. Firstly, there are concerns that 'over-patentability' in the biotechnology industry may have the potential to stifle innovation in the private and public sector rather than promote it. This is linked to the scope of the claims that can be made in the field of agro-biotechnology. The perception is often that broad claims are necessary to provide the industry with sufficient incentives to innovate but that IPRs claims should not extend to the primary material for research because this tends to stifle scientific and technological innovation. This constitutes a difficult debate in the present environment. Generally, scientific innovation benefits from free access to all primary materials for research. However, current scientific research often requires access to patented technologies beyond the primary biological material. Further, the products of scientific research are increasingly often patented. From a policy-making point of view, it is necessary to determine whether the primary holders of biological material and knowledge should avail their resources and knowledge free to the whole of humankind for the greater common good. It is noteworthy in this context that the introduction of plant breeders' rights, as distinguished from patents, was partly based on the premise that innovations by breeders could only be sustained if the primary and protected material remained freely available for further research. Secondly, an other point concerns the extent to which it is reasonable to expect the research agenda to be geared towards the needs of individuals below the poverty line as long as most of the research is carried out with a view to develop commercially valuable products. In fact, it is noteworthy that the first generation of genetically modified crops have generally not been bred for raising yield potential, and that any gains in yields and production have come primarily from reduced losses to pests. This tends to indicate that the introduction of IPRs in agriculture in developing countries should be accompanied by further measures to ensure that research is also geared towards the needs of the poor. This concern leads the FAO to suggest that public sector research will have a strong role to play, in particular with regard to the need to raise productivity of the poor in the agro-ecological and socio-economic environments where they practise agriculture and earn their living.

The introduction of IPRs in agriculture must also be examined in its broader context which includes, for instance, the impacts of IPRs in agriculture on biodiversity management. Biodiversity and agricultural-biodiversity in particular, is of primary importance for the sustainability of agricultural systems in the long term. Agro-biodiversity is of special importance because it directly contributes to feeding people. Agriculture and biodiversity management are inextricably intertwined because biological resources constitute a primary input to agricultural production systems and the majority of existing agricultural products have evolved through selection and collection of plant and animal species. In this context, landraces which are geographically or ecologically distinct crops or animals selected by farmers for their overall economic value are of special importance. IPRs in agriculture have an inherent tendency to displace landraces because protected varieties generally offer higher yields than local counterparts. This process of displacement tends to promote homogenisation in agricultural fields (or in other words monocultures) which leads to a loss in diversity and generally reduces crops' resilience to pests and diseases. Other elements that must be taken into account include problems related to the development of resistance by pests to biopesticides. Further, there are some specific concerns with regard to the potential harmful impacts of transgenic plant varieties on specific species. While a number of the impacts of the introduction of transgenic plant varieties can be compared from an environmental point of view to the impacts of the introduction of Green Revolution varieties and may not be specific to the context of this discussion, they should nevertheless be fully considered.

INTERNATIONAL LAW AND FOOD SECURITY

The international legal framework for food security is found in a number of different treaties and instruments which belong to completely different areas of international law. Firstly, some treaties and institutions deal with food security from the point of view of agriculture. Secondly, IPRs treaties only deal indirectly with food security but their implementation has significant impacts for food security in developing countries. Thirdly, several environment-related treaties have important implications for food security. Finally, human rights treaties focusing on the right to food or related rights also have a central place in the overall framework.

Agriculture Related Legal and Institutional Framework

Legal instruments sponsored by the FAO

The FAO, in keeping with its role as the central UN organisation dealing with agriculture, has logically played an important role in defining the food security related legal framework. In fact, the two main instruments adopted in the FAO context, the 1983 International Undertaking for Plant Genetic Resources (International Undertaking) and the 2001 International Treaty on Plant Genetic Resources for Food and Agriculture (PGRFA Treaty) clearly reflect the evolution of the overall legal system in this area. The importance of the International Undertaking and the PGRFA Treaty derives from their focus on the legal status of agricultural plant genetic resources, the focus on farmers' rights and at least an attempt to provide a coherent system taking into account the different interests at stake, from the imperative of access to food to agro-biodiversity management and the granting of incentives to commercial breeders through IPRs.

The international legal regime for the conservation and use of agricultural plant genetic resources has been marked by significant changes over the past few decades. Traditionally, plant genetic resources for food and agriculture (PGRFA) were freely exchanged on the understanding that PGRFA constituted a common heritage of humankind. As a result, rights over PGRFA could not be appropriated by private entities. These principles were embodied in the 1983 International Undertaking. It affirms the principle that plant genetic resources are a heritage of humankind which should be made available without restriction to anyone. This covers not only traditional cultivars and wild species but also varieties developed by scientists in laboratories. The International Undertaking was adopted as a non-binding conference resolution. However, the emphasis on the free availability of PGRFA proved to be unacceptable to some developed countries which already had interests in genetic engineering. Broader acceptance of the International Undertaking was only achieved after the FAO Conference passed interpretative resolutions in 1989 and 1991. These resolutions affirm the need to balance the rights of formal innovators as breeders of commercial varieties and breeders' lines on the one hand, with the rights of informal innovators of farmers' varieties on the other. Resolution 4/89 recognises that plant breeders' rights, as provided for in the UPOV Convention, are not inconsistent with the Undertaking, and simultaneously recognises farmers' rights as defined in Resolution 5/89. Resolution 3/91 further recognises the sovereign rights of nations over their own genetic resources.

Further revision of the International Undertaking was prompted by the growing importance of biological and genetic resources at the international level. In 1992, Agenda 21 called for the strengthening of the FAO Global System on Plant Genetic Resources, and its adjustment in accordance with the outcome of negotiations on the Biodiversity Convention. Negotiations for the revision of the Undertaking in harmony with the Convention began with the First Extraordinary Session of the Commission on Plant Genetic Resources in November 1994 and continued until November 2001.

The new Undertaking is now a binding treaty, the PGRFA Treaty. The Treaty was the object of arduous negotiations which led to a final consensus text which was acceptable to all the states present apart from the United States and Japan which abstained from voting. The overall objectives of the PGRFA Treaty are significantly different from those of the 1983 Undertaking. The Treaty, reflecting the new orientation given by the Biodiversity Convention, emphasises the conservation of PGRFA, their sustainable use and benefit sharing. The guiding principles for these three objectives are the promotion of sustainable agriculture and food security. The PGRFA Treaty focuses on issues not addressed in other international treaties such as farmers' rights but it does not address directly patents or plant breeders' rights covered in other treaties. The PGRFA Treaty has a number of unique characteristics.

Firstly, it is the first treaty providing a legal framework which not only recognises the need for conservation and sustainable use of PGRFA but also delineates a regime for access and benefit sharing, and in this process provides direct and indirect links to IPRs instruments.

Secondly, it directly links plant genetic resource conservation, IPRs, sustainable agriculture and food security.

Thirdly, the element which remains the distinguishing feature of the PGRFA Treaty in the field of plant variety protection is its focus on farmers' rights.

In fact, the term farmers' rights is slightly misleading. The PGRFA Treaty gives recognition to farmers' contribution to conserving and enhancing PGRFA. It further gives broad guidelines to states concerning the scope of the rights to be protected under this heading but overall devolves the responsibility for realising farmers' rights to member states. This includes the protection of traditional knowledge, farmers' entitlement to a part of benefit sharing arrangements and the right to participate in decision-making regarding the management of plant genetic resources. However, the treaty is silent with regard to farmers' rights over their landraces. In fact, the 'recognition' of farmers' contribution to plant genetic resource conservation and enhancement does not include any property rights. In this context, the only rights that are recognised are the residual rights to save, use, exchange and sell farm-saved seeds.

On important aspect of the PGRFA Treaty is the novel scheme devised to regulate access and benefit sharing of PGRFA covered under the Treaty. The underlying reason for the inclusion of a system of facilitated access is that the sovereign rights of states over their PGRFA are qualified by the recognition that these resources are a common concern of humankind and that all countries depend largely on PGRFA that originated in other countries. As a result, donor countries have full control over their PGRFA but there are strict limitations on their ability to restrict access to other states. As per the PGRFA Treaty, access is to be provided only for the purpose of utilisation and conservation for research, breeding and training for food and agriculture. As a result of the recognition of PGRFA as a common concern, access has to be accorded expeditiously. Concerning material which is under development by farmers or breeders at the time when access is requested, the Treaty gives the country of origin the right to delay access during the period of development.

One of the most difficult part of the Treaty negotiations related to the treatment of IPRs. The compromise solution is that recipients of PGRFA cannot claim IPRs that limit the facilitated access to the PGRFA, or their genetic parts or components, in the form received from the Multilateral System. Further, PGRFA accessed under the Multilateral System must also be made available to other interested parties by the recipient under the conditions laid out by the Treaty. This provision which stops the appropriation of isolated components from material accessed under the Multilateral System was strongly opposed by some countries which argued that this would stifle innovation. On the other hand, when the PGRFA in question are already protected by intellectual property or other property rights, access can only take place in conformity with the treaties regulating the particular kind of property rights. As is the case with some other treaties like the Biosafety Protocol, the PGRFA Treaty refuses to establish a hierarchy between itself and other related treaties, such as IPRs treaties. This leaves the door open for divergent interpretation at the time of implementation. The question of access is closely related to that of benefit sharing. In fact, the benefit sharing regime constitutes another part of the bargaining process which seeks to make PGRFA a common concern of humankind. The rationale for benefit sharing is that countries providing facilitated access to their PGRFA are granted in return the right to receive some forms of benefits. Different types of benefit sharing mechanisms are provided for under the Treaty: These include the exchange of information, access to and transfer of technology, capacity building, and the sharing of the benefits arising from commercialisation. With regard to the sharing of information, the Treaty envisages that member states will, for instance, provide catalogues and inventories, information on technologies, and the results of technical, scientific and socio-economic research. Concerning technology transfer, the Treaty provides only a general obligation to facilitate access to technologies for the conservation, characterisation, evaluation and use of PGRFA which is further qualified by the fact that access to such technologies is subject to applicable property rights. In the case of developing countries, specific mention is made of the fact that even technologies protected by IPRs should be transferred under 'fair and most favourable terms', in particular in the case of technologies for use in conservation as well as technologies for the benefit of farmers in developing countries. Finally, the Treaty provides for the sharing of monetary benefits.

These include, for instance, the involvement of the private sector in developing countries in research and technology development. Further, the standard Material Transfer Agreement, through which facilitated access will be implemented, will include a requirement that an equitable share of the benefits arising from the commercialisation of products that incorporates material accessed through the Multilateral System will have to be paid to the trust account set up under the Treaty. The benefits that arise under the benefit sharing arrangements must be primarily directed to farmers who conserve and sustainably use PGRFA. Overall, the Treaty which constitutes the outcome of many years of negotiations is noteworthy for linking the conservation of PGRFA, their use, the rights of farmers over resources and knowledge and finally the IPRs system. It provides an interesting, though inconclusive, attempt to link these different elements. The provisions concerning access and benefit sharing typically seek to build a bridge between the different forms of property rights recognised under the PGRFA Treaty and in other relevant treaties such as the TRIPS Agreement. They, however, largely lack in specificity, partly because they reflect the difficult balancing of interests that the negotiators had to achieve between the interests of developed and developing countries, big private seed companies and small farmers and a number of other actors in between.

Consultative group on international agricultural research

Since its inception in 1971, the CGIAR has played an important role in the management of genetic resources used to meet food needs and in defining property rights policies in this regard. The CGIAR brings together a network of IARCs which have important *ex situ* germplasm collections. The CGIAR aims at alleviating poverty, achieving food security and assuring sustainable use of natural resources. It has traditionally sought to fulfil its mandate through the development of freely accessible *ex situ* collections and the production of freely available improved varieties. However, in keeping with the progressive move towards the establishment of sovereign and private property rights over biological and genetic resources, the CGIAR has gradually modified its stance concerning real and intellectual property rights.

In the past decade, a number of important developments have taken place. Firstly, starting in 1994, the Centres have signed agreements that place their collections held in trust for humankind under the auspices of the FAO and that restrict them from claiming IPRs over designated germplasm or related information. Secondly, the CGIAR and the IARCs progressively developed new guiding principles on intellectual property with a view to harmonise the CGIAR's core principles that designated germplasm is held in trust for the world community with the recognition of various forms of property rights, including sovereign rights, farmers' rights and IPRs. To-date, the Centres do not normally apply intellectual property protection to their designated germplasm and require recipients to observe the same conditions. They also refrain from asserting IPRs over the products of their research. An exception to this rule is made in case the assertion of IPRs facilitates technology transfer or otherwise protects developing countries' interests. The CGIAR also imposes that any IPRs on the IARCs' output should be assigned to the Centre and not an individual. While the guiding principles on intellectual property generally seek to contain to an extent the monopoly elements of IPRs such as patents, plant breeders' rights are specifically welcomed. Recipients of germplasm can apply for plant breeders' rights as long as this does not prevent others from using the original materials in their own breeding programmes.

Thirdly, the PGRFA Treaty will further change the conditions under which the CGIAR operates. In future, guidance concerning the management of CGIAR collections will come from the Treaty's Governing Body. In fact, the Centres having signed agreements with the FAO are now invited to sign new agreements with the Treaty's Governing Body. These agreements will provide that the collections of the Centres will be governed by the access provisions of the PGRFA Treaty. This will, however, only cover materials collected after the entry into force of the Treaty and that fall within its scope. The Centres are also put under an obligation to provide preferential treatment to countries that provided material to their gene banks and are not to request any material transfer agreement if a country of origin wants to access its own material. Generally, the Centres will have to recognise the authority of the PGRFA Treaty's Governing Body to provide policy guidance relating to their *ex situ* collections. Overall, the PGRFA Treaty will foster more coordination between the FAO and the CGIAR. This will, in particular, have significant impacts in terms of their outlook on IPRs which will have to be broadly similar, at least with regard to the CGIAR collections falling in the scope of the PGRFA Treaty.

The CGIAR has long benefited from its hybrid institutional status among international institutions which contributed in part to making possible its contribution to the alleviation of food insecurity in developing countries. In recent times, however, the CGIAR has found it increasingly difficult to reconcile its original mission with the changing legal and policy framework in which it operates. Thus, the decision to accept the Syngenta Foundation for Sustainable Agriculture as a new CGIAR member has been criticised as sign that the CGIAR is moving away from its public sector research mission. Further, the CGIAR has also found it difficult to adjust to some of the challenges of biotechnology. The case of the controversy over the introduction of genetically modified maize in Mexico – the primary centre of diversity for maize – illustrates the challenges that lie ahead for an organisation which is striving to maintain its significant collections of germplasm while endorsing at the same time biotechnology as 'one of the critical tools for providing food security for the poor'.

Intellectual Property Rights Related Legal and Institutional Framework

Developments in the agricultural field are of central importance because they directly concern food security. However, with the large-scale development of genetic engineering, IPRs standards have become increasingly important in their own right and because they influence the development of the legal and policy framework in agriculture and other fields.

This section does not attempt to provide an exhaustive analysis of the IPRs framework in the field of food security but focuses on some of the most important treaties and institutions from the point of view of developing countries. Further, it only covers under the heading of IPRs, rights that have generally been considered as falling within the subject matter of intellectual property protection.

TRIPS agreement

The TRIPS agreement is today the most important intellectual property treaty for all WTO member states. The TRIPS Agreement is only indirectly concerned with agriculture and environmental management but the IPRs standards it sets have wide-ranging impacts on agricultural management.

The TRIPS Agreement is a general treaty which covers different types of IPRs, such as patents, copyright and geographical indications. It seeks to introduce minimum standards of IPRs in all member states. In practice, this generally has the effect of extending the application of IPRs standards already in use in most OECD countries to all WTO member states and thus imposes a significant burden of adjustment on developing country member states. The framework provided by the TRIPS Agreement must be understood in the context of the interpretative clauses that are part of the treaty. Article 7 recalls that IPRs protection must both contribute to the promotion of technological innovation and at the same time to the transfer and dissemination of technology in a manner conducive to social and economic welfare, and to a balance of rights and obligations. Further, Article 8 concedes that in implementing TRIPS obligations at the domestic level, states have the possibility to adopt measures to protect nutrition and to promote the public interest in sectors of vital importance to their socio-economic and technological development.

Among the types of IPRs protected under the TRIPS Agreement, patent rights stand out in the context of food security. The Agreement uniformly provides that patents must be available for inventions, whether products or processes, in all fields of technology. Some general exceptions are granted and states can, for instance, exclude patentability where this is necessary to protect human, animal or plant life or health, or to avoid serious prejudice to the environment. They can also exclude from patentability plants and animals other than micro-organisms.

International convention for the protection of new varieties of plants

The International Convention for the Protection of New Varieties of Plants (UPOV Convention) is the only intellectual property treaty which directly focuses on agriculture. It was adopted in 1961 by a group of western European countries which sought to introduce IPRs in agriculture but were not prepared to accept the introduction of patents in this field. As a result, the UPOV Convention proposes the adoption of plant breeders' rights. The UPOV Convention's main aim is to protect new varieties of plants in the interests of both agricultural development and commercial plant breeders.

Plant breeders' rights differ from patent rights but they also share a number of basic characteristics with them. Plant breeders' rights provide exclusive commercial rights to rights holders, reward an inventive process, and are granted for a limited period of time after which they pass into the public domain. More specifically, UPOV recognises the exclusive rights of individual plant breeders to produce or reproduce protected varieties, to condition them for the purpose of propagation, to offer them for sale, to commercialise them, including exporting and importing them, and to stock them for production or commercialisation. Protection under UPOV is granted for developed or discovered plant varieties which are new, distinct, uniform and stable. While novelty is a criterion shared with patent law, UPOV adopts a different approach. Under UPOV, a variety is novel if it has not been sold or otherwise disposed of for purposes of exploitation of the variety. Novelty is thus defined in relation to commercialisation and not by the fact that the variety did not exist previously. UPOV gives a specific time frame for the application of novelty. To be novel, a variety must not have been commercialised in the country where the application is filed for more than a year before the application and in other member countries for more than four years. The criterion of distinctness requires that the protected variety should be clearly distinguishable from any other variety whose existence is a matter of common knowledge at the time of the filing of the application. Stability is obtained if the variety remains true to its description after repeated reproduction or propagation. Finally, uniformity implies that the variety remains true to the original in its relevant characteristics when propagated.

One of the main distinguishing features of the UPOV regime is that the recognition of plant breeders' rights is circumscribed by two main exceptions. Firstly, under the 1978 version of the Convention, the so-called 'farmer's privilege' allows farmers to re-use propagating material from the previous year's harvest and to freely exchange seeds of protected varieties with other farmers. Secondly, plant breeders' rights do not extend to acts done privately and for non-commercial purposes or for experimental purposes and do not extend to the use of the protected variety for the purpose of breeding other varieties and the right to commercialise such other varieties. The 1991 version of the Convention, by strengthening plant breeders' rights, has conversely limited existing exceptions. The remaining exceptions include acts done privately and for non-commercial purposes, experiments, and for the breeding and exploitation of other varieties. Breeders are now granted exclusive rights to harvested materials and the distinction between discovery and development of varieties has been eliminated. Further, the right to save seed is no longer guaranteed as the farmer's privilege has been made optional.

Environment Related Legal Framework

International environmental legal instruments have increasingly taken a broad perspective of the environment over time. This is in keeping with the shift of international environmental law towards an international law of sustainable development. As a result of the broader perspective of environmental treaties, environmental management is seen in a broader light which includes for instance links with agricultural management, human rights and IPRs. Among the different treaties with food security links, the regime for biodiversity management is noteworthy because it provides the general legal framework for biological resource management.

The Convention on Biological Diversity (Biodiversity Convention) is a framework treaty which seeks to regulate the conservation and use of biological resources. Its three main goals are the conservation of biological diversity, the sustainable use of its components, and the fair and equitable sharing of the benefits derived from the use of genetic resources. In the context of food security and IPRs, the Biodiversity Convention makes several distinct contributions.

Firstly, the specific role and importance of agro-biodiversity has been recognised by the Conference of the Parties and a special programme on agro-biodiversity was established in 1996. It generally aims to promote the positive effects and mitigating the negative impacts of agricultural practices on biological diversity in agricultural ecosystems and their interface with other ecosystems. Further, it seeks to promote the conservation and sustainable use of genetic resources of actual or potential value for food and agriculture. Over time, the agro-biodiversity programme has taken up specific challenges, deepened its cooperation with the FAO and examined cross-sectorial issues such as the potential impacts of patented genetic use restriction technologies on farmers.

Secondly, the Biodiversity Convention provides one of the few existing statements on the relationship between the management of biological and genetic resources and IPRs. Article 16 clearly indicates that IPRs should not undermine the working of the Convention. The actual relationship of the Biodiversity Convention with the TRIPS Agreement is an issue which has not been solved. This is partly due to the fact that a clear statement on the matter would have significant repercussions for the development of international law in these two fields.

Thirdly, the Biodiversity Convention has also made its own contribution to the development of access and benefit sharing schemes, effort supplemented with the adoption by the Conference of the Parties of the Bonn Guidelines on access and benefit sharing. The Convention attempts to provide a framework which respects donor countries' sovereign rights over their biological and genetic resources while facilitating access by users. Access must therefore be provided on 'mutually agreed terms' and is subject to the 'prior informed consent' of the country of origin. Further, the Biodiversity Convention provides that donor countries of micro-organisms, plants or animals used commercially have the right to obtain a fair share of the benefits derived from use. Benefit sharing as conceived under the Convention and the Bonn Guidelines can take the form of monetary benefits or non-monetary benefits such as the sharing of research and development results, collaboration in scientific research and access to scientific information relevant to conservation and sustainable use of biological diversity. Overall, the contribution of the Biodiversity Convention and the PGRFA Treaty concerning access and benefit sharing are complementary even though the latter's framework goes further insofar as it constitutes an integral part of the treaty while the Bonn Guidelines remain at present purely voluntary.

Fourthly, the Biodiversity Convention also provides in general terms for the conservation of traditional knowledge, a question that is closely linked to the fulfilment of basic food needs and to the protection of agro-biotechnology through IPRs. The Convention provides under Article 8(j) a general duty for all member states to respect, preserve and maintain knowledge, innovations and practices of indigenous and local communities pertaining to the management of biological resources, promote their wider application with prior informed consent and encourage the equitable sharing of the benefits arising from such utilisation. This provision has been supplemented with the setting up of a working group mandated with the task of giving advice on legal and other means of protection of traditional knowledge. While the Convention has addressed the conservation of traditional knowledge and the issue of access and benefit sharing, it has not really tackled questions surrounding the ownership of biodiversity-related traditional knowledge, an area which remains generally unsettled in international law.

While the Biodiversity Convention plays a dominant role in the international environmental law field, a great number of other treaties are also significant in the context of this study. Of particular relevance is the Desertification Convention. This Convention is noteworthy because it directly recognises the links between desertification as an environmental problem and socio-economic problems such as food security. It also specifically indicates that national action programmes to be developed by state parties must include among the measures to mitigate the effects of drought the establishment and strengthening of food security measures, including storage and marketing facilities. Further, the Desertification Convention is more specific than most treaties with regard to the protection of traditional knowledge insofar as it directs states not only to respect it but also to provide 'adequate protection'.

Human Rights Related Legal Famework

The realisation of food security at the level of each and every individual level can be broadly equated with the realisation of the human right to food. While the realisation of the right to food can be analysed separately from the concerns examined in this study, it provides the underlying guiding framework for analysing the relationship between IPRs and food security. Further, even though human rights and IPRs operate largely independently, some specific links need to be analysed.

The human right to food is recognised, for instance, in the Covenant on Economic Social and Cultural Rights (ESCR Covenant) which provides a right to adequate food and a right to be free from hunger. The right to food, like other socio-economic requires the state to take measures to progressively realise this right through positive steps which include the improvement of production methods and output, the improvement of food distribution networks and at the international level a better distribution of world food supplies in relation to the needs of each country. In practical terms, the right to food is realised when all individuals have physical and economic access at all times to adequate food or means for its procurement. Adequate food under the Covenant does not just imply a minimum package of calories and nutrients but takes into account a much broader set of factors to determine whether particular foods or diets that are accessible can be considered the most appropriate under given circumstances. As expounded by the Committee on Economic Social and Cultural Rights, the realisation of the right to food requires the availability of food in a quantity and quality that is sufficient to satisfy the dietary needs of individuals and that is free from adverse substances. It also implies that the accessibility of food must be sustainable and should not interfere with the enjoyment of other human rights.

The link between IPRs and human rights surfaces at different levels. The ESCR Covenant recognises everyone's right to take part in cultural life and the right 'to enjoy the benefits of scientific progress and its application'. This general entitlement promoting the sharing of knowledge is supplemented by another provision which recognises everyone's right 'to benefit from the protection of the moral and material interests resulting from any scientific, literary or artistic production of which he is the author'. The interpretation of these two provisions together may be interpreted as indicating that the recognition of the material interests of an individual IPRs holder does not prevail over everyone's right to the enjoyment of scientific and technological development.

IPRS AND FOOD SECURITY — GENERAL TRENDS AND IMPLEMENTATION

The international legal regime outlined above has evolved in response to different challenges and changes, such as the development of genetic engineering over the past couple of decades. In turn, the international legal regime has also had – and is having – a significant influence on the development of national legal frameworks in developing countries. This section first examines the broad trends that have marked the international legal regime in recent years and then goes on to analyse in more detail the situation in India, a country which has adopted significant changes in its domestic legal framework in recent years, partly with a view to implement its international obligations.

FOOD SECURITY AND INTELLECTUAL PROPERTY RIGHTS IN THE SOUTH: LESSONS FROM RECENT DEVELOPMENTS IN INDIA

A number of countries have attempted or are in the process of implementing their different international obligations concerning both IPRs and food security. In nearly all cases and even in the case of India which has moved far towards the implementation of its international commitments, there remain a number of areas that have not yet been addressed. Further, the adoption of the PGRFA Treaty in 2001 has added a new layer of international obligations which will have to be taken into account by all PGRFA Treaty member states.

Given that a number of developed countries introduced IPRs in agriculture a long time before developing countries, it may seem appropriate to examine the impacts that this had to understand the likely impacts of the introduction of agriculture-related IPRs in the South. This comparison would not yield significant insights, in part because the socio-economic conditions of developing countries are too different from the situation of developed countries, even a few decades ago. To take but one example, while the percentage of people engaged in the agricultural sector in the European Union in 1961 was 20% when the UPOV Convention was adopted, the population active in the agricultural sector in developing countries today amounts to 86 per cent of the rural population and 52 per cent of the total population in developing countries.

Indian Situation

India is an interesting case study because it has been through different shifts in policy over food security policies in the context of IPRs since independence. India inherited at independence a patent law which was deemed inappropriate to realise the economic development goals of the country because the colonial act had failed to stimulate invention by Indian citizens and to encourage the development and exploitation of new inventions for industrial purposes in the country so as to secure benefits to the largest section of the people. Patent law was thus overhauled in the decades following independence in an attempt to make it fit the developmental priorities of the country. The resulting Patents Act, 1970 retained the western model of intellectual property but provided a number of exception with a view to foster the fulfilment of basic needs. In particular, the Act excluded the patentability of life forms and specifically precluded the patentability of methods of agriculture or horticulture. Further, while allowing process patents on substances intended for use as food, medicine or drug, the Act rejected the possibility of granting patents in respect of the substances themselves. Insofar as the duration of the rights conferred was concerned, the normal 14-year term was reduced to 7 years with respect to processes of manufacture for substances intended for use as food, medicine or drug. The Patents Act, 1970 also introduced a series of measures restricting the rights of patent holders, in particular to encourage use of the invention in India. The rationale for the introduction of limiting clauses in the Act was in part to foster the growth of local industries and in part to foster the availability of essential items such as food and medicine by keeping the prices as low as possible in areas related to the fulfilment of basic needs.

The absence of patents in agriculture contributed to the development of a system of agricultural management based on the sharing of genetic material and related knowledge. At the same time, it did not provide significant incentives for the development of a private seed industry. As a result of these policies, the public sector has until recently been a major force in agricultural management.

The ratification of the TRIPS Agreement by India has been the trigger for significant changes in the IPRs related national legal framework. This has included in particular the adoption of a Plant Variety Act, a series of significant changes to the Patents Act, 1970 and the adoption of IPRs-related clauses in the recently adopted Biodiversity Act. These three main legislative instruments are examined in turn.

Historically, the protection of plant varieties through IPRs was barred, as reflected in the Patents Act, 1970. The introduction of plant variety protection thus constitutes a step in a completely different direction. As noted, TRIPS imposes the introduction of plant variety protection but leaves member states to choose the specific form of protection they want to adopt. It does not privilege plant breeders' rights (or in other words, the UPOV Convention) over alternatives such as farmers' rights. The Indian legislation was first introduced in Parliament in December 1999, just before the TRIPS Agreement's compliance deadline. The main characteristic of the first draft was to propose a plant variety protection model largely fashioned after the UPOV Convention. This first draft was referred to a Parliamentary Committee which conducted further hearings in 2000 and put forward a substantially revised Bill. This second draft was adopted by Parliament in 2001 and is now the Protection of Plant Varieties and Farmers' Rights Act (Plant Variety Act). Generally, the Act differs from the first draft of the bill insofar as it clearly seeks to establish both plant breeders' rights and farmers' rights. The proposed regime for plant breeders' rights largely follows the model provided by the UPOV Convention. It introduces rights which are meant to provide incentives for the further development of a commercial seed industry in the country. The criteria for registration are thus the same as those found in UPOV, namely novelty, distinctiveness, uniformity and stability. The Act incorporates a number of elements from the 1978 version of UPOV and also includes some elements of the more stringent 1991 version, like the possibility of registering essentially derived varieties. The Act now seeks to put farmers' rights on par with breeders' rights. It provides, for instance, that farmers are entitled, like commercial breeders, to apply to have a variety registered. Farmers are generally to be treated like commercial breeders and are to receive the same kind of protection for the varieties they develop. However, it is unsure whether these provisions will have a significant impact in practice since the Act accepts the registration criteria of the UPOV Convention which cannot easily be used for the registration of farmers' varieties. The Act incorporates other provisions which are directly related to food security concerns. These include, for instance, a section which specifically bars the registration of plant varieties with genetic restriction use technologies.

Overall, the Act is noteworthy for making a real attempt at balancing breeders' and farmers' rights. However, two main facts are likely to hamper the effectiveness of the provisions for farmers' rights. Firstly, since farmers' rights were only added as an afterthought without changing the criteria for registration of varieties, the existing regime exclusively reflect the registration needs of commercial breeders and is therefore heavily tilted against farmers. Secondly, even though India intended to provide a *sui generis* response to the need to provide plant variety protection under the TRIPS Agreement, it is now in the process of formally joining UPOV, a move which will tilt the balance further away from farmers.

Apart from adopting plant variety legislation, India has passed substantial amendments to its patent legislation. The modifications to the Patents Act required to fulfil TRIPS obligations have resulted in the dismantling of most of the specificities that were introduced by the 1970 Act in view of the explicit recommendations concerning the working of the earlier colonial patent act. Among the major changes required is an increase in the general patent term from 14 years to 20 years, and from 7 years to 20 years in the case of process patents on food related inventions. Certain control mechanisms restricting the scope of the rights granted to patent holders such as the existence of licences of right, and more specifically automatic licences of right in the case of process patents relating to substances used as food, have been removed from the Act. In general, the 2002 amendments to the Patents Act, 1970 will contribute to the development of agro-biotechnology. However, the Amendment Act takes into account some of the concerns that have been voiced in recent times, in particular with regard to 'biopiracy' or the unwarranted use of traditional knowledge. It now obliges inventors to disclose the geographical origin of any biological material used in an invention. Further, there is a specific exclusion on patents that are anticipated in traditional knowledge.

Besides the plant variety and patents legislation, the Biodiversity Act is also important because the regulation of biodiversity management has direct impacts on food security and because the Act directly links biodiversity management and IPRs. The main focus of the Act is on the question of access to resources. Its response to current challenges is to assert the country's sovereign rights over natural resources. It therefore proposes to put stringent limits on access to biological resources or related knowledge for all foreigners. The Act's insistence on sovereign rights reflects current attempts by various countries to assert control over the resources or knowledge they control. While the Act focuses on preserving India's interests vis-à-vis other states in rather strong terms, its main impact within the country will be to concentrate power in the hands of the government. Indeed, Indian citizens and legal persons must give prior intimation of their intention to obtain biological resources to the state biodiversity boards. The Act is even more stringent in terms of IPRs since it requires that all inventors obtain the consent of the National Biodiversity Authority before applying for such rights. The impact of this clause is, however, likely to be limited since patent applications are covered by a separate clause. Further, the Authority has no extra-territorial authority.

The Biodiversity Act implicitly takes the position that India cannot do more than regulate access by foreigners to its knowledge base. It does, however, attempt to discipline the IPRs system in some respects. As noted, it requires inventors who want to apply for IPRs to seek the National Biodiversity Authority's permission. It also authorises the Authority to allocate a monopoly right to more than one actor. Further the Authority is also entitled to oppose the grant of intellectual property rights outside India. The Act also seeks to address the question of the rights of holders of local knowledge by setting up a system of benefit sharing. The benefit sharing scheme is innovative insofar as it provides that the Authority can decide to grant joint ownership of a monopoly intellectual right to the inventor and the Authority or the actual contributors if they can be identified.

However, the sharing of IPRs is only one of the avenues that the Authority can choose to fulfil its obligation to determine benefit sharing. It is also in the Authority's power to allocate rights solely to itself or a contributor such as a farmer contributor. Other forms of benefit sharing include technology transfers, the association of benefit claimers in research and development or the location of production, research and development units in areas where this will facilitate better living standards to the benefit claimers. On the whole, the Biodiversity Act effectively condones the introduction of IPRs in the management of biological resources provided for in the TRIPS Agreement but does not specifically seek to ensure that IPRs are supportive of the goals of the Biodiversity Convention.

The different legislative changes introduced in India will have profound impacts on the development of IPRs based industries such as agro-biotechnology and on food security. From a legal point of view, the adopted regime is noteworthy for attempting to reconcile to a certain extent India's international obligations with its domestic priorities. However, on the whole, it is unsure whether India has managed to provide a balance which puts food security concerns at the forefront and serves its interests. This is, for instance, illustrated by the apparent tension in the Biodiversity Act between the emphasis on India's claim over its biological resources and an acknowledgment that India cannot control the use that is made of related knowledge because it cannot control patent applications in other parts of the world. Further, with regard to the development of agro-biotechnology, existing studies seem to indicate that neither the public nor the private domestic sector have been until now in a position to take advantage of the opportunities to appropriate benefits of the new IPRs regime. With regard to food security at the individual level, the Plant Variety Act makes a determined attempt to adopt a balanced legal regime which gives incentives to the private sector seed industry but also protects individual farmers and farming communities. In practice, however, the proposed farmers' rights regime is unlikely to be effective. Further, the effectiveness of the adopted regime is likely to be hampered by the lack of coordination between the three acts. Potential problems range from the lack of institutional coordination to the definition of different benefit sharing schemes under the Plant Variety and Biodiversity Acts. Finally, the adopted legal regime fails to take into account a significant proposal by the Indian Law Commission linking biodiversity management, food security and plant variety protection. The Commission proposed its own draft Biodiversity Bill in which it introduced a provision which stated that no IPRs should be granted on species used for alimentary or medicinal purposes. This was meant as an attempt to integrate the right to food with the exceptions allowed in the TRIPS Agreement, a proposal which was not maintained subsequently.

On the whole, the Indian legal framework constitutes a good starting point for a regime seeking to comply with all relevant international obligations in the field of food security and IPRs. However, it remains inadequate in important areas like farmers' rights and the protection of traditional knowledge. This may be explained to an extent by the fact that these are new areas and that the development of appropriate legal frameworks is a lengthy exercise. In the context of long-term policy objectives, including the ratification of the PGRFA Treaty and discussions taking place in WIPO on the protection of traditional knowledge, it seems important to further pursue the development of the legal framework even in a country like India which has gone through substantial legislative effort in recent years. In any case, the current legal regime needs at the very least adjustments to make the different pieces of the puzzle work together harmoniously. This is a challenge that many other countries face because most countries tend to give authority for the implementation of different acts with different focuses to different ministries even if there are strong links between them, such as in the case of the Biodiversity Act, the Plant Variety Act and the Patents Act in India.

Finally, the capacity of the Indian legal regime to provide a model for other developing countries is limited. Even though many countries face a number of similar structural constraints and similar socio-economic conditions, the protection of farmers' rights and traditional knowledge should be tailored to the specific conditions of individual countries. The last section of this study examines some of the general options that developing countries may consider to implement their international obligations. It also examines some avenues that may go beyond the generally accepted interpretation of existing treaties but could nevertheless be considered to foster individual countries' food security, environmental and economic interests.

TOWARDS SUI GENERIS INTELLECTUAL PROPERTY PROTECTION

As noted above, Article 27(3)b of the TRIPS Agreement provides an opportunity for developing states to develop their own IPRs framework in the field of plant varieties, taking into account such concerns as food security at the individual and national levels. This flexibility can be used in the narrow context of an intellectual property treaty such as the TRIPS Agreement. However, given that the introduction of IPRs in agriculture has broader implications beyond the strict field of intellectual property, it appears opportune to pursue a broader strategy whereby the legal framework introduced in the context of plant variety takes into account a number of other goals. These include elements covered by other international treaties such as the introduction of farmers' rights, the protection of traditional knowledge and benefit-sharing regimes. It also includes other links such as the relationship between the introduction of IPRs in agriculture and the realisation of the human right to food, a dimension which is often unjustifiably sidelined.

There are further reasons for developing countries to devise their own legal framework in the area of food security and IPRs. The current and evolving international legal regime in relevant areas increasingly promotes the appropriation of biological and genetic resources, the appropriation of knowledge related to biological and genetic resources, and trade in resources and knowledge. The international legal system has until now generally protected developing countries' interests in this area by constantly reaffirming their sovereignty over their natural resources. New developments in genetic engineering are increasingly making access to physical resources much less important than the control over knowledge. At present, the IPRs system only offers one type of protection, namely protection for state-of-the-art inventions granted in exclusivity to the rights holders. In general, the existing system has not been conceived with the situation of developing countries in mind. As a result, while developing countries can benefit to a certain extent from the existing system, this must be supplemented with other measures destined to take into account their specificities. This includes, for instance, the need to provide legal frameworks which provide strong property rights to all relevant actors in the field. This is not due to the fact that property rights are better able to promote food security than existing systems based on exchange and free flows of information but to the fact that in a world where the scope of appropriation is rapidly increasing, it is especially important to make sure the weaker actors such as farmers and traditional knowledge holders are well protected.

To sum up, the challenge of enhancing food security for each individual and each country around the world will require tremendous efforts on the part of all actors involved if malnutrition is ever to be eradicated. Food insecurity in developing countries has been a concern for long and is associated with a number of general and specific policy challenges. The development of genetically modified plant varieties and the introduction of IPRs in agriculture constitute two related and significant changes in the policy environment for addressing food security.

Overall, the need to develop a legal framework that goes beyond traditionally recognised IPRs regimes is based on a number of reasons. At a basic level, the introduction of IPRs in agriculture can only be justified if IPRs foster food security, or in other words the realisation of the human right to food. There are a number of ways to foster food security. One of them includes the appropriation of knowledge related to plant varieties through property rights. In this scheme which is promoted today at the international level, control over knowledge is only offered to state-of-the-art inventions. In fact, the introduction of property rights in agriculture should benefit all actors involved in agricultural management. This is the gap that developing countries must fill given that their agricultural systems are often overwhelmingly dependent on the contributions of a significant number of small individual farmers, local farming communities and public sector institutions rather than private actors. In this situation, the development of positive farmers' rights is necessary not only for the benefit of farmers but also their countries. In fact, appropriately designed farmers' rights should provide benefits to farmers and farming communities, should foster sustainable agro-biodiversity management, should provide tools for governments to fight biopiracy and overall should provide a set of incentives to tackle food insecurity. Such farmers' rights need not be envisaged as opposed to existing IPRs. They should be complementary, possibly overlapping forms of property rights, and on the whole they should foster, like patents and plant breeders' rights, further incentives towards the realisation of the human right to food.

Glossary

Absorption	:	Absorption, in chemistry, is a physical or chemical phenomenon or a process in which atoms, molecules or ions enter some bulk phase—gas, liquid and solid material.
Acetogenesis	:	Acetogenesis is the formation of acetate from CO_2 as the result of the metabolic processes of certain bacteria, in contrast to methanogenesis, three recognised processes are the acetyl-CoA pathway, the glycine synthase-dependent pathway, and the reductive citric acid cycle.
Actinomycetes	:	Bacteria in which species are characterised by the formation of branching and/or true filaments.
Activated sludge process	:	Biological waste-water treatment process in which a mixture of the waste-water and activated sludge is aerated in a reactor basin or aeration tank. Active biological solids bio-oxidise the waste matter and the biological solids are removed by secondary clarification or final settling.
Acute bronchitis	:	Disease that presents with cough productive of sputum that is often secondary to a viral or bacterial infection and the inhalation route of exposure to high concentrations of bioaerosols in agriculture facilities.
Aerated lagoon	:	Waste-water treatment pond in which mechanical or diffused air aeration is used to supplement oxygen supply.
Aerobes	:	Group of organisms that require air or oxygen for their survival and growth.
Aerobic	:	Presence of free molecular oxygen is required.
Aerobic digestion	:	Digestion of suspended organic matter by aerobic microbes.
Agar	:	A derivative of marine seaweed used as a solidifying agent in many microbiological media.
Anaerobes	:	Group of organisms that cannot tolerate the presence of air or oxygen or survive in the absence of air or oxygen.
Anaerobic bacteria	:	Bacteria that require combined oxygen and the absence of free molecular oxygen.
Anaerobic digestion	:	Digestion of suspended organic matter by anaerobic microbial action.
Anthropogenic	:	Originating in human activity.
Archaebacteria	:	Most primitive type of micro-organism among prokaryotes.
Auxotroph	:	A cell that requires nutritional supplements for growth.
Bacillus thuringiensis	:	Bacterium used in crop protection, and in the generation of *Bt* plants that are resistant to insect attack. The bacterium produces a toxin that affects the insect.

Bacteria	:	These are ubiquitous, single celled prokaryotic micro-organisms, the domain level of taxonomy of micro-organisms that are distinctly different from prokaryotes in the domain Archaea and the eukaryotes in the domain Eukarya.
Bacteriocins	:	A group of bacterial proteins toxic to other bacteria.
Batch reactor	:	A reactor that does not have continuous streams entering or leaving. The reactants are added, reaction occurs, then the products are discharged.
Binary fission	:	The manner in which most bacteria multiply. The parent cell divides, usually into two daughter cells.
Biochemical oxidation	:	Oxidation caused by biological activity resulting in a chemical combination of oxygen with organic matter to produce relatively stable end-products.
Biodegradation	:	The process of chemical breakdown of a substance to smaller products caused by micro-organisms or their enzymes.
Biodeterioration	:	The chemical or physical alteration of a product that decreases the usefulness of that product for its intended purpose, caused by micro-organisms or their enzymes.
Biological oxidation	:	An oxidation caused by biological activity resulting in a chemical combination of oxygen with organic matter to produce stable end-products known as both biochemical oxidation and bio-oxidation.
Biopreservation	:	The use of micro-organisms or microbial by-products to prevent spoilage and extend the shelf-life of foods.
Bioremediation	:	Process of using organisms to consume or otherwise help remove pollutants from the environment.
Biotransformation	:	Enzyme-catalysed conversion of one chemical, other than the normal body constituents of live organisms, into another. Normal metabolism refers to such conversions restricted to carbohydrates, fats, proteins, etc. taking place inside the body.
Brownian movement	:	Random zig-zag movement of microscopic particles in a gaseous system or suspended in a liquid medium.
Buffer action	:	Action of certain ions in solution to oppose a change in pH.
Carboxydotrophic bacteria	:	Bacteria that aerobically utilise carbon monoxide as both a carbon source and as an energy source, for example, *Pseudomonas carboxidoflava* and *Pseudomonas carboxidohydrogena*.
Catabolism	:	Breakdown of complex biological molecules into simpler ones, usually accompanied with the release of energy in the form of ATP.
Chemical oxygen demand (COD)	:	The amount of oxygen required to chemically oxidise the organic and sometimes inorganic matter in water or waste-water. Usually expressed in mg/l. COD test does not measure the oxygen required to convert ammonia to nitrites and nitrites to nitrates. COD is frequently assumed to be equal to the ultimate first-stage biochemical oxygen demand.
Chemical sludge	:	Sludge produced by chemical coagulation or chemical precipitation.
Coagulation	:	In water or waste-water treatment, the destabilisation and initial aggregation of colloidal and finely divided suspended solids by the addition of floc-forming chemicals.

Completely mixed activated sludge	:	An activated sludge process with a completely mixed reactor basin. Usual basin is square, circular or slightly rectangular in plan view, and the influent, on entering, is almost immediately dispersed throughout the reactor basin.
Cosmid	:	Phage-plasmid artificial hybrids.
Demineralisation	:	Removal of all salts from a water.
Digested sludge	:	Sludge digested by aerobic or anaerobic action to the degree that the volatile content is low enough for the sludge to be stable.
Dispersed plug-flow activated sludge	:	Activated sludge process with a dispersed plug-flow reactor basin. The basin is rectangular in plan view and has significant longitudinal or axial dispersion of fluid elements throughout its length.
Dispersed plug-flow reactor	:	Reactor that is rectangular in plan view and has significant longitudinal mixing of fluid elements throughout its length.
Downstream processing	:	Refers to the procedures used to purify products (usually proteins) after they have been expressed in bacterial, fungal or mammalian cells.
Ecosystem	:	A community of organisms in their natural environment.
Electrophoresis	:	Technique for separating molecules based on the differential mobility in an electric field.
Enzymes	:	Proteins specialised to trigger biological reactions, e.g. the conversion of certain organic substances into different ones.
Exotoxin	:	A metabolic poison produced chiefly by Gram-positive bacteria, exotoxins are released to the environment, they are composed of protein and affect various organs and systems of the body.
Extended aeration activated sludge process	:	Activated sludge process with a detention time long enough to allow the amount of cells synthesised to be endogenously decayed.
Fermentation	:	Process by which enzymes, usually coming from micro-organisms, cause desired changes in taste, smell and texture.
Filtration	:	Unit operation that consists of passing a liquid through a granular medium for the removal of suspended and colloidal matter.
Fixed-bed	:	In carbon adsorption or ion exchange treatments using columns or open beds, this refers to a bed that is stationary in the column or in the structure for the open bed.
Flocculation	:	Slow stirring of a coagulated water or waste-water to aggregate the destabilised particles and form a rapid-settling floc. In biological waste-water treatment where a coagulant is not used, aggregation may be accomplished biologically.
Fluidised bed	:	Refers to a bed in which the particles are not in continuous contact due to the upward flow of the water or waste-water.
Genetic engineering	:	The *in vitro* manipulation of gene sequences.
Green bacteria	:	Anoxygenic phototrophic bacteria that conduct photophosphorylation using Bchl *c* and Bchl *d* chlorophyll pigments that are located within chlorosomes, in contrast to the purple bacteria.
Growth curve	:	Depiction of the cycle of a microbial population in culture in which the population increases, stabilises, and decreases, defined for each microbial population with a lag phase, an exponential phase, a stationary phase, and a death phase.
Halophiles	:	Organisms requiring NaCl for growth, extreme halophiles grow in concentrated brines.

Hepatotoxin	:	A compound that is toxic to the liver.
Immobilised enzyme	:	An enzyme bound to a solid support.
Isozymes (isoenzymes)	:	Multiple forms of an enzyme that differ in properties such as substrate specificity and maximum activity.
Lag phase	:	Initial period of time in the growth curve of micro-organisms in which growth does not occur immediately, the period to time prior to exponential growth.
Latent virus	:	A virus whose genome is integrated into the host's genome, but is not expressed, upon activation (e.g. by stress or exposure to ultraviolet irradiation), infective virus particles are produced and symptoms of infection appear.
Leaching	:	The removal of metal from ore by chemical or microbial activity, the transport of dissolved materials from upper soil layers deeper into the subsurface.
Legionellaceae	:	Bacterial family, characterised as intracellular parasites or endosymbionts of free-living parasites.
Ligation	:	Formation of a phosphodiester bond to link two adjacent bases separated by a nick in one strand of the double helix of DNA.
Linker	:	A synthetic self-complementary oligonucleotide that contains a restriction enzyme recognition site. Used to add cohesive ends to DNA molecules that have blunt ends.
Lipase	:	Enzyme that hydrolyses fats (lipids).
Lye dip	:	Soaking fruits or vegetables in a lye solution, which makes the product easier to dry and makes the peel easier to remove.
Lyophilisation	:	Process in which cold temperature and air evacuation are used for preservation of micro-organisms, also termed freeze-drying.
MA	:	Muramic acid.
Macromonas	:	Bacterial genus, cylindrical to bean-shaped cells that oxidise sulphur and sulphur compounds, may accumulate calcium carbonate with sulphur globules, found in seawater.
Macromolecules	:	Large molecules, contained within a cell, with molecular weights ranging from a few thousand to hundreds of millions.
Magnetospirillum	:	Bacterial genus, magnetotactic organisms that participate in biomineralisation of magnetosomes.
MCYSTs	:	Microcystins.
Mechanical aeration	:	Transfer of oxygen from the atmosphere into a liquid by the mechanical action of a turbine or other mechanisms. Mixing by mechanical means of the mixed liquor in the reactor basin or aeration tank of an activated sludge treatment plant.
Mesophilic digestion	:	Anaerobic digestion by biological oxidation by anaerobic action at or below 45°C (110°F).
Micro (μ)	:	SI prefix, 10^{-6}.
Microbial activity	:	Chemical changes resulting from biochemical action, the metabolism of living organisms.
Microbial pest control agent (MPCA)	:	A nonpathogenic micro-organism applied to agricultural crops to minimise the colonisation of a microbial phytopathogen.
Microconidia	:	Plural of microconidium.
Microcosm	:	A small-scale experimental model that is designed to reproduce the environmental conditions of interest as closely as possible, used in laboratory experiments to define environmental conditions and test biological populations.

Microenvironment	:	The physical and chemical conditions in the area immediately surrounding an organism.
Microinjection	:	Introduction of DNA into the nucleus or cytoplasm of a cell by insertion of a microcapillary and direct injection.
Micro-organism	:	A microscopic organism that exists as a single cell or in an aggregate of cells or as an acellular entity (i.e. virus).
Microtubule	:	A structural entity of eukaryotic flagella.
Mixed culture	:	Microbial culture consisting of two or more species.
Monosaccharides	:	Chemical building blocks of carbohydrates with the empirical formula $(CH_2O)_n$.
MPA	:	Microscopic particulate analysis.
MPCA	:	Microbial pest control agent.
Multimedia filtration	:	Filtration of water or waste-water through a granular bed containing two or more filter media.
Mutagenesis	:	The process of inducing mutations in DNA.
Mutant	:	An organism (or gene) carrying a genetic mutation.
Mutation	:	An alteration to the sequence of bases in DNA. May be caused by insertion, deletion or modification of bases.
Neutralisation	:	A type of antigen-antibody reaction in which the activity taking place between reactants is not visible.
Neutralism	:	A relationship of microbial populations in which there is no interaction between the populations, occurs when populations of organisms have different metabolic capabilities, populations are spatially distant from each other, when environmental conditions are unfavourable for active growth or when organisms are in a resting state.
Nitrifying bacteria	:	A group of bacteria that oxidises ammonia to nitrite or nitrite to nitrate.
Node	:	A joint, point of origin of fungal hyphae or an enlarged area on a fungal hypha.
Nod genes	:	Genetic sequences in nitrogenic fixing bacteria that direct specific steps in the formation of a root nodule.
Nuclease	:	An enzyme that hydrolyses phosphodiester bonds.
Nucleoside	:	A nitrogenous base bound to a sugar.
Nucleotide	:	A nucleoside bound to a phosphate group.
Nucleus	:	Membrane-bound region in a eukaryotic cell that contains the genetic material.
Oligonucleotide	:	A short sequence of nucleotides.
Outlier	:	A value that is inconsistent with the other data obtained during statistical analysis.
PAB	:	Propionic acid bacterium.
PADs	:	Phenolic acid decarboxylases.
Papovavirus	:	Any virus in the family Papovaviridae.
Parasite	:	An organism that lives on or in a host.
Parasitic bacteria	:	Bacteria that require living host organism but do not harm the host.
Passive sampling	:	Collection of material without the use of a mechanical device, gravitational sampling.
Pasteurisation	:	Preservation method in which bottled or canned food is heated at a maximum temperature of 100°C. This process kills most micro-organisms and thereby increases the product's shelf-life up to several weeks, but it is not as effective as sterilisation.
Pathogen/Pathogenic	:	An organism that infects a host and is capable of causing disease.

PBBs	:	Polybrominated biphenyls.
PCA	:	Principal component analysis.
PCBs	:	Polychlorinated biphenyls.
PCP	:	Pentachlorophenol.
Repressor protein	:	A protein that inhibits the activity of certain genes, lysogeny is established when repressor protein is produced under direction of a virus.
Retrovirus	:	Any virus in the family Retroviridae, viruses are icosahedral in shape, surrounded by a lipid envelope, contain single-stranded RNA, human pathogens in this family cause different types of cancer, and the human immunodeficiency viruses (causative agent of AIDS) are members of the lentivirus genus in this family.
Reverse transcriptase	:	An enzyme that synthesises a DNA molecule from the code supplied by a RNA molecule.
Rhizobiaceae	:	Bacterial family, characterised by their ability to fix atmospheric nitrogen.
Ribonuclease (RNase)	:	An enzyme that hydrolyses RNA.
Sake	:	A type of rice beer produced primarily in the orient.
Salmonella enterica	:	Bacterial species, foodborne pathogen transmitted via the ingestion route of exposure through the consumption of raw tomatoes.
Salmonella paratyphi	:	Bacterial species, waterborne and foodborne pathogen transmitted via the ingestion route of exposure.
Sanitary landfill	:	Landfill for disposing of solid wastes.
Scotochromogenesis	:	Formation of pigment only when the micro-organism is cultured in dark, used in the classification of some *Mycobacterium* spp., in contrast of photochromogenesis.
Secondary effluent	:	Effluent leaving the secondary or final clarifier at a waste-water treatment plant.
Secondary sludge	:	Sludge from the final clarifier at waste-water treatment plant. For the activated sludge process, it is the sludge to be recycled. For the trickling filter process, it is the trickle filter growths that have sloughed off—that is, the trickling filter humus.
Sedimentation	:	Removal of settleable suspended solids from water or waste-water by gravity settling in a quiescent tank or basin. Also called clarification or settling.
Selection bias	:	The introduction of a systematic error due to the manner in which the test and control populations are selected in contrast to surveillance bias and misclassfication bias.
Settled waste-water	:	Waste-water that has been treated by sedimentation. Also called clarified waste-water.
Silencing	:	The process whereby an organism shuts down the expression of a gene.
Sludge conditioning	:	Treatment of sludge, usually by chemical means, to enhance its dewatering characteristics.
Sludge digestion	:	Biological oxidation of organic or volatile matter in sludges to produce more stable substances.
Sludge digestion tank	:	Tank used for the anaerobic digestion of organic sludges.
Somatic cell	:	Body cell, as opposed to germ-line cell.
SOP	:	Standard operating procedure.
Sperm	:	The mature male gamete.
Spore-formers	:	Type of bacteria that carry a certain type of seed that can withstand high temperatures and that grow into bacteria at low temperatures.
Sterilisation	:	Preservation method in which bottled or canned food is heated at a temperature of 100°–121°C. This process kills all micro-organisms, and extends the product's

		shelf-life up to a maximum of one year, but it does not kill the spores, which can grow into bacteria once the container is reopened.
Structural gene	:	A gene that encodes a protein product.
Suspended matter	:	Solids in suspension in water or waste-water that can be removed by laboratory filtration techniques, such as membrane filtration.
TEM	:	Transmission electron microscopy.
Temperate	:	Refers to bacteriophages that can undergo lysogenic infection of the host cell.
Tertiary treatment	:	Use of physical, chemical or biological means to upgrade a secondary effluent.
Thermal death point	:	The temperature required to kill an organism in a given length of time.
Toxin	:	A poisonous substance produced by a species of micro-organism, bacterial toxins are classified as exotoxins or endotoxins.
Transcription	:	The synthesis of RNA using a DNA template.
Transduction	:	Genetic recombination process mediated by virus, DNA is incorporated from one cell to another with the assistance of virus by generalised transduction or specialised transduction.
Transfection	:	Introduction of purified phage or virus DNA into cells.
Transformant	:	A cell that has been transformed by exogenous DNA.
Transformation	:	The process of introducing DNA (usually plasmid DNA) into cells. Also used to describe the change in growth characteristics when a cell becomes cancerous.
Transgene	:	The target gene involved in the generation of a transgenic organism.
Transgenic micro-organism	:	A micro-organism with a cloned DNA sequence from another organism.
Trickling filter	:	Biological filter consisting of a bed of coarse material, such as stone, over which waste-water is distributed by a spray from a moving distributor or other device. The waste-water trickles through the bed to the underdrains, giving the microbial slimes an opportunity to absorbed the organic material and clarify the waste-water.
Turbidity	:	Suspended matter in water or waste-water that causes the scattering or absorption of light rays.
Vaccinia	:	The alternative name for cow-pox.
Vacuole	:	A large, fluid-filled sac located in the cytoplasm.
Vector	:	A DNA molecule that is capable of replication in a host organism, and can act as a carrier molecule for the construction of recombinant DNA.
Virus	:	An infectious agent that cannot replicate without a host cell.
VNA	:	Viral nucleic acid.
Waste stabilisation	:	Process of reducing the BOD or COD of organic wastes to render them harmless.
Waste-water analysis	:	The determination of the physical, chemical, and biological characteristics of a waste-water or treatment plant effluent.
Waterborne disease	:	Disease caused by organisms or toxic materials transported by water. The most common waterborne diseases are typhoid fever, cholera, dysentery and other intestinal disturbances.
Water moulds	:	Aquatic fungi that are members of the Oomycetes.
Waterwashed disease	:	Illness caused by organisms that originate in feces and are transmitted through contact because of inadequate sanitation.
WHC	:	Water holding capacity.

Whey	:	A waste liquid of the dairy industry containing lactose and minerals that is used in industrial processes as supplemental carbon.
Xanthomonadins	:	Yellow, membrane-bound, halogenated aryl polyene pigments that are produced by *Xanthomonas* spp. and may provide some protection against photodamage.
Xanthomonas	:	Bacterial genus, member of the family Pseudomonaceae, chemoorganotrophic, Gram-negative, straight, obligate aerobic, bacillus that is motile by a single polar flagellum, many species are phytopathogens.
Xanthomonas campestris	:	Bacterial species, phytopathogen, causative agent of black rot of crucifers.
Xanthomonas oryzae	:	Bacterial species, phytopathogen, causative agent of blight of rice.
Xanthomonas vascularum	:	Bacterial species, phytopathogen, causative agent of gumming of sugar cane.
Xenobiotic	:	Group of chemicals unfamiliar or foreign to micro-organisms and thus not easily degradable by them.
XPS	:	X-ray photoemission spectroscopy.
Yeast	:	Unicellular fungus that reproduces by budding, most are members of the Ascomycetes, some have a filamentous phase, for example, *Saccharomyces cerevisiae* are used extensively in food production for leavening of bread and in beer and wine fermentation.
Yersinia	:	Bacterial genus, member of the family Enterobacteriaceae, facultative, Gram-negative, nonsporulating bacilli with 10 established species, most with simple nutritional requirements, previously classified as the genus *Pasteurella*.
Zonate	:	Arranged in zones or rings radiating from the centre.
Zone of inhibition	:	The area in which an antimicrobial substance prevents growth.
Zooglea	:	The gelatinous material resulting from the attrition of bacterial slime layers. An important constituent of activated sludge floc and trickling filter growths.
Zooglea ramigera	:	Bacterial species, produces extracellular polysaccharide slime matrix during sewage treatment.
Zygomycetes	:	Fungal class, rapid growing non-septate fungi with sporangiospores produced in sporangia and some species have rhizoids and stolons, sexual reproduction produces a dark thick-walled zygospore.
Zymocide	:	A factor present in some yeast cells that is toxic to other yeasts.
Zymomonas	:	Bacterial genus, tolerant of low pH and ethanol concentrations up to 10 per cent, large, Gram-negative bacillus that ferments sugars to ethanol, active in fermentation of plant sap for industrial production (e.g. fermentation of agave in Mexico to produce tequila and palm sap in tropical areas), responsible for spoilage of fruit juices and production of an odour of rotten apples in spoiled beer.

References

Allen, J.L., *Basic Concepts of Microbiology*, Butterworths, London.

Alvarez, A.J., *Microbiology of Food and Allied Products*, Reinhold Publication Corporation, New York.

Anke, K.T., *Chemistry and Technology of Food and Food Products*, Butterworths, London.

Bradley, R.S., *Engineering Aspects of Food Biotechnology*, Academic Press, London.

Brown, N.H., *Food Science and Technology*, Harcourt Brace Jovanovich, New York.

Budyko, N.I., *Introduction to Food Biotechnology*, Butterworths, London

Cambell, K.E. and Lemer, H.A., *Food Biochemistry*, Academic Press, London.

Coolingwood, S.W., *Food Science and Food Biotechnology*, D. Van Nostrand, New York.

Daniel, G.L., *Industrial Drying*, John Wiley & Sons, New York.

Desmond, S.T., *Fundamentals of Food Biotechnology*, Academic Press, London.

Downe, S.A., *Biochemistry*, Pergamon Press, Oxford.

Goldman, M., *Food Quality Safety and Technology*, Gordon and Breach, Science Publishers, New York.

Gould, G.W., *Food Biochemistry*, D. Van Nostrand, New York.

Harding, G., *Food Chemistry*, Prentice-Hall, London.

Jackson, M.L., *Fundamental of Biotechnology*, Marcel Dekker, New York.

Kim, C.K., *Food Sanitation*, Butterworths, London.

Krieg, G.M., *Food Engineering and Process Applications,* Heinemann, London.

Lechevallier, M., *Industrial Microbiology*, Academic Press, London.

Lewis, B.B., *Food Microbiology*, Elsevier Scientific Publishing Co., Amsterdam.

McCaull, J. and Crossland, J., *General Microbiology*, Harcourt Brace Jovanovich, New York.

Miller, B.M., *Industrial Microbiology*, McGraw-Hill Book Company, New York.

Miller, M.S., *Micro-organisms in Food*, John Wiley & Sons, New York.

Nuan, E., *Food Engineering Operations*, Reinhold Publication Corporation, New York.

Odum, S.K., *Introduction to Microbiology*, W.B. Saunders and Co., New York.

Pearsons, K.S., *Practical Food Microbiology and Technology*, Williams and Wilkins Company, Baltimore.

Phillips, D.J.H., *Microbes and their Importance*, Applied Science Publishers, London.

Rainbow, C., *Food Microbiology*, Academic Press, New York.

Reid, G.K., *Technology of Food Preservation*, Reinhold Publication Corporation, New York.

Riemann, D., *Encyclopedia of Biotechnology and Industrial Microbiology*, Academic Press, London.

Robert, B. and Evison, L., *Principles of Food Science*, D. Van Nostrand, New York.

Sengner, J., *Micro-organisms in Food*, Leonard Hill Books, London.

Smith, H.S., *Chemistry of Micro-organisms*, Chilton Book Company, Radnor, Pennsylvania.

Smith, P., *Encyclopedia of Environmental Microbiology*, Cambridge University Press, Cambridge.

Tanaka, S.K., *Biotechnology of Enzymes*, Gordon and Breach, Science Publishers, New York.

Taylor, K.C. and Borrel, N., *Basic Concepts of Biochemistry*, McGraw-Hill, New York.

Thatcher, F.S., *Micro-organisms in Food*, University of Toronto Press, Toronto.

Vollenweider, R.A., *Industrial Biotechnology*, Blackwell Scientific Publications, New York.

Whittaker, R.H., *Basic Concepts of Environmental Biotechnology*, John Wiley & Sons, New York.

Wilson, W.L., *Biochemistry of Industrial Micro-organisms*, Academic Press, London.

Index

Food Processing and Preservation: Volume I

Contents

Food Processing and Preservation: Volume II

Contents